PROBABILITY, STATISTICS AND MODELLING IN PUBLIC HEALTH

PROBABILITY, STATISTICS AND MODELLING IN PUBLIC HEALTH

Edited by

MIKHAIL NIKULIN
Université Victor Segalin Bordeaux 2, France
V. Steklov Mathematical Institute, Saint Petersburg, Russia

DANIEL COMMENGES
Université Victor Segalin Bordeaux 2, France

CATHERINE HUBER
Université René Descartes, Paris, France

Springer

Library of Congress Control Number: 2005052019

ISBN-10: 0-387-26022-6 e-ISBN: 0-387-26023-4

ISBN-13: 978-0387-26022-8

Printed on acid-free paper.

Printed in the United States of America.

9 8 7 6 5 4 3 2 1

springeronline.com

Dedicated to Marvin ZELEN

Preface

On September 23, 2003 Marvin Zelen was awarded the title of Docteur Honoris Causa de l'Université Victor Segalen Bordeaux 2, Bordeaux, France. Professor Zelen was the third biostatistician to receive this title after David Cox (1999) and Norman Breslow (2001). To mark the occasion and the importance of the contribution of Professor Zelen in development of biostatistics in public health and especially in the *War on Cancer*, a special symposium, *Probabilités, Statistics and Modelling in Public Health*, was organized in Marvin's honor by Daniel Commenges and Mikhail Nikulin. This workshop took place on September 22-23, 2003, in Bordeaux. Several well known biostatisticians from Europe and America were invited. A special issue of *Lifetime Data Analysis* was published (Volume 10, No 4), gathering some of the works discussed at this symposium. This volume gathers a larger number of papers, some of them being extended versions of papers published in the *Lifetime Data Analysis* issue, others being new. We present below several details of the biography of Professor Zelen.

Marvin Zelen is Professor of Statistics at the Harvard School of Public Health in Boston. He is one of the major researchers in the field of statistical methods in public health.

Since 1960, Professor Zelen constantly worked in several fields of applied statistics, specifically in biology and epidemiology of cancer. He is very well known for his work on clinical trials in oncology, on survival analysis, reliability and planning of experiments and prevention. His papers have now become classics among epidemiologists and biostatisticians who work in the field of cancer.

Since 1967, Professor Zelen was involved in different scientific groups such as the Eastern Cooperative Oncology Group, the Veteran's Administration Lung Cancer Group, the Gastrointestinal Tumor Study Group, and the Radiation Therapy Oncology Group to do statistical research in cancer clinical trials in the USA. Professor Zelen made also significant contributions to reliability theory and random processes, mainly Markov and semi-Markov processes, in biostatistics and epidemiology. Professor Zelen is famous all over the world for the development of the Biostatistics Department in the Harvard School of Public Health. He received several awards for his contributions to statistical methodology in the biomedical field. Among them, in 1967, the Annual Award, Washington Academy of Science, for Distinguished Work in Mathematics, in 1992, the *Statistician of the Year* award of Boston Chapter of the American Statistical Association, and, in 1996, the *Morse Award for Cancer Research.*

We thank all participants of the workshop in Bordeaux and all colleagues and friends of Marvin for supporting us in the organization of the meeting in Bordeaux and for their contributions in preparation of this volume. Especially we thank Thelma Zelen, Mei-Ling Ting Lee, Stephen Lagakos, Dave Harrington, Bernard Begaud, Roger Salamon, Valia Nikouline, Elizabeth Cure and the participants of the European Seminar *Mathematical Methods for Reliability, Survival Analysis and Quality of Life* for their help in organization of the meeting and preparation of the proceedings. We thank also l'IFR-99 "Santé Publique" for financial support of our project.

We sincerely hope that this volume will serve as a valuable reference for statisticians.

Mikhail Nikulin, Daniel Commenges and *Catherine Huber*, editors
March, 2005, Bordeaux

Contents

List of Contributors

K. G. Arbeev Center for Demographic Studies, Duke
University, 2117 Campus Drive, Box 90408, Durham, NC
27708-0408,USA
arbeev@cds.duke.edu

L. S. Arbeeva Ulyanovsk State University, Leo Tolstoi St.
42, 432700 Ulyanovsk, Russia
arbeev@mail.ru

V.N. Ardashev Burdenko Main Military Clinical Hospital,
Moscow, Russia

V.Bagdonavičius
Department of Mathematical Statistics, Vilnius University,
Lithuania
vilius@sm.u-bordeaux2.fr

P. Barberger-Gateau IFR
99 Santé Publique, Université Victor Segalen Bordeaux 2,
France
nikou@sm.u-bordeaux2.fr

Vlad Barbu
Université de Technologie de Compiègne, Laboratoire
de Mathématiques Appliquées de Compiègne, BP 20529, 60205
Compiègne, France
barbu@dma.utc.fr

Barsov M.I.
Burdenko Main Military Clinical Hospital, Moscow,
Russia

A. Bikelis
Vilnius University, Naugarduko 24, Vilnius,
Lithuania
marius@post.omnitel.net

D. Commenges INSERM E0338, Université
Victor Segalen Bordeaux 2,
146 rue Léo Saignat, 33076 Bordeaux Cedex, FRANCE.
daniel.commenges@isped.u-bordeaux2.fr

V. Couallier
Equipe Statistique Mathematique et ses Applications
U.F.R. Sciences et Modelisation, Universite Victor Segalen
Bordeaux 2
146 rue Leo Saignat
33076 Bordeaux cedex FRANCE
couallier@sm.u-bordeaux2.fr

D. M. Dabrowska
Department of Biostatistics,
University of California, Los
Angeles, CA 90095-1772
dmdabrowska@yahoo.com

J.P.Daurès
Laboratoire de Biostatistique,
Institut Universitaire de
Recherche Clinique, 641 avenue de
Doyen Gaston Giraud, 34093
Montpellier, France.

P. Deheuvels
L.S.T.A., Université Paris VI, 7
avenue du Château, F
92340 Bourg-la-Reine, France
pd@ccr.jussieu.fr

G. Derzko
Sanofi-Synthélabo Recherche, 371
rue du Professeur Joseph Blayac,
34184 Montpellier Cedex 04, France
Gerard.Derzko@sanofi-aventis.com

J-F. Dupuy
Laboratoire de Statistique et
Probabilités, Université Paul
Sabatier,
118, route de Narbonne, 31062
Toulouse cedex 4, France
dupuy@math.ups-tlse.fr

R. M. Elashoff
Department of Biostatistics,
University of California, Los
Angeles, CA 90095-1772

M.S. Finkelstein
Department of Mathematical
Statistics
University of the Free State
PO Box 339, 9300 Bloemfontein,
Republic of South Africa
and Max Planck Institute for
Demographic Research
Konrad-Zuse-Strasse 1
18057 Rostock, Germany
FinkelM.SCI@mail.uovs.ac.za

S.I. Glukhova
Burdenko Main Military Clinical
Hospital, Moscow,
Russia

C. Gras
Laboratoire de Biostatistique,
Institut Universitaire de
Recherche Clinique, 641 avenue de
Doyen Gaston Giraud, 34093
Montpellier, France.
claudine.gras@iurc.montp.inserm.fr

S. Gulati
Department of Statistics, The Honors College, Florida International
University,
Miami, FL 33199,USA
gulati@fiu.edu

B.P. Harlamov
Institute of Problems of Mechanical
Engineering,
Russian Academy of Sciences,
Saint-Petersburg,
harlamov@random.ipme.ru

D. Harrington
Department of Biostatistics,
Harvard School of Public Health,
and Department of Biostatistical
Science, Dana-Farber Cancer
Institute, 44 Binney Street, Boston,
Massachusetts 02115, U.S.A.
dph@jimmy.harvard.edu

N. Heutte
IUT de Caen, Antenne de
Lisieux, Statistique et Traitement
Informatique des Données. 11,
boulevard Jules Ferry 14100 Lisieux,
France
N.Heutte@lisieux.iutcaen.
unicaen.fr

J. Huang
Department of Preventive Medicine, Feinberg School of Medicine, Northwestern University, 680 N. Lake Shore Drive Suite 1102, Chicago, Illinois 60611, U.S.A.
jjhuang@northwestern.edu

C. Huber-Carol
University Paris 5, 45 rue des Saints-Pères, 75270 Paris Cedex 06, France and U 472 INSERM, 16bis avenue P-V Couturier, 94 800, Villejuif, France
catherine.huber@univ-paris5.fr

W. Kahle
Otto-von-Guericke-University, Faculty of Mathematics, D-39016 Magdeburg, Germany
waltraud.kahle@mathematik.uni-magdeburg.de

V. Kazakevičius
Vilnius University, Naugarduko 24, Vilnius, Lithuania
Vytautas.kazakevicius.maf.vu.lt

N. Keiding
Department of Biostatistics, University of Copenhagen, Blegdamsvej 3, DK-2200 Copenhagen N, Denmark
nk@biostat.ku.dk

V.M. Klyuzhev
Burdenko Main Military Clinical Hospital, Moscow, Russia

S. W. Lagakos
Department of Biostatistics, Harvard School of Public Health
655 Huntington Avenue, Boston MA 02115
lagakos@hsph.harvard.edu

H. Läuter
Institute of Mathematics, University of Potsdam
laeuter@rz.uni-potsdam.de

A. Lehmann
Otto–von–Guericke–University Magdeburg
Institute of Mathematical Stochastics
PF 4120, D-39016 Magdeburg, Germany
axel.lehmann@mathematik.uni-magdeburg.de

Q. H. Li
Food and Drug Administration Center for Drug and Evaluation Research, HFD-705
7500 Standish Place, Metro Park North (MPN) II, Rockville, MD 20855
liq@cder.fda.gov

H. Liero
Institute of Mathematics, University of Potsdam
liero@rz.uni-potsdam.de

N. Limnios
Université de Technologie de Compiègne, Laboratoire de Mathématiques Appliquées de Compiègne
Nikolaos.Limnios@utc.fr

B. Liquet
Laboratoire de Statistique et Analyse des Données,
BHSM, 1251 avenue centrale BP 47
38040 Grenoble Cedex 09, FRANCE

N.G. Mamchich
Burdenko Main Military Clinical Hospital, Moscow,
Russia

M. Mesbah
Laboratoire de Statistique Théorique et Appliquée (LSTA),
Université Pierre et Marie Curie - Paris VI, Boîte 158, - Bureau 8A25 - Plateau A. 175 rue du Chevaleret,
75013 Paris, France
mesbah@ccr.jussieu.fr

D. L. Morton
John Wayne Cancer Institute,
Santa Monica, CA 90404

U. U. Müller
Fachbereich 3, Universität Bremen,
Postfach 330 440, 28334 Bremen,
Germany
uschi@math.uni-bremen.de

M. Newby
Centre for Risk Management,
Reliability and Maintenance
City University
LONDON EC1V 0HB

M.Nikulin
99 Santé Publique, Université Victor Segalen Bordeaux 2,
France
nikou@sm.u-bordeaux2.fr

G. Parmigiani
Departments of Oncology,
Biostatistics and Pathology
Johns Hopkins University,
Baltimore, MD 21205
gp@jhu.edu

C. Pinçon
EA 3614 - Laboratoire de Biomathématiques
3, rue du Professeur Laguesse - 59006 Lille cédex - France.
cpincon@pharma.univ-lille2.fr

O. Pons
Département MIA - INRA
Domaine de Vilvert - 78352
Jouy-en-Josas cédex - France.
Odile.Pons@jouy.inra.fr

V.P. Pop
Burdenko Main Military Clinical Hospital, Moscow,
Russia

S. Rosthøj
Department of Biostatistics,
University of Copenhagen,
Blegdamsvej 3, DK-2200 Copenhagen N, Denmark
S.Rosthoej@biostat.ku.dk

O.A. Rukavicin
Burdenko Main Military Clinical Hospital, Moscow,
Russia

A. Schick
Department of Mathematical Sciences, Binghamton University,
Binghamton, NY 13902-6000, USA
anton@math.binghamton.edu

V. Sebille
Laboratoire de Biostatistiques,
Faculté de Pharmacie, Université de Nantes, 1
rue Gaston Veil, BP 53508, 44035
Nantes Cedex 1, France.
veronique.sebille@univ-nantes.fr

S.V. Shamansky
Burdenko Main Military Clinical Hospital, Moscow,
Russia

Yu Shen
Department of Biostatistics and Applied Mathematics
M. D. Anderson Cancer Center, University of Texas
Houston, TX 77030
yshen@mdanderson.org

G.I. Sidorovich
Burdenko Main Military Clinical Hospital, Moscow,
Russia

V. Solev
Steklov Institute of Mathematics at St. Petersburg, nab.
Fontanki, 27 St.Petersburg 191023
Russia,
solev@pdmi.ras.ru

B.Tretarre
Registre des Tumeurs de l'Hérault, bâtiment
recherche, rue des Apothicaires
B.P. 4111,
34091 Montpellier Cedex 5.

S. V. Ukraintseva
Center for Demographic Studies, Duke University, 2117 Campus Drive, Box 90408, Durham, NC 27708-0408, USA
ukraintseva@cds.duke.edu

R. Viertl
Vienna University of Technology,
1040 Wien, Austria
R.Viertl@tuwien.ac.at

F. Vonta
Department of
Mathematics and Statistics, University of Cyprus P.O. Box 20537,
CY-1678, Nicosia, Cyprus,
vonta@ucy.ac.cy

W. Wefelmeyer
Mathematisches Institut, Universität zu Köln,
Weyertal 86–90, 50931 Köln,
Germany
wefelmeyer@math.uni-koeln.de

H. Wendt
Otto-von-Guericke-University,
Faculty of Mathematics,
D-39016 Magdeburg, Germany

H.-D. I. Wu
School of Public Health, China Medical University,
91 Hsueh-Shih Rd., Taichung 404,
TAIWAN.
honda@mail.cmu.edu.tw

A. I. Yashin
Center for Demographic Studies, Duke University, 2117 Campus Drive, Box 90408, Durham, NC 27708-0408, USA
yashin@cds.duke.edu

S. Zacks
Department of Mathematical Sciences
Binghamton University
shelly@math.binghamton.edu

M. Zelen
Harvard School of Public Health and the Dana-Farber Cancer Institute
Boston, MA 02115, U.S.A.

O. Zdorova-Cheminade
Université Victor Segalen
Bordeaux 2, France

Forward and Backward Recurrence Times and Length Biased Sampling: Age Specific Models

Marvin Zelen[1]

Harvard School of Public Health and the Dana-Farber Cancer Institute
Boston, MA 02115, U.S.A. name@email.address

Summary. Consider a chronic disease process which is beginning to be observed at a point in chronological time. The backward recurrence and forward recurrence times are defined for prevalent cases as the time with disease and the time to leave the disease state respectively, where the reference point is the point in time at which the disease process is being observed. In this setting the incidence of disease affects the recurrence time distributions. In addition, the survival of prevalent cases will tend to be greater than the population with disease due to length biased sampling. A similar problem arises in models for the early detection of disease. In this case the backward recurrence time is how long an individual has had disease before detection and the forward recurrence time is the time gained by early diagnosis; i.e. until the disease becomes clinical by exhibiting signs or symptoms. In these examples the incidence of disease may be age related resulting in a non-stationary process . The resulting recurrence time distributions are derived as well as some generalization of length-biased sampling.

1 Introduction

Consider a sequence of events occuring over time in which the probability distribution between events is stationary. Consider a randomly chosen interval having endpoints which are events and select at random a time point in the interval. The forward recurrence time is defined as the time from the random time point to the next event; the backward recurrence time is the time from the time point to the previous event; cf. Cox and Miller [CM65].

An example illustrating these recurrence times is the so-called "waiting time paradox"; cf. Feller [FEL71]. Suppose the events are defined as bus arrivals at a particular location. A person arriving at the bus stop has a waiting time until the next bus arrives. The waiting time is the forward recurrence time. The backward recurrence time is how long the person missed the previous bus.

Backward and forward recurrence times play an important role in several biomedical applications. However in many instances the distribution of events may have a distribution which changes with time. Furthermore time may be chronological or age. In some applications it may be necessary to consider two time scales incorporating both chronological time and age.

In addition, a closely related topic is length biased sampling . Referring to the bus waiting problem, when the individual arrives at the bus stop, she is intersecting a time interval having endpoints consisting of the previous bus arrival and the next arrival. Implicitly these intervals are chosen so that the larger the interval, the greater the probability of selecting it. The selection phenomena is called length bias sampling.

We will consider two motivating examples for generalizing the recurrence time distributions and length biased sampling. One example deals with a model of the natural history of a chronic disease . The other example refers to modeling the early detection of disease . The mathematics of the examples are the same. However, they are both important in applications and we use both to motivate our investigation. This paper is organized as follows. Section 2 describes the two motivating examples and summarizes results for stationary processes. Section 3 develops the model for the chronic disease example; section 4 indicates the necessary changes for the early detection example. The paper concludes with a discussion in section 5.

2 Motivating Problems and Preliminary Results

2.1 Chronic Disease Modeling

Consider a population and a chronic disease such that at any point in time a person may be disease free (S_0), alive with disease (S_a) or may have died of the specific disease (S_d). The natural history of the disease will be $S_0 \to S_a \to S_d$. The transitions $S_0 \to S_a$ corresponds to the (point) incidence of the disease and $S_a \to S_d$ describes the (point) mortality.

Of course an individual may die of other causes or may be cured by treatment. Our interest is in disease specific mortality. Hence an individual who dies of other causes while in S_a is regarded as being censored for the particular disease. An individual who is cured of a disease will still be regarded as being in S_a and eventual death due to other causes will be viewed as a censored observation. This model is a progressive disease model and is especially applicable for many chronic diseases — especially some cancers, cardiovascular disease and diabetes.

Consider a study where at some point in time, say, t_0 this population will be studied. At this point in time some individuals will be disease free (S_0) while others will be alive with disease (S_a). Those in S_a are prevalent cases. The backward recurrence time is how long a prevalent case has had disease up to the time t_0. The forward recurrence time refers to the eventual time of death of the prevalent cases using t_0 as the origin. The sum of the backward and forward recurrence times is the total survival of prevalent cases.

2.2 Early Detection Modeling

Consider a population in which at any point in time a person may be in one of three states: disease free (S_0), pre-clinical (S_p), or clinical (S_c). The pre-clinical state refers to individuals who have disease, but there are no signs or symptoms. The individual is unaware of having disease. The clinical state refers to the clinical diagnosis of the disease when the disease interferes with the functioning of an organ system or causes pain resulting in the individual seeking medical help leading to the clinical diagnosis of the disease. The natural history of the disease is assumed to be $S_0 \to S_p \to S_c$. Note that the transition from $S_0 \to S_p$ is never observed. The transition $S_p \to S_c$ describes the disease incidence. The aim of an early detection program is to diagnose individuals in the pre-clinical state using a special examination. If indeed, the early detection special examination does diagnose disease in the pre-clinical state, the disease will be treated and the natural history of the disease will be interrupted. As a result, the transition $S_p \to S_c$ will never be observed. The time gained by earlier diagnosis is the forward recurrence time and the time a person has been in the pre-clinical state before early diagnosis is the backward recurrence time. If t_0 is the time (either age or chronological time) in which the disease is detected, we then have an almost identical model as the chronic disease model simply by renaming the states.

2.3 Preliminary Results

Consider a non-negative random variable T having the probability density function $q(t)$. A length biased sampling process chooses units with a probability proportional to t $(t < T \leq t + dt)$. Samples of T are drawn from a length biased process. Suppose the random variable is randomly split into two parts (U, V) so that $T = U + V$. The random variable U and V are the backward and forward recurrence times. The model assumes that for fixed $T = t$ $(t < T \leq t + dt)$ a point u is chosen according to a uniform distribution over the interval $(0, t)$. Then if $q_f(v)$ and $q_b(u)$ are the probability density functions of the forward and backward recurrence times it is well known that with length biased sampling for selecting T; cf. Cox and Miller [CM65].

$$q_f(t) = q_b(t) = Q(t)/m, \qquad t > 0 \tag{1}$$

where $Q(t) = \int_t^\infty q(x)dx$ and $m = \int_0^\infty Q(x)dx$.
Also the p.d.f. of T is

$$f(t) = tq(t)/m. \tag{2}$$

Note that the first moments of these distributions are:

$$\int_0^\infty \frac{tQ(t)}{m}dt = \tfrac{m}{2}(1+C^2),$$

$$\int_0^\infty \frac{t^2q(t)}{m}dt = m(1+C^2) \tag{3}$$

where $C = \sigma/m$ is the coefficient of variation associated with $q(t)$. If $q(t)$ is the exponential distribution with mean m, the forward and backward recurrence times have the same exponential distribution as $q(t)$ and $C = 1$.

A reviewer suggested that a simpler way to discuss these results is to initially assume that the joint distribution of (U, V) is $f(u, v) = q(u+v)I(u \geq 0, v \geq 0)/m$.. Then all the results above are readily derived. Implication in this assumption is $f(u/T) = 1/t$ and length biased sampling.

3 Development of the Chronic Disease Model

In this section we will investigate generalizations of the distribution of the backward and forward recurrence times using the chronic disease model as a motivating example. We remark that for the chronic disease model, the process may have been going on for a long time before being observed at time t_0.

Suppose at chronological time t_0 the disease process is being observed. The prevalent cases at time t_0 will have an age distribution denoted by $b(z|t_0)$. We will initially consider the prevalent cases who have age z. Later by weighting by the age distribution for the whole population we will derive properties of the prevalent cases for the population. The prevalent cases could be regarded as conditional on the time t_0 when observations began. Another model is that the prevalent cases could be assumed to have arisen by sampling the population at a random point in time which is t_0. We shall consider both situations.

Define

$$a(z|t_0) = \begin{cases} 1 \text{ if individual of age } z \text{ is in } S_a \text{ at time } t_0. \\ 0 \text{ if individual of age } z \text{ is not in } S_a \text{ at time } t_0, \\ \quad \text{but was incident with disease before age } z. \end{cases}$$

$$a(t_0) = \begin{cases} 1 \text{ if individual is in } S_a \text{ at time } t_0. \\ 0 \text{ if individual is not in } S_a \text{ at time } t_0, \\ \quad \text{but was incident with disease before time } t_0. \end{cases}$$

$$P(z|t_0) = P\{a(z|t_0) = 1\}, \quad P_0 = P\{a(t_0) = 1\} = \int_0^{t_0} P(z|t_0)b(z|t_0)dz \tag{4}$$

Note that someone with disease at time t_0 having age z was born in the year $v = t_0 - z$. Hence the probability distribution of ages at time t_0 is equivalent to the distribution of birth cohorts at time t_0.

3.1 Forward Recurrence Time Distribution

Define

$$\begin{aligned}
T_f &= \text{Forward recurrence time random variable} \\
q_f(t|z)dt &= P\{t < T_f \le t + dt \mid a(z|t_0) = 1\} \\
Q_f(t|z) &= P\{T_f > t \mid a(z|t_0) = 1\} \\
I(\tau)d\tau &= P\{S_0 \to S_a \text{ during } \tau, \tau + d\tau\}
\end{aligned}$$

where τ refers to the age of incidence. Consider the probability of being in S_a at time t_0 and having age z. If an individual becomes incident at age τ, then $P\{a(z|t_0) = 1|\tau\} = P\{T > z - \tau\} = Q(z - \tau)$. Multiplying by $I(\tau)d\tau$ and integrating over the possible values of τ $(0 < \tau \le z)$ results in

$$P\{a(z|t_0) = 1\} = \int_0^z I(\tau)Q(z - \tau)d\tau \tag{5}$$

This probability applies to the birth cohort year $v = t_0 - z$; i.e. an individual born in year v who is prevalent at time t_0 having age z.

Consider the joint distribution of an individual having age z at time t_0 and staying in S_a for at least an additional t time units. If τ is the age of entering S_a, then

$$P(z|t_0, \tau)Q_f(t|z, \tau) = P\{T > z - \tau + t\} = Q(z - \tau + t)$$

and multiplying by $I(\tau)d\tau$ and integrating over $(0, z)$ gives

$$P(z|t_0)Q_f(t|z) = \int_0^z I(\tau)Q(z - \tau + t)d\tau \tag{6}$$

In the above it is assumed that the time entering S_a (τ) is not known, requiring integration over possible values of (τ). Consequently the p.d.f. of the forward recurrence time is

$$q_f(t|z) = -\frac{d}{dt}Q_f(t|z) = \int_0^z I(\tau)q(z - \tau + t)d\tau / P(z|t_0) \tag{7}$$

Suppose the incidence is constant, $I(\tau) = I$ then

$$q_f(t|z) = [Q(t) - Q(t + z)] / \int_0^z Q(y)dy. \tag{8}$$

If $Q(z)$ is negligible, then

$$q_f(t|z) \sim Q(t)/m$$

which is the usual forward recurrence time distribution for a stationary process.

Define $q_f(t|t_0)$ as the forward recurrence time averaged over the population. By definition we can write

$$P(a(t_0) = 1)q_f(t|t_0) = \int_0^{t_0} P(z|t_0)q_f(b|z)b(z|t_0)dz \tag{9}$$

When the age distribution is uniform so that $b(z|t_0) = b$ then it can be shown, cf. Zelen and Feinleib [ZF69]

$$\int_0^{\infty} q_f(t|t_0)P(a|t_0) = 1)dt_0 / \int_0^{\infty} P(a(t_0) = 1)dt_0 = Q(t)/m.$$

Thus if the sampling point is regarded as a random point in time, the forward recurrence time distribution as $t_0 \to \infty$ is the same as the stationary forward recurrence time distribution.

3.2 Backward Recurrence Time Distribution

The backward recurrence time refers to the time in S_a up to time t_0 (or age z). Let T_b be the backward recurrence time random variable and $q_b(t|z)$ be the conditional p.d.f. with $Q_b(t|z) = \int_t^z q_b(y|z)dy$. Note that $0 < t \leq z$. Then using the same reasoning as in deriving the forward recurrence time distribution we have

$$P\{T_b > t, a(z|t_0) = 1\} = P(z|t_0)Q_b(t|z) = \int_0^{z-t} I(\tau)Q(z-\tau)d\tau \tag{10}$$

which allows the calculation of $q_b(t|z)$; i.e.,

$$q_b(t|z) = I(z-t)Q(t)/P(z|t_0), \quad 0 < t \leq z \tag{11}$$

When $I(\tau) = I$, $q_b(t|z) = Q(t)/\int_0^z Q(y)dy$.

Finally the average backward recurrence time distribution is

$$q_b(t|t_0) = Q(t)\int_t^{t_0} I(z-t)b(z|t_0)dz/P_0 \tag{12}$$

Note the distinction between $q_b(t|z)$ and $q_b(t|t_0)$. The former refers to individuals having age z at time t_0 whereas the latter refers to the weighted average over age for prevalent cases at time t_0. When $b(z|t_0) = b$, we can integrate over t_0 and show that the backward recurrence time averaged over t_0 is $Q(t)/m$.

3.3 Length Biased Sampling and the Survival of Prevalent Cases

As pointed out earlier, the prevalence cases are not a random sample of cases, but represent a length biased sample. In this section, we investigate the consequences of length biased sampling when disease incidence is age-related. We also derive the survival of prevalent cases.

Define $T = T_b + T_f$ which is the time in which prevalent cases are in S_a. This is the survival of prevalent cases from the time when they become incident with disease. We will derive $f(t|z)$, the *pdf* of the time in S_a for prevalent cases who have age z at chronological time t_0. Since the age z is fixed at time t_0, it is necessary to consider $t > z$ and $t \le z$ separately. If t is fixed and $t > z$, then $P\{a(z|t_0) = 1 \mid t > z\} = \int_0^z I(\tau)d\tau$. Similarly, if t is fixed and $t < z$, in order to be prevalent at time t_0 and be of age z, it is necessary that $z - t < \tau < z$. Thus, we have for fixed t $(t < T \le t + dt)$

$$P\{a(z|t_0) = 1 \mid t < T \le t + dt\} = \begin{cases} \int_0^z I(\tau)d\tau, & \text{if } \ t > z \\ \int_{z-t}^z I(\tau)d\tau, & \text{if } \ t \le z \end{cases} \tag{13}$$

Note that $\int_{z-t}^t I(\tau)d\tau$ is an increasing function of t. Consequently, individuals with long sojourn times in S_a have a greater probability of being in S_a at time t_0. Our development is a generalization of the usual considerations of length biased sampling as we have shown how length biased sampling is affected by the transition into S_a. The usual specification of length biased sampling is to assume $P\{a(z) = 1 \mid t < T \le t + dt\} \propto t$, which in our case would be true if $I(\tau) = I$ and $t \le z$. We also remark that $P\{a(z|t_0) = 0 \mid t < T \le t + dt\} = \int_0^{z-t} I(\tau)d\tau$ refers to individuals, conditional on having survival $t < T \le t + dt$, who entered S_a and died before time t_0, but would have been age z at time t_0 if they had lived. Another interpretation of this probability is that a birth cohort born in $v = z - t$ was incident with disease but died before reaching age z. Using (13) the joint distribution of $a(z|t_0)$ and T is

$$P\{a(z|t_0) = 1, \ t < T \le t + dt\} = \begin{cases} q(t)dt \int_0^z I(\tau)d\tau, & \text{if } \ t > z \\ q(t)dt \int_{z-t}^z I(\tau)d\tau, & \text{if } \ t \le z. \end{cases} \tag{14}$$

Therefore, the time in S_a for cases prevalent at t_0 and having age z is

$$f(t|z)dt = \frac{P\{a(z|t_0) = 1, \ t < T \le t + dt\}}{P(z)}. \tag{15}$$

Some simplifications occur if $I(\tau) = I$. Then

$$f(t|z) = \begin{cases} zq(t)/\int_0^z Q(x)dx & \text{if } \ t > z \\ tq(t)/\int_0^z Q(x)dx & \text{if } \ t \le z \end{cases} \tag{16}$$

If $q(t)$ is negligible in the neighborhood of z, and $t \le z$, then $f(t|z) \simeq tq(t)/m$ which is the usual distribution for the sum if the forward and backward recurrence time random variables.

Using the same development, we can calculate $f(t|a(z|t_0)=0)$ which refers to the survival of individuals who died before t_0, but would have been age z at time t_0. Since

$$P\{a(z|t_0)=0, \quad t<T\le t+dt\} = \left[\int_0^{z-t} I(\tau)d\tau\right] q(t)dt, \; t\le z$$

and

$$P(a(z|t_0)=0) = \int_0^z \left[\int_0^{z-t} I(\tau)d\tau\right] q(t)dt$$

we have

$$f(t|a(z|t_0)=0) = \frac{\left[\int_0^{z-t} I(\tau)d\tau\right] q(t)}{P(a(z|t_0)=0)} \quad \text{if } t\le z \tag{17}$$

which is the distribution of those who died before time t_0, but would have been age z at t_0 if they had lived. If $I(z)=I$, the distribution is

$$f(t|a(z|t_0)=0) = \frac{(1-\frac{t}{z})q(t)}{\int_0^z(1-\frac{t}{z})q(t)dt} \quad \text{for } t\le z. \tag{18}$$

Note that if $z\to\infty$, then

$$f(t|a(z|t_0)=0) = q(t)$$

which is the population survival pdf.

3.4 Chronological Time Modeling

Suppose that the incidence is a function of chronological time rather than age. Also, in some cases, t_0 may be regarded as far removed from the origin as the disease process has been going on a long time. Then the equations for the forward and backward times may be modified by replacing z by t_0. Therefore, we have

$$\begin{aligned}
q_f(t|t_0) &= \int_0^{t_0} I(\tau)q(t_0-\tau+t)dt/P(t_0)\\
q_b(t|t_0) &= I(t_0-t)Q(t)/P(t_0)\\
f(t|t_0) &= \begin{cases} q(t)\int_0^{t_0} I(\tau)d\tau & \text{if } t>t_0\\ q(t)\int_{t_0-t}^{t_0} I(\tau)d\tau & \text{if } t\le t_0\end{cases}\\
f(t|a(t_0)=0) &= q(t)\int_0^{t_0-t} I(\tau)d\tau/P\{a(t_0)=0\} \quad \text{for } t\le t_0
\end{aligned} \tag{19}$$

with $P(t_0)=P\{a(t_0)=1\}=\int_0^{t_0} I(\tau)Q(t_0-\tau)d\tau$.

If $I(\tau)=I$ then

$$\begin{aligned} q_f(t|t_0) &= [Q(t) - Q(t+t_0)] / \int_0^{t_0} Q(y)dy \\ q_b(t|t_0) &= Q(t) / \int_0^{t_0} Q(y)dy, \quad 0 < t \leq t_0 \\ f(t|t_0) &= \begin{cases} t_0 q(t) / \int_0^{t_0} Q(y)dy & \text{if } t > t_0 \\ t q(t) / \int_0^{t_0} Q(y)dy & \text{if } t \leq t_0 \end{cases} \\ f(t|a(t_0)=0) &= q(t)(1-\frac{t}{t_0}) / \int_0^{t_0} (1-\frac{y}{t_0}) q(y)dy, \quad t \leq t_0. \end{aligned} \tag{20}$$

Consequently, if $t_0 \to \infty$

$$\begin{aligned} \lim_{t_0 \to \infty} q_f(t|t_0) &= \lim_{t_0 \to \infty} q_b(t|t_0) = Q(t)/m \\ \lim_{t_0 \to \infty} f(t|t_0) &= t q(t)/m \\ \lim_{t_0 \to \infty} f(t|a(t_0)=0) &= q(t). \end{aligned} \tag{21}$$

4 Early Detection Disease Model

In this section, the results of the chronic disease model will be adapted to the early detection model. The states for this model are S_0, S_p and S_c having the natural history $S_0 \to S_p \to S_c$. Define the probability of the transition $S_0 \to S_p$ during $(\tau, \tau + d\tau)$ by $w(\tau)d\tau$ where τ refers to age. The point incidence refers to the transition $S_p \to S_c$. The relation between $w(z)$ and $I(z)$ is

$$I(z) = \int_0^z w(\tau) q(z-\tau) d\tau \tag{22}$$

where $q(t)$ is the pdf of the sojourn time in the pre-clinical state. Assume that at time t_0 the disease is detected early and the age at which the disease is detected is z. Then the expression for the forward and backward recurrence time distribution are

$$\begin{aligned} P(z)Q_f(t|z) &= \int_0^z w(\tau) Q(z-\tau+t) d\tau \\ P(z)Q_b(t|z) &= \int_0^{z-t} w(\tau) Q(z-\tau) d\tau, 0 < t \leq z \end{aligned} \tag{23}$$

where $P(z) = \int_0^z w(\tau) Q(z-\tau) d\tau$.

Equation (23) enables the calculation of the pdf's of the forward and backward recurrence time distribution. Note that aside from definitions the only change is the substitution of $w(\tau)$ for $I(\tau)$ in (4) and (6).

Similarly, the distribution of the time in the pre-clinical state $T = T_b + T_f$ is

$$f(t|z) = \begin{cases} q(t) \int_0^z w(\tau)d\tau / P(z) & \text{if } t > z \\ q(t) \int_{z-t}^z w(\tau)d\tau / P(z) & \text{if } t \leq z \end{cases} \tag{24}$$

If $w(\tau) = w$,

$$\begin{aligned} Q_f(t|z) &= \int_t^{z+t} Q(y) / \int_0^z Q(y) \\ Q_b(t|z) &= \int_t^z Q(y)dy / \int_0^z Q(y)dy, \quad 0 < t \leq z \\ f(t|z) &= \begin{cases} zq(t) / \int_0^z Q(y)dy & \text{if } t > z \\ tq(t) / \int_0^z Q(y)dy & \text{if } t \leq z \end{cases} \end{aligned} \tag{25}$$

and as $z \to \infty$, the usual results for stationary processes hold.

5 Discussion

Backward and forward recurrence time distributions play an important role in modeling human disease. Two examples are presented which make use of recurrence times; i.e. modeling both the progressive chronic disease and the early disease detection processes. In these examples, the process is influenced by disease incidence which is usually age related. As a result, the recurrence time distributions are generalized to be functions of disease incidence. A characteristic of the progressive chronic disease model is that the process may have been going on for a long time relative to when the process is beginning to be studied at time t_0. As a result, it may be convenient to consider t_0 to be far removed from the origin.

A closely related topic is the role of disease incidence in length biased sampling. We have generalized the length biased process to take account of incidence. In our general development, we have also derived the distribution of the "complement" of length biased sampling which may be called "anti-length biased sampling." Our setting is that at time t_0 the process is being observed and individuals who are alive with disease are characterized by being length biased and tend to have longer time in the state in which they are observed at this time. However, individuals who entered and exited the state before time t_0 will tend to have a shorter stay in that particular state. In the progressive disease modeling example, those who died of disease before time t_0 will tend to have shorter survival times than those in the disease state at t_0.

One can envision a study in which observations are taken within the chronological time period (t_0, t_1). We have considered the situation where t_0 is the beginning of a study. However, if t_1 represents the time point at which the study stops, observations may be right censored at that time. The conventional approach is to have a model which incorporates a censoring distribution and the observation is the minimum of the censoring distribution and the distribution in that state. However, another formulation is that the time in the state up to t_1 is a backward recurrence time. Consequently, the likelihood function will be different than the usual formulation for right censored observations.

More generally, the results presented here are applicable when a stochastic process has discrete (or countable) states and is initially being observed. The sojourn times for those observations in the initially observed states are forward recurrence times. Consequently, any modeling of the process using the initial observations must incorporate these forward recurrence times.

References

[CM65] Cox, D.R. and Miller, H.D. : The theory of stochastic processes. Chapman, London (1965)

[FEL71] Feller, W. : An introduction to probability theory and its application. **Volume II, 3rd edition**. Wiley, New York (1971)

[ZF69] Zelen, M. and Feinleib, M. : On the theory of screening for chronic diseases. Biometrika , **56**, 601–14 (1969)

This article has appeared in the December 2004 issue of Lifetime Data Analysis

Difference between Male and Female Cancer Incidence Rates: How Can It Be Explained?

Konstantin G. Arbeev[1], Svetlana V. Ukraintseva[2], Lyubov S. Arbeeva[3], and Anatoli I. Yashin[4]

[1] Center for Demographic Studies, Duke University, 2117 Campus Drive, Box 90408, Durham, NC 27708-0408, USA arbeev@cds.duke.edu
[2] Center for Demographic Studies, Duke University, 2117 Campus Drive, Box 90408, Durham, NC 27708-0408, USA ukraintseva@cds.duke.edu
[3] Ulyanovsk State University, Leo Tolstoi St. 42, 432700 Ulyanovsk, Russia arbeev@mail.ru
[4] Center for Demographic Studies, Duke University, 2117 Campus Drive, Box 90408, Durham, NC 27708-0408, USA yashin@cds.duke.edu

Summary. Age patterns of male and female cancer incidence rate do not look similar. This is because of the biologically based difference in susceptibility to cancer of different sites. This argument, however, does not clarify how age patterns of male and female cancer incidence rate must look like. The analysis of epidemiological data on cancer in different countries and in different years shows that male and female cancer incidence rates intersect around the age of female climacteric. We explain the observed pattern using the difference in ontogenetic components of aging between males and females. The explanation requires a new model of carcinogenesis, which takes this difference into account. Application to data on cancer incidence in Japan (Miyagi prefecture) illustrates the model.

Key words: cancer , model, incidence , ontogenesis

1 Introduction

The analysis of epidemiological data on cancer in different countries and at different time periods reveals common age patterns and universal time-trends in cancer incidence rates. Some of these features have been observed and discussed before. These include an increase of cancer incidence rate over time, both for males and females, and an increase, a leveling-off, and then a decline of the age pattern of this rate. Note that a consensus among cancer epidemiologists has not been arrived at concerning the explanation of these phenomena.

The new interesting feature is related to the joint pattern of cancer incidence rate for males and females. The data show strange regularity in relative behavior of male and female cancer incidence rates. In all countries and time periods, these curves intersect at the interval of ages near female climacteric. For all investigated countries at different time periods, the total cancer incidence rates for females are higher than those for males up to middle age (near the age of female climacteric). After that the incidence rates for females become lower than for males. The growth of incidence rates over age is much more rapid for males than for females. In the latter case the growth is nearly linear (Fig. 1). Similar effects are also observed in cohort data (Fig. 2).

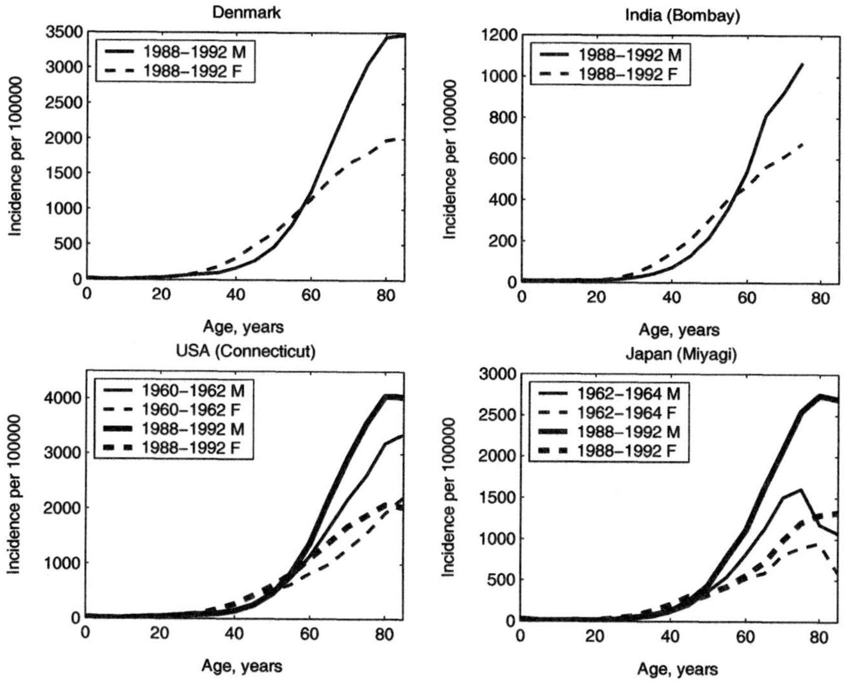

Fig. 1. Typical patterns of intersection between male and female overall cancer incidence rates: Denmark (1988–1992), India (Bombay, 1988–1992), USA (Connecticut, 1960–1962 and 1988–1992), and Japan (Miyagi prefecture, 1962–1964 and 1988–1992). 'M' – males, 'F' – females; data source: [3]–[9].

Common sense suggests that male and female age patterns of overall cancer incidence rate must differ because of biologically-based differences in specific cancer sites (such as breast, ovarian cancers for females and prostate cancer for males). However, the differing age-pattern, and its relative stability over time and place, cannot be predicted from such a consideration.

The differences in mechanisms involved in cancer initiation and development for males and females would be better understood if one could explain forces shaping the age-trajectories of cancer incidence rates, evaluate the role

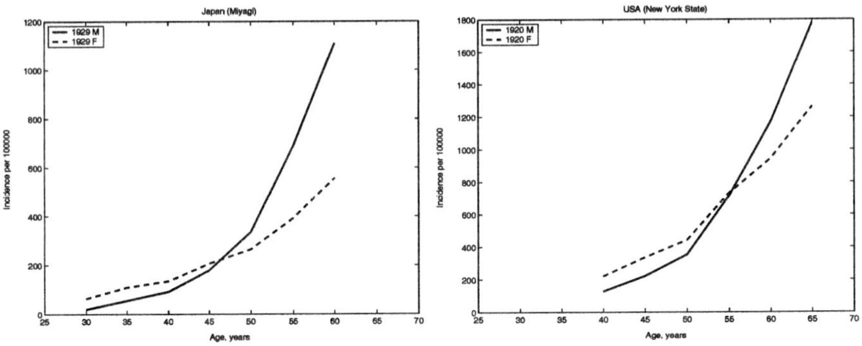

Fig. 2. Female and male "cohort" cancer incidence rates in Japan (Miyagi prefecture), 1929 "cohort" and in USA (New York State), 1920 "cohort". Data source: [3]–[9].

of gender in this process, as well as factors responsible for observed time-trends of these rates. Below, we describe the approach, which has the capability to explain the relative stability of the age pattern of cancer incidence and mortality rates for males and females, as well as their change over time. The approach explores the possibility to represent cancer incidence rate in terms of age-related processes. This involves a new mathematical model of carcinogenesis . This model represents cancer incidence rate as a sum of two components reflecting basic types of age-related changes in an organism (see [15]). We show that in contrast to traditional models of carcinogenesis, the new model, which we call the *ontogenetic* model, captures main features of the age pattern and time-trend of cancer incidence rates. It also explains the relative stability of the intersection pattern of male and female cancer incidence rates. We illustrate this model by the application to data on overall cancer incidence rates in Japan (Miyagi prefecture) (data source: [3]–[9]).

2 Data

We apply our model to data on female and male cancer incidence rates in Japan (Miyagi prefecture). The International Agency for Research on Cancer (IARC) provides the data on cancer incidence in different countries, in seven volumes ([3]–[9]). Each volume covers a time period of several years (usually 3–5 years) for each country (or province and/or ethnic group). The periods vary for different countries. In each volume, female and male average annual cancer incidence per 100000 over the corresponding time period are given for the specific country (province and/or ethnic group), in five-year age groups up to age 85+ (for some countries the first group 0–4 is separated into 0 and 1–4). The data are given for separate sites and for all sites combined. Not all countries are presented in each volume. The longest time series are available for Japan (Miyagi prefecture). Each of the seven volumes contains the data on cancer incidence in this territory over different time periods. This data set is the foremost one to analyze time trends in cancer incidence rates over time, and is used in this study.

3 Three Components of the Individual Aging Process

Ukraintseva and Yashin [15] suggested studying individual aging by analyzing three internal biological processes that have different age-related dynamics. These include *basal*, *ontogenetic*, and *exposure-related* components. These processes also affect the shape of cancer incidence rate. We assume that any observed age pattern of this rate is the result of the combined influence of these three age-related processes.

The main characteristic of the *basal* component is the age-related decline in the individual rate of living (i.e., in the metabolic and information processing rates). This component is responsible for the deceleration of change in many physiological parameters of an organism with advanced age. It can be responsible for the leveling-off of the morbidity rate at old ages, observed for many chronic diseases (see [15]). This component may also contribute towards the acceleration in rates of onset of *acute* health disorders leading to death (due to deceleration in the potency to recover, and hence due to the progressive decline in individual stress resistance at old ages).

The term *ontogenetic* refers to the developmental history of an organism. The *ontogenetic component* of aging represents effects of metabolic switches accompanying changes in stages of ontogenesis during life (e.g., in infancy, in the reproductive period and at the climacteric). This component of individual aging can be responsible for non-monotonic change in vulnerability of an organism to stress and diseases due to a variation in hormonal balance in an organism. The *exposure-related component* is responsible for long-term accumulation of specific lesions in an organism, which contribute to an increase in the morbidity rate.

A properly balanced combination of all these components may be used for an explanation of age-specific morbidity and mortality patterns in human populations, including cancer morbidity. The obvious advantage of such an approach is that by dividing individual aging into the processes with different age-related dynamics, one has an opportunity to use information from different studies focused on specific aspects of individual aging. For example, the age pattern of ontogenetic vulnerability used in the respective component of cancer incidence rate in our study was obtained from asthma studies (see [12]). A similar pattern is also produced in the studies of other chronic diseases, as shown in [11]. The limitations of this approach are associated with the large amounts of data required for identification of model parameters.

4 The Incorporated Ontogenetic Model of Cancer

To capture the age pattern, time-trends, as well as the intersection of age-specific incidence rates for males and females, we incorporate the three-component model of individual aging [15] into the tumor latency model of carcinogenesis [18]. We specify patterns of age-dependence for different components in the oldest cohort, and set a rule of changing these components from one cohort to the next to construct the corresponding period rates. Following this idea we define cancer incidence rates as

$$\mu_i(x) = \int_0^x h_i(x-t)\,dF(t), \tag{1}$$

where $i = 1 \dots n$ stands for a cohort, $h_i(x)$ is an age-specific intensity of unrepaired lesion formation in i^{th} cohort, and $F(t)$ is a cumulative probability distribution function of progression times. We suppose that progression times are gamma distributed with fixed shape and scale parameters k and λ and the functions $F(t)$ are the same for all cohorts.

We also assume that the age-specific intensity of unrepaired lesion formation $h_i(x)$ is a result of the combined influence of age-related processes in an organism which are represented by the *basal*, *ontogenetic* and *exposure-related* components described above.

The part of the hazard rate, associated with the *basal* component, should be increasing with the declining rate with age. Respectively, the part of the hazard rate, associated with the *exposure-related* component, should exhibit accelerated increase with age by definition of this component. For the sake of simplicity, we combine the *exposure-related* and *basal* effects and specify one general pattern of hazard rate for these components (referred to as *time-component*). We denote this general component $h_i^{\text{time}}(x)$, where index i is associated with the birth year of the cohort, and x is an individual's age. Thus, the exposure-related lesions in an organism accumulate with age, on the grounds of a basal deceleration in the individual rate of living (e.g., due to general deceleration in information processing) in an organism.

The *ontogenetic* component has a wave-like shape for both males and females, with peaks at early ages and around ages of climacterics for females, and between ages 55 and 65 for males. The peaks correspond to the ages of hormonal imbalance where this component largely influences risks of morbidity and mortality. A similar pattern of morbidity is observed for many human chronic diseases (see [11], [12], [13], [14]). In principle, one can use these patterns to model the *ontogenetic* component. However, these rates, in essence, reflect not only the ontogenetic changes, but also the other factors responsible for the manifestation of the disease. Thus, to model the ontogenetic changes at advanced ages influencing unrepaired lesion formation, we use the function with a pronounced peak around some specific age, and zero otherwise. The peak is around the age of menopause for females, and the pattern is shifted to the right for males (see Fig. 3). We ignore the peak at early ages for the sake of simplicity. This component is the same for all cohorts.

Denote $h^{\text{ont}}(x)$ the value of the *ontogenetic* component of hazard rate at age x and let $h_0^{\text{time}}(x)$ be the value of the general component (combined from *exposure-related* and *basal* effects) at age x for the oldest cohort. We suppose that the last component may change for different cohorts due to an increasing influence of harmful factors on an organism. The dynamics of this component for i^{th} cohort, $i = 1 \ldots n$ is described as

$$h_i^{\text{time}}(x) = (1 + i\ d)\, h_0^{\text{time}}(x), \tag{2}$$

where parameter d characterizes the growth rate of the hazard rate over time. The introduced values $h^{\text{ont}}(x)$ and $h_i^{\text{time}}(x)$ are used to define the age-specific intensity of unrepaired lesion formation for i^{th} cohort, $i = 1 \ldots n$, as a sum of these two components,

$$h_i(x) = h^{\text{ont}}(x) + h_i^{\text{time}}(x). \tag{3}$$

5 Application of the Ontogenetic Model to Data on Cancer Incidence Rate by Sex

We apply the model to data on cancer incidence in Japan (Miyagi prefecture) (data source: [3]–[9]). The parameters of the model are fixed at $d = 0.2$, $k = 25$, and $\lambda = 1$. The patterns of the *ontogenetic* component ($h^{\text{ont}}(x)$) and the *time-dependent* component in the oldest cohort ($h_0^{\text{time}}(x)$), for both males and females, are shown in Fig. 3. The trajectories of $h_0^{\text{time}}(x)$ were assumed piecewise constant and were estimated using Matlab's least-square routine.

The observed and estimated male and female incidence rates are shown in Fig. 3.2. Table 1 illustrates the fit of the model. Note that for the sake of simplicity, we used a rather straightforward pattern of $h_i^{\text{time}}(x)$ and a number of fixed parameters of the model. A more elaborated specification of $h_i^{\text{time}}(x)$ and an estimation of all the parameters would likely provide a better fit to the data. However, the message here is that this model captures all the features of the observed cancer incidence rates mentioned above. It describes an increase of the rates over time, the deceleration and decline of the rates at the oldest old ages, and the intersection of male and female incidence rate curves near the age of female climacteric. Increasing $h_i^{\text{time}}(x)$ in cohorts gives an increase of the period incidence rates over time. The specification of a cumulative probability distribution function of progression times and difference in ontogenetic component for males and females produces a decline of the rates at oldest old ages and the intersection of the male and female rates.

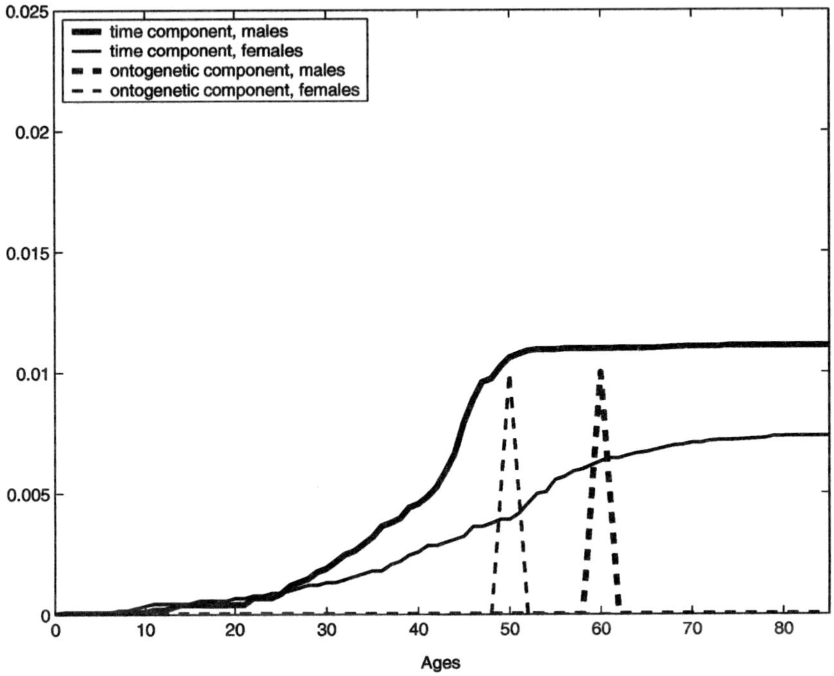

Fig. 3. The ontogenetic model of cancer applied to data on overall cancer incidence rates in Japan (Miyagi prefecture): curves of the combined exposure-related and basal component ("time component") in the oldest cohort and the ontogenetic component for males and females. Data source: [3]–[9].

Table 1. The ontogenetic model of cancer applied to data on overall female and male cancer incidence rates in Japan (Miyagi prefecture): norm of differences (columns 'Norm') and correlation (columns 'Corr') between modeled and observed incidence rates. Data source: [3]–[9].

Period	Norm (Females)	Corr (Females)	Norm (Males)	Corr (Males)
1959–1960	436.105	0.972	384.487	0.990
1962–1964	285.894	0.982	614.039	0.973
1968–1971	211.032	0.993	198.371	0.998
1973–1977	200.417	0.995	502.704	0.994
1978–1981	233.626	0.998	452.061	0.999
1983–1987	94.588	0.999	196.311	0.999
1988–1992	165.258	0.999	201.993	0.999

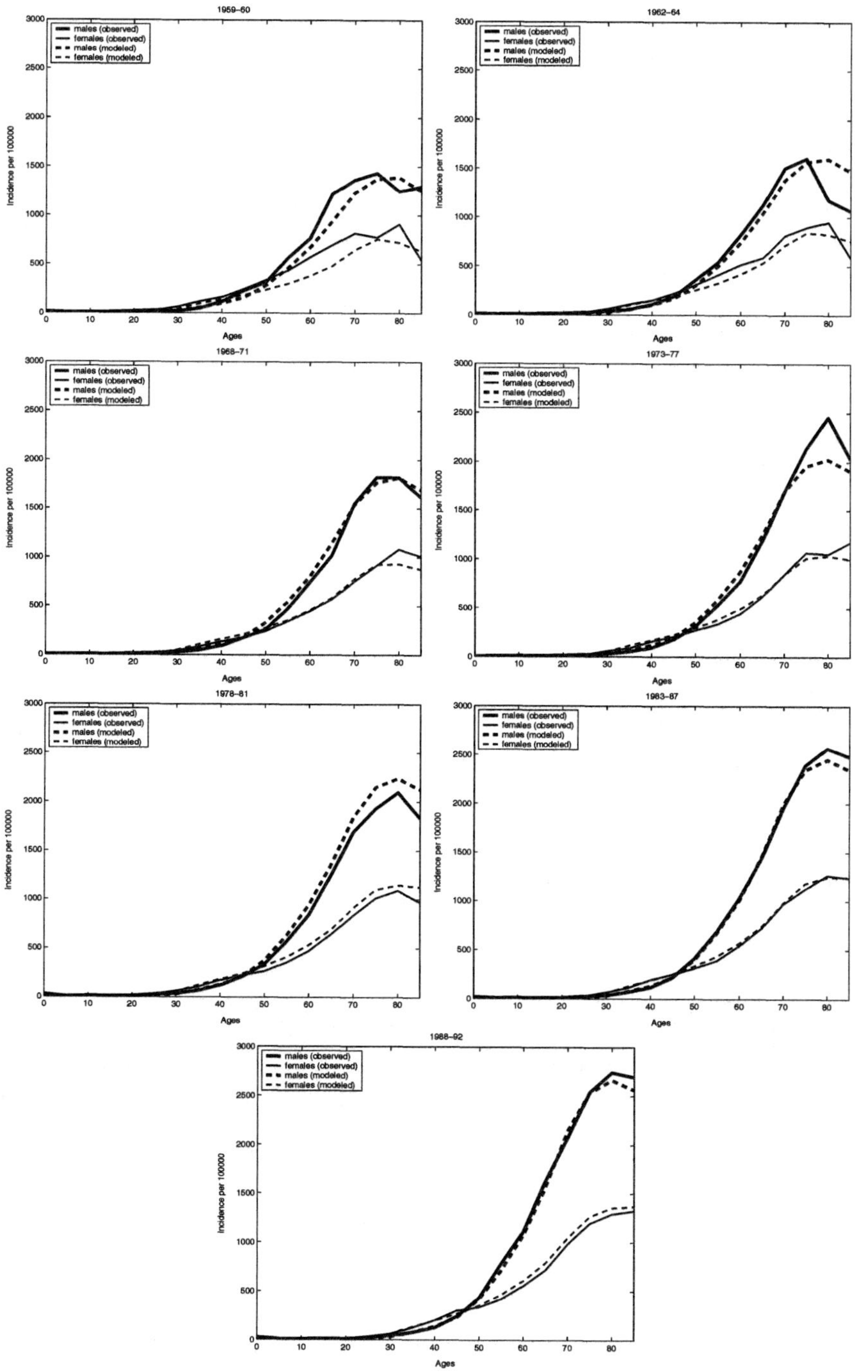

Fig. 4. The ontogenetic model of cancer applied to data on overall cancer incidence rates in Japan (Miyagi prefecture): male and female observed and modeled rates for different time periods. Data source: [3]–[9].

6 Conclusion

The analysis of epidemiological data on cancer shows that cancer became the leading cause of death in most productive ages of human life. The range of ages where cancer maintains its leading role tends to increase with years. In many developed countries the overall cancer incidence rate still tends to increase. Many factors associated with the economic progress could be mainly responsible for the increase in cancer incidence rate. Among those are the improved cancer diagnostics, elevated exposure to external carcinogens such as car exhaust pollution, and factors associated with a Western-like life style (such as dietary habits, new medicines and home-use chemicals). This increase is not likely to be explained by the improvement in cancer diagnostics alone. The survival of cancer patients differs in different countries, despite continuing efforts in sharing medical information on efficiency of cancer treatment procedures and respective drugs.

Different models of carcinogenesis can explain some of the observed phenomena of human cancer incidence rates. The literature on cancer modeling is extensive. The list of classical models includes the multistage model of cancer by Armitage-Doll (AD model), the two-event model by Moolgavkar-Venzon-Knudson (MVK model), and the tumor latency model by Yakovlev and Tsodikov. These models describe biological mechanisms involved in cancer initiation and development, and derive mathematical representation for cancer incidence rate. This representation can then be used in the statistical estimation procedures to test hypotheses about regularities of respective mechanisms and the validity of basic assumptions. The multi-stage model of carcinogenesis [2] explains the increase of the rates over age, but does not describe the entire age-trajectory of cancer incidence rate and does not explain the intersection of male and female incidence rates. The two-mutation model [10], as well as the tumor latency model (see [16], [17]), is capable of describing the entire age-trajectory of cancer incidence rate. However, they cannot explain the stable intersection pattern of male and female cancer incidence rates.

It is clear that the overall cancer incidence rates for males and females do have different age patterns. This conclusion stems from the basic biological knowledge about the difference between male and female organisms. This difference is responsible for the different susceptibility to cancer of certain sites (e.g., breast cancer). The exposure to hazardous materials can also be different for males and females because of their difference in social and economic life. There is, however, neither a theory nor a mathematical model that predicts how age-trajectories of cancer incidence rates will behave, and to what extent these trajectories are affected by environmental and living conditions experienced by populations in different countries.

In this paper we show that the relative difference in age patterns of male and female cancer incidence rates may be explained by the difference in ontogenetic curves of age-dependent susceptibility to cancer for males and females.

This is because the peak of hormonal imbalance in females is between ages 45 and 55, when the reproductive system ultimately stops functioning. In males this peak is shifted to the right (between 55 and 65). The age pattern of cancer incidence rate reflects the contribution of the ontogenetic component of age-related processes in an organism. The heterogeneity in individual frailty may also have a substantial contribution. The ontogenetic model is capable of describing the time trends and the stable pattern of intersection in the male and female incidence rates. In our recent paper [1], we pointed out that the universal pattern of male/female cancer incidence rates might also be a result of different strategies of resource allocation between "fighting" against external stresses and "fighting" against physiological aging used by male and female organisms. This effect needs further explanation, from both biological and mathematical perspectives. The availability of molecular-biological and epidemiological data on stress resistance (e.g., cellular sensitivity to oxidative stress) would allow for the development of more sophisticated mathematical models of such mechanisms. New models are also needed to explain age pattern and time-trends in male/female cancer mortality rates. These models should include information on cancer incidence rates as well as on survival of cancer patients.

Acknowledgements

The authors wish to thank Prof. James W. Vaupel for the opportunity to complete this work at the Max Planck Institute for Demographic Research, Germany.

References

[1] Arbeev, K.G., Ukraintseva, S.V., Arbeeva, L.S., Yashin, A.I.: Mathematical Models for Human Cancer Incidence Rates. Demographic Research, **12**, in press, (2005)

[2] Armitage, P., Doll, R.: The age distribution of cancer and a multistage theory of carcinogenesis. Br. J. Cancer, **8**, 1–12 (1954)

[3] IARC: Cancer Incidence in Five Continents. Volume I. International Agency for Research on Cancer, Lyon (1965)

[4] IARC: Cancer Incidence in Five Continents. Volume II. International Agency for Research on Cancer, Lyon (1970)

[5] IARC: Cancer Incidence in Five Continents. Volume III. IARC Sci Publ, **15** (1976)

[6] IARC: Cancer Incidence in Five Continents. Volume IV. IARC Sci Publ, **42** (1982)

[7] IARC: Cancer Incidence in Five Continents. Volume V. IARC Sci Publ, **88** (1987)

[8] IARC: Cancer Incidence in Five Continents. Volume VI. IARC Sci Publ, **120** (1992)
[9] IARC: Cancer Incidence in Five Continents. Volume VII. IARC Sci Publ, **143** (1997)
[10] Moolgavkar, S.H., Luebeck, E.G.: Two-event model for carcinogenesis: biological, mathematical and statistical considerations. Risk Anal., **10**, 323–341 (1990)
[11] Sankaranarayanan, K., Chakraborty, R., Boerwinkle, E.: Ionizing radiation and genetic risks – VI. Chronic multifactorial diseases: a review of epidemiological and genetical aspects of coronary heart disease, essential hypertension and diabetes mellitus. Mutat. Res., **436**, 21–57 (1999)
[12] Ukraintseva, S.V.: Genetic-Epidemiological Analysis of Predisposition to Asthma. Ph.D. Thesis, Research Center for Medical Genetics, Russian Academy of Medical Sciences, Moscow (1998)
[13] Ukraintseva, S.V.: On the role of age in asthma morbidity. Clinical Gerontology, **6**, 29–33 (2000)
[14] Ukraintseva, S.V., Sergeev, A.: Analysis of genetic heterogeneity of bronchial asthma as related to the age of onset. Russian Journal of Genetics, **36**, 201–205 (2000)
[15] Ukraintseva, S.V., Yashin, A.I.: How individual aging may influence human morbidity and mortality patterns. Mech. Aging and Dev., **122**, 1447–1460 (2001)
[16] Yakovlev, A.Yu., Asselain, B., Bardou, V.–J., Fourquet, A., Hoang, T., Rochefodiere, A., Tsodikov, A.D.: A simple stochastic model of tumor recurrence and its application to data on pre-menopausal breast cancer. In: Asselain, B., Boniface, M., Duby, C., Lopez, C., Masson, J.P., Tranchefort, J., (ed) Biometrie et Analyse de Donnees Spatio-Temporelles, **12**. Societe Francaise de Biometrie. ENSA, Rennes (1993)
[17] Yakovlev, A.Yu., Tsodikov, A.D., Bass, L.: A stochastic model of hormesis. Math. Biosci., **116**, 197–219 (1993)
[18] Yakovlev, A.Yu., Tsodikov, A.D.: Stochastic Models of Tumor Latency and their Bio-statistical Applications. World Scientific, Singapore (1996)

Non-parametric estimation in degradation-renewal-failure models

V. Bagdonavičius[1], A. Bikelis[1], V. Kazakevičius[1] and M. Nikulin[2]

[1] Vilnius University, Naugarduko 24, Vilnius, Lithuania
`Vilijandasbag@techas.lt` `marius@post.omnitel.net`
`Vytautas.kazakevicius.maf.vu.lt`
[2] Bordeaux Victor Segalen University, France `nikou@sm.u-bordeaux2.fr`

1 Introduction

Classical reliability theory and survival analysis parts give methods for analysis of failure time data.

An important part of modern reliability theory and survival analysis is modelling and statistical analysis of ageing , wearing, damage accumulation, degradation processes of technical units or systems, living organisms ([ME98], [WK04], [BN02]).

Lately, methods for simultaneous degradation-failure time data analysis are being developed ([BN01], [WT97], [BBK04]).

Some degradation processes are not non-reversible and degradation processes may be renewed. For example, the degradation of pancreas and thyroid can be defined by the quantities of secreted insulin and thyroidal hormone, respectively. By injection of insulin the degradation process of pancreas is (indirectly) renewed. By injection of thyroxine (case of hyperthyroid) or carbimazole (case of hypothyroid) the degradation process of thyroid is (indirectly) renewed. If the value of hormone (for example, insulin) approaches a critical value, the risk of failure increases quickly.

In reliability, a simple example is a tire with renewable protector. The risk of failure depends on the level of protector wear .

We consider relatively simple linear (or loglinear) degradation process. On the other side, we consider rather complicated situation of non-parametric estimation when units are renewable and the failure intensities depend on degradation level of the unit.

We consider nonparametric estimation of degradation and failure process characteristics using degradation and failure time data with renewals. See also the paper [L04] who considers parametric estimation in similar context.

2 Model

For $j \geq 1$, let S_j denote the moment of the jth renewal (we assume $S_1 = 0$) and A_j be the inverse to the degradation rate in the interval $(S_j; S_{j+1}]$.

Assume that the random variables $A_1, A_2, \ldots$ are independent and identically distributed according to some cumulative distribution function π (or only independent with the cumulative distribution functions $\pi_1, \pi_2, \ldots$).

Denote by $Z(t)$ the value of the degradation process at the moment t.

Degradation process model:

$$Z(t) = (t - S_j)/A_j \quad \text{for } S_j < t \leq S_{j+1}, \tag{1}$$

where $S_{j+1} = \sum_{i=1}^{j} A_i z_0$.

Let T denote the moment of a traumatic failure.

Failure model:

$$P(T > t \mid Z(s), 0 \leq s \leq t) = \exp\left\{ - \int_0^t \lambda\big(Z(s)\big) ds \right\}, \tag{2}$$

λ being a positive function.

Denote

$$m(t) = j, \quad \text{if} \quad t \in (S_j; S_{j+1}] \quad (j \geq 1), \quad m = m(T). \tag{3}$$

The failure occurs in the interval $(S_m, S_{m+1}]$.

The data (for one unit) can be defined as the following vector of a random length:

$$\big(S_1, \ldots, S_m, T, Z(T)\big).$$

Remark 1. The conditional distribution of T (with respect to the σ-algebra $\mathcal{A}$ generated by the random variables $A_1, A_2, \ldots$) can be defined in another way, which is more convenient for computer simulations.

Firstly, define recursively conditionally independent random variables ΔT_1, $\Delta T_2, \ldots$ such that

$$P_{\mathcal{A}}(\Delta T_j > t) = e^{-\int_0^t \lambda(s/A_j) ds}$$

(here $P_{\mathcal{A}}$ denotes the conditional probability with respect to $\mathcal{A}$). Secondly, set

$$\tilde{T} = \begin{cases} S_1 + \Delta T_1, & \text{if } \Delta T_1 \leq A_1 z_0; \\ S_2 + \Delta T_2, & \text{if } \Delta T_1 > A_1 z_0,\ \Delta T_2 \leq A_2 z_0; \end{cases}$$

Then conditional distribution of $\tilde{T}$ coincides with that of T.

Indeed, if $t \in (S_j, S_{j+1}]$ then

$$P_{\mathcal{A}}\{\tilde{T} > t\} = P_{\mathcal{A}}\{\Delta T_1 > A_1 z_0, \ldots, \Delta T_{j-1} > A_{j-1} z_0, \Delta T_j > t - S_j\}$$

$$= P_{\mathcal{A}}\{\Delta T_1 > A_1 z_0\} \cdots P_{\mathcal{A}}\{\Delta T_{j-1} > A_{j-1} z_0\} P_{\mathcal{A}}\{\Delta T_j > t - S_j\}$$

$$= e^{-\int_0^{A_1 z_0} \lambda(s/A_1)ds} \dots e^{-\int_0^{A_{j-1} z_0} \lambda(s/A_{j-1})ds} e^{-\int_0^{t-S_j} \lambda(s/A_j)ds}$$

$$= e^{-\int_{S_1}^{S_2} \lambda((s-S_1)/A_1)ds} \dots e^{-\int_{S_{j-1}}^{S_j} \lambda((s-S_{j-1})/A_{j-1})ds} e^{-\int_{S_j}^{t} \lambda((s-S_j)/A_j)ds}$$

$$= e^{-\int_0^t \tilde{\lambda}(s)ds}.$$

Set

$$\Lambda(z) = \int_0^z \lambda(y)dy.$$

We suppose that the distribution function π and the cumulative intensity function Λ are completely unknown.

We are interested in the probability

$$p_j(z) = P(T > S_j + zA_j \mid T > S_j)$$

to attain the level of degradation z $(0 \le z \le z_0)$ before a failure occurs given that an unit had been renewed $j-1$ times $(j = 1, 2, \dots)$.

The considered model implies that

$$p_j(z) = \int_0^\infty \exp\{-a\Lambda(z)\}\, d\pi_j(a).$$

3 Decomposition of a counting process associated with $Z(T)$

For $z \in [0; z_0)$ set

$$N(z) = 1_{\{Z(T) \le z\}} \tag{4}$$

and let $\mathcal{F}_z$ denote the σ-algebra generated by the following collections of events:

$$\{A_1 \le a_1, \dots, A_j \le a_j\} \cap \{m = j\} \tag{5}$$

and

$$\{A_1 \le a_1, \dots, A_j \le a_j\} \cap \{m = j\} \cap \{Z(T) \le y\}; \tag{6}$$

here $j \ge 1$, $a_1, \dots, a_j > 0$ and $y \le z$.

Theorem 1. *The process $N(z)$ can be written as the sum*

$$N(z) = \int_0^z Y(y)d\Lambda(y) + M(z), \tag{7}$$

where $M(z)$ is a martingale with respect to the filtration $(\mathcal{F}_z \mid 0 \le z < z_0)$ and

$$Y(y) = \frac{A_m 1_{\{Z(T) \ge y\}}}{1 - e^{-A_m(\Lambda(z_0) - \Lambda(y))}}. \tag{8}$$

Proof. Fix $y < z$, $j \geq 1$ and denote by $X = 1_{\{A_1 \leq a_1, \ldots, A_j \leq a_j\}}$. Then

$$E\Big[X 1_{\{m=j\}}\big(N(z) - N(y)\big)\Big] = E\big[X 1_{\{S_j + A_j y < T \leq S_j + A_j z\}}\big]$$

$$= E\Big[X\big(e^{-\int_0^{S_j+A_j z} \lambda(Z(s))ds} - e^{-\int_0^{S_j+A_j y} \lambda(Z(s))ds}\big)\Big] = E\Big[X \int_y^z e^{-\int_0^{S_j+A_j x} \lambda(Z(s))ds} A_j\Big] d\Lambda(x)$$

$$= \int_y^z E\Big[X e^{-\int_0^{S_j+A_j x} \lambda(Z(s))ds} A_j d\Lambda(x)\Big] = \int_y^z E\Big[X A_j \frac{1_{\{S_j+A_j x < T \leq S_{j+1}\}}}{1 - e^{-\int_{S_j+A_j x}^{S_{j+1}} \lambda(Z(s))ds}}\Big] d\Lambda(x)$$

$$= \int_y^z E\Big[X 1_{\{m=j, Z(T) \geq x\}} \frac{A_j}{1 - e^{-A_j(\Lambda(z_0) - \Lambda(x))}}\Big] d\Lambda(x) = E\Big[X 1_{\{m=j\}} \int_y^z Y(x) d\Lambda(x)\Big].$$

Moreover, for each $y' \leq y$,

$$E\Big[X 1_{\{m=j, Z(T) \leq y'\}}\big(N(z) - \tilde{N}(y)\big)\Big] = 0$$

and

$$E\Big[X 1_{\{m=j, Z(T) \leq y'\}} \int_y^z Y(x) d\Lambda(x)\Big] = 0$$

(because $1_{\{Z(T) \leq y'\}} Y(x) = 0$ for all $x > y$). Since the union of collections (5) and (6) is closed with respect to finite intersections of events, obtained equalities mean, by the Monotone Class Theorem, that

$$E\Big[N(z) - \int_0^z Y(x) d\Lambda(x) \mid \mathcal{F}_y\Big] = N(y) - \int_0^y Y(x) d\Lambda(x).$$

Hence the process $M(z)$ is a martingale.

The proof is complete.

Note that we can not use the Nelson–Aalen estimator based on the obtained decomposition because the function $Y(y)$ depends on the values of Λ in the point $z_0 > y$. On the other hand, the decomposition is useful for demonstration of asymptotic properties of estimators.

Let us consider another decomposition of the process $N(z)$. Set

$$N^*(t) = 1_{\{T \leq t\}}, \quad Y^*(t) = 1_{\{T \geq t\}}.$$

Denote by $\mathcal{F}_t^*$ the σ-algebra generated by $N^*(s), Y^*(s), 0 \leq s \leq t$. Then

$$N^*(t) = \int_0^t \lambda(Z(u)) Y^*(u) du + M^*(u), \tag{9}$$

where $M^*(u)$ is a martingale with respect to the filtration $(\mathcal{F}_t^* \mid t \geq 0)$.

Set

$$Z = Z(T), \quad Z_j = \begin{cases} z_0, \text{if } j < m; \\ Z, \text{ if } j = m. \end{cases} \tag{10}$$

Theorem 2. *The process $N(z)$ can be written as the sum*

$$N(z) = \int_0^z Y^{**}(y) d\Lambda(y) + M^{**}(z), \tag{11}$$

where

$$Y^{**}(y) = \sum_{j=1}^{m} A_j 1_{\{Z_j \geq y\}}, \quad M^{**}(z) = \int_0^\infty 1_{\{Z(u) \leq z\}} dM^*(u). \tag{12}$$

Proof.

$$N(z) = 1_{\{Z(T) \leq z\}} = \int_0^\infty 1_{\{Z(u) \leq z\}} dN^*(u) = \int_0^\infty 1_{\{Z(u) \leq z\}} \lambda(Z(u)) Y^*(u) du$$

$$+ \int_0^\infty 1_{\{Z(u) \leq z\}} dM^*(u) = \int_0^T 1_{\{Z(u) \leq z\}} \sum_{j=1}^{\infty} \lambda\left(\frac{u - S_j}{A_j}\right) 1_{\{S_j < u \leq S_{j+1}\}} du + M^{**}$$

$$= \sum_{j=1}^{m-1} \int_{S_j}^{S_{j+1}} 1_{\left\{\frac{u-S_j}{A_j} \leq z\right\}} \lambda\left(\frac{u - S_j}{A_j}\right) du + \int_{S_m}^{T} 1_{\left\{\frac{u-S_m}{A_m} \leq z\right\}} \lambda\left(\frac{u - S_m}{A_m}\right) du + M^{**} =$$

$$\sum_{j=1}^{m-1} \int_0^{\frac{S_{j+1}-S_j}{A_j}} 1_{\{v \leq z\}} \lambda(v) A_j dv + \int_0^{\frac{T-S_m}{A_m}} 1_{\{v \leq z\}} \lambda(v) A_m dv + M^{**} =$$

$$\sum_{j=1}^{m} \int_0^{Z_j} 1_{\{v \leq z\}} A_j d\Lambda(v) + M^{**} = \sum_{j=1}^{m} A_j \int_0^z 1_{\{v \leq Z_j\}} d\Lambda(v) + M^{**}.$$

4 Estimation

4.1 The data

Suppose that n units are on test and, for ith unit, denote by S_{ij} $(j \geq 1)$ the moment of jth renewal, by A_{ij} the inverse to the degradation rate in the interval $(S_{ij}; S_{i,j+1}]$, by $Z_i(t)$ the degradation process and by T_i the moment of its failure.

Set $\Delta S_{ij} = S_{i,j+1} - S_{ij}$ and define m_i as in the case of one unit writing (for the ith unit) S_{ij} and T_i instead of S_j and T. Denote $Z_i = Z_i(T_i)$. Then the data can be defined as the following collection of vectors of a random length:

$$(S_{i1}, \ldots, S_{im_i}, T_i, Z_i), \quad i = 1, \ldots, n.$$

Define the processes $N_i(z)$, $Y_i(y)$ and $M_i(z)$ as in Theorem 1, with $Z(T)$, A_m, S_m replaced by $Z_i(T_i)$, S_{im_i}, A_{im_i}. Set

$$\bar{N}(z) = \sum_{i=1}^{n} N_i(z), \quad \bar{Y}(y) = \sum_{i=1}^{n} Y_i(y), \quad \bar{M}(z) = \sum_{i=1}^{n} M_i(z)$$

and let $\bar{\mathcal{F}}_z$ denote the σ-algebra, generated by the events of the form given just before Theorem 1 with A_j, m, $Z(T)$ replaced by A_{ij}, m_i, $Z_i(T_i)$ $(i = 1, \ldots, n)$. Then Theorem 1 implies that

$$\bar{N}(z) = \int_0^z \bar{Y}(y) d\Lambda(y) + \bar{M}(z), \tag{13}$$

and $\bar{M}(z)$ is a martingale with respect to the filtration $(\bar{\mathcal{F}}_z)$.

Theorem 2 implies another decomposition:

$$\bar{N}(z) = \int_0^z \tilde{Y}(y) d\Lambda(y) + \tilde{M}(z), \tag{14}$$

where

$$\tilde{Y}(y) = \sum_{i=1}^{n} \sum_{j=1}^{m_i} A_{ij} 1_{\{Z_{ij} \geq y\}},$$

$$\tilde{M}(z) = \sum_{i=1}^{n} \int_0^{\infty} 1_{\{Z_i(u) \leq z\}} dM_i^*(u);$$

here M_i^* is a martingale with respect to the filtration $(\mathcal{F}_{it}^* \mid t \geq 0)$, where

$$\mathcal{F}_{it}^* = \sigma(N_i^*(s), Y_i^*(s), 0 \leq s \leq t), \quad N_i^*(t) = 1_{\{T_i \leq t\}}, Y_i^*(t) = 1_{\{T_i \geq t\}}.$$

4.2 Estimation of Λ

Consider the problem of non-parametric estimation of Λ. Note that we can not use the Nelson–Aalen estimator based on the decomposition (13) because the function $\bar{Y}(y)$ depends on the values of the function Λ in the interval $[y, z_0]$.

The decomposition (14) implies the estimator

$$\hat{\Lambda}(z) = \int_0^z \frac{d\bar{N}(y)}{\tilde{Y}(y)}. \tag{15}$$

We shall show that this estimator can be obtained by other way, considering the non-parametric model as the limit of a sequence of parametric models.

Consider some parametric family $(\lambda_\theta \mid \theta \in \Theta \subset \mathbf{R}^p)$ of intensity functions λ and find the maximum likelihood estimators $\hat{\theta}$ of unknown parameter θ.

Let $\Theta = (0; \infty)^p$ and for $\theta = (\theta_1, \ldots, \theta_p)$ set

$$\lambda_\theta(z) = \begin{cases} \theta_1, \text{for } z_{(0)} < z \leq z_{(1)}; \\ \vdots \\ \theta_p, \text{for } z_{(p-1)} < z \leq z_{(p)}; \end{cases}$$

here $0 = z_{(0)} < z_{(1)} < \cdots < z_{(p-1)} < z_{(p)} = z_0$ are fixed cut points.

A natural container of the data vectors

$$U_i = (A_{i1}, \ldots, A_{im_i}, Z_i(T_i)) \tag{16}$$

is the space $E = \coprod_{j=1}^{\infty} E_j$, where $E_j = (0;\infty)^{j+1}$ and $\coprod$ stands for the direct sum of topological spaces. Define the Borel measure ν on E by setting, for each Borel subset $C \subset E$,

$$\nu(C) = \nu_j(C \cap E_j);$$

here $d\nu_j(a_1, \ldots, a_j, z) = d\pi(a_1)\cdots d\pi(a_j)dz$. Then the probability density function $p(u)$ of the random element (16) (with respect to the measure ν) is given by the equalities

$$p(u) = e^{-\tilde{\Lambda}(s_j + a_j z)} a_j \tilde{\lambda}(s_j + a_j z) \quad \text{for } u = (a_1, \ldots, a_j, z) \in E_j;$$

here $s_j = (a_1 + \cdots + a_{j-1})z_0$, $\tilde{\lambda}(t) = \lambda(Z(t))$, $\tilde{\Lambda}(t) = \int_0^t \tilde{\lambda}(s)ds$. Indeed, if $C \cap E_j = (0; a_1^*] \times \cdots \times (0; a_j^*] \times (0; z^*]$, then

$$P\{U_i \in C\} = \sum_{j=1}^{\infty} P\{U_i \in C \cap E_j\}$$

$$= \sum_{j=1}^{\infty} P\{m_i = j, A_{i1} \le a_1^*, \ldots, A_{ij} \le a_j^*, Z_i(T_i) \le z^*\}$$

$$= \sum_{j=1}^{\infty} P\{A_{i1} \le a_1^*, \ldots, A_{ij} \le a_j^*, S_{ij} < T_i \le S_{ij} + A_{ij}z^*\}$$

$$= \sum_{j=1}^{\infty} \int_0^{a_1^*} d\pi(a_1) \cdots \int_0^{a_j^*} d\pi(a_j) \big[e^{-\tilde{\Lambda}(s_j)} - e^{-\tilde{\Lambda}(s_j + a_j z^*)}\big]$$

$$= \sum_{j=1}^{\infty} \int_0^{a_1^*} d\pi(a_1) \cdots \int_0^{a_j^*} d\pi(a_j) \int_0^{z^*} e^{-\tilde{\Lambda}(s_j + a_j z)} a_j \tilde{\lambda}(s_j + a_j z) dz$$

$$= \int_C p(u) d\nu(u).$$

The log-likelihood equals

$$L(\theta) = \sum_{i=1}^{n} \Big[-\tilde{\Lambda}(S_{im_i} + A_{im_i} Z_{im_i}) + \log(A_{im_i}) + \log \tilde{\lambda}(S_{im_i} + A_{im_i} Z_{im_i})\Big]$$

$$= \sum_{i=1}^{n} \Big[-\sum_{j=1}^{m_i} \int_{S_{ij}}^{S_{ij} + A_{ij} Z_{ij}} \lambda_\theta((s - S_{ij})/A_{ij}) ds + \log \lambda_\theta(Z_{im_i}) + const\Big]$$

$$= \sum_{i=1}^{n} \Big[-\sum_{j=1}^{m_i} A_{ij} \int_0^{Z_{ij}} \lambda_\theta(z)dz + \log \lambda_\theta(Z_{im_i}) \Big] + const$$

$$= -\sum_{i=1}^{n}\sum_{j=1}^{m_i}\sum_{k=1}^{p} A_{ij} \int_{z_{(k-1)}}^{z_{(k)}} \theta_k 1_{\{Z_{ij}\ge z\}} dz + \sum_{i=1}^{n}\sum_{k=1}^{p} \log \theta_k 1_{\{z_{(k-1)}<Z_{im_i}\le z_{(k)}\}} + const.$$

The maximum likelihood estimators satisfy the equations

$$-\sum_{i=1}^{n}\sum_{j=1}^{m_i} A_{ij} \int_{z_{(k-1)}}^{z_{(k)}} 1_{\{Z_{ij}\ge z\}} dz + \frac{1}{\theta_k}\sum_{i=1}^{n} 1_{\{z_{(k-1)}<Z_{im_i}\le z_{(k)}\}} = 0,$$

i.e.

$$\hat\theta_k = \frac{\sum_{i=1}^{n} 1_{\{z_{(k-1)}<Z_{im_i}\le z_{(k)}\}}}{\sum_{i=1}^{n}\sum_{j=1}^{m_i} A_{ij}\int_{z_{(k-1)}}^{z_{(k)}} 1_{\{Z_{ij}\ge z\}}dz}.$$

Now formally take $p = n$ and $z_{(k)} = Z_{(k)}$, where $Z_{(1)}, \dots, Z_{(n)}$ are the values Z_{im_i} in ascending order. Then $1_{\{Z_{ij}\ge z\}} = 1_{\{Z_{ij}\ge Z_{(k)}\}}$ for each $z \in (Z_{(k-1)}; Z_{(k)}]$ and therefore

$$\hat\lambda(Z_{(k)}) = \frac{1}{\sum_{i=1}^{n}\sum_{j=1}^{m_i} A_{ij} 1_{\{Z_{ij}\ge Z_{(k)}\}} (Z_{(k)} - Z_{(k-1)})}.$$

The cumulative intensity function at point $Z_{(k)}$ then can be estimated by

$$\sum_{l=1}^{k} (Z_{(l)} - Z_{(l-1)}) \hat\lambda(Z_{(l)}).$$

We get the following estimator:

$$\hat\Lambda(z) = \sum_{Z_{(l)}\le z} \frac{1}{\tilde Y(Z_{(l)})} = \int_0^z \frac{d\bar N(y)}{\tilde Y(y)}.$$

4.3 Large sample properties of $\hat\Lambda$

Proposition 1. *The process $\hat\Lambda$ is a semi-martingale with the characteristics (B_h, C_h, ν), where*

$$B_h(z) = \int_0^z h(\tilde Y^{-1}(y)) \bar Y(y) d\Lambda(y),$$

$$C_h(z) = \int_0^z h^2(\tilde Y^{-1}(y)) \bar Y(y) d\Lambda(y),$$

$$\nu(dy, du) = \bar Y(y) d\Lambda(y) \epsilon_{\tilde Y^{-1}(y)}(du),$$

where ϵ_u denotes the Dirac measure concentrated at point u, $h : \mathbf{R} \to \mathbf{R}$ is a continuous function with compact support, which equals u for u in some neighborhood of 0.

Let us find the first characteristic, B_h, of the process $\hat{\Lambda}$. We have

$$\Delta\hat{\Lambda}(z) = \hat{\Lambda}(z) - \hat{\Lambda}(z-) = \frac{\Delta\bar{N}(z)}{\tilde{Y}(z)} = \sum_{i=1}^{n} \frac{1_{\{Z_{im_i}=z\}}}{\tilde{Y}(Z_{im_i})},$$

$$\sum_{y\le z} \Delta\hat{\Lambda}(y) = \sum_{i=1}^{n}\sum_{y\le z} \frac{1_{\{Z_{im_i}=z\}}}{\tilde{Y}(Z_{im_i})} = \sum_{i=1}^{n} \frac{1_{\{Z_{im_i}\le z\}}}{\tilde{Y}(Z_{im_i})} = \int_0^z \frac{d\bar{N}(y)}{\tilde{Y}(y)} = \hat{\Lambda}(z),$$

$$\sum_{y\le z} h(\Delta\hat{\Lambda}(y)) = \sum_{i=1}^{n} 1_{\{Z_{im_i}\le z\}} h(\tilde{Y}^{-1}(Z_{im_i})).$$

So

$$\hat{\Lambda}_h(z) = \Lambda(z) + \sum_{y\le z}[h(\Delta\Lambda(y)) - \Delta\Lambda(y)]$$

$$= \sum_{i=1}^{n} 1_{\{Z_{im_i}\le z\}} h(\tilde{Y}^{-1}(Z_{im_i})) = \int_0^z h(\tilde{Y}^{-1}(y))d\bar{N}(y).$$

By (13),

$$\hat{\Lambda}_h(z) = \int_0^z h(\tilde{Y}^{-1}(y))\bar{Y}(y)d\Lambda(y) + \int_0^z h(\tilde{Y}^{-1}(y))d\bar{M}(y).$$

Since the process $\tilde{Y}$ is left-continuous, the second term in the right-hand side is a martingale. The first term is continuous and therefore predictable. Hence it equals $B_h(z)$.

The second characteristic, C_h, is a compensator of the process $(\hat{\Lambda}_h - B_h)^2$. By the well-known formula for predictable variation of stochastic integrals,

$$C_h(z) = \left\langle \int h(\tilde{Y}^{-1})d\bar{M} \right\rangle(z)$$

$$= \int_0^z h^2(\tilde{Y}^{-1}(y))d\langle\bar{M}\rangle(y) = \int_0^z h^2(\tilde{Y}^{-1}(y))\bar{Y}(y)d\Lambda(y).$$

The third characteristic, ν, is a compensator of the jump measure ρ of the process $\hat{\Lambda}$. Obviously,

$$\rho(dy, du) = \sum_{y\ge 0} 1_{\{\Delta\hat{\Lambda}(y)\ne 0\}} \epsilon_{(y,\Delta\tilde{\Lambda}(y))}(dy, dx)$$

$$= \sum_{i=1}^{n} \epsilon_{(Z_{i\cdot},\tilde{Y}^{-1}(Z_{im_i}))}(dy, dx).$$

If $U(y)$ is a continuous adapted process and $f(u)$ is a deterministic continuous function, then

$$\int_0^z \int_{\mathbf{R}} U(y) f(u) \rho(dy, du) = \sum_{i=1}^n U(Z_{im_i}) f(\tilde{Y}^{-1}(Z_{im_i})) 1_{\{Z_{im_i} \le z\}}$$

$$= \int_0^z U(y) f(\tilde{Y}^{-1}) d\bar{N}(y) = \int_0^z U(y) f(\tilde{Y}^{-1}) \bar{Y}(y) d\Lambda(y) + \int_0^z U(y) f(\tilde{Y}^{-1}) d\bar{M}(y).$$

The first term in the right-hand side is continuous and therefore predictable. The second term is a martingale. The first term can be written

$$\int_0^z \int_{\mathbf{R}} U(y) f(x) \bar{Y}(y) d\Lambda(y) \epsilon_{\tilde{Y}^{-1}}(dx) = \int_0^z \int_{\mathbf{R}} U(y) f(x) \nu(dy, du).$$

The proof is completed.

To formulate the conditions under which the estimator $\hat{\Lambda}$ is consistent, we need the following notation:

$$b(z) = n^{-1} E\bar{Y}(z) = E\Big[\frac{A_m 1_{\{Z(T) \ge z\}}}{1 - e^{-A_m(\Lambda(z_0) - \Lambda(y))}}\Big].$$

Direct calculations show that $E\tilde{Y}(z)$ also equals $nb(z)$. Indeed, $Z(T) = (T - S_j)/A_j$ for $S_j < T \le S_{j+1}$, so

$$b(z) = \sum_{k=1}^{\infty} E\Big[\frac{A_k 1_{\{S_k + A_k z < T \le S_{k+1}\}}}{1 - e^{-A_m(\Lambda(z_0) - \Lambda(y))}}\Big]$$

$$= \sum_{k=1}^{\infty} E\Big[\frac{A_k}{1 - e^{-A_m(\Lambda(z_0) - \Lambda(y))}} \Big(e^{-\tilde{\Lambda}(S_k + A_k z)} - e^{-\tilde{\Lambda}(S_{k+1})}\Big)\Big]$$

$$= \sum_{k=1}^{\infty} E\Big[A_k e^{-\tilde{\Lambda}(S_k + A_k z)}\Big] = \sum_{k=1}^{\infty} E\Big[A_k 1_{\{T \ge S_k + A_k z\}}\Big] = E \sum_{k=1}^{m} \Big[A_k 1_{\{T \ge S_k + A_k z\}}\Big]$$

$$= E \sum_{k=1}^{m} \Big[A_k 1_{\{Z_k \ge z\}}\Big] = n^{-1} E\tilde{Y}(z).$$

Theorem 3. *Suppose that*

(i) $\inf_{z \le z_0} b(z) > 0$, $\sup_{z \le z_0} b(z) < \infty$,

(ii) $\sup_{z \le z_0} \mid n^{-1}\bar{Y}(z) - b(z) \mid \xrightarrow{P} 0$, $\sup_{z \le z_0} \mid n^{-1}\tilde{Y}(z) - b(z) \mid \xrightarrow{P} 0$, *as* $n \to \infty$.

Then the estimator $\hat{\Lambda}$ *is uniformly consistent, i.e.*

$$\sup_{z \le z_0} \mid \hat{\Lambda}(z) - \Lambda(z) \mid \xrightarrow{P} 0,$$

as $n \to \infty$.

Proof. By Theorem VIII.2.17 of [1] and Proposition1 it suffices to prove the following:

1) $\sup_{z\le z_0} \mid B_h(y) - \Lambda(z) \mid = \sup_{z\le z_0} \mid \int_0^z h(\tilde{Y}^{-1}(y))\bar{Y}(y)d\Lambda(y) - \Lambda(z) \mid \xrightarrow{P} 0$;

2) $\sup_{z\le z_0} \mid C_h(y) \mid = \int_0^{z_0} h^2(\tilde{Y}^{-1}(z))\bar{Y}(z)d\Lambda(z) \xrightarrow{P} 0$;

3) for each bounded continuous non-negative function g, which equals 0 in some neighborhood of 0,

$$\sup_{z\le z_0} \mid \int_0^z \int_{\mathbf{R}} g(x)\nu(dy,dx) \mid = \int_0^{z_0} g(\tilde{Y}^{-1}(z))\bar{Y}(z)d\Lambda(z) \xrightarrow{P} 0.$$

Set $c_1 = \inf_{z\le z_0} b(z)$, $c_2 = \sup_{z\le z_0} b(z)$ and suppose $g(u) = 0$ for $\mid u \mid \le c$. Then, for $n \ge 2/(cc_1)$,

$$P\Big\{\int_0^{z_0} g(\tilde{Y}^{-1}(z))\bar{Y}(z)d\Lambda(z) > \epsilon\Big\}$$

$$= P\Big\{\exists z \le z_0 : \tilde{Y}^{-1}(z) > c, \int_{g(\tilde{Y}^{-1}(z))>c} g(\tilde{Y}^{-1}(z))\bar{Y}(z)d\Lambda(z) > \epsilon\Big\}$$

$$\le P\{\exists z \le z_0 : \tilde{Y}^{-1}(z) > c\} \le P\{\exists z \le z_0 : n^{-1}\tilde{Y}(z) < \frac{1}{nc}\}$$

$$\le P\{\exists z \le z_0 : n^{-1}\tilde{Y}(z) < c_1/2\} \le P\{\exists z \le z_0 : n^{-1}\tilde{Y}(z) < c_1/2 + b(z) - c_1\}$$

$$\le P\{\sup_{z\le z_0} \mid n^{-1}\tilde{Y}(z) - b(z) \mid > c_1/2\} \to 0,$$

which gives 3).

Now suppose that $h(u) = u$ for $\mid u \mid \le c$. Similarly as above

$$P\Big\{\int_0^{z_0} h^2(\tilde{Y}^{-1}(z))\bar{Y}(z)1_{\{\tilde{Y}^{-1}(z)>c\}}d\Lambda(z) > \epsilon\Big\} \le P\{\exists z \le z_0 : \tilde{Y}^{-1}(z) > c\} \to 0.$$

Moreover, for n sufficiently large,

$$P\Big\{\int_0^{z_0} h^2(\tilde{Y}^{-1}(z))\bar{Y}(z)1_{\{\tilde{Y}^{-1}(z)\le c\}}d\Lambda(z) > \epsilon\Big\} \le P\Big\{\int_0^{z_0} \tilde{Y}^{-2}(z)\bar{Y}(z)d\Lambda(z) > \epsilon\Big\}$$

$$\le P\{\exists z \le z_0 : \tilde{Y}^{-2}(z)\bar{Y}(z) > \epsilon/\Lambda(z_0)\} \le P\left\{\frac{\sup_z n^{-1}\bar{Y}(z)}{\inf_z n^{-2}\tilde{Y}^2(z)} > \frac{n^{1/2}\epsilon}{n^{-1/2}\Lambda(z_0)}\right\}$$

$$\le P(\sup_z n^{-1}\bar{Y}(z) > n^{1/2}\epsilon) + P(\inf_z n^{-2}\tilde{Y}^2(z) < n^{-1/2}\Lambda(z_0))$$

$$= P\{\exists z \le z_0 : n^{-1}\bar{Y}(z) > n^{1/2}\epsilon\} + P\{\exists z \le z_0 : n^{-2}\tilde{Y}^2(z) < n^{-1/2}\Lambda(z_0)\}$$

$$\le P\{\exists z \le z_0 : n^{-1}\bar{Y}(z) > 2c_2) + P\{\exists z \le z_0 : n^{-1}\tilde{Y}(z) < c_1/2\}\}$$

$$\le P\{\sup_{z\le z_0} \mid n^{-1}\bar{Y}(z) - b(z) \mid > c_2\} + P\{\sup_{z\le z_0} \mid n^{-1}\tilde{Y}(z) - b(z) \mid > c_1/2\} \to 0.$$

This yields 2).

Relation 1) will be proved if we show that

$$\sup_{z\le z_0} \mid h\big(\tilde Y^{-1}(z)\big)\bar Y(z) - 1 \mid \xrightarrow{P} 0.$$

Similarly as above we get

$$P\Big\{\sup_{z\le z_0} \mid h\big(\tilde Y^{-1}(z)\big)\bar Y(z) - 1 \mid 1_{\{n^{-1}\tilde Y(z)<c_1/2\}} > \epsilon\Big\}$$

$$\le P\{\exists z \le z_0 : n^{-1}\tilde Y(z) < c_1/2\} \to 0.$$

On the other hand, for n sufficiently large

$$\sup_{z\le z_0} \mid h\big(\tilde Y^{-1}(z)\big)\bar Y(z) - 1 \mid 1_{\{n^{-1}\tilde Y(z)>c_1/2\}} =$$

$$\sup_{z\le z_0} \mid \tilde Y^{-1}(z)\bar Y(z) - 1 \mid 1_{\{n^{-1}\tilde Y(z)>c_1/2\}} = \sup_{z\le z_0} \mid \frac{\bar Y(z) - \tilde Y(z)}{\tilde Y(z)} \mid 1_{\{\frac{1}{\tilde Y(z)}<\frac{2}{nc_1}\}}$$

$$\le \frac{2}{nc_1}\sup_{z\le z_0} \mid \bar Y(z) - \tilde Y(z) \mid = \frac{2}{c_1}\sup_{z\le z_0} \mid n^{-1}\bar Y(z) - n^{-1}\tilde Y(z) \mid$$

$$\le \frac{2}{c_1}\sup_{z\le z_0} \mid n^{-1}\bar Y(z) - b(z) \mid + \frac{2}{c_1}\sup_{z\le z_0} \mid n^{-1}\tilde Y(z) - b(z) \mid \to 0.$$

The proof is complete.

Theorem 4. *Suppose that the conditions of Theorem 3 are satisfied,* $E(A) < \infty$. *Then the random function*

$$\sqrt{n}(\hat\Lambda - \Lambda)$$

tends in distribution in the space $D[0, z_0]$ *to the mean zero Gaussian process* V *with the covariance function*

$$\sigma(z_1, z_2) = cov(V(z_1), V(z_2)) =$$

$$E\int_0^{z_1}\int_0^{z_2}\sum_{k=1}^m \frac{A_k^2}{e^{\tilde\Lambda(S_{k+1})} - e^{\tilde\Lambda(S_k + A_k(u\wedge v))}}\frac{d\Lambda(u)d\Lambda(v)}{b(u)b(v)} + \int_0^{z_1\wedge z_2}\frac{d\Lambda(y)}{b(y)}.$$

Proof. The asymptotic distribution of the estimator $\hat\Lambda$ can be found using the martingale decomposition of $\bar N$, i.e. using the equality

$$n^{1/2}(\hat\Lambda(z) - \Lambda(z)) = n^{1/2}\int_0^z \left(\frac{\bar Y(y)}{\tilde Y(y)} - 1\right) d\Lambda(y) + n^{1/2}\int_0^z \frac{d\bar M(y)}{\tilde Y(y)}$$

$$= n^{-1/2}\int_0^z \frac{\bar{Y}(y)-\tilde{Y}(y)}{b(y)}\,d\Lambda(y) + n^{-1/2}\int_0^z \left(\frac{n}{\tilde{Y}(y)}-\frac{1}{b(y)}\right)(\bar{Y}(y)-\tilde{Y}(y))\,d\Lambda(y)$$

$$+n^{-1/2}\int_0^z \frac{d\bar{M}(y)}{\tilde{Y}(y)} = \Delta_1(z)+\Delta_2(z)+\Delta_3(z).$$

Let us find the mean and the covariance function of the first term. The equality $E\tilde{Y}(y) = E\bar{Y}(y)$ implies that

$$E\Delta_1(z) = En^{-1/2}\int_0^z \frac{\bar{Y}(y)-\tilde{Y}(y)}{b(y)}\,d\Lambda(y) = 0.$$

Let us find the covariance function: for $z_1 \le z_2 \le z_0$

$$cov(\Delta_1(z_1), \Delta_1(z_2))$$

$$= n^{-1}\int_0^{z_1}\int_0^{z_2} \frac{1}{b(u)}\frac{1}{b(v)} E(\bar{Y}(u)-\tilde{Y}(u))(\bar{Y}(v)-\tilde{Y}(v))d\Lambda(u)d\Lambda(v),$$

where

$$n^{-1}E(\bar{Y}(u)-\tilde{Y}(u))(\bar{Y}(v)-\tilde{Y}(v))$$

$$= E\sum_{k=1}^{m} A_k^2\left[\frac{1_{\{S_k+A_k u<T\le S_{k+1}\}}}{1-e^{-A_m(\Lambda(z_0)-\Lambda(u))}} - 1_{\{T_i\ge S_{ik}+A_{ik}u\}}\right]\times$$

$$\left[\frac{1_{\{S_k+A_k v<T\le S_{k+1}\}}}{1-e^{-A_m(\Lambda(z_0)-\Lambda(v))}} - 1_{\{T_i\ge S_{ik}+A_{ik}v\}}\right] = E\sum_{k=1}^{m}\frac{A_k^2}{e^{\tilde{A}(S_{k+1})}-e^{\tilde{A}(S_k+A_k(u\wedge v))}}.$$

The second term converges in probability to zero uniformly on $[0, z_0]$. The first and the third terms are asymptotically independent. The asymptotic distribution of the third term is obtained similarly as the asymptotic distribution of the Nelson-Aalen estimator.

It tends in distribution in the space $D[0; z_0]$ to the zero mean Gaussian process W with the covariance function

$$E\{W(z)W(z')\} = \sigma_k^2(z\wedge z'), \tag{17}$$

where

$$\sigma_k^2(z) = \int_0^z \frac{d\Lambda(y)}{b(y)}. \tag{18}$$

4.4 Estimation of the probability $\mathbf{p_j(z)}$

If $\pi_i = \pi$ then the probability

$$p_j(z) = P(T > S_j + zA_j \mid T > S_j)$$

to attain the level of degradation z $(0 \le z \le z_0)$ before a failure occurs given that an unit had been renewed $j-1$ times $(j = 1, 2, \dots)$ is estimated by the statistic

$$\hat{p}_j(z) = \int_0^\infty \exp\left\{-a\hat{\Lambda}(z)\right\} d\hat{\pi}(a) = \frac{1}{m}\sum_{i=1}^{n}\sum_{j=1}^{m_i} \exp\left\{-A_{ij} \sum_{Z_{k,m_k}\leq z} 1/\tilde{Y}(Z_{k,m_k})\right\},$$

because

$$\hat{\pi}(a) = \frac{1}{m}\sum_{i=1}^{n}\sum_{j=1}^{m_i} 1_{\{A_{ij}\leq a\}}, \quad m = \sum_{i=1}^{n} m_i.$$

Otherwise, the estimator is

$$\hat{p}_j(z) = \int_0^\infty \exp\left\{-a\hat{\Lambda}(z)\right\} d\hat{\pi}_j(a) = \frac{1}{m(j)}\sum_{m_i\geq j} \exp\left\{-A_{ij} \sum_{z_{k,m_k}\leq z} 1/\tilde{Y}(Z_{k,m_k})\right\},$$

where

$$\hat{\pi}_j(a) = \frac{\sum_{i=1}^{n} 1_{\{A_{ij}\leq a, m_i\geq j\}}}{m(j)}, \quad m(j) = \sum_{i=1}^{n} 1_{\{j\leq m_i\}}.$$

References

[BN02] Bagdonavičius, V. and Nikulin, M. Accelerated Life Models. Chapman and Hall/CRC: Boca Raton(2002)

[BN01] Bagdonavičius, V., Nikulin, M.: Estimation in Degradation Models with Explanatory variables. Lifetime Data Analysis, **7**, 85–103 (2001)

[BBK04] Bagdonavičius, V., Bikelis, A., Kazakevičius, V. : Statistical Analysis of Linear Degradation and Failure Time Data with Multiple Failure Modes. Lifetime Data Analysis, **10**, 65–81(2004)

[1] Jacod, J. and Shyriayev, A. N.: Limit theorems for stochastic processes. Springer, New York (1987)

[ME98] Meeker, W.Q. and Escobar L.: Statistical Methods for Reliability Data. J.Wiley and Sons (1988)

[WT97] Wulfsohn M. and Tsiatis A. A Joint Model for Survival and Longitudinal Data Measured with Error. Biometrics, **53**, 330–339 (1997)

[L04] Lehmann, A. On a Degradation-Failure Models for Repairable Items. In: Nikulin, M., Balakrishnan, N, Mesbah, M., Limnios, N. (ed) Parametric and Semiparametric Models with Applications to Reliability, Survival Analysis, and Quality of Life. Birkhauser, Boston, 65–80 (2004)

[WK04] Wendt, H., Kahle, W. On Parametric Estimation for a Position-Dependent Marking of a Doubly Stochastic Poisson Process. In: Nikulin, M., Balakrishnan, N, Mesbah, M., Limnios, N. (ed) Parametric and Semiparametric Models with Applications to Reliability, Survival Analysis, and Quality of Life. Birkhauser, Boston, 473–486 (2004)

The Impact of Dementia and Sex on the Disablement in the Elderly

P.Barberger-Gateau[1], V.Bagdonavičius[2], M.Nikulin[1], O.Zdorova-Cheminade[1],

[1] IFR 99 Santé Publique, Université Victor Segalen Bordeaux 2, France
nikou@sm.u-bordeaux2.fr

[2] Department of Mathematical Statistics, Vilnius University, Lithuania
vilius@sm.u-bordeaux2.fr

Summary. The paper considers the analysis of disablement process of the elderly using the general path model with noise. The impact of dementia and sex on degradation is analysed. These joint model for survival and longitudinal data is discussed.

Key words: aging, censored data , conjoint model , degradation process , dementia, disability , failure , elderly , noise , path model , Wulfsohn-Tsiatis model .

1 Introduction

Aging of the French population, which included 21.3% persons aged 60 and over at the 1999 census, is the result of the decrease of the number of births and the increase of life expectancy. This phenomenon raises the problem of the management by the health care system and the society of age-related diseases and their consequences.

The World Health Organisation (WHO) proposed the International Classification of Impairment, Disability and Handicap to conceptualise the consequences of disease [WHO80]. Disability is defined as the reduction of the capacity to accomplish daily activities, in a way normal for a given age and gender. These activities include basic Activities of Daily Living (ADL), such as bathing, dressing or eating, and more complex activities including household activities called Instrumental Activities of Daily Living (IADL). Activities performed outside, often referred to "mobility", correspond to an even higher level of difficulty. There is a hierarchical relationship between these three domains of disability. We showed that an indicator combining mobility disability, assessed by the Rosow scale [RB66], IADL disability assessed by the Lawton scale [LB69] and ADL disability assessed by five items of the Katz scale [KDCG70] was an almost perfect four grade Guttman scale

[B-GRLD00]. These four grades correspond to four degrees of increasing disability, from full functional independence to severe disability. Transitions observed over time between these disability grades may be progressions to a more severe grade (called increasing degradation) but also regressions to a less severe grade (called decreasing degradation) are possible. Transition to death may be observed from any disability grade. Thus the degradation process may be modelled by a five state model, including an absorbing state: death. Markov models have been used to estimate the transition intensities between these states, function of covariates [B-GVP01]. Other models may be used, which model the degradation process as a continuum from full independence to the most severe disability, including phases of recovery and different slopes of decline, function of the characteristics of the subject. Disability at older ages results from lifelong disabling diseases and living conditions, in addition to specific age-associated diseases such as dementia . Thus the degradation process will be modified by covariates, in particular socio-demographic factors. Age is a major factor to be taken into account since the risk of death strongly increases with age, but also because age is the main risk factor of age-associated diseases. In particular, oldest old persons, those aged 80 and over, often suffer from several pathologies in addition to the proper effect of physiological aging. Women have a longer life expectancy than men, and they experience living conditions and disease different from those of men. Dementia is a major disabling disease in the elderly. Thus all these factors are expected to impact the degradation process and the risk of death. The objective of this research was to describe the degradation process in elderly persons aged 65 and over function of their socio-demographic characteristics and the diagnosis of dementia, using the general path model of degradation with noise.

1.1 Data

The data come from the PAQUID (Personnes Agées QUID) epidemiological study which aims to study cerebral aging and disability in elderly people. PAQUID is a prospective cohort study in which 3777 community dwellers aged 65 and over were included in 1988-89. The participants were randomly selected from electoral rolls of 75 parishes in Gironde and Dordogne, in south-western France. The initial participation rate was 68% and the sample was representative in terms of age and sex of the local aged population. The participants were visited at home by a psychologist for the baseline interview, and then visited again one, three, five, eight, ten and thirteen years afterwards in the same manner. Disability was recorded at each follow-up with the following instruments: - Five activities from the Katz ADL scale [KDCG70]: bathing, dressing, toileting, transferring, and feeding. For each activity the subject was rated on a three grade scale : independent, needs partial help, dependent. We considered a subject as dependent for ADL if he was dependent for at least one of the five activities according to the threshold defined by Katz for each activity. - Five IADL from the Lawton scale [LB69] common to both sexes: using the telephone, means of transportation, shopping, responsibility for medication and budget management. Three activities were added when assessing women: meals preparation, housekeeping and doing the laundry. For each item a threshold for dependence was defined by Lawton and we considered that a subject was dependent for IADL if he was dependent for at least one in five (for men) or in eight (for women) of these activities. - Mobility was assessed on the Rosow and Breslau scale which includes three activities : doing heavy housework, climbing stairs, and walking between 500 m and 1 km. Subjects were considered as dependent if they were unable to perform at least one of these activities.

A four grade hierarchical disability indicator was built as follows [B-GRLD00] :

Grade 0: fully independent subject for the three domains of disability. Grade 1: subjects dependent only for the mobility scale, but independent for IADL and ADL.

Grade 2: subjects dependent for mobility and IADL but independent for ADL.

Grade 3: subjects dependent for each of the three domains. This indicator classified 99.3% of the subjects at baseline with a scalability coefficient of 0.98 [B-GRLD00]. In a second step a hierarchy was also identified within each of the three disability categories. The mobility disabled subjects (grade 1 disability) were divided into two subgroups : those dependent only for doing heavy housework and those also dependent for climbing stairs or walking. For IADL disability three groups of progressively increasing disability were identified : disability for shopping and/or using means of transportation, disability for managing medication and/or budget in addition to the previous category, and disability for using the telephone in addition to the two previous categories. For ADL disability three subgroups were identified : those dependent only for bathing and/or dressing, those also dependent for toileting and/or transferring, and those also dependent for feeding. Thus the number of disability items was reduced to eight levels of increasing disability. A score was built as follows.

Among the 3777 participants in the PAQUID cohort at baseline, 3642 had all the relevant disability variables recorded and at least one follow-up visit or deceased. This sample included 1530 men (42%) and 2112 women (58%). The distribution of the score in this sample is given in table 1. Older subjects had higher disability scores. In each age group women tended to have higher disability scores than men. Only 2864 subjects (1183 men and1681 women) with at least two measures of the score could be used for modelisation. In this sample 403 subjects (14.1%) had a diagnosis of dementia at any time of the follow-up. The sample included 929 subjects (32.4%) who had not achieved the "Cetrificat d'Etudes Primaires (CEP) corresponding to about seven years of schooling. These subjects were considered as "low educated", in opposition to the 1935 "high educated" who had at least reached this level.

Zdorova-Cheminade [Z-C03] studied by simulation the considered model as statistical degradation model of disablement in the elderly to verify that a hierarchical relationship exists between the concepts of Activities Daily Living, Instrumental Activities of Daily Living and mobility and to use this model to study the evolution of disability. The cumulative disability scale was used to describe the degradation process in time. In longitudinal analysis an additional level was considered to the disability index to take death into account. It is evident that this approach can be used in many other medical studies where a degradation is observed, especially in oncology.

Each of the 13 initial activities was given a rating between 0 and 1 :

- - 0 corresponds to the ability to perform the activity without help;
- - 1 corresponds to full dependency for this activity;
- - a step $1/(m-1)$ was added for each intermediate level of ability, m being the number of degrees on each activity in its original version [RB66] – [KDCG70]. m varied between 3 and 5, function of the activity.

Each of the eight disability levels was then given a rating between 0 and 1 :

- - if the level corresponded to a single activity, the rating was that of the activity;
- - if the level was a combination of several activities, the rating was the mean of the ratings of each activity.

A score was built by summing up the eight ratings corresponding to each disability level. This score varies between 0 and 8. The score increases with increasing disability.

The time scale was age, starting at age 65 to model the process

$$Z(t) = \text{score} + 1$$

2 Degradation model

Let us consider the *degradation model with the noise*:

$$Z(t) = g(t, A)\, U(t), \quad t \geq 0,$$

here $A = (A_1, A_2)$ is a random vector with the distribution function F_A, g is a continuously differentiable function, $U(t)$, $t > 0$ is the noise. We suppose that

$$V(t) = \ln U(t) = \sigma W(\ln(1+t)),$$

W is the standard Wiener process independent on A. The component $g(t, A)$ explains the interior degradation process, different for each individual, the noise U explains the complementary influence on the obtained disability score by such factors as temporary disability, desease, low spirits, breaked leg, etc. Note that for any $t > 0$ the median of the the random variable $U(t)$ is 1.

If $\sigma = 0$ then we have the *General Path Model* [MEL98].

3 Estimation of the mean degradation

Assume that the c.d.f. F_A of A and the function g are unknown.

Fix the degradation measurement moments $t_{i,1}, \ldots, t_{i,m_i}$ of the i th individual ($i = 1 \ldots, n$). If the death time τ_i of this individual occurs in the interval $[t_{i,j_i}, t_{i,j_i+1})$ ($j_i = 1, \ldots, m_i; t_{m_i+1} = \infty$) then the values $Z_{i1}, \ldots, Z_{i,j_i}$ of the degradation process Z_i of the ith individual are observed at the time moments $t_{i1}, \ldots, t_{i,j_i}$.

The PAQUID data has the following properties:

1. number of measurements per individual is small;
2. the time of follow-up $t_{ij_i} - t_{i1}$ is short for each individual i;
3. there are important differences in intervals $[t_{i1}; t_{ij_i}]$ for different individuals, for example $[60; 65]$ and $[90, 95]$, etc.

In each short time interval $[t_{i1}; t_{ij_i}]$ we model the real degradation $Z_r(t)$ by the loglinear model

$$g(t, A_i) = e^{A_{i1}} (1 + t)^{A_{i2}},$$

where $A_i = (A_{i1}, A_{i1})$, and $A_1, \ldots, A_n$ are n independent replicates of the random vector A.

To have stable estimators of the mean degradation attained at the moment t we use the degradation values of individuals with indices s such that

$$s \in D(t) = \{i \,:\, t \in [t_{i1}; t_{ij_i}],\ j_i \geq 2\}.$$

Set

$$Y_{ij} = \ln Z_{ij}, \quad Y_i = (Y_{i1}, \ldots, Y_{ij_i})^T.$$

Then given $A_i = a_i$, j_i

$$Y_i \sim N(\mu_i, \sigma^2 \Sigma_i),$$

where

$$\mu_i = (\mu_{i1}, \ldots, \mu_{i,j_i})^T, \quad \mu_{ij} = \mu_{ij}(a_i) = \ln g(t_{ij}, a_i), \quad \Sigma_i = \| s_{ikl} \|_{j_i \times j_i},$$

$$s_{ikl} = c_{ik} \wedge c_{il}, \quad c_{ij} = \ln(1 + t_{ij}).$$

Denote by b_{ikl} the elements of the inverse matrix Σ_i^{-1}, and by N the number of individuals such that $j_i \geq 2$.

The predictors $\hat{A}_i$ of the random vectors A_i for the ith individual are found minimizing with respect to $a_1, \ldots, a_N$ the quadratic form

$$Q(Y, a) = \sum_{i=1}^{N} (Y_i - \mu_i(a_i))^T \; \Sigma_i^{-1} \; (Y_i - \mu_i(a_i)). \tag{1}$$

Denote by $m = \sum_{i=1}^{N} j_i$ the total number of measures over all of individuals. The estimator $\hat{\sigma}^2$ of the parameter σ^2 is found maximizing with respect to σ^2 the conditional likelihood function

$$L(\sigma^2 \mid \hat{A}_1, \ldots, \hat{A}_N) =$$

$$\frac{1}{(2\pi)^{m/2}\sigma^m} \prod_{i=1}^{N} \mid \Sigma_i \mid^{-1/2} \exp\left\{ -\frac{1}{2\sigma^2} \sum_{i=1}^{N} (Y_i - \mu_i(\hat{A}_i))^T \, \Sigma_i^{-1} \, (Y_i - \mu_i(\hat{A}_i)) \right\}.$$

The minimization of the quadratic form (1) and the maximization of the conditional likelihood function gives the following equations to compute $\hat{\sigma}^2$ and $\hat{A}_i$:

$$\hat{A}_{i1} = \frac{c_i d_i - e_i f_i}{c_i^2 - b_i e_i}, \quad \hat{A}_{i2} = \frac{c_i f_i - b_i d_i}{c_i^2 - b_i e_i}$$

where

$$b_i = \mathbf{1}^T \Sigma_i^{-1} \mathbf{1}, \quad c_i = C_i^T \Sigma_i^{-1} \mathbf{1}, \quad d_i = Y_i^T \Sigma_i^{-1} C_i,$$
$$e_i = C_i^T \Sigma_i^{-1} C_i, \quad f_i = Y_i^T \Sigma_i^{-1} \mathbf{1}, \quad g_i = Y_i^T \Sigma_i^{-1} Y_i m$$
$$\mathbf{1} = (1, \ldots, 1)_{j_i}^T, \quad C_i = (c_{i1}, \ldots, c_{i,j_i})^T,$$

and

$$\hat{\sigma}^2 = \frac{\hat{c}}{m},$$

where

$$\hat{c} = Q(Y, \hat{A}) = \sum_{i=1}^{N} (g_i + b_i \hat{A}_{1i}^2 + e_i \hat{A}_{2i}^2 + 2c_i \hat{A}_{1i} \hat{A}_{2i} - 2f_i \hat{A}_{1i} - 2d_i \hat{A}_{2i}).$$

The conditional mean of the estimator $\hat{\sigma}^2$ given N and j_i is:

$$E(\hat{\sigma}^2 \mid j_i, i = 1, \ldots, N) = \frac{m+2}{m} \sigma^2,$$

so $\hat{\sigma}^2$ is a consistent estimator of σ^2.

A consistent estimator of the mean degradation $m(t) = E(Z(t))$ is

$$\hat{m}(t) = e^{\frac{1}{2}\hat{\sigma}^2 \ln(1+t)} \frac{\sum_{i \in D(t)} g(t, \hat{A}_i)}{\sum_{i \in D(t)}^{n} \exp\{\frac{1}{2}[\hat{\sigma}_{i1}^2 + 2\hat{\sigma}_{i1}\hat{\sigma}_{i2}\hat{\rho}_i \ln(1+t) + \hat{\sigma}_{i2}^2 \ln^2(1+t)]\}},$$

where

$$\hat{\sigma}_{i1}^2 = \frac{e_i}{b_i e_i - c_i^2} \hat{\sigma}^2, \quad \hat{\sigma}_{i2}^2 = \frac{b_i}{b_i e_i - c_i^2} \hat{\sigma}^2, \quad \hat{\rho}_i = -\frac{c_i}{\sqrt{b_i e_i}}.$$

4 Application to the PAQUID data

4.1 The estimated mean of the disablement process in men and women

The estimator of the parameter σ^2 is $\hat{\sigma}^2 = 0.57$ for women and $\hat{\sigma}^2 = 0.69$ for men. So the noise is large.

The mean disablement processes are represented in figure 1 for women and men. The score is always higher in women, and the magnitude of the difference increases after age 75.

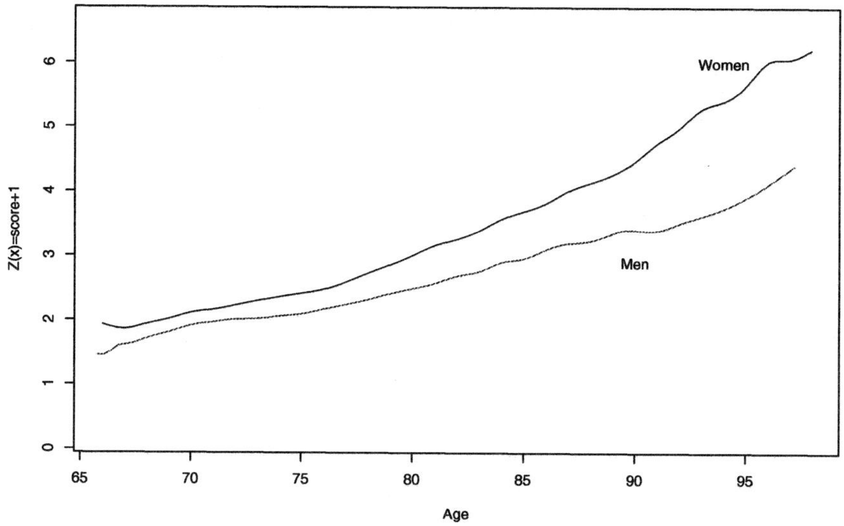

Figure 1

Many studies have found a similar difference between older women and men, women being generally more disabled than men of the same age [SKCC92]–[REG02]. In particular, Verret [VER99] showed the influence of sex on mild and moderate disability in the PAQUID cohort. Using a five state Markov model with piecewise constant transition intensities on the same data, Regnault [REG02] confirmed that women were at higher risk at the beginning of the disablement process. The model of degradation shows that women are more disabled than men, but also that the difference increases with aging. The degradation process is faster in women.

4.2 The estimated mean of the disablement process in demented and non-demented subjects

The estimator of the parameter σ^2 is $\hat{\sigma}^2 = 0.79$ for demented subjects and $\hat{\sigma}^2 = 0.51$ for the non demented. The noise is large.

The mean degradation processes in demented and non-demented subjects are represented in figure 2.

The difference between these two groups of subjects is large. Subjects who will be diagnosed as demented at follow-up were more disabled at any age, even at entry in the cohort at age 65 before the diagnosis of dementia was made. This study confirms the strong impact of dementia on the disablement process. Future demented subjects were more disabled even before the clinical diagnosis of dementia was made and they had a higher speed of degradation. The model with time dependent covariates developed by Bagdonavicius and Nikulin [BN04] could be used to take the pre-clinical phase into account. Dementia is major cause of disablement in older persons [REG02], [DB-GG91]–[DGM91]. So *the individual degradation curve of a non-demented individual is an important predictive factor of dementia in the future.*

The five state Markov model with piecewise constant transition intensities used by Regnault [REG02] on the same data showed similar results : dementia was associated with progression from mild to moderate disability and then to severe disability.

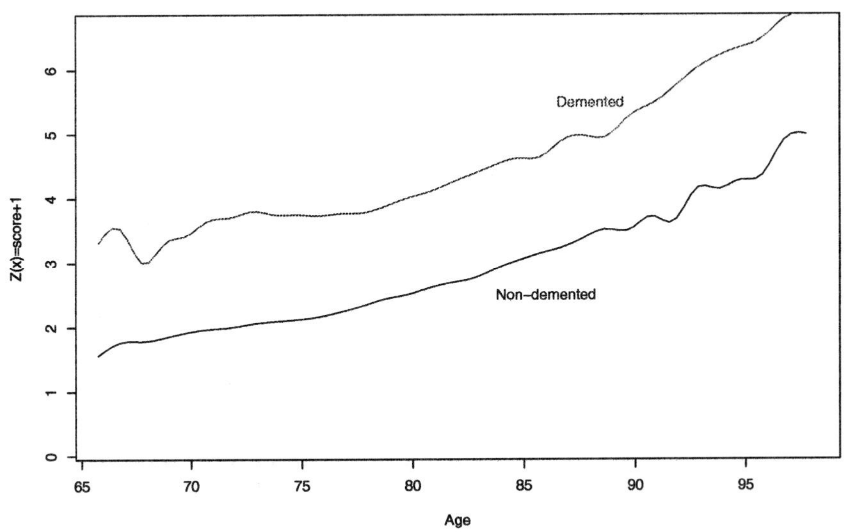

Figure 2

4.3 The estimated mean of the disablement process in demented and non-demented men

The mean degradation processes in demented and non-demented men are represented in figure 3.

The difference between these two groups of men is large. Men who will be diagnosed as demented at follow-up were more disabled at any age, even at entry in the cohort at age 65 before the diagnosis of dementia was made. Future demented subjects were more disabled even before the clinical diagnosis of dementia was made and they had a higher speed of degradation.

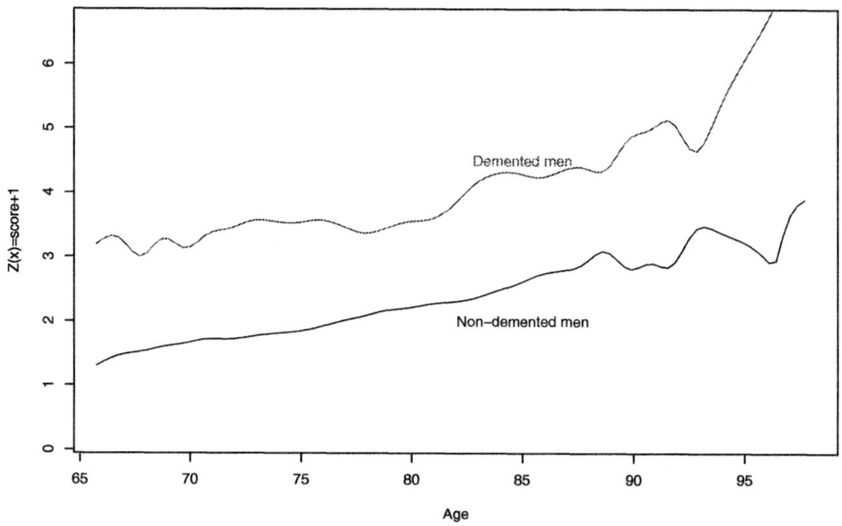

Figure 3

4.4 The estimated mean of the disablement process in demented and non-demented women

The mean degradation processes in demented and non-demented men are represented in figure 4.

As in the case of men, the difference between demented and non-demented women is large. Future demented women were more disabled even before the clinical diagnosis of dementia was made and they had a higher speed of degradation.

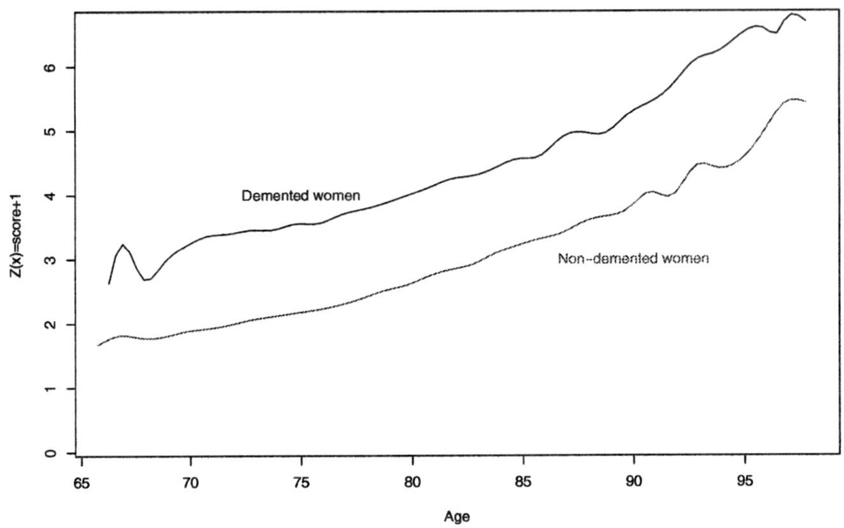

Figure 4

4.5 The estimated mean of the disablement process in demented men and women

The mean degradation processes in demented men and women are represented in figure 5.

For the men and women who will be diagnosed as demented at follow-up the degradation process develops similarly. So the difference between the degradation of older women and men is observed only for non-demented individuals.

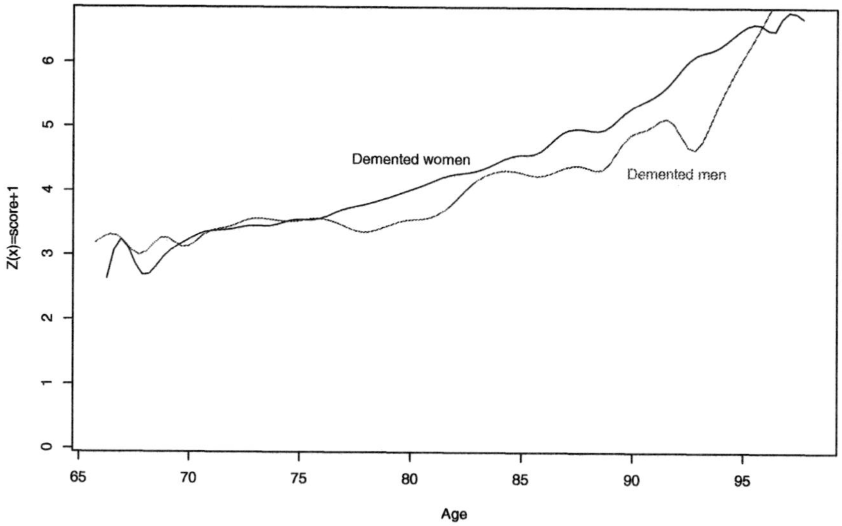

Figure 5

4.6 The estimated mean of the disablement process in non-demented men and women

The mean degradation processes in non-demented men and women are represented in figure 6.

For the men and women who will be diagnosed as demented at follow-up the degradation process develops differently. The disablement process develops quicker in women then in men.

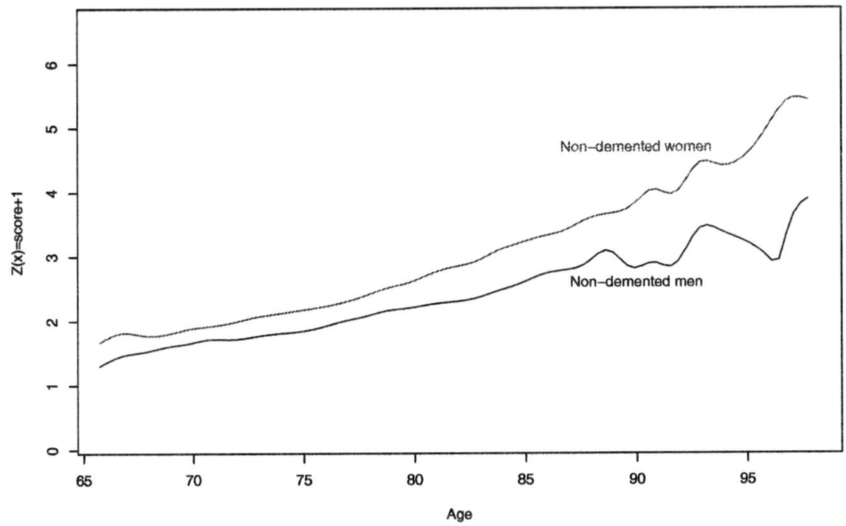

Figure 6

4.7 The estimated mean of the disablement process in high and low educated subjects

The mean degradation processes in high (primary education present)and low educated (primary education absent) subjects are represented in figure 7.

The disablement process develops quicker in low educated subjects.

More about the influence of the level of education on the aging-degradation process one can see in Barberger-Gateau et al [B-GVP01], Dartigues et al [DB-GG91], etc.

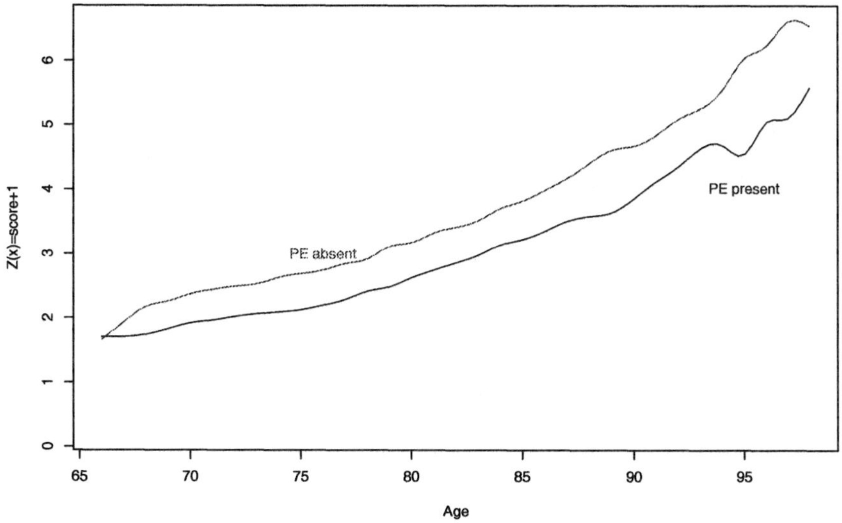

Figure 7

5 Joint model for degradation-failure time data

Joint degradation and failure time data may be analized using the following model. We call a failure of an individual *natural* if the degradation process attains a *critical level* z_0. Denote by T^0 the moment of non-traumatic failure, i.e. the moment when the degradation attains some critical value z_0:

$$T^0 = \sup\{t : Z(t) \leq z < z_0\}$$

We denote T the time to death. In this case the observed failure moment is

$$\tau = T^0 \wedge T.$$

We shall consider the model when the hazard rate depends on degradation.

Following [BN04] let us consider the joint degradation model according to which the conditional survival of T given the degradation process has the form:

$$S_T(t \mid A) = P\{T > t \mid Z(s)), 0 \leq s \leq t\} = \exp\Big\{-\int_0^t \lambda_0(s,\alpha)\lambda\big(g(s,A)\big)ds\Big\},$$

where λ is the unknown intensity function, $\lambda_0(s,\alpha)$ being from a parametric family of hazard functions.

Note that the function λ is defined on the set of degradation values, not on the time scale.

The model states [BN04] that the conditional hazard rate $\lambda_T(t \mid A)$ at the moment t given the degradation $g(s, A), 0 \leq s \leq t$, has the form

$$\lambda_T(t \mid A) = \lambda_0(t, \alpha)\lambda(g(t, A)).$$

The term $\lambda(g(t, A))$ shows the influence of degradation on the hazard rate, the term $\lambda_0(t, \alpha)$ shows the-influence of time on the hazard rate not explained by degradation. If, for example,

$$\lambda_0(t, \alpha) = (1 + t)^{\alpha}, e^{\alpha t},$$

then $\alpha = 0$ corresponds to the case when the hazard rate at any moment t is a function of the degradation level at this moment.

Wulfsohn and Tsiatis [WT97] considered the so called joint model for survival and longitudinal data measured with error, given by

$$\lambda_T(t \mid A) = \lambda_0(t)e^{\beta(A_1 + A_2 t)}$$

with bivariate normal distribution of of (A_1, A_2). The difference: in our model the function λ, characterizing the influence of degradation on the hazard rate, is non-parametric, in the *Wulfsohn-Tsiatis model* this function is parametric. On the other hand, the baseline hazard rate λ_0 (it is proportional to the hazard rate which should be observed if the degradation would be absent) is parametric in our model and non-parametric in Wulfsohn-Tsiatis model.

The analysis of the PAQUID data using the joint model see in Zdorova-Cheminade [Z-C03].

References

[WHO80] WHO : World Health Organization . The International Classification of Impairments, Disabilities, and Handicaps - a manual relating to the consequences of disease. Geneva, WHO (1980)

[RB66] Rosow, I. and Breslau, N. : A Guttman health scale for the aged. J Gerontol., **21**, 556–9 (1966)

[LB69] Lawton, M. P. and Brody, E. M. : Assessment of older people : self-maintaining and instrumental activities of daily living. The Gerontologist, **9**, 179–86 (1969)

[KDCG70] Katz, S., Downs, T. D., Cash, H. R. and Grotz, R. C.: Progress in development of the index of ADL. The Gerontologist, **10**, 20–30 (1970)

[B-GRLD00] Barberger-Gateau, P., Rainville, C., Letenneur, L. and Dartigues, J.-F.: A hierarchical model of domains of disablement in the elderly: a longitudinal approach. Disability and Rehabilitation, **22**, 308–17 (2000)

[B-GVP01] Barberger-Gateau, P., Verret, C. and Peres, K.: Factors associated with progression and regression through states of disability in elderly people. Gerontology, **47**, 372 (2001)

[Z-C03] Zdorova-Cheminade, O. : Modélisation du processus d'évolution de l'incapacité chez les personnes agées. Mémoire de DEA "Epidémiologie et Intervention en Santé Publique", Université Bordeaux 2, Bordeaux (2003)

[MEL98] Meeker, W., Escobar, L. and Lu, C. : Accelerated degradation tests : modeling and analysis. Technometrics, **40**, 89–99 (1998)

[SKCC92] Strawbridge, W., Kaplan, G., Camacho, T., Cohen, R.: The dynamics of disability and functional change in an elderly cohort: results from the Almeda county study. J AM Geriatr. Soc., **40**, 789–806 (1992)

[SBGDLD94] Sauvel, C., Barbeger-Gateau, P., Dequae, L., Letenneur, L., Dartigues, J. : Facteurs associés à l'évolution à un an de l'autonomie fonctionnelle des personnes âgées vivant à leur domicile. Rev. Epidemiol. et Santé Publ., **42**, 13–23 (1994)

[CRJ-F87] Colvez, A., Robine, J., Jouan-Flahault, C. : Risque et facteur de risque d'incapacité aux âges éleés. Rev. Epidemiol. et Santé Publ., **35**, 257–69 (1987)

[GLA93] Guralnik, J., Lacroix, A., Abbott, R., et al. : Maintain in mobility in late life. I. Demographic characteristics and chronic conditions. Am. J. Epidemiol., **137**, 845–57 (1993)

[VER99] Verret, C. : Etude de l'évolution de la dépendance et de ses facteurs associś chez les personnes âgés en Aquitaine, 1988-98. Mémoire de DEA "Epidémiologie et Intervention en Santé Publique", Université Bordeaux 2, Bordeaux (1999)

[REG02] Regnault, A. : Modélisation de l'évolution de la dépendance des personnes âgées : Rôlle de la démence et facteurs associés. Mémoire de DESS "SA3S", Université Bordeaux 2, Bordeaux (2002)

[BN04] Bagdonavicius, V., Nikulin, M.S. : Semiparametric analysis of degradation and failure time data with covariates. In: Nikulin, M.S., Balakrishnan, N., Mesbah, M., Limnios, N. (eds) Parametric and Semiparametric Models with Applications to Reliability, Survival Analysis, and Quality of Life. Birkhäuser, Boston Basel Berlin (2004)

[DB-GG91] Dartigues, J., Barbeger-Gateau, P., Gagnon, M., et al.: PAQUID : Etude épidémiologique du vieillissement normal et pathologique. Rev. Geriatr., **16**, 5–11 (1991)

[DGM91] Dartigues, J., Gagnon, M., Michel, P., et al. : Le programme de recherche PAQUID sur l'épidémiologie de la démence. Méthodes et résultats initiaux. Rev. Neurol., **147(3)**, 225–30 (1991)

[WT97] Wulfsohn, M. and Tsiatis, A. : A Joint Model for Survival and Longitudinal Data Measured with Error. Biometrics, **53**, 330–339 (1997)

Nonparametric Estimation for Failure Rate Functions of Discrete Time semi-Markov Processes

Vlad Barbu[1] and Nikolaos Limnios[2]

[1] Université de Technologie de Compiègne, Laboratoire de Mathématiques Appliquées de Compiègne, BP 20529, 60205 Compiègne, France
barbu@dma.utc.fr

[2] Université de Technologie de Compiègne, Laboratoire de Mathématiques Appliquées de Compiègne Nikolaos.Limnios@utc.fr

Summary. We consider a semi-Markov chain, with a finite state space. Taking a censored history, we obtain empirical estimators for the discrete semi-Markov kernel, renewal function and semi-Markov transition function. We propose estimators for two different failure rate functions: the usual failure rate, BMP-failure rate, defined by [BMP63], and the new introduced failure rate, RG-failure rate, proposed by [RG92]. We study the strong consistency and the asymptotic normality for each estimator and we construct the confidence intervals. We illustrate our results by a numerical example.

Key words: semi-Markov chain, discrete time semi-Markov kernel , reliability , BMP-failure rate , RG-failure rate , nonparametric estimation , asymptotic properties .

1 Introduction

Continuous time semi-Markov systems are important framework for reliability studies, cf. [LO99, LO01]. Estimation of reliability and related measures for semi-Markov systems can be found in [LO03], [OL99].

As compared to the attention given to the continuous time semi-Markov processes and related reliability matters, the discrete time semi-Markov processes ($DTSMP$) are less studied. This is rather surprising, because considering a discrete time semi-Markov system instead of a continuous one offers important advantages, especially for applications. Indeed, a semi-Markov chain makes only a finite number of transitions in a finite time interval and the Markov renewal function can be expressed as a finite sum, which is not the case for a continuous semi-Markov process. Consequently, the related numerical computations for a discrete time semi-Markov process are much faster and more accurate than for a continuous one.

An introduction to discrete time Markov renewal processes can be found in [How71], [MS00], [BBL04]. For the discrete time semi-Markov model for reliability see [Cse02], [BBL04]. Estimators of the kernel, transition matrix, renewal function, reliability, availability of a discrete time semi-Markov system and their asymptotic properties are given in [BL04a] and [BL04b].

The aim of this paper is to construct estimators for failure rate functions of a discrete time semi-Markov process and to give their properties. We consider two different failure rate functions: the usual failure rate, BMP-failure rate, defined by [BMP63, BP75], and the new introduced failure rate, RG-failure rate, proposed by [RG92]. The reasons for introducing this new definition of a failure rate for discrete time models are given in [Bra01, BGX01].

The estimator of the classical failure rate for a continuous time semi-Markov system have been studied in [OL98]. Statistical estimation and asymptotic properties for RG-failure rate and BMP-failure rate of a discrete time homogeneous Markov system are presented in [SL02].

The present paper is organized as follows. In Section 2 we give some definitions and recall some previously obtained results concerning discrete time semi-Markov processes and the associated reliability metrics (see [BBL04], [BL04a, BL04b]). In Section 3 we construct empirical estimators for failure rates and we study the strong consistency and the asymptotic normality for the proposed estimators. Asymptotic confidence intervals are also given. All the above results are proved in Section 4. In Section 5 we illustrate our results by a numerical example of a three state discrete time semi-Markov system.

2 Preliminaries

2.1 The Discrete Time semi-Markov Model

Let us consider:

- E, the state space. We suppose E to be finite, say $E = \{1, \ldots, s\}$.
- The stochastic process $J = (J_n;\ n \in \mathbb{N})$ with state space E for the system state at the n-th jump.
- The stochastic process $S = (S_n;\ n \in \mathbb{N})$ with state space $\mathbb{N}$ for the n-th jump of the process. We suppose $S_0 = 0$ and $0 < S_1 < S_2 < \ldots < S_n < S_{n+1} < \ldots$.
- The stochastic process $X = (X_n;\ n \in \mathbb{N}^*)$ with state space $\mathbb{N}^*$ for the sojourn time X_n in state J_{n-1} before the n-th jump. Thus, we have for all $n \in \mathbb{N}^*$
$$X_n = S_n - S_{n-1}.$$

We denote by $\mathcal{M}_E$ the set of non negative matrices on $E \times E$ and by $\mathcal{M}_E(\mathbb{N})$, the set of matrix-valued functions : $\mathbb{N} \to \mathcal{M}_E$.

Definition 1. *The stochastic process $(J, S) = ((J_n, S_n);\ n \in \mathbb{N})$ is said to be a discrete time Markov renewal process (DTMRP) if for all $n \in \mathbb{N}$, for all $i, j \in E$ and for all $k \in \mathbb{N}$ it almost surely satisfies*

$$\mathbb{P}(J_{n+1} = j, S_{n+1} - S_n = k \mid J_0, \dots, J_n; S_0, \dots, S_n) = \mathbb{P}(J_{n+1} = j, S_{n+1} - S_n = k \mid J_n). \tag{1}$$

Moreover, if (1) is independent of n, (J, S) is said to be homogeneous, with discrete semi-Markov kernel $q(k) = (q_{ij}(k);\ i, j \in E) \in \mathcal{M}_E$ defined by

$$q_{ij}(k) := \mathbb{P}(J_{n+1} = j, X_{n+1} = k \mid J_n = i).$$

Definition 2. *The transition function of the embedded Markov chain $(J_n;\ n \in \mathbb{N})$ is the matrix-valued function $V \in \mathcal{M}_E$ defined by*

$$V = (p_{ij})_{i,j \in E}, \quad p_{ij} := \mathbb{P}(J_{n+1} = j \mid J_n = i), \quad i, j \in E,\ n \in \mathbb{N}. \tag{2}$$

Definition 3. *For all $i, j \in E$ such that $p_{ij} \neq 0$, let us denote by:*

1. *$f_{ij}(\cdot)$, the conditional distribution of the sojourn time in state i before going to state j :*

$$f_{ij}(k) = \mathbb{P}(X_{n+1} = k \mid J_n = i, J_{n+1} = j), \quad k \in \mathbb{N}, \tag{3}$$

2. *$h_i(\cdot)$, the sojourn time distribution in state i:*

$$h_i(k) = \mathbb{P}(X_{n+1} = k \mid J_n = i) = \sum_{l \in E} q_{il}(k), \quad k \in \mathbb{N}^*,$$

3. *$H_i(\cdot)$, the sojourn time cumulative distribution function in state i :*

$$H_i(k) = \mathbb{P}(X_{n+1} \leq k \mid J_n = i) = \sum_{l \geq 1}^{k} h_i(l), \quad k \in \mathbb{N}^*.$$

Obviously, for all $i, j \in E$ such that $p_{ij} \neq 0$ and for all $k \in \mathbb{N}$, we have

$$q_{ij}(k) = p_{ij} f_{ij}(k). \tag{4}$$

Let us give some definitions and results from [BBL04], which will be useful for the estimation presented in this paper.

Definition 4. *(discrete time convolution product) Let $A, B \in \mathcal{M}_E(\mathbb{N})$ be two matrix-valued functions. The matrix convolution product $A * B$ is the matrix-valued function $C \in \mathcal{M}_E(\mathbb{N})$ defined by*

$$C_{ij}(k) := \sum_{k \in E} \sum_{l=0}^{k} A_{ik}(k-l)\, B_{kj}(l), \quad i, j \in E, \quad k \in \mathbb{N}.$$

Lemma 1. *Let $\delta I = (d_{ij}(k);\ i,j \in E) \in \mathcal{M}_E(\mathbb{N})$ be the matrix-valued function defined by*

$$d_{ij}(k) := \begin{cases} 1 \text{ if } i = j \text{ and } k = 0 \\ 0 \text{ elsewhere.} \end{cases}$$

Then, δI is the neutral element for the discrete time matrix convolution product, i.e., δI satisfies

$$\delta I * A = A * \delta I = A, \quad A \in \mathcal{M}_E(\mathbb{N}).$$

Definition 5. *Let $A \in \mathcal{M}_E(\mathbb{N})$ be a matrix-valued function. If there exists some $B \in \mathcal{M}_E(\mathbb{N})$ such that*

$$B * A = \delta I, \tag{5}$$

then B is called the left inverse of A in the convolution sense and it is denoted by $A^{(-1)}$.

We stress the fact that the left inverse of A is not always defined. The next proposition gives a sufficient condition for the existence and uniqueness of the left inverse.

Proposition 1. *Let $A \in \mathcal{M}_E(\mathbb{N})$ be a matrix-valued function. If* $\det A(0) \neq 0$*, then the left inverse of A exists and is unique.*

Definition 6. *(discrete time n-fold convolution) Let $A \in \mathcal{M}_E(\mathbb{N})$ be a matrix-valued function and $n \in \mathbb{N}$. The n-fold convolution $A^{(n)}$ is the matrix function $C \in \mathcal{M}_E(\mathbb{N})$ defined recursively by:*

$$A_{ij}^{(0)}(k) := \begin{cases} 1 \text{ if } \ k = 0 \text{ and } \ i = j \\ 0 \text{ else} \end{cases},$$

$$A_{ij}^{(1)}(k) := A_{ij}(k)$$

and

$$A_{ij}^{(n)}(k) := \sum_{l \in E} \sum_{s=0}^{k} A_{lj}^{(n-1)}(k-s)\, A_{il}(s), \quad n \geq 2, k \geq 1.$$

For a DTMRP (J, S), the n-fold convolution of the semi-Markov kernel can be expressed as follows.

Proposition 2. *For all $i, j \in E$, for all n and $k \in \mathbb{N}$, we have*

$$q_{ij}^{(n)}(k) = \mathbb{P}(J_n = j, S_n = k \mid J_0 = i). \tag{6}$$

Let us consider the matrix-valued functions $Q = (Q(k);\ k \in \mathbb{N}) \in \mathcal{M}_E(\mathbb{N})$, defined by

$$Q_{ij}(k) := \mathbb{P}(J_{n+1} = j, X_{n+1} \leq k \mid J_n = i) = \sum_{l=1}^{k} q_{ij}(l), \quad i,j \in E,\ k \in \mathbb{N} \tag{7}$$

and $\psi = (\psi(k);\ k \in \mathbb{N}) \in \mathcal{M}_E(\mathbb{N})$, defined by

$$\psi_{ij}(k) := \sum_{n=0}^{k} q_{ij}^{(n)}(k), \quad i,j \in E,\ k \in \mathbb{N}. \tag{8}$$

Proposition 3. *The matrix-valued function $\psi = (\psi(k);\ k \in \mathbb{N})$ is given by:*

$$\psi(k) = (\delta I - q)^{(-1)}(k), \tag{9}$$

where $(\delta I - q)^{(-1)}$ denotes the left convolution inverse of the matrix function $(\delta I - q)$ and is computed using the following forward algorithm

$$\begin{cases} (\delta I - q)^{(-1)}(k) = I_E & \text{if } k = 0 \\ (\delta I - q)^{(-1)}(k) = -\sum_{s=0}^{k-1} (\delta I - q)^{(-1)}(s)\,(\delta I - q)(k-s) & \text{if } k \in \mathbb{N}^*. \end{cases} \tag{10}$$

Definition 7. *The matrix renewal function $\Psi = (\Psi(k);\ k \in \mathbb{N}) \in \mathcal{M}_E(\mathbb{N})$ of the DTMRP is defined by*

$$\Psi_{ij}(k) := \mathbb{E}_i[N_j(k)], \quad i,j \in E,\ k \in \mathbb{N}, \tag{11}$$

where $N_j(k)$ is the number of visits to state j before time k.

The matrix renewal function can be expressed in the following form:

$$\Psi_{ij}(k) = \sum_{n=0}^{k} Q_{ij}^{(n)}(k) = \sum_{l=0}^{k} \psi_{ij}(l), \quad i,j \in E,\ k \in \mathbb{N}. \tag{12}$$

Definition 8. *A stochastic process $Z = (Z_k;\ k \in \mathbb{N})$ is called the discrete time semi-Markov process associated with the DTMRP (J,S), if*

$$Z_k = J_{N(k)}, \quad k \in \mathbb{N},$$

where $N(k) := \max\{n \geq 0; S_n \leq k\}$ is the discrete time counting process of the number of jumps in $[1,k] \subset \mathbb{N}$.

Thus, Z_k gives the state of the process at time k. We have also $J_n = Z_{S_n}, n \in \mathbb{N}$.

Let us now define the discrete time semi-Markov transition matrix and propose a computation procedure.

Definition 9. *The transition matrix of the semi-Markov process Z is the matrix-valued function $P \in \mathcal{M}_E(\mathbb{N})$ defined by*

$$P_{ij}(k) := \mathbb{P}(Z_k = j \mid Z_0 = i), \quad i,j \in E,\ k \in \mathbb{N}.$$

The Markov renewal equation for the semi-Markov transition function P is (see [BBL04])

$$P = I - H + q * P, \tag{13}$$

where $H(k) := diag(H_i(k);\ i \in E)$.

Solving the Markov renewal equation (13) (see [BBL04]) we obtain that the unique solution is

$$P(k) = \Big[(\delta I - q)^{(-1)} * (I - H)\Big](k) = \Big[\psi * (I - diag(Q \ \cdot \mathbf{1}))\Big](k), \tag{14}$$

where $\mathbf{1}$ denotes the s-column vector whose all elements are 1.

2.2 Basic Results on semi-Markov Chains Estimation

We will give the following results for a MRP which satisfies some conditions.

Assumptions

1. The Markov chain $(J_n;\ n \in \mathbb{N})$ is irreducible;
2. The mean sojourn times are finite, i.e., $\sum_{k \geq 0} k h_i(k) < \infty$ for any state $i \in E$.
3. The $DTMRP$ $((J_n, S_n);\ n \in \mathbb{N})$ is aperiodic.

Let us consider a history $\boldsymbol{H}(M)$ of the MRP $((J_n, S_n);\ n \in \mathbb{N})$, censored at time $M \in \mathbb{N}$,

$$\boldsymbol{H}(M) := (J_0, X_1, \ldots, J_{N(M)-1}, X_{N(M)}, J_{N(M)}, U_M),$$

where we set $N(M) := \max\{n \mid S_n \leq M\}$ and $U_M := M - S_{N(M)}$.

Definition 10. *For all $i, j \in E$ and $k \leq M$, we define:*

1. $N_i(M) := \sum_{n=0}^{N(M)-1} \mathbf{1}_{\{J_n = i\}}$, *the number of visits to state i, up to time M;*
2. $N_{ij}(M) := \sum_{n=1}^{N(M)} \mathbf{1}_{\{J_{n-1} = i, J_n = j\}}$, *the number of transitions from i to j, up to time M;*
3. $N_{ij}(k, M) := \sum_{n=1}^{N(M)} \mathbf{1}_{\{J_{n-1} = i, J_n = j, X_n = k\}}$, *the number of transitions from i to j, up to time M, with sojourn time in state i equal to $k, 1 \leq k \leq M$.*

Taking a history $\boldsymbol{H}(M)$ of a discrete time MRP, for all $i, j \in E$ and $k \in \mathbb{N}, k \leq M$, we define the empirical estimators of the probability transition function p_{ij}, sojourn conditioned time $f_{ij}(k)$ and discrete semi-Markov kernel $q_{ij}(k)$ by

$$\hat{p}_{ij}(M) := \frac{N_{ij}(M)}{N_i(M)}, \tag{15}$$

$$\hat{f}_{ij}(k, M) := \frac{N_{ij}(k, M)}{N_{ij}(M)} = \frac{1}{N_{ij}(M)} \sum_{n=1}^{N(M)} \mathbf{1}_{\{J_{n-1} = i, J_n = j, X_n = k\}}, \tag{16}$$

$$\hat{q}_{ij}(k, M) := \frac{N_{ij}(k, M)}{N_i(M)} = \frac{1}{N_i(M)} \sum_{n=1}^{N(M)} \mathbf{1}_{\{J_{n-1} = i, J_n = j, X_n = k\}}. \tag{17}$$

Let us also set $\hat{q}(k,M) := (\hat{q}_{ij}(k,M);\ i,j \in E)$.

Remark The above empirical estimators are approached nonparametric maximum likelihood estimators, i.e., they maximize an approached likelihood function, obtained by neglecting the part corresponding to $U_M := M - S_{N(M)}$.

Replacing q by its estimator in the expressions of Q, ψ, Ψ and P we obtain the corresponding estimators:

$$\hat{Q}(k,M) := \sum_{l=1}^{k} \hat{q}(l,M), \quad \hat{\psi}(k,M) := \sum_{n=0}^{k} \hat{q}^{(n)}(k,M) \tag{18}$$

$$\hat{\Psi}(k,M) := \sum_{l=0}^{k} \hat{\psi}(l,M) = \sum_{l=0}^{k}\sum_{n=0}^{l} \hat{q}^{(n)}(l,M) \tag{19}$$

$$\hat{P}(k,M) := \Big[\Big(\delta I - \hat{q}(\cdot,M)\Big)^{(-1)} * \Big(I - diag(\hat{Q}(\cdot,M)\cdot \mathbf{1})\Big)\Big](k) \tag{20}$$

$$= \Big[\hat{\psi}(\cdot,M) * \Big(I - diag(\hat{Q}(\cdot,M)\cdot \mathbf{1})\Big)\Big](k), \tag{21}$$

where $\hat{q}^{(n)}(k,M)$ is the n-fold convolution of $\hat{q}(k,M)$ (see Definition 6).

We can prove the strong consistency for the proposed estimators and the asymptotic normality of $\hat{q}_{ij}(k,M)$, $\hat{\psi}_{ij}(k,M)$, $\hat{P}_{ij}(k,M)$ (see [BL04b]).

3 Failure Rates Estimation

The objective of this section is to construct empirical estimators for two failure rate functions of a discrete time semi-Markov system.
Firstly, we give a method for computing the reliability of a discrete time semi-Markov system and we propose an empirical estimator.

Let E be partitioned into two subsets U and D, respectively for the up states and for the down states, where $E = U \cup D$ and $U \cap D = \emptyset$. Without loss of generality, we can suppose that

$$U = \{1, \ldots, s_1\} \text{ and } D = \{s_1+1, \ldots, s\}, \text{ with } 0 < s_1 < s.$$

We will partition all vectors and matrix-valued functions according to this partition. For instance, the transition matrix P of the semi-Markov process and the initial distribution vector α can be written as follows:

$$P(k) = \begin{array}{c} \begin{array}{cc} U & D \end{array} \\ \begin{pmatrix} P_{11}(k) & P_{12}(k) \\ P_{21}(k) & P_{22}(k) \end{pmatrix} \end{array} \begin{array}{c} U \\ D \end{array} \qquad \alpha = \begin{array}{c} \begin{array}{cc} U & D \end{array} \\ (\alpha_1 \ \ \alpha_2) \end{array}$$

For $m, n \in \mathbb{N}^*$ such that $m > n$, let $\mathbf{1}_{m,n}$ denote the m-column vector whose first n elements are 1 and the last $m-n$ elements are 0; for $m \in \mathbb{N}^*$, let $\mathbf{1}_m$ denote the m-column vector whose all elements are 1.

Let T_D denotes the first passage time in the subset D, i.e., $T_D := \inf\{n \in \mathbb{N};\ \ Z_n \in D\}$. The reliability at time $k \in \mathbb{N}$ is given by

$$R(k) := \mathbb{P}(T_D > k) = \mathbb{P}(Z_n \in U, n \in \{0, \ldots, k\}).$$

We define a new semi-Markov process $Y = (Y_n;\ n \in \mathbb{N})$ with state space $E_Y = U \cup \{\Delta\}$, where Δ is an absorbing state

$$Y_n := \begin{cases} Z_n \text{ if } n < T_D \\ \Delta \text{ if } n \geq T_D \end{cases}, \ n \in \mathbb{N}.$$

The semi-Markov kernel of the process Y is

$$q_Y(k) = \begin{bmatrix} q_{11}(k) & q_{12}(k)\,\mathbf{1}_{s-m} \\ \mathbf{0}_{1m} & 0 \end{bmatrix}, \ k \in \mathbb{N},$$

where $\mathbf{0}_{1m}$ is an m-dimensional row vector whose all elements are 0. Let P_Y denote the transition function of Y. Thus, for all $k \in \mathbb{N}$, the reliability is given by

$$\begin{aligned} R(k) &= \mathbb{P}(Y_k \in U) = \sum_{j \in U} \sum_{i \in U} \mathbb{P}(Y_k = j \mid Y_0 = i)\, \mathbb{P}(Y_0 = i) \\ &= [\alpha_1, 0]\, P_Y(k)\, \mathbf{1}_{m+1,m} = \alpha_1 \cdot P_{11}(k) \cdot \mathbf{1}_{s_1} \\ &= \alpha_1\, \psi_{11} * (I - H_1)(k)\, \mathbf{1}_{s_1} = \alpha_1\, \psi_{11} * (I - diag(Q \cdot \mathbf{1})_{11}) \mathbf{1}_{s_1}. \end{aligned}$$

We propose the following estimator for the system reliability:

$$\begin{aligned} \hat{R}(k, M) &:= \alpha_1 \cdot \hat{P}_{11}(k, M) \cdot \mathbf{1}_{s_1} \\ &= \alpha_1 \Big[\hat{\psi}_{11}(\cdot, M) * \Big(I - diag(\hat{Q}(\cdot, M) \cdot \mathbf{1})_{11} \Big) \Big](k) \mathbf{1}_{s_1}, \end{aligned} \tag{22}$$

where the estimators $\hat{\psi}$ and $\hat{Q}$ are defined in (18).

All the results which follow are proved under **Assumptions** (1), (2) and (3).

Theorem 1. *(see [BL04a]) The estimator of the reliability of a discrete time semi-Markov system is strongly consistent, i.e.,*

$$\max_{0 \leq k \leq M} \mid \hat{R}(k, M) - R(k) \mid \xrightarrow[M \to \infty]{a.s.} 0.$$

Let us denote by T the discrete random variable describing the lifetime of the system. We consider two different definitions of the failure rate function.

• BMP-failure rate function $\lambda(k)$

It is the usual failure rate, defined by [BMP63] as the conditional probability that the failure of the system occurs at time k, given that the system has worked until time $k - 1$ (see also [BP75] and [SL02] in the case of Markov chains).

It is worth noticing that the failure rate in discrete case is a probability function and not a general positive function as in continuous time case.

For any $k \geq 1$

$$\lambda(k) := \mathbb{P}(T = k \mid T \geq k) = \begin{cases} \frac{\mathbb{P}(T=k)}{\mathbb{P}(T\geq k)}, & R(k-1) \neq 0 \\ 0, & otherwise. \end{cases} = \begin{cases} 1 - \frac{R(k)}{R(k-1)}, & R(k-1) \neq 0 \\ 0, & otherwise \end{cases}$$

and

$\lambda(0) := 1 - R(0).$

- RG-failure rate function $r(k)$

A new definition of the discrete failure rate function is proposed in [RG92] for solving some of the problems raised by the use of the usual failure rate function $\lambda(k)$ in discrete time. A detailed argument for the introduction of the new definition of the failure rate function is given in [Bra01, BGX01].

$$r(k) := \begin{cases} \ln \frac{R(k-1)}{R(k)}, & k \geq 1 \\ -\ln R(0), & \mathrm{k} = 0 \end{cases}.$$

The two failure rate functions are related by

$$r(k) = -\ln(1 - \lambda(k)).$$

We propose the following estimators for the failure rates:

$$\hat{\lambda}(k, M) := \begin{cases} 1 - \frac{\hat{R}(k,M)}{\hat{R}(k-1,M)}, & \hat{R}(k-1, M) \neq 0 \\ 0, & otherwise \end{cases}$$

$$\hat{\lambda}(0, M) := 1 - \hat{R}(0, M)$$

and

$$\hat{r}(k, M) := \begin{cases} \ln \frac{\hat{R}(k-1,M)}{\hat{R}(k,M)}, & k \geq 1 \\ -\ln \hat{R}(0, M), & \mathrm{k} = 0 \end{cases}.$$

The following results concern the uniform strong consistence and the asymptotic normality of the empirical estimators of the failure rates.

Theorem 2. *The estimators of the failure rates of a discrete time semi-Markov system are strongly consistent, in the sense that*

$$\max_{0 \leq k \leq M} \mid \hat{\lambda}(k, M) - \lambda(k) \mid \xrightarrow[M \to \infty]{a.s.} 0$$

$$\max_{0 \leq k \leq M} \mid \hat{r}(k, M) - r(k) \mid \xrightarrow[M \to \infty]{a.s.} 0.$$

Notation: For a matrix function $A \in \mathcal{M}_E(\mathbb{N})$, we denote by $A^+ \in \mathcal{M}_E(\mathbb{N})$ the matrix function defined by $A^+(k) := A(k+1),\ k \in \mathbb{N}$.

Theorem 3. *For any fixed $k \in \mathbb{N}$ we have the following convergence in distribution*

$$\sqrt{M}[\hat{\lambda}(k,M) - \lambda(k)] \xrightarrow[M\to\infty]{\mathcal{D}} \mathcal{N}(0, \sigma_\lambda^2(k)),$$

where

$$\sigma_\lambda^2(k) = \frac{1}{R^4(k-1)}\sigma_1^2(k),$$

$$\begin{aligned}\sigma_1^2(k) = \frac{\mu_{jj}^*}{\mu_{jj}}\sum_{i=1}^{s}\frac{\mu_{ii}^2}{\mu_{ii}^*}\Big\{ & R^2(k)\sum_{j=1}^{s}\Big[D_{ij}^U - \mathbf{1}_{\{i\in U\}}\sum_{t\in U}\alpha(t)\Psi_{ti}\Big]^2 * q_{ij}(k-1) \\ & + R^2(k-1)\sum_{j=1}^{s}\Big[D_{ij}^U - \mathbf{1}_{\{i\in U\}}\sum_{t\in U}\alpha(t)\Psi_{ti}\Big]^2 * q_{ij}(k) - T_i^2(k) \\ & + 2R(k-1)R(k)\sum_{j=1}^{s}\Big[\mathbf{1}_{\{i\in U\}}D_{ij}^U\sum_{t\in U}\alpha(t)\Psi_{ti}^+ + \mathbf{1}_{\{i\in U\}}(D_{ij}^U)^+\sum_{t\in U}\alpha(t)\Psi_{ti} \\ & - (D_{ij}^U)^+ D_{ij}^U - \mathbf{1}_{\{i\in U\}}\Big(\sum_{t\in U}\alpha(t)\Psi_{ti}\Big)\Big(\sum_{t\in U}\alpha(t)\Psi_{ti}^+\Big)\Big] * q_{ij}(k-1)\Big\},\end{aligned} \quad (23)$$

where

$$\begin{aligned}T_i(k) := \sum_{j=1}^{s}\Big[& R(k)D_{ij}^U * q_{ij}(k-1) - R(k-1)D_{ij}^U * q_{ij}(k) \\ & - R(k)\mathbf{1}_{\{i\in U\}}\sum_{t\in U}\alpha(t)\psi_{ti} * Q_{ij}(k-1) + R(k-1)\mathbf{1}_{\{i\in U\}}\sum_{t\in U}\alpha(t)\psi_{ti} * Q_{ij}(k)\Big]\end{aligned}$$

$$D_{ij}^U := \sum_{n\in U}\sum_{r\in U}\alpha(n)\psi_{ni} * \psi_{jr} * \Big(I - diag(Q\cdot\mathbf{1})\Big)_{rr},$$

μ_{ii}^* is the mean recurrence time of state i for the embedded Markov chain $((J_n);\ n \in \mathbb{N})$ and μ_{ii} is the mean recurrence time of the state i for the $DTMRP$ (J,S).

Corollary 1. *For any fixed $k \in \mathbb{N}$, $\sqrt{M}[\hat{r}(k,M) - r(k)]$ converges in distribution to a zero mean normal random variable with variance*

$$\sigma_r^2(k) = \frac{1}{(1-\lambda(k))^2}\sigma_\lambda^2(k) = \frac{1}{R^2(k-1)R^2(k)}\sigma_1^2(k),$$

where $\sigma_1^2(k)$ is given in Equation (23).

Asymptotic Confidence Intervals for Failure Rates

Let us now give the asymptotic confidence intervals for failure rates, using the above asymptotic results.

For $k \in \mathbb{N}, k \leq M$, replacing $q(k), Q(k), \psi(k), \Psi(k)$ respectevely by $\hat{q}(k,M)$, $\hat{Q}(k,M)$, $\hat{\psi}(k,M)$, $\hat{\Psi}(k,M)$ in Equation (23), we obtain an estimator $\hat{\sigma}_\lambda^2(k)$ of the variance $\sigma_\lambda^2(k)$. From the strong consistency of the estimators $\hat{q}(k,M), \hat{Q}(k,M)$, $\hat{\psi}(k,M)$ and $\hat{\Psi}(k,M)$ (see [BL04b]), we obtain that $\hat{\sigma}_\lambda^2(k)$ converges almost surely to $\sigma_\lambda^2(k)$, as M tends to infinity.

For $k \in \mathbb{N}, k \leq M$, the estimated asymptotic confidence interval of BMP-failure rate function $\lambda(k)$ at level $100(1-\gamma)\%$, $\gamma \in (0,1)$, is given by

$$\hat{\lambda}(k,M) - u_{1-\gamma/2}\frac{\hat{\sigma}_\lambda(k)}{\sqrt{M}} \leq \lambda(k) \leq \hat{\lambda}(k,M) + u_{1-\gamma/2}\frac{\hat{\sigma}_\lambda(k)}{\sqrt{M}}, \tag{24}$$

where u_γ is the $\gamma-$ quantile of an $N(0,1)$- distributed variable. In the same way, we obtain the asymptotic confidence interval of RG-failure rate function.

4 Proofs

In order to prove the above results, we need the following lemmas.

Lemma 2. *Let $A \in \mathcal{M}_E(\mathbb{N})$ be a matrix function and let q be the semi-Markov kernel of the DTMRP (J,S). For any fixed $k \in \mathbb{N}$, i, j, l and $r \in E$, we have*

1. $(A_{ij} * \mathbf{1}_{\{x=\cdot\}})(k) = \begin{cases} A_{ij}(k-x), & x \leq k \\ 0, & \textit{otherwise}. \end{cases}$
2. $(A_{ij} * \mathbf{1}_{\{x=\cdot\}})^2(k) = \begin{cases} A_{ij}^2(k-x), & x \leq k \\ 0, & \textit{otherwise}. \end{cases}$
3. $\sum_{x=0}^{\infty}(A_{lr} * \mathbf{1}_{\{x=\cdot\}})(k) q_{ij}(x) = (A_{lr} * q_{ij})(k).$
4. $\sum_{x=0}^{\infty}(A_{lr} * \mathbf{1}_{\{x=\cdot\}})^2(k) q_{ij}(x) = (A_{lr}^2 * q_{ij})(k).$

Lemma 3. *Let $A, B \in \mathcal{M}_E(\mathbb{N})$ be two matrix functions. For any fixed $k, y \in \mathbb{N}$, i, j, l, r, u and $v \in E$, we have*

1. $A_{uv} * \mathbf{1}_{\{y\leq\cdot\}}(k) = \begin{cases} \sum_{t=0}^{k-y} A_{uv}(t), & y \leq k \\ 0, & otherwise \end{cases}.$

2. $\sum_{x=0}^{\infty} A_{uv} * \mathbf{1}_{\{x\leq\cdot\}}(k) q_{ij}(x) = A_{uv} * Q_{ij}(k).$

3. $\sum_{x=0}^{\infty} \Big(A_{uv} * \mathbf{1}_{\{x\leq\cdot\}}(k)\Big)^2 q_{ij}(x) = \Big(\sum_{t=0}^{\cdot} A_{uv}(t)\Big)^2 * q_{ij}(k).$

4. $\sum_{x=0}^{\infty} A_{uv} * \mathbf{1}_{\{x=\cdot\}}(k) B_{lr} * \mathbf{1}_{\{x\leq\cdot\}}(k) q_{ij}(x) = \Big[A_{uv}(\cdot) \sum_{t=0}^{\cdot} B_{lr}(t)\Big] * q_{ij}(k).$

Lemma 4. *Let $A, B \in \mathcal{M}_E(\mathbb{N})$ be two matrix functions. For any fixed $k \in \mathbb{N}$, i, j, l, r, u and $v \in E$, we have*

1. $\sum_{x=0}^{\infty} (A_{uv} * \mathbf{1}_{\{x=\cdot\}})(k)(B_{lr} * \mathbf{1}_{\{x=\cdot\}})(k-1) q_{ij}(x) = (A_{uv}^{+} B_{lr}) * q_{ij}(k-1),$

2. $\sum_{x=0}^{\infty} (A_{uv} * \mathbf{1}_{\{x=\cdot\}})(k)(B_{lr} * \mathbf{1}_{\{x\leq\cdot\}})(k-1) q_{ij}(x)$

   $= \Big[A_{uv}^{+}(\cdot) \sum_{t=0}^{\cdot} B_{lr}(t)\Big] * q_{ij}(k-1),$

3. $\sum_{x=0}^{\infty} (A_{uv} * \mathbf{1}_{\{x=\cdot\}})(k-1)(B_{lr} * \mathbf{1}_{\{x\leq\cdot\}})(k) q_{ij}(x)$

   $= \Big[A_{uv}(\cdot) \Big(\sum_{t=0}^{\cdot} B_{lr}(t)\Big)^{+}\Big] * q_{ij}(k-1),$

4. $\sum_{x=0}^{\infty} (A_{uv} * \mathbf{1}_{\{x\leq\cdot\}})(k-1)(B_{lr} * \mathbf{1}_{\{x\leq\cdot\}})(k) q_{ij}(x)$

   $= \Big[\Big(\sum_{t=0}^{\cdot} A_{uv}(t)\Big) \Big(\sum_{t=0}^{\cdot} B_{lr}(t)\Big)^{+}\Big] * q_{ij}(k-1).$

In the sequel, we give the proofs of Theorem 2, 3 and Corollary 1. All along this section, we will use the notation $\Delta q_{ij}(k, M) := \hat{q}_{ij}(k, M) - q_{ij}(k)$, $\Delta P_{ij}(k, M) := \hat{P}_{ij}(k, M) - P_{ij}(k)$, etc. We will also omit the censoring time M as an argument of the estimators; for instance, we write $\hat{q}_{ij}(k)$ instead of $\hat{q}_{ij}(k, M)$.

Proof of Theorem 2. We have

$$\max_{0\leq k\leq M} | \hat{\lambda}(k, M) - \lambda(k) |$$

$$\leq \max_{0\leq k\leq M} \frac{| \hat{R}(k, M) - R(k) |}{\hat{R}(k-1, M)} + \max_{0\leq k\leq M} \frac{R(k)}{R(k-1)} \frac{| \hat{R}(k-1, M) - R(k-1) |}{\hat{R}(k-1, M)}.$$

>From the uniform strong consistency of the reliability estimator we conclude that the right-hand side term converges almost surely to zero, when M tends to infinity.

Using the relation between the BMP-failure rate and the RG-failure rate

$$r(k) = -\ln(1 - \lambda(k)),$$

we infer the consistency of the RG-failure rate. □

Proof of Theorem 3.

$$\begin{aligned}&\sqrt{M}[\hat{\lambda}(k, M) - \lambda(k)] \\ &= \frac{\sqrt{M}[R(k)(\hat{R}(k-1, M) - R(k-1)) - (\hat{R}(k, M) - R(k))R(k-1)]}{\hat{R}(k-1, M)R(k-1)}.\end{aligned}$$

>From the consistency of the reliability estimator (see Theorem 1), we have $\hat{R}(k-1, M) \xrightarrow[M\to\infty]{a.s.} R(k-1)$, so, in order to obtain the asymptotic normality for the BMP-failure rate, we need only to prove that

$$\sqrt{M}\Big[R(k)\Big(\hat{R}(k-1, M) - R(k-1)\Big) - \Big(\hat{R}(k, M) - R(k)\Big)R(k-1)\Big] \xrightarrow[M\to\infty]{\mathcal{D}} \mathcal{N}(0, \sigma_1^2(k)).$$

We obtain that

$$\sqrt{M}\Big[R(k)\Big(\hat{R}(k-1, M) - R(k-1)\Big) - \Big(\hat{R}(k, M) - R(k)\Big)R(k-1)\Big]$$

has the same limit in distribution as

$$\begin{aligned}&\sqrt{M}\sum_{l,r=1}^{s} R(k)D_{rl}^{U} * \Delta q_{rl}(k-1) - \sqrt{M}\sum_{r\in U}\sum_{l=1}^{s} R(k)\Big(\sum_{n\in U}\alpha(n)\psi_{nr}\Big) * \Delta Q_{rl}(k-1) \\ &-\sqrt{M}\sum_{l,r=1}^{s} R(k-1)D_{rl}^{U} * \Delta q_{rl}(k) + \sqrt{M}\sum_{r\in U}\sum_{l=1}^{s} R(k-1)\Big(\sum_{n\in U}\alpha(n)\psi_{nr}\Big) * \Delta Q_{rl}(k) \\ &= \frac{1}{\sqrt{M}}\sum_{n=1}^{N(M)}\sum_{l,r=1}^{s}\frac{M}{N_r(M)}\Big[R(k)D_{rl}^{U} * \Big(\mathbf{1}_{\{J_{n-1}=r, J_n=l, X_n=\cdot\}} - q_{rl}(\cdot)\mathbf{1}_{\{J_{n-1}=r\}}\Big)(k-1) \\ &-R(k)\mathbf{1}_{\{r\in U\}}\Big(\sum_{t\in U}\alpha(t)\psi_{tr}\Big) * \Big(\mathbf{1}_{\{J_{n-1}=r, J_n=l, X_n\leq\cdot\}} - Q_{rl}(\cdot)\mathbf{1}_{\{J_{n-1}=r\}}\Big)(k-1) \\ &-R(k-1)D_{rl}^{U} * \Big(\mathbf{1}_{\{J_{n-1}=r, J_n=l, X_n=\cdot\}} - q_{rl}(\cdot)\mathbf{1}_{\{J_{n-1}=r\}}\Big)(k) \\ &-R(k-1)\mathbf{1}_{\{r\in U\}}\Big(\sum_{t\in U}\alpha(t)\psi_{tr}\Big) * \Big(\mathbf{1}_{\{J_{n-1}=r, J_n=l, X_n\leq\cdot\}} - Q_{rl}(\cdot)\mathbf{1}_{\{J_{n-1}=r\}}\Big)(k)\Big] \\ &= \frac{1}{\sqrt{M}}\sum_{l=1}^{N(M)} f(J_{l-1}, J_l, X_l),\end{aligned}$$

where we have defined the function $f : E \times E \times \mathbb{N} \to \mathbb{R}$ by

$$f(i,j,x) = \frac{M}{N_i(M)}\Big[R(k)D_{ij}^U * \mathbf{1}_{\{x=\cdot\}}(k-1) - R(k-1)D_{ij}^U * \mathbf{1}_{\{x=\cdot\}}(k)$$
$$-R(k)\mathbf{1}_{\{i\in U\}}\Big(\sum_{t\in U}\alpha(t)\psi_{ti}\Big) * \mathbf{1}_{\{x\leq\cdot\}}(k-1)$$
$$+R(k-1)\mathbf{1}_{\{i\in U\}}\Big(\sum_{t\in U}\alpha(t)\psi_{ti}\Big) * \mathbf{1}_{\{x\leq\cdot\}}(k) - T_i(k)\Big].$$

We will obtain the desired result from the Central limit theorem for discrete time Markov renewal processes (see [PS64] and [MP68]).

Using Lemmas 2, 3 and 4, we obtain:

$$A_{ij} := \sum_{x=1}^{\infty} f(i,j,x)q_{ij}(x)$$
$$= \frac{M}{N_i(M)}\Big[R(k)D_{ij}^U * q_{ij}(k-1) - R(k-1)D_{ij}^U * q_{ij}(k)$$
$$-R(k)\mathbf{1}_{\{i\in U\}}\Big(\sum_{t\in U}\alpha(t)\psi_{ti}\Big) * q_{ij}(k-1)$$
$$+R(k-1)\mathbf{1}_{\{i\in U\}}\Big(\sum_{t\in U}\alpha(t)\psi_{ti}\Big) * q_{ij}(k) - T_i(k)p_{ij}\Big].$$

$$A_i := \sum_{j=1}^{s} A_{ij} = \frac{M}{N_i(M)}\Big[T_i(k) - T_i(k)\sum_{j=1}^{s} p_{ij}\Big] = 0.$$

>From Lemmas 3 and 4 we get

$$
\begin{aligned}
B_{ij} &:= \sum_{x=1}^{\infty} f^2(i,j,x) q_{ij}(x) \\
&= \Big(\frac{M}{N_i(M)}\Big)^2 \Big\{ R^2(k)(D_{ij}^U)^2 * q_{ij}(k-1) + R^2(k-1)(D_{ij}^U)^2 * q_{ij}(k) \\
&\quad + R^2(k)\mathbf{1}_{\{i\in U\}} \Big(\sum_{l=0}^{\cdot}\sum_{t\in U} \alpha(t)\psi_{ti}(l)\Big)^2 * q_{ij}(k-1) \\
&\quad + R^2(k-1)\mathbf{1}_{\{i\in U\}} \Big(\sum_{l=0}^{\cdot}\sum_{t\in U} \alpha(t)\psi_{ti}(l)\Big)^2 * q_{ij}(k) + T_i^2(k) p_{ij} \\
&\quad - 2R(k-1)R(k)\Big(D_{ij}^U (D_{ij}^U)^+\Big) * q_{ij}(k-1) \\
&\quad - 2R^2(k)\Big[D_{ij}^U(\cdot)\mathbf{1}_{\{i\in U\}} \Big(\sum_{l=0}^{\cdot}\sum_{t\in U} \alpha(t)\psi_{ti}(l)\Big)\Big] * q_{ij}(k-1) \\
&\quad + 2R(k-1)R(k)\Big[D_{ij}^U(\cdot)\mathbf{1}_{\{i\in U\}} \Big(\sum_{l=0}^{\cdot}\sum_{t\in U} \alpha(t)\psi_{ti}(l)\Big)^+\Big] * q_{ij}(k-1) \\
&\quad + 2R(k-1)R(k)\Big[(D_{ij}^U)^+(\cdot)\mathbf{1}_{\{i\in U\}} \Big(\sum_{l=0}^{\cdot}\sum_{t\in U} \alpha(t)\psi_{ti}(l)\Big)\Big] * q_{ij}(k-1) \\
&\quad - 2R^2(k-1)\Big[D_{ij}^U(\cdot)\mathbf{1}_{\{i\in U\}} \Big(\sum_{l=0}^{\cdot}\sum_{t\in U} \alpha(t)\psi_{ti}(l)\Big)\Big] * q_{ij}(k) \\
&\quad - 2R(k-1)R(k)\mathbf{1}_{\{i\in U\}}\Big[\Big(\sum_{l=0}^{\cdot}\sum_{t\in U} \alpha(t)\psi_{ti}(l)\Big)\Big(\sum_{l=0}^{\cdot}\sum_{t\in U} \alpha(t)\psi_{ti}(l)\Big)^+\Big] * q_{ij}(k-1) \\
&\quad - 2T_i(k)\Big[R(k)D_{ij}^U * q_{ij}(k-1) - R(k-1)D_{ij}^U * q_{ij}(k) \\
&\qquad - R(k)\mathbf{1}_{\{i\in U\}}\Big(\sum_{t\in U}\alpha(t)\psi_{ti}\Big) * Q_{ij}(k-1) \\
&\qquad + R(k-1)\mathbf{1}_{\{i\in U\}}\Big(\sum_{t\in U}\alpha(t)\psi_{ti}\Big) * Q_{ij}(k)\Big]\Big\}.
\end{aligned}
$$

$$
\begin{aligned}
B_i &:= \sum_{j=1}^{s} B_{ij} \\
&= \left(\frac{M}{N_i(M)}\right)^2 \Big\{ R^2(k) \sum_{j=1}^{s} \Big[D_{ij}^U - \mathbf{1}_{\{i\in U\}} \sum_{t\in U} \alpha(t)\Psi_{ti} \Big]^2 * q_{ij}(k-1) \\
&+ R^2(k-1) \sum_{j=1}^{s} \Big[D_{ij}^U - \mathbf{1}_{\{i\in U\}} \sum_{t\in U} \alpha(t)\Psi_{ti} \Big]^2 * q_{ij}(k) - T_i^2(k) \\
&+ 2R(k-1)R(k) \sum_{j=1}^{s} \Big[\mathbf{1}_{\{i\in U\}} D_{ij}^U \sum_{t\in U} \alpha(t)\Psi_{ti}^+ + \mathbf{1}_{\{i\in U\}} (D_{ij}^U)^+ \sum_{t\in U} \alpha(t)\Psi_{ti} \\
&- (D_{ij}^U)^+ D_{ij}^U - \mathbf{1}_{\{i\in U\}} \Big(\sum_{t\in U} \alpha(t)\Psi_{ti}\Big)\Big(\sum_{t\in U} \alpha(t)\Psi_{ti}^+\Big) \Big] * q_{ij}(k-1) \Big\}.
\end{aligned}
$$

Since $N_i(M)/M \xrightarrow[M\to\infty]{a.s.} 1/\mu_{ii}$ (see, e.g., [LO01]), applying the central limit theorem, we obtain the desired result. □

Proof of Corollary 1. The relation between the BMP-failure rate and the RG-failure rate can be written in the form $r(k) = \phi(\lambda(k))$, where ϕ is the function defined by $\phi(x) := -\ln(1-x)$. Using delta method and the asymptotic normality of the BMP-failure rate (see Theorem 3), we obtain that the RG-failure rate converges in distribution to a zero mean normal random variable with the variance $\sigma_r^2(k)$, given by

$$
\sigma_r^2(k) = \Big(\phi'(\lambda(k))\Big)^2 \sigma_\lambda^2(k) = \frac{1}{(1-\lambda(k))^2}\sigma_\lambda^2(k).
$$

□

5 Numerical Example

In this section we apply the previous results to a three-state discrete time semi-Markov process described in Figure 1.

Let us consider that the state space $E = \{1, 2, 3\}$ is partitioned into the up-state set $U = \{1, 2\}$ and the down-state set $D = \{3\}$. The system is defined by the initial distribution $\mu := (1\ 0\ 0)$, by the transition probability matrix V of the embedded Markov chain $(J_n;\ n \in \mathbb{N})$

$$
V := \begin{pmatrix} 0 & 1 & 0 \\ 0.95 & 0 & 0.05 \\ 1 & 0 & 0 \end{pmatrix}
$$

and by the conditional distributions of the sojourn time

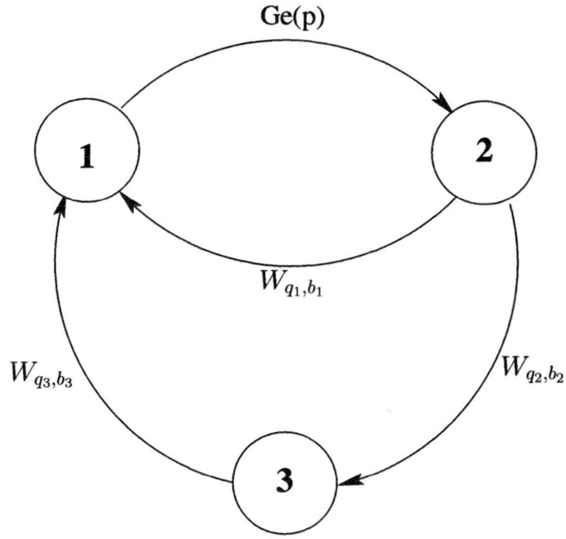

Fig. 1. A three-state discrete time semi-Markov system

$$f(k) := \begin{pmatrix} 0 & f_{12}(k) & 0 \\ f_{21}(k) & 0 & f_{23}(k) \\ f_{31}(k) & 0 & 0 \end{pmatrix}, \; k \in \mathbb{N}.$$

We consider the following distributions for the conditional sojourn time:

- f_{12} is a geometric distribution defined by

$$f_{12}(0) := 0,$$
$$f_{12}(k) := p(1-p)^{k-1}, k \geq 1,$$

  where we take $p = 0.8$.
- $f_{21} := W_{q_1,b_1}$, $f_{23} := W_{q_2,b_2}$ and $f_{31} := W_{q_3,b_3}$ are discrete time, first type Weibull distributions, defined by

$$W_{q,b}(0) := 0,$$
$$W_{q,b}(k) := q^{(k-1)^b} - q^{k^b}, k \geq 1,$$

  where we take $q_1 = 0.5, b_1 = 0.7, q_2 = 0.6, b_2 = 0.9, q_3 = 0.5, b_3 = 2$ (for discrete time Weibull distribution, see, e.g., [Bra01], [BGX01]).

The empirical estimator and confidence interval at level 95% for the BMP-failure rate and the RG-failure rate of the system, for the total time of observation $M = 5000$, are given in Figure 2.

Figure 3 gives a comparison between failure rates estimators obtained for different sample sizes. We see that, as M increases, the estimators approach the true value. We also notice that the failure rates become constant as time

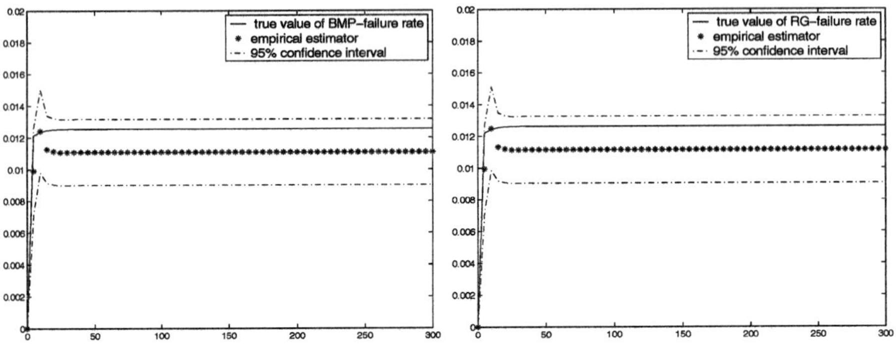

Fig. 2. Failure rates estimators and confidence interval at level 95%

increases, that is the semi-Markov system approaches a Markov one as time increases.

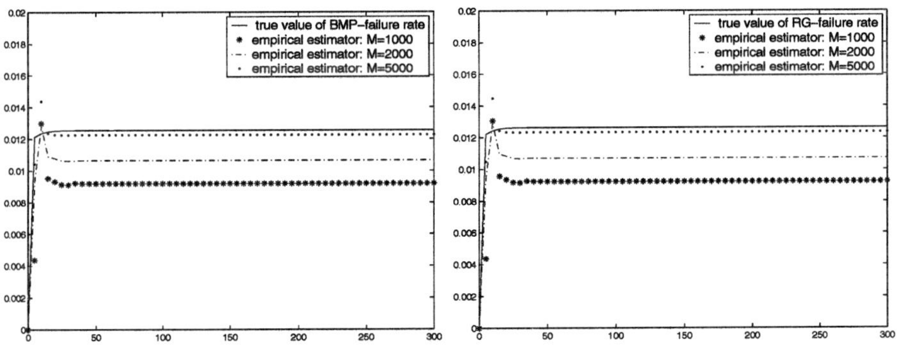

Fig. 3. Failure rates estimators consistency

Figure 4 presents the failure rates variances estimators for some different values of the sample size. As in Figure 3, we can note the consistency of variances estimators.

References

[BBL04] Barbu, V., Boussemart, M., Limnios, N.: Discrete time semi-Markov model for reliability and survival analysis. To appear in: Communications in Statistics-Theory and Methods, **33** (11) (2004)

[BL04a] Barbu, V., Limnios, N.: Discrete Time Semi-Markov Processes for Reliability and Survival Analysis - A Nonparametric Estimation Approach. In Nikulin, M., Balakrishnan, N., Meshbah, M., Limnios, N. (eds) Parametric and Semiparametric Models with

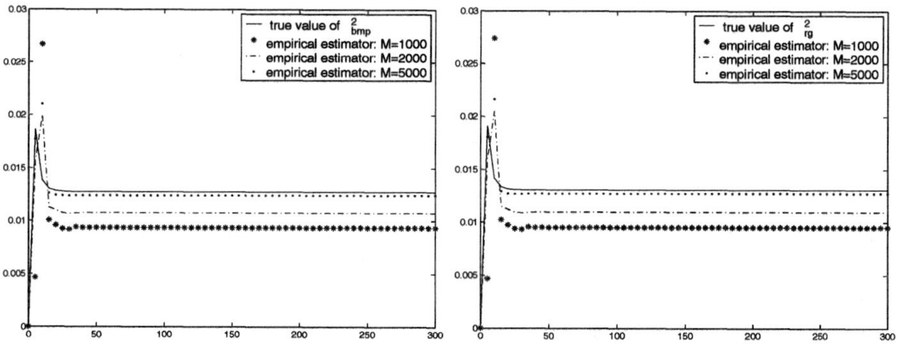

Fig. 4. Consistency of failure rates variances estimators

Applications to Reliability, Survival Analysis and Quality of Life. Birkhäuser, Boston, 487–502 (2004)

[BL04b] Barbu, V., Limnios, N.: Nonparametric estimation for discrete time semi-Markov processes with applications in reliability. Submitted for publication (2004)

[BMP63] Barlow, R.E., Marshall, A.W., Prochan, F.: Properties of probability distributions with monotone hazard rate. Annals of Mathematical Statistics, **34**, 375–389 (1963)

[BP75] Barlow, R.E., Prochan, F.: Statistical theory of reliability and life testing: probability models. Holt, Rinehart and Winston, New York (1975)

[Bra01] Bracquemond, C.:Modélisation stochastique du vieillissement en temps discret. PhD Thesis, Laboratoire de Modélisation et Calcul, Grenoble (2001)

[BGX01] Bracquemond, C., Gaudoin, O., Xie, M.: Towards a new definition of failure rate for discrete distributions. 10th International Symposium on Applied Stochastic Models and Data Analysis, Vol. 2, 266–270 (2001)

[Cse02] Csenki, A.: Transition analysis of semi-Markov reliability models - A tutorial review with emphasis on discrete-parameter approaches. In: Osaki, S. (ed) Stochastic Models in Reliability and Maintenance. Springer, Berlin, 219–251 (2002)

[How71] Howard, R.: Dynamic Probabilistic systems. vol. II, Wiley, New York (1971)

[LO99] Limnios, N., Oprişan, G.: A unified approach for reliability and performability evaluation of semi-Markov systems. Appl. Stochastic Models Bus. Ind., **15**, 353–368 (1999)

[LO01] Limnios, N., Oprişan, G.: Semi-Markov Processes and Reliability. Birkhäuser, Boston (2001)

[LO03] Limnios, N., Ouhbi, B.: Empirical estimators of reliability and related functions for semi-Markov systems. In: Lindqvist, B., Doksum, K.A. (eds) Mathematical and Statistical Methods in Reliability. World Scientific Publishing, **7** (2003)

[MS00] Mode, C.J., Sleeman, C.K.: Stochastic processes in epidemiology. World Scientific, New Jersey (2000)

[MP68] Moore, E.H., Pyke, R.: Estimation of the transition distributions of a Markov renewal process. Ann. Inst. Statist. Math., **20**, 411–424 (1968)

[NO75] Nakagawa, T., Osaki, S.: The discrete Weibull distribution. IEEE Trans. On Reliability, **R-24**, 300–301 (1975)

[OL99] Ouhbi, B. Limnios, N.: Nonparametric estimation for semi-Markov processes based on its hazard rate functions. Statistical Inference for Stochastic Processes, **2** (2), 151–173 (1999)

[OL98] Ouhbi, B., Limnios, N.: Non-parametric failure rate estimation of semi-Markov systems. In: Janssen, J., Limnios, N. (eds) Semi-Markov models and applications. Kluwer, Dordrecht, 207–218 (1998)

[PS64] Pyke, R., Schaufele, R.: Limit theorems for Markov renewal processes. Annals of Mathematical Statistics, **35**, 1746–1764 (1964)

[RG92] Roy, D., Gupta, R.: Classification of discrete lives. Microelectron. Reliab., **32** (10), 1459–1473 (1992)

[SL02] Sadek, A., Limnios, N.:Asymptotic properties for maximum likelihood estimators for reliability and failure rates of Markov chains. Communications in Statistics-Theory and Methods, **31** (10), 1837–1861 (2002)

Some recent results on joint degradation and failure time modeling

Vincent Couallier

Equipe Statistique Mathematique et ses Applications
U.F.R. Sciences et Modelisation, Universite Victor Segalen Bordeaux 2
146 rue Leo Saignat 33076 Bordeaux cedex FRANCE
couallier@sm.u-bordeaux2.fr

Key words: degradation process, hazard rate, hitting time, regression, correlated errors, generalized least squares estimation, Nelson-Aalen estimate, semiparametric estimation

1 Introduction

Analyzing survival data is historically and classically based on a sample of n real and non negative random variables $(T_1, ..., T_n)$ each measuring a individual time to event. This event can reflect a time of diagnosis, a death, a failure time, a breakdown or every change in the status of an individual but whatever the field of applications is, biostatistics, industrial statistics or econometrics, modelling or making inference on the distribution of the random variable T often requires additional data and precise definition of the sampling process. Some well known fields of research are devoted to failure time regression data, frailty models or competing risk models. Usually, one attempts to conditionally define the reliability characteristics such that the survival function, the hazard rate or the cumulative hazard rate given some random covariate describing either the frailty of the item, or the environmental conditions.

For instance, conditionally on the random vector A, the famous Proportional Hazard rate model by Cox specifies that the hazard rate verifies $\lambda_T(t|A) = \lambda_o(t) \times e^{\beta^T A}$ where λ_o is a baseline hazard function and β is a vector of parameters. This model was first defined for constant in time covariate but it is possible to allow a time-varying effect of the environment on the survival of the item. [WT97] study the model $\lambda_T(t|A) = \lambda_o(t) \times e^{\beta(A_1+A_2t)}$ where $A = (A_1, A_2)$ is some random but fixed in time vector of coefficients. Also, this unit-to-unit variability in the definition of the hazard rate can be interpreted as an individual frailty, modelled by a stochastic process and reflecting an internal accumulation of wear called aging or degradation process.

In fact an increasing hazard rate usually describes degradation in engineering applications [FIN03]. [BN04] define the conditional hazard rate given the random vector A as $\lambda(t|A) = \lambda_o(t) \times \lambda(g(t, A))$ where g is a given non decreasing function. Is is strongly related to the model of Wulfsohn and Tsiatis but the assumptions made for estimation and inference are completely different. If we consider that $g(., A)$ is the degradation function with random coefficient A, we obtain a model where the degradation is varying unit-to-unit and influences the hazard rate of a so called traumatic failure time.

2 Joint models for degradation and failure time modeling

In this section, we are interested in defining the failure time due to wear or aging. This point has been an interesting field of research for the last twenty years. The most important probabilistic issue being to jointly model the degradation process and the associated failure times. Some recent results use the fact that when longitudinal data of degradation or markers of it can be measured before the failure occurs, it is possible to estimate reliability characteristics of the item. Numerous models exist now and we will present some of the most important ones.

Two main joint models exist. The first one considers a failure time which is directly defined by the degradation process, the second one considers that the degradation process influences the hazard rate. Obviously some joints models include both by considering multiple failure modes.

Let us assume that the degradation of an item is given by a real-valued right continuous and left hand limited stochastic process $Z(t)$, $t \in I$. Some well known reliability characteristics will be defined conditionally on Z but even without any information on Z, we assume that the life time T_0 is in fact the first time of crossing a ultimate threshold z_0 (which can be random) for Z(t)

$$T_0 = \inf\{t \in I, Z(t) \geq z_0\}, \tag{1}$$

if the degradation process tends to increase with time (the definition can obviously be reversed if Z models an anti-aging process). We shall define as well the failure time T with the conditional survival function given the past degradation process as

$$P(T > t|Z(s), 0 \leq s \leq t) = \exp\left(-\int_0^t \lambda_T(s|Z(u), 0 \leq s \leq u)ds\right). \tag{2}$$

The failure time T_0 is sometimes called soft failure (or failure directly due to wear) because in most of industrial applications, z_0 is fixed and the experiment is voluntarily ceased at the time the degradation process reaches the level z_0. Even if the degradation process is completely unknown, it can be useful and meaningful to analyze the links between the assumed distribution function of T_0 and the definition of the underlying degradation process [AG01]. This issue is strongly related to the theory of hitting time of stochastic processes.

In the following section, we recall some well known results dealing with hitting times such that T_0. The analysis of the traumatic failure time T is related to accelerated life models and is postponed to Sect. 2.2.

2.1 Failure time as hitting times of stochastic processes

stochastic degradation defined as diffusion

The degradation process is often assumed to be defined as the solution of a stochastic diffusion, for instance

$$dX(t) = \mu(X(t))dt + \sigma X(t)d\gamma(t),$$

where γ is often a Wiener process. The degradation path can also been given directly as the sample path of a given stochastic process. The gaussian process with positive drift

$$X(t) = a + bt + W(t),$$

can bc casily defined in both manners. In these cases, it is possible to link the distribution of the failure time T_0 to the diffusion process and most of the parametric choices for survival distributions have a stochastic justification in terms of an unknown degradation process suitably defined. For instance, [SIN95] recalls that if Z_0 is fixed then a Wiener process which starts out at a deterministic X_0 will have a time to absorption with an inverse gaussian distribution. According to [COX99] this fact has sometimes been used to suggest the Inverse-Gaussian distribution as a good one for parametric analysis of failure data but this simple case has deeper generalizations and we refer to [SIN95] and [AG01] for further details.

Estimation procedures for degradation data based on Wiener processes (eventually measured with errors) have extensively been studied by [WHI95], [DN95], [WS97], [WCL98] and [PT04] among others. The mathematical property of independency of the increments of degradation data is obviously used to derive the likelihood functions needed for the estimation in such parametric models.

Stochastic processes such that general Levy processes, gamma processes or shock processes have also been studied as accumulative degradation processes. All of these processes have non-decreasing sample paths what is meaningful for the accumulation of wear.

A gamma process as degradation process

Following [BN01], the family of Gamma processes with right-continuous trajectories ensures the growth of the paths and the fact that, denoting

$$Z(t) = \sigma^2 \gamma(t) \quad \text{with } \gamma(t) \sim G(1, \nu(t)) = G(1, \frac{m(t)}{\sigma^2}),$$

where $G(1, \nu(t))$ is a Gamma distribution with parameters 1 and $\nu(t)$, the increment from t to $t + \Delta t$ for $\Delta t > 0$ lies in the same family of distributions since $Z(t + \Delta t) - Z(t) = \sigma^2 \Delta\gamma(t)$ and

$$\Delta\gamma(t) = \gamma(t + \Delta t) - \gamma(t) \sim G(1, \frac{\Delta m(t)}{\sigma^2}).$$

Then

$$\mathrm{E}Z(t) = m(t) \quad \text{and} \quad \mathrm{E}\Delta Z(t) = \Delta m(t).$$

A useful property for estimation procedures is the independency of increments. Parametric models include functions $m(t) = m(t, \theta)$ where θ is a random coefficient but semiparametric methods (where m is nonparametric) have also been developed in [BN01]. Note that covariates describing some environmental stress are also considered and influence the degradation function as in an accelerated failure time model. [COU04] has studied the numerical properties in both models for constant in time stresses. See Also [LC04] who have recently used these methods to degradation data in a stressed environment with random effects.

A marked point process as degradation

When the degradation consists in an accumulation of shocks, each of them leading to a random increment of degradation, it is possible to model it by a marked point process $\Phi = (T_n, X_n)_{n \geq 1}$ where T_n is the time of the n-th shock and X_n is the size of the corresponding increment of degradation. Thus

$$Z(t) = \sum_{i=1}^{+\infty} I(t_n \leq t) X_n = \sum_{i=1}^{N(t)} X_i$$

is the sample path of the degradation and $\Phi(t) = \sum_{i=1}^{+\infty} I(t_n \leq t)$ is the counting process associated to the marked point process.

In these models also, a failure time can be defined as first crossing time of a level z_0. Due to the piecewise constant shape of the degradation, the failure time T_0 is necessarily one of the times of shock. For a fixed level z_0, some parametric estimations and large sample results have recently been provided in [WK04] under the multiplicative intensity assumption of Aalen. It assumes that the stochastic intensity function of the point process $(T_n)_{n \geq 1}$ is the product of a random variable Y and a deterministic function $\eta(t), t \in \mathbb{R}^+)$. Hence, given $Y = y$, the point process Φ is a Poisson process with deterministic intensity function $y.\eta(.)$. The process Φ is called a doubly stochastic Poisson process. It includes for different choices of the distribution function of Y and η the mixed Poisson process and the non homogeneous Poisson process.

In order to describe the distribution function of the soft failure T_0, it is necessary to jointly define the intensity function of the point process and the distribution function of the marks. The simpler hypothesis is the independency between T_n and X_n for all $n \geq 1$. A dependent position distribution is easily defined if we let $X_n = U_n e^{\delta T_n}$ where $(U_n)_{n \geq 1}$ is a i.i.d. sequence of random variables. The condition $\delta > 0$ allows increasing effects of shocks on the degradation level and $\delta < 0$ leads to absorbed shocks. [WK04] and [KW04] discuss estimation procedures for inverse gaussian and gamma-type distribution of Y and different shapes of the function η (see also [BN02] chap.13).

A mixed regression as degradation process : the general path model

The real degradation is defined by the sample path of the process

$$Z(t) = g(t, \theta), \tag{3}$$

where g is a deterministic non decreasing continuous function of the time and of the random coefficient $\theta \in \mathbb{R}^p$. The distribution function F_θ of θ is unknown. In this case, the hitting time of a given threshold z_0 is easily found by inverting g in its first parameter. There exists h such that

$$Z(T_0, \theta) = z_0 \leftrightarrow T_0 = h(z_0, \theta)$$

Thus, the distribution function of the failure time T_0 can be written in terms of z_0, h and of the distribution function F_θ of θ.

For instance, [OC04] study linear degradation $g(t, \beta_1, \beta_2) = \beta_1 + \beta_2 t$ with fixed β_1 and random β_2 following Weibull(α, γ) or log-normal(μ, σ^2) distributions. Hence T_0 follows a reciprocal Weibull or log-normal distribution respectively (see also [ME98]). In the case of noised degradation values, random coefficients are estimated unit-to-unit and a pseudo-failure time $\hat{T_0}$ is deduced from the degradation data. Finally classical survival analysis is applied to these pseudo failure times to analyze and estimate the real distribution function of the failure time T_0.

The observed degradation of the i-th item is often a partially and erroneous version of the real degradation. It is defined by the the vector of the m_i measurements of the real degradation at given times $t_{i1} < .. < t_{im_i}$

$$Z_{ij} = Z(t_{ij}) = g(t_{ij}, \theta_i) + \epsilon_{ij}, \tag{4}$$

where the ϵ_i are zero mean random variables. [BN04] have also assumed multiplicative errors in (4). In this case, we assume

$$Z_{ij} = Z(t_{ij}) = g(t_{ij}, \theta_i)\epsilon_{ij} \tag{5}$$

and using $Y_{ij} = \ln Z_{ij} = \ln Z(t_{ij}) = \ln g(t_{ij}, \theta_i) + \ln \epsilon_{ij} = \tilde{g}(t_{ij}, \theta_i) + \tilde{\epsilon}_{ij}$, as transformed degradation data, estimation procedures with additive noise are available (with suitable assumptions on $\tilde{\epsilon}_{ij}$).

Numerous models are based on (3) and (4) or (5) and differ only by assumptions made on the function g (or $\tilde{g}$) and the distributions of θ and ϵ (or $\tilde{\epsilon}$). [LM93] and [ME98] for instance use non linear mixed effect models in which g has a specified shape, θ follows a multivariate gaussian distribution and unknown parameters are estimated by maximizing the global likelihood function. The distribution function of T_0 is numerically estimated with Monte-Carlo method. The method is illustrated on the famous Fatigue Crack Growth Data by [BK85]. In the same framework and on the same data set, [RC00] use bayesian approach to estimate the unknown reliability characteristics such the failure time distribution of T_0 and [GLJ03] propose an estimation procedure based on artificial neural network.

A slightly different approach consists in supposing that the unit-to-unit variability of the degradation paths has a part explained by the environmental conditions, that is θ is a random parameter whose distribution depends on a external covariate describing the stress. In a purely parametric model, [YU03] and [YU04] study optimal designs of accelerated degradation experiments where a transformed degradation function is $g(t) = -\beta t^{\alpha}$, α is fixed and known, β follows a Log-Normal(μ, σ^2) or a reciprocal-Weibull distribution (with parameters varying with the stress) and the errors are gaussian.

Also [BBK04] study a linear degradation model with multiple failure modes. The degradation process is $Z(t) = t/\theta$ and is observed without error. This is a very simple path model but the complexity relies on the fact that no parametric assumption is made on the distribution of θ and some competing failure times are censoring the degradation process.

2.2 Failure times with degradation-dependent hazard rate

The failure time T_0 defined above and named soft failure is directly due to the evolution of the degradation because it is a crossing time. Another way to model a link between the degradation and a failure time is to consider that the degradation level is an internal covariate influencing the survival function of a traumatic failure T, through the conditional definition

$$P(T > t|Z(s), 0 \leq s \leq t) = \exp\left(-\int_0^t \lambda_T(s|Z(u), 0 \leq s \leq u)ds\right).$$

[COX99] notes that the essential point is that given the degradation history $\{Z(s), 0 < s < t\}$, the hazard rate λ_T does not depend on time and considers as an example that $\lambda_T(t|Z(u), 0 \leq u \leq t) = \alpha + \beta Z(t)$.

Remark 1 : This definition is equivalent to assuming that T is the first time a doubly stochastic Poisson process with intensity $\lambda_T(s|Z(u), 0 \leq s \leq u)$ jumps (see [BN02] chap.3).

Remark 2 : This definition has an analogy with the hitting-time definition of lifetime seen in Sect. 2.1. In fact for T_0 defined as the first time the degradation process reaches the threshold z_0, we have

$$\begin{aligned} P(T_0 > t|Z(s), 0 \le s \le t) &= P(z_0 > Z(t)|Z(s), 0 \le s \le t) \\ &= G_0(Z(t)) \end{aligned}$$

where G_0 is the survival function of z_0. Hence, if z_0 is fixed then

$$G_0(Z(t)) = 1_{\{Z(t)<z_0\}}$$

because G_0 is the survival function of the dirac random variable whose realizations give z_0 almost surely.

Definition (2) is related to some well known accelerated failure time model with external time-varying covariate $X(t), t > 0$ satisfying

$$P(T > t|X(s), 0 < s < t) = G\Big(\int_0^t \psi(X(s), \beta)ds\Big)$$

where G is a survival function and ψ is a positive function in the space of covariate values sometimes called transfer function or ideal time scale ([BN97], [DL00], [DL02]). Such conditional survival functions have also been studied by [YM97].

2.3 The joint model : a mixed regression model with traumatic censoring

Two failure mode are considered here. The failure time T_0 is the first time the unknown real degradation $Z(t) = g(t, \theta)$ reaches a given threshold z_0. As in Sect. 2.2, the traumatic failure time T is defined through its conditional survival function given the past degradation

$$P(T > t|Z(s), 0 \le s \le t) = \exp\left(-\int_0^t \lambda_T(s|Z(u), 0 \le s \le u)ds\right)$$

Noting that Z is a parameterized degradation function with random parameter θ, we restrict ourself here to the following assumption for conditional hazard rate λ_T.

$$\lambda_T(t|Z(u), 0 \le u \le t) = \lambda_T(t|\theta) = \lambda(g(t, \theta)) \qquad (6)$$

where λ is a nonparametric function in the degradation domain.

For the item i, define T_0^i the random variable defined as the hitting time of the threshold degradation z_0 and T^i the failure time whose conditional hazard rate is defined in (6). The last values of degradation can be censored either by the failure time T_0 due to degradation or the traumatic failure time T. If the failure is due to degradation, we define $\tau^i \in \{t_{ij}, i = 1..m_i\}$ such that the degradation $Z^{obs}(\tau^i)$ at time τ^i is the first degradation value strictly greater than z_0. f_i is the number of degradation values and is such that $t_{if_i} = \tau_i$. If the failure is due to traumatism, we define $\tau^i \in \{t_{ij}, i = 1..m_i\}$ such that τ^i is the last time of measurement before T. Then the data for the item i are the date of failure $U^i = \min(\tau^i, T^i, t_{im_i})$, a discrete variable δ^i giving the cause of failure and the $f_i \leq m_i$ noised measures of degradation $0 < Z_{i1} < ... < Z^{obs}(\tau_i) = Z_{if_i}$. These degradation data can follow a regression model like (4) or (5). In the following, we develop such models. Note that [HA03] use the same data to fit a purely parametric model (using Weibull distribution) in a competing risks framework but they do not infer on the degradation curve. The degradation distributions are only inferred at times $t_1, ..., t_m$ but the longitudinal aspect of the data is not considered.

3 Some recent results in semiparametric estimation in the general path model

We assume here that the degradation evolves in the same manner an unknown function g growths with time. The function g is a continuous parametrized function with unit-to-unit dependent parameters. Hence the *real unobserved* degradation of the i-th item is

$$Z_{real}(t) = g(t, \theta_i) \tag{7}$$

where $\theta_i \in \mathbb{R}^p$ is random vector of parameters with unknown distribution.

In order to get explicit formulae, we first restrict ourself to functions $g(., \theta)$ in (7) leading to a linearized problem of estimation. Nonlinear regression models are obviously useful in some applications but need numerical optimizations which do not provide estimates in closed form.

3.1 Linear estimation

For each item i, $i = 1..n$, we assumed that we have at our disposal the survival time U_i, the indicator δ_i and the degradation data as described in Sect. 2.3. The model considered here assumes that there exists a function $\mathcal{F}$ such that the f_i available measurements $Z_{i1}, ..Z_{if_i}$ at times $t^i = (t^i_1, .., t^i_{f_i})'$ of the degradation level of the i-th item satisfy

$$Y_{ij} = \mathcal{F}(Z_{ij}) = \tilde{t}^i \tilde{\theta}_i + \tilde{\epsilon}_{ij}, \quad j = 1..f_i, \tag{8}$$

where $\tilde{\theta}_i \in \mathbb{R}^p$ is a function of the random vector $\theta_i \in \mathbb{R}^p$ and $\tilde{t}^i = (\tilde{t}^i_1, .., \tilde{t}^i_{f_i})'$ is a $f_i \times p$-design matrix whose j-th raw is the raw-vector $\tilde{t}^i_j \in \mathbb{R}^p$ function of t_{ij}. Finally $(\tilde{\epsilon}^i_j)_{j=1,..,f_i}$ is the vector of additive noises of the i-th item. Thus the function $\mathcal{F}$ permits to get a linear expression of the degradation

in a transformed scale of time and we denote by Y^i the vector of measures for the item i. For instance in [BN04], the real degradation of the i-th item, i=1,..,n, is $Z(t) = e^{a_1^i}(1+t)^{a_2^i}$, $t \in \mathbb{R}^+$ where $(a_1^i, a_2^i) = \theta_i$ is a random vector with unknown distribution function F_θ, and the data consist in $Z_{ij} = Z(t_{ij})U_{ij}, j = 1,..,f_i, \quad i = 1,..,n$. In fact, taking $Y_{ij} = \ln Z_{ij}$, we get that $Y_{ij} = a_1^i + a_2^i \ln(1+t_{ij}) + \ln \epsilon_{ij}$ and thus in that case $\tilde{\theta}^i = \theta_i$, $\tilde{t}_j^i = (1, \ln(1+t_{ij}))$ and $\tilde{\epsilon}_j^i = \ln \epsilon_{ij}$.
Three main assumptions are made for the correlation structure of the noises namely

H1 $(\tilde{\epsilon}_j^i)_{j=1..f_i}$ are i.i.d. random variables with

$$\mathrm{E}\tilde{\epsilon}_j^i = 0, \ \mathrm{cov}[\tilde{\epsilon}_{j_1}^i, \tilde{\epsilon}_{j_2}^i] = \sigma_i^2 \mathbf{1}_{\{j_1=j_2\}}$$

H2 $(\tilde{\epsilon}_j^i)_{j=1..f_i}$ are identically distributed random variables with

$$\mathrm{E}\tilde{\epsilon}_j^i = 0, \ \mathrm{cov}[\tilde{\epsilon}_{j_1}^i, \tilde{\epsilon}_{j_2}^i] = \sigma_i^2 \left(c(t_{j_1}^i) \wedge c(t_{j_2}^i)\right)$$

where c is a positive nondecreasing function.

H3 $(\tilde{\epsilon}_j^i)_{j=1..f_i}$ are identically distributed random variables with

$$\mathrm{E}\tilde{\epsilon}_j^i = 0, \ \mathrm{cov}[\tilde{\epsilon}_{j_1}^i, \tilde{\epsilon}_{j_2}^i] = \sigma_i^2 \phi^{|t_{j_1}^i - t_{j_2}^i|}$$

where $0 \le \phi < 1$.

If the errors are gaussian, under H3, $\tilde{\epsilon}_j^i = \sigma^2 W(c(t_j^i))$ where W is a brownian motion. Under H2, if the times $(t_j^i)_{(j)}$ are equidistant, the $\tilde{\epsilon}_j^i$ is a AR(1) sequence of noises.

For item i, denote by $\sigma_i^2 \Sigma_i$ the $f_i \times f_i$-variance matrix of $(\tilde{\epsilon}_j^i)_{(j)}$. Hence a predictor for θ_i is found by generalized least square estimation

$$\hat{\theta}_i = \mathrm{argmin}_{a \in \mathbb{R}^p} (Y^i - \tilde{t}^i a)' \Sigma_i^{-1} (Y^i - \tilde{t}^i a) \tag{9}$$

denoting X' for the transposition of matrix X. We recall the following well known results of generalized least square estimation

Proposition 1 : *The predictor*

$$\hat{\theta}_i = (\tilde{t}^{i'} \Sigma_i^{-1} \tilde{t}^i)^{-1} \ \tilde{t}^i \ \Sigma_i^{-1} Y^i$$

is also the ordinary least square estimate of the transformed data

$$\hat{\theta}_i = argmin_{a \in \mathbb{R}^p} \| R^i Y^i - R^i \tilde{t}^i a \|,$$

where $R^i = (U^{i'})^{-1}$ *and* U^i *satisfies* $\Sigma^i = U^{i'} U^i$

Proposition 2 :

under H1 we have $R = Id$,
under H2 $diag(R) = \left(1, 1/\sqrt{c(t_2^i) - c(t_1^i)}, ..., \sqrt{c(t_{f_i}^i) - c(t_{f_i-1}^i)}\right)$, $R_{i+1,i} = -R_{i+1,i+1}$ *and* $R_{ij} = 0$ *elsewhere,*
under H3 with equally spaced measures, $diag(R) = (\sqrt{1-\phi^2}, 1.., 1)$, $R_{i+1,i} = -\phi$ *and* $R_{i,j} = 0$ *elsewhere.*

These results use the Cholesky decomposition of matrices and are valid if the nuisance parameter Φ and c are known. If ϕ is unknown, it can be estimated with two-step procedures or iterative estimation ([SW03], p. 279). Interestingly, under H2 the estimation procedure involves only the standardized increments $(Y_j^i - Y_{j-1}^i)/\sqrt{c(t_j^i) - c(t_{j-1}^i)}$ and we have

Proposition 3 : *In the model*

$$Y_{ij} = \theta_1^i + \theta_2^i f(t_j^i) + \tilde{\epsilon}_j^i,$$

where f is any increasing function, under H2 we obtain

$$\theta_1^i = \frac{c(t_1^i)}{\gamma - c(t_1^i)}(\gamma Y_{i1} - c(t_1^i)\Delta' Y^i) \quad \theta_2^i = \frac{c(t_1^i)}{\gamma - c(t_1^i)}(\Delta' Y^i - Y_{i1})$$

where, if we denote $dl(t_j^i) = l(t_{j+1}^i) - l(t_j^i)$ *for any function* l *with* $dl(t_0^i) = dl(t_{f_i}^i) = 0$,

$$\gamma = \sum_{j=0}^{f_i-1} df(t_j^i)^2/dc(t_j^i)$$

and

$$\Delta = \left(df(t_{j-1}^i)/c(t_{j-1}^i) - df(t_j^i)/c(t_j^i)\right)_{j=1..f_i}$$

The simple case $f = c$ in found in [BN04] where $f(t) = c(t) = ln(1+t)$. In this case the calculations reduce to

$$\begin{pmatrix} \hat{\theta}_1^i \\ \hat{\theta}_2^i \end{pmatrix} = \begin{pmatrix} Y_1^i - \hat{\theta}_2^i c(t_1^i) \\ (Y_{f_i}^i - Y_1^i)/(f(t_{f_i}^i) - f(t_1^i)) \end{pmatrix},$$

and thus the estimate is not consistent in this case.

3.2 Nonlinear estimation

In the nonlinear case, whatever the hypothesis about the correlation of the noises is, the predictor $\hat{\theta}^i$ of θ^i is found by least square minimization

$$\hat{\theta}^i = \text{argmin}_{a \in \mathbb{R}^p}(Y^i - g(t^i, a))' \Sigma_i^{-1}(Y^i - g(t^i, a)) \quad i = 1..n$$

where $g(t^i, a)$ is the vector in $\mathbb{R}^{f_i}$ of the values $(g(t_{ij}, a))_{j=1..f_i}$. Under H1, [LM93] provide direct estimation of the distribution of the θ_i via a parametric assumption $\theta_i \sim \mathcal{N}(\beta, \Sigma)$, $i = 1..n$ and estimate β and Σ in a nonlinear mixed effect model with maximum likelihood estimators. Here we do not make such assumption. Thus we have to construct some predictor $\hat{\theta}_i$ of the random coefficient θ_i and shall plug these predictors in some non parametric estimate of F_θ in the next section. Closed form for these predictors are not available but numerous numerical optimization procedures exist. We illustrate two well known algorithms on the Fatigue crack size propagation Alloy-A data ([LM93]). These data represent the growth of cracks in metal for 21 test units with a maximum of twelve equally spaced measurements. Testing was stopped if the crack length exceeded 1.60 inches which defines the threshold z_0. No traumatic failure mode is defined in this example. The Paris growth curve

$$g(t, m, C) = \left(0.9^{\frac{2-m}{2}} + \frac{2-m}{2} C \sqrt{\pi}^m t\right)^{\frac{2}{2-m}}$$

with unit-to-unit coefficients $\theta^i = (m_i, C_i)$ is fitted on each item. The aim is thus to estimate the distribution function $F_{(m,C)}$ with the noised measurements of degradation under hypothesis H1.

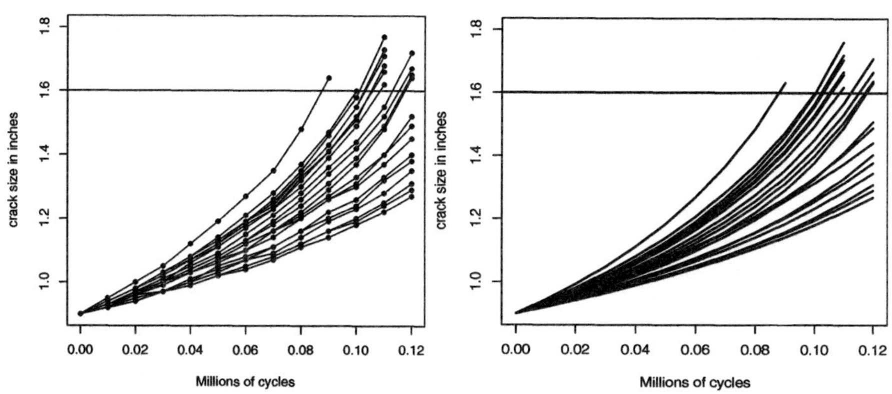

Fig. 1. Fatigue Crack Size for Alloy-A and Nonlinear fitting of Paris curves with nls and gnls

Under H1, We used a Gauss-Newton algorithm implemented by Bates and DebRoy in *nls* function of Splus and *R* softwares ([BC92]). A similar algorithm minimizing generalized least squares and allowing correlated errors is obviously useful under H2 and H3. Thus we tested also the *gnls* function by Pinheiro and Bates in *R*. Both methods give a very well fit to the data but the predictions present some little difference as it is shown in figure 2.

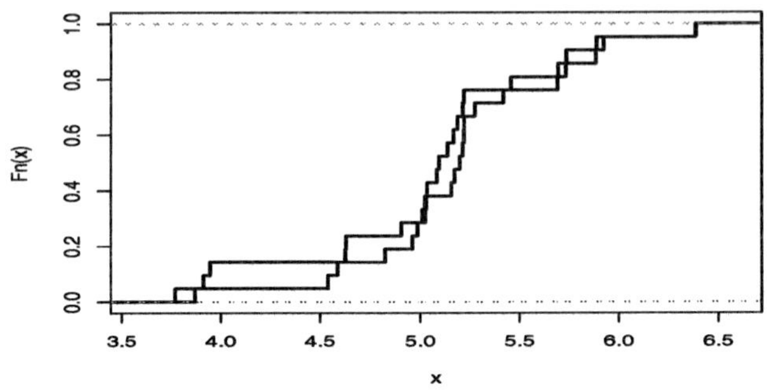

Fig. 2. E.c.d.f. of m with both methods

In fact we could compare the methods by calculating the empirical mean square error. In this case, the mean square error of method nls in the lowest ($MSE_{GNLS} = 0.0175$, $MSE_{NLS} = 0.0105$).

3.3 Estimation of the reliability functions

If the θ_i where known, an estimate of the distribution function F_θ would be the classical empirical distribution function $\hat{F}_\theta$. If the predictions $\hat{\theta}^i$ of the random coefficients θ^i are consistent then a nonparametric estimation of F_θ is

$$\hat{\hat{F}}_\theta(a) = \frac{1}{n}\sum_{i=1}^{n} \mathbf{1}_{\{\hat{\theta}^i \leq a\}}, \quad a \in \mathbb{R}^p$$

and when a traumatic failure mode exists we can plug the $\hat{\theta}^i$'s in a Nelson-Aalen type estimate of the cumulative hazard function in the degradation space $\Lambda(z) = \int_0^s \lambda(u)du$ to get

$$\hat{\Lambda}(z) = \sum_{Z^i_{f_i} \leq z, \delta_i = 1} \frac{1}{\sum_{j, Z^j_{f_j} \geq Z^i_{f_i}} h'(Z^i_{f_i}, \hat{\theta}^i)}$$

For further details we refer to [BN04]. The overall survival function S of the failure time $U = min(T_0, T)$ is estimated by

$$\begin{aligned}\hat{S}(t) &= \int \Big[\exp - \int_0^{g(t,a)} h'(z,a)d\hat{\Lambda}(z)\Big] 1_{t<h(z_0,a)} d\hat{\hat{F}}_\theta(a)\\ &= \frac{1}{n}\sum_{i=1}^{n} \exp\Big[- \int_0^{g(t,\hat{\theta}^i)} h'(z,\hat{\theta}^i)d\hat{\Lambda}(z)\Big] 1_{t<h(z_0,\hat{\theta}^i)}.\end{aligned}$$

Example : As an illustration we consider three simulations of n=100 degradation curves $Z(t, \theta_1, \theta_2) = e^{\theta_1}(1+t)^{\theta_2}$, $t \in [0, 12]$ with multiplicative noise and traumatic failure times with a hazard rate in the degradation space of Weibull-type $\lambda(t) = \beta/\alpha(x/\alpha)^{\beta-1}$, $\alpha = 5$, $\beta = 2.5$ and (θ_1, θ_2) is a gaussian vector with mean (-2,2) and Varθ_1 =Varθ_2 $= 0.1^2$, Corr$(\theta_1, \theta_2) = -0.7$. In this case the results of section 3.1 hold with Y_{ij} =lnZ_{ij} and the additive errors $\tilde{\epsilon}$ follow H1 or H2 with $\Phi = 0.9$ or H3 with c(t)=ln(1+t) and $\sigma_i = 0.05$ for all i. The path i is censored by the minimum $min(T_0^i, T^i, 12)$ and the estimation of the coefficients are carried out according to section 3.1.

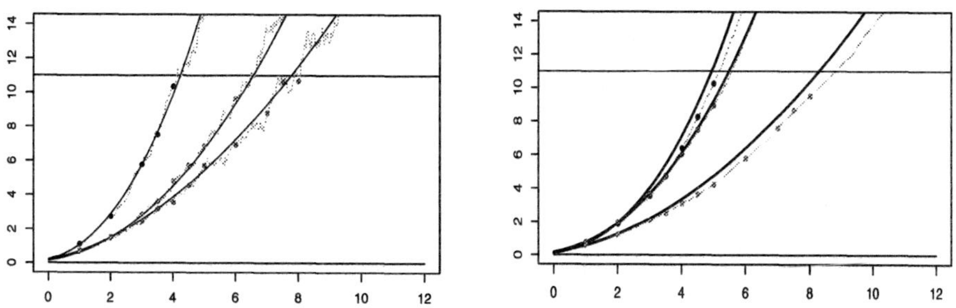

Fig. 3. 3 simulated path with CAR(1) and Wiener-type noises

The estimation behaves well under H1 and H2 but is less efficient under H3 (see the right hand side of figure 4). In fact, figure 5 shows that under H1 the distribution function of the random variable θ_1 is well estimated both by $\hat{F}_{\theta_1}$ and $\hat{\hat{F}}_{\theta_1}$ but not under H3.

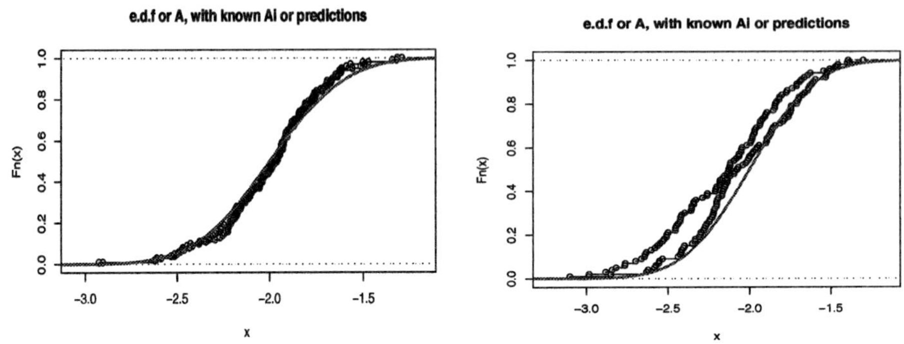

Fig. 4. Estimation of F_{θ_1} under H1 and H3

Finally, we present an estimation of the cumulative hazard rate functuion Λ in the degradation space under H1 and a Monte Carlo simulation giving a 95% empirical confidence band under H2.

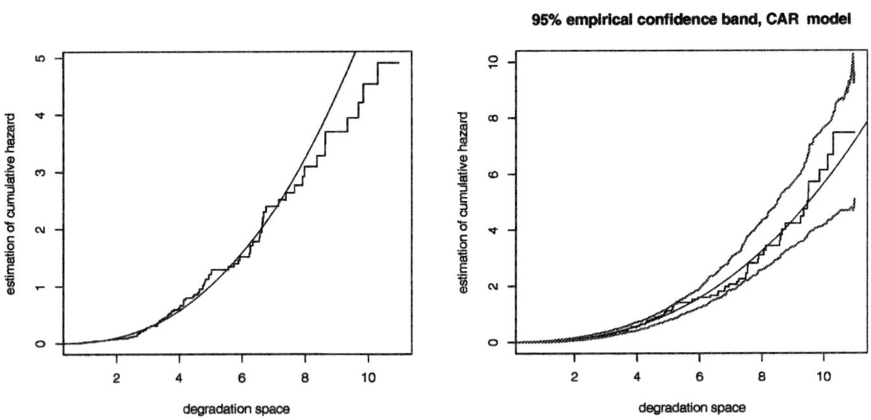

Fig. 5. Estimation of Λ under H1 and 95 % empirical confidence band of Λ under H2

References

[AG01] Aalen, O.O., Gjessing, H.K.: Understanding the shape of the hazard rate: a process point of view. Statist. Sci. **16**(1) 1–22 (2001)

[BBK04] Bagdonavicius V, Bikelis A, Kazakevicius V.: Statistical analysis of linear degradation and failure time data with multiple failure modes. Lifetime Data Anal., **10**(1), 65-81 (2004)

[BC92] Bates, D. M., Chambers, J. M.: Nonlinear models, Chapter 10 of Statistical Models in S, eds J.M. Chambers and T.J. Hastie, Wadsworth & Brooks/Cole (1992)

[BK85] Bogdanoff, J. L., Kozin F.: Probability Models of Cumulative Damage. Wiley, New York (1985)

[BN97] Bagdonavicius V., Nikulin M.S.: Transfer functionals and semiparametric regression models. Biometrika, **84** 365–78 (1997)

[BN01] Bagdonavicius V., Nikulin M.S.: Estimation in degradation models with explanatory variables. Lifetime Data Anal., **7**(1), 85–103 (2001)

[BN02] Bagdonavicius V., Nikulin M.S.: Accelerated Life Models : Modeling and Statistical Analysis. Chapman & Hall / CRC, Boca Raton (2002)

[BN04] Bagdonavicius, V., Nikulin, M., Semiparametric analysis of degradation and failure times data with covariates, in *Parametric and Semiparametric Models with Applications to Reliability, Survival Analysis, and Quality of Life* Series : Statistics for Industry and Technology Nikulin, M.S.; Balakrishnan, N.; Mesbah, M.; Limnios, N. (Eds.). Birkauser (2004)

[COU04] Couallier, V., Comparison of parametric and semiparametric estimates in a degradation model with covariates and traumatic censoring in *Parametric and Semiparametric Models with Applications to Reliability, Survival Analysis, and Quality of Life* Series : Statistics for Industry and Technology Nikulin, M.S.; Balakrishnan, N.; Mesbah, M.; Limnios, N. (Eds.). Birkauser (2004)

[COX99] Cox DR.: Some remarks on failure-times, surrogate markers, degradation, wear, and the quality of life. Lifetime Data Anal., **5**(4), 307–14 (1999)

[DN95] Doksum K.A., Normand S.L.: Gaussian models for degradation processes-Part I: Methods for the analysis of biomarker data. Lifetime Data Anal., **1**(2), 131–44 (1995)

[DL00] Duchesne T, Lawless J.: Alternative time scales and failure time models. Lifetime Data Anal., **6**(2),157–79 (2000)

[DL02] Duchesne T, Lawless J.: Semiparametric inference methods for general time scale models. Lifetime Data Anal., **8**(3),263-76 (2002)

[FIN03] Finkelstein MS.: A model of aging and a shape of the observed force of mortality. Lifetime Data Anal., **9**(1),93-109 (2003)

[GLJ03] Girish, T., Lam, S.W.: Jayaram, S.J.: Reliability Prediction Using Degradation Data - A Preliminary Study Using Neural Network-based Approach. Safety and Reliability - Bedford & van Gelder(eds): Proceedings of ESREL 2003, European Safety and Reliability Conference, 15-18 June, Maastricht, The Netherlands. 681–88. (c) Swets & Zeitlinger, Lisse (2003)

[HA03] Huang, W., Askin, R.G.: Reliability analysis of electronic devices with multiple competing failure modes involving performance aging degradation. Quality and Reliability Engineering International **19**(3), 241–54 (2003)

[KW04] Kahle, W., Wendt, H.: On a cumulative damage process and resulting first passages times. Applied Stochastic Models in Business and Industry, **20**(1) 17–26 (2004)

[LC04] Lawless, J., Crowder, M.: Covariates and random effects in a gamma process model with application to degradation and failure. Lifetime Data Anal., **10**(3), 213–27 (2004)

[LM93] Lu, C. J. , Meeker, W. Q.: Using degradation measures to estimate a time-to-failure distribution. Technometrics, 35, 161-174 (1993)

[ME98] Meeker, W.Q., Escobar, L.: Statistical Analysis for Reliability Data. John Wiley and Sons, New York (1998)

[OC04] Branco de Oliveira, V.R., Colosimo E.A.: Comparison of Methods to Estimate the Time-to-failure Distribution in Degradation Tests. Quality and Reliability Engineering International, **20**(4), 363–73 (2004)

[PT04] Padgett, W.J., Tomlinson, M.A.: Inference from Accelerated Degradation and Failure Data Based on Gaussian Process Models. Lifetime Data Anal., **10**(2) 191–206 (2004)

[RC00] Robinson, M.E., Crowder, M.J.: Bayesian methods for a growth-curve degradation model with repeated measures. Lifetime Data Anal., **6**(4), 357–74 (2000)

[SW03] Seber, G.A.F., Wild, C.J.: Nonlinear regression. John Wiley & Sons, New Jersey (2003).

[SIN95] Singpurwalla, N.D.: Survival in dynamic environments. Statist. Sci. **10**(1), 86–103 (1995)

[WK04] Wendt, H., Kahle, W.: On parameter stochastic poisson process, in Parametric and Semiparametric Models with Applications to Reliability, Survival Analysis, and Quality of Life, Series : Statistics for Industry and Technology, Nikulin, M.S.; Balakrishnan, N.; Mesbah, M.; Limnios, N. (Eds.), Birkauser, 473–86 (2004)

[WHI95] Whitmore, G.A.: Estimating degradation by a Wiener diffusion process subject to measurement error. Lifetime Data Anal., **1**(3), 307–19 (1995)

[WS97] Whitmore, G.A., Schenkelberg F.: Modelling accelerated degradation data using Wiener diffusion with a time scale transformation. Lifetime Data Anal., 3(1), 27-45 (1997)

[WCL98] Whitmore, G.A., Crowder, M.J., Lawless, J.F.: Failure inference from a marker process based on a bivariate Wiener model. Lifetime Data Anal., **4**(3), 229–51 (1998)

[WT97] Wulfsohn M.S., Tsiatis A.A.: A joint model for survival and longitudinal data measured with error. Biometrics **53** 330–39 (1997)

[YM97] Yashin, A.I., Manton, K.G.: Effects of unobserved and partially observed covariate processes on system failure: a review of models and estimation strategies. Statist. Sci., **12**(1), 20–34 (1997)

[YU03] Yu, H.F.: Designing an accelerated degradation experiment by optimizing the estimation of the percentile. Quality and Reliability Engineering International, **19**(3), 197-214 (2003)

[YU04] Yu, H.F.: Designing an accelerated degradation experiment with a reciprocal Weibull degradation rate. Journal of Statistical Planning and Inference, In press, corrected proof (2004)

Estimation in a Markov chain regression model with missing covariates

Dorota M. Dabrowska[1], Robert M. Elashoff[1] and Donald L. Morton[2]

[1] Department of Biostatistics, University of California, Los Angeles, CA 90095-1772
[2] John Wayne Cancer Institute, Santa Monica, CA 90404

Summary. Markov chain proportional hazard regression model provides a powerful tool for analysis of multiple event times. We discuss estimation in absorbing Markov chains with missing covariates . We consider a MAR model assuming that the missing data mechanism depends on the observed covariates, as well as the number of events observed in a given time period, their types and times of their occurrence. For estimation purposes we use a piecewise constant intensity regression model.

1 Introduction

Missing covariate measurements arise frequently in regression analyses of survival time data. The most common approach to handling such measurements corresponds to the case deletion method. It consists of exclusion of subjects with missing covariates and analysis of the data based on information collected on the remaining subjects. This method can be highly inefficient, and can lead to biased estimates, if complete cases do not form a random sample of the original data (Little and Rubin, 1987).

Several authors have proposed methods for analysis of the proportional hazard model with missing covariates. In particular, Zhou and Pepe (1995) and Lin and Ying (1993) suggested methods for regression analysis in the case of covariates missing completely at random (MCAR). This model assumes that the distribution of the missing data mechanism does not depend on the outcome variables. The approach taken by Zhou and Pepe and Lin and Ying corresponds to estimation of regression coefficients based on modified partial likelihoods obtained by approximating the conditional expectation of a covariate $Z(t)$ given the risk process based on subjects who have complete measurements and remain at risk for failure at time t. Martinussen (1999) and Chen and Little (1999) considered the more parsimonious model assuming that covariates are missing at random (MAR). Under assumptions of this model, the missing data mechanism may depend the observed data, but not on the values of the missing covariates. Several methods for handling missing covariates in both proportional hazard model and parametric survival analysis models were also proposed by Lipsitz and Ibrahim (1996, 1998), Chen and Ibrahim (2001).

In this paper we consider estimation in a multivariate counting process corresponding to a finite state proportional hazard Markov chain model (Andersen, et al. 1993, Andersen, Hansen and Keiding, 1992). As opposed to single endpoint models, here inferences refer to a stochastic process $\{J(t), t \in [0, \tau]\}$ such that at time t, $J(t)$ takes on values in a finite set $E = \{1, \ldots, k\}$ representing possible events in the evolution of a disease. Along with a possibly censored realization of the process $J(t)$, we also observe a vector of time independent covariates Z. The model assumes that conditionally on Z, the process $J(t)$ forms an inhomogeneous Markov chain and intensities of transitions among adjacent states have proportional hazard form. In Section 2 we allow some components of the vector Z to be missing. We define a MAR model, assuming that the missing data mechanism depends on the observed covariates as well as the number of events observed during a specified period of time, their types and times of the occurrence. We consider censoring corresponding to the termination of the study at a fixed time point τ, and random censoring representing an absorbing state of the observed model. For purposes of estimation of the parameters of the Markov chain we use a modification of Freedman's (1982) approach to analysis of the proportional hazard model with piecewise constant hazard rates. The method uses histogram approximation to the hazard rates and a piecewise linear approximation to cumulative intensities.

In their analysis of MCAR and MAR models, Little and Rubin (1987, p. 90) showed that the likelihoods for estimation of the parameters of interest are the same under assumptions of both models. More precisely, the likelihoods differ only in the proportionality factors depending on the parameters describing the missing data mechanism at hand. In the present setting, the MCAR model allows for estimation of the unknown parameters as well as estimation of a modified matrix of transition probabilities (Section 2.1). On the other hand, the MAR condition depends on the sequence of states visited and times of entrances into these states. Similarly to Little and Rubin (1987, p. 90), the likelihood for estimation of the regression coefficients is the same under the MAR and the MCAR model (up to proportionality factors), however, we show that the MAR model does not allow in general for estimation of the transition probabilities among different states of the model.

For illustrative purposes in Section 3 we use data on 4141 patients diagnosed with malignant melanoma and treated at the John Wayne Cancer Institute, Santa Monica (JWCI) and UCLA. The JWCI/UCLA database was initiated over 30 years ago. This clinio-pathological demographic database has been established in late 1970's and identified as a national resource for melanoma studies. The database has expanded in the number and type of variables included as well as addition of clinical trials data developed by JWCI. Data analyses of this database encountered the not uncommon situation where observations on some important variables are missing. Thus for example, depth of invasion of the primary tumor has not been observed in a moderate number of cases.

2 The model and estimation

2.1 The model

Throughout we consider estimation in a finite state Markov chain process. We assume that the chain $\{J(t) : t \in [0, \tau]\}$ is observed over a finite time period ($\tau < \infty$) and its state space $E = \{1, \dots, k\}$ can be partitioned into two disjoint sets $\mathcal{T} \cup \mathcal{A} = E, \mathcal{T} \cap \mathcal{A} = \emptyset$, representing transient ($\mathcal{T}$) and absorbing ($\mathcal{A}$) states. A pair of distinct states $(i, j) \in E \times E, i \neq j$ is called adjacent if transition from state i to state j is possible in one step. The collection of all such adjacent pairs is denoted by $E_0, E_0 \subset E \times E$.

A Markov chain regression model can be specified in terms of two parameters. They correspond to (i) the joint marginal distribution of the initial state J_0 and the covariates $Z = (Z_1, \dots, Z_d)$; and (ii) the conditional cumulative intensity matrix $\mathcal{A}(t; z) = [A_{ij}(t; z)]_{i,j \in E}$. The entries of the matrix $\mathcal{A}(t; z)$ are given by

$$A_{ij}(t; z) = \int_0^t \alpha_{ij}(u; z) du \quad \text{if} \quad i \neq j \; ,$$
$$A_{ii}(t; z) = -\sum_{j \neq i} A_{ij}(t; z) \quad \text{if} \quad i = j \; .$$

For $i \neq j$, the functions $\alpha_{ij}(u, z)$ represent conditional hazard rates of one-step transitions among adjacent states of the model. The negative on-diagonal entries form cumulative hazard functions accounting for the sojourn time in each state of the model. The (i, j) entry of the conditional transition probability matrix

$$\mathcal{P}(s, t; z) = [\mathcal{P}_{ij}(s, t; z)]_{i,j=1,\dots,k} = [Pr(J(t) = j | J(s) = i, Z = z)]_{i,j=1,\dots,k}$$

provides the conditional probability that the process occupies state j at time t, $J(t) = j$, given $Z = z$ and given that that at time $s, s < t$ the process is in state i. The matrix $\mathcal{P}(s, t; z)$ forms solution to Kolmogorov equations

$$\mathcal{P}(s, t; z) = I + \int_s^t \mathcal{P}(s, u-; z) \mathcal{A}(du; z) = I + \int_s^t \mathcal{A}(du; z) \mathcal{P}(u, t; z) \; ,$$

where I is the identity matrix. Methods for its computation are discussed in Chiang (1968), Aalen and Johansen (1978) and Andersen et al (1993), among others.

Associated with the pair $(Z, \{J(t), t \in [0, \tau]\})$ is a marked point process $\{Z, (T_m, J_m)_{m \geq 0}\}$, where $0 = T_0 < T_1 < \ldots < T_m < \ldots$ are times of consecutive entrances into the possible states of the model, and $J_0, J_1, \ldots, J_m, \ldots$ are states visited at these times. Let $W_m = (T_\ell, J_\ell : \ell = 0, \ldots, m), m \geq 0$, be the first m pairs in the sequence $(T_m, J_m)_{m \geq 0}$. The assumption that, conditionally on the covariates, the process $J(t)$ forms a Markov chain entails that

$$\Pr(T_m \leq t, J_m = j_m | W_{m-1}, Z) = \int_{(T_{m-1}, t]} f(u, j_m | T_{m-1}, J_{m-1}, Z) ,$$

where

$$f(u, j_m | t_{m-1}, j_{m-1}, Z) = 1(u > t_{m-1}) F(u | t_{m-1}, j_{m-1}, z) \alpha_{j_{m-1}, j_m}(u; z)$$

and

$$\begin{aligned} F(t | t_{m-1}, j_{m-1}, z) &= \exp[- \sum_{l:(j_{m-1}, l) \in E_0} \int_{(t_{m-1}, t]} \alpha_{j_{m-1}, l}(u; z) du] \\ &\qquad \text{if} \quad t > t_{m-1} , \\ &= 1 \qquad \text{otherwise} . \end{aligned}$$

The function $F(\cdot | t_{m-1}, j_{m-1}, z)$ represents the conditional survival function in state j_{m-1}, that is

$$Pr(T_m > t | W_{m-1}, Z) = F(t | T_{m-1}, J_{m-1}, Z) .$$

Finally, the probability of one-step transition into state j at time T_m is given by

$$\Pr(J_m = j | Z, T_m = u, W_{m-1}) = \frac{\alpha_{j_{m-1}, j}(u; z)}{\sum_{l:(j_{m-1}, \ell) \in E_0} \alpha_{j_{m-1}, l}(u; z)} .$$

From this it also follows that

$$\begin{aligned} &\lim_{s \downarrow 0} \frac{1}{s} Pr(T_m \in [t, t+s], J_m = j | Z, T_m \geq t, W_{m-1}) = \\ &\qquad 1(T_m > t > T_{m-1}) 1(J_{m-1} = j_{m-1}) \alpha_{j_{m-1}, j}(t; z) . \end{aligned}$$

If we denote by $N_h(t)$ and $Y_h(t), h \in E_0$, the processes

$$N_h(t) = \sum_{m \geq 1} N_{hm}(t) , \quad Y_h(t) = \sum_{m \geq 1} Y_{hm}(t) ,$$

where

$$N_{hm}(t) = 1(T_{m-1} < T_m \leq t, J_{m-1} = i, J_m = j) ,$$
$$Y_{hm}(t) = 1(T_m \geq t > T_{m-1}, J_{m-1} = i) ,$$

then $N(t) = \{N_h(t) : t \in [0,\tau], h = (i,j) \in E_0\}$ is a multivariate counting process whose components record transitions among adjacent states occurring during the time interval $[0,t]$. Assuming that the process $N(t)$ is defined on a complete probability space $(\Omega, \mathcal{F}, \Pr)$, its compensator $\Lambda(t) = \{\Lambda_h(t) : t \in [0,\tau], h \in E_0\}$, relative to the self-exciting filtration $\{\mathcal{G} \otimes \mathcal{F}_t\}_{t \leq \tau}, \mathcal{G} = \sigma(Z), \mathcal{F}_t = \sigma\{J_0, N_h(s), Y_h(s+) : s \leq t, h \in E_0\}$ satisfies

$$\Lambda_h(dt) = E[N_h(dt)|\mathcal{G} \times \mathcal{F}_{t-}] = Y_h(t)\alpha_h(t; Z)dt .$$

The assumption that the process forms a proportional hazard Markov chain corresponds to the choice

$$\Lambda_h(dt) = Y_h(t)e^{\beta^T Z_h}\alpha_h(t)dt ,$$

where $\alpha = [\alpha_h : h \in E_0]$ are unknown baseline hazards, $[Z_h : h \in E_0]$ is a vector of transition specific covariates, and $\beta = [\beta_h : h \in E_0]$ is a conformal vector of regression coefficients. For any pair of adjacent states, $h \in E_0$, the vector Z_h is either equal to the covariate Z, or else it represents a function $Z_h = \Theta_h(Z)$ derived from the covariate Z.

We assume now that the covariate Z can be partitioned into two non-empty blocks, $Z = (Z_0, Z_1)$ such that $Z_0 = (Z_{01}, \ldots, Z_{0q})$ and $Z_1 = (Z_{11}, \ldots, Z_{1,d-q})$. We shall use the following regularity conditions.

Condition 2.1

(i) The conditional distribution of Z_0 given (Z_1, J_0) has density $g_\theta(z_0|z_1, j_0)$ with respect to a product dominating measure $\otimes_{i=1}^q \mu_i$ and dependent on a parameter $\theta \in \Theta \subset R^q$.

(ii) Conditionally on (Z_0, Z_1, J_0), the sequence $(T_m, J_m)_{m \geq 0}$ forms a proportional hazard Markov chain model with parameters $\alpha = [\alpha_h : h \in E_0]$ and $\beta = [\beta_h : h \in E_0]$.

(ii) The parameter θ is noninformative on (α, β).

We denote by $\psi = (\theta, \alpha, \beta)$ the unknown parameters. In Appendix 2, we give a recurrent formula for the conditional density of the the covariate Z_0 given the vector V, $V = [N_.(\tau), (J_\ell, T_\ell)_{\ell=0}^{N_.(\tau)}, Z_1]$. We also show that the "marginal" model, obtained by omitting the covariate Z_0 forms a non-Markovian counting process.

Here we assume that some components of this vector may be missing. Let $R = (R_1, \ldots, R_q)$ be a binary vector defined by

$$R_j = 1 \quad \text{if} \quad Z_{0j} \quad \text{is observed}$$
$$R_j = 0 \quad \text{if} \quad Z_{0j} \quad \text{is unobserved} .$$

Then for a subject whose missing data indicator R is equal to $r = (r_1, \ldots, r_q)$, we observe the vector $(V, Z_0(r))$, where $V = [N_.(\tau), (T_\ell, J_\ell)_{\ell=1}^{N_.(\tau)}, Z_1]$ and

$$Z_0(r) = (Z_{0j} : r_j = 1) .$$

We also denote by $z_0(\overline{r}) = [z_{0j} : r_j = 0]$ a potential realization of the missing covariate.

We shall treat the variable R as an extra covariate taking values in the set $\mathcal{R} = \{0,1\}^q$. Denoting the sample space of covariates Z_0 and Z_1 by $\mathcal{Z}_0$ and $\mathcal{Z}_1$, respectively, the unobserved model is defined on the probability space $(\Omega' \times \Omega, \{\mathcal{G}' \otimes \mathcal{F}_t\}_{t\leq\tau}, Pr)$, where $\Omega' = \mathcal{R} \times \mathcal{Z}_0 \times \mathcal{Z}_1$, $\mathcal{G}'$ is the Borel σ–field of Ω' and "Pr" is defined in the condition (2.1) below. The observable model corresponds to the transformation of this space according to the assignment $X(r, z_0, z_1, \omega) = (r, z_0(r), z_1, \omega)$, where r is the realization of the missing data indicator, $(z_0(r), z_1)$ is the realization of the observed covariate and ω is the sequence of states visited and times of entrances into these states. Since the number of events $N_{.}(\tau) = \sum_{h\in E_0} N_h(\tau)$ observed during the time period $[0, \tau]$ is random, we specify the MAR assumption by conditioning on the number of events $N_{.}(\tau)$ and type and time of their occurrence.

Condition 2.2.

(i) The conditional distribution of the missing data indicator satisfies

$$Pr(R = r|V, Z_0) = \nu(V, Z_0(r)) ,$$

where $V = [N_{.}(\tau), (J_l, T_l)_{l=0}^{N_{.}(\tau)}, Z_1]$ and ν is a proper conditional probability measure not dependent on missing covariates.

(ii) The parameters of the conditional distribution ν of the missing data indicators are noninformative on the parameter $\psi = (\theta, \alpha, \beta)$.

The MAR condition (i) is a type of conditional independence assumption. It is satisfied for example, if the vectors R and Z_0 are conditionally independent given V. In the latter case, the probability distribution ν depends only on the sequence V. The stronger MCAR condition assumes that the function ν depends only on the pair (Z_1, J_0), but not on the vector $V_1 = [N_{.}(\tau), (J_l, T_l)_{l=0}^{N_{.}(\tau)}]$ or the observed covariates $Z_0(R)$. The MCAR model is satisfied for example, if R and the sequence $[Z_0, (T_\ell, J_\ell)_{\ell\geq 0}]$ are conditionally independent given (Z_1, J_0).

The difference between these two models can be better understood in the context of prediction. If parameters of the missing data mechanism are noninformative on ψ and not estimated from the data, then the MAR and the MCAR model lead to the same likelihoods and the same estimates of the parameter ψ. In the case of the MCAR model, the resulting parameters can be also used to estimate some parameters related to the prediction of the survival status of a new patient based on his/her observed covariates and follow-up history. For example, we can define an analogue of the transition probability matrix by setting

$$\begin{aligned} &\Pr(J(t) = j|J(s) = i, Z_0(R), Z_1, J_0 = \ell, R) = \qquad (1) \\ &= \int \mathcal{P}_{ij}(t, s|Z_0(R), z_0(\overline{r}), Z_1)\pi(dz_0(\overline{r})|i, \ell, s, Z_1, Z_0(R)) , \end{aligned}$$

where $\pi(\cdot|i,\ell,s,Z_1,Z_0(R))$ is the conditional density of the missing covariate $Z_0(\overline{R})$ given the observed covariate $(Z_1, Z_0(R))$ and given that the states occupied at times 0 and s are $J(0) = \ell$ and $J(s) = i$, respectively. This posterior density is given by

$$\pi(dz_0(\overline{r})|i,\ell,s,Z_1,Z_0(R)) =$$
$$= \frac{\mathcal{P}_{\ell i}(0,s|Z_0(R),z_0(\overline{r}),Z_1)g_\theta(Z_0(r),z_0(\overline{r})|Z_1,\ell)\mu_{\overline{r}}(dz_0(\overline{r}))}{\int \mathcal{P}_{\ell i}(0,s|Z_0(R),z_0(\overline{r}),Z_1)g_\theta(Z_0(r),z_0(\overline{r})|Z_1,\ell)\mu_{\overline{r}}(dz_0(\overline{r}))}$$

where $\mu_{\overline{r}} = \otimes_{i:r_i=0}\mu_i$. Under the MCAR model, the matrix (1) can be estimated using plug-in method.

However, under the weaker assumption of the MAR model, the transition probabilities (1) depend in general on the conditional distribution of the missing data indicator, and hence its parameters must from the data. A "partial" MAR model, assuming

$$\Pr(R = r|V, Z_0) = \nu(J_0, Z_1, Z_0(r)) \tag{2}$$

instead of the condition 2.2 (i), is sufficient to ensure ignorability of the missing data mechanism for the modified transition probability matrix (1). In sum, although the MAR model forms an ignorable missing data mechanism for estimation of the parameter ψ, this is not the case for estimation of transition probabilities or other parameters related to prediction. At the same time, the transition probabilities derived under the MCAR condition, or the partial MAR model (2), are in general biased.

2.2 Example

Here we consider a four state illness model assuming that a healthy person (state 0) can develop two forms of a disease: D1 or D2 (state 1 or state 2) and subsequently die (state 3), or else he/she dies without developing the disease.

The matrix of baseline intensities is of the form

$$\alpha(t) = \begin{pmatrix} -\sum_{j=1}^{3}\alpha_{0j}(t) & \alpha_{01}(t) & \alpha_{02}(t) & \alpha_{03}(t) \\ 0 & -\alpha_{13}(t) & 0 & \alpha_{13}(t) \\ 0 & 0 & -\alpha_{23}(t) & \alpha_{23}(t) \\ 0 & 0 & 0 & 0 \end{pmatrix}$$

The corresponding diagram of transitions is presented in Figure 2.1.

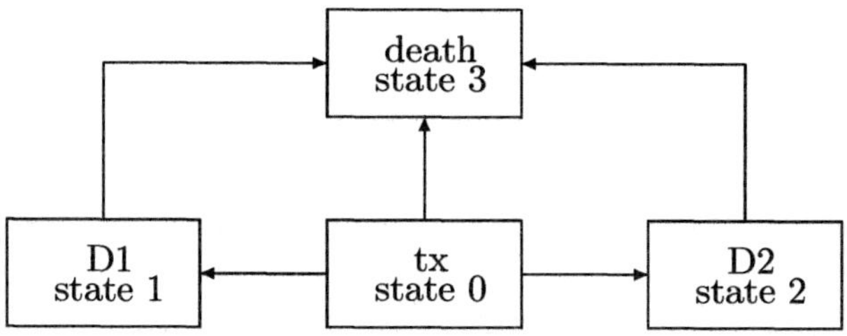

Figure 2.1. A four state illness model

The matrix of conditional transition probabilities $\mathcal{P}(s,t;z) = [\mathcal{P}_{ij}(s,t;z)]_{i,j=0,\ldots,3}$ has entries

$$\begin{aligned}
\mathcal{P}_{00}(s,t;z) &= \Pr(T_1 > t | T_1 > s, Z = z) = F(t|s,0,z)\ , \\
\mathcal{P}_{ii}(s,t;z) &= \Pr(T_2 > t | T_2 > s > T_1, J_1 = i, Z = z) = F(t|s,i,z)\ , \quad i = 1,2\ , \\
\mathcal{P}_{0i}(s,t;z) &= \Pr(T_1 \leq t < T_2, J_1 = 1 | T_1 > s, Z = z) = \\
&= \int_s^t \mathcal{P}_{00}(s,u-;z)\alpha_{0i}(du;z)\mathcal{P}_{ii}(u,t;z)\ , \quad i = 1,2\ , \\
\mathcal{P}_{i3}(s,t;z) &= \Pr(T_2 \leq t <, J_2 = 3 | T_1 < s < T_2, J_1 = i, Z = z) = \\
&= \int_s^t \mathcal{P}_{ii}(s,u-;z)\alpha_{i3}(du;z) = \int_s^t f(u,3|s,i,z)du\ , \quad i = 1,2\ , \\
\mathcal{P}_{03}(s,t;Z) &= \Pr(T_1 \leq t, J_1 = 3 | T_1 > s, Z = z) + \Pr(T_2 \leq t, J_2 = 2 | T_1 > s, Z = z) \\
&= \sum_{i=0}^{2} \int_s^t \mathcal{P}_{0i}(s,u-;z)\alpha_{i3}(du;z)\ .
\end{aligned}$$

In addition, $\mathcal{P}_{33}(s,t;z) = 1$ and the remaining entries are 0.

In the case of complete covariates, the assumption that observations are censored at a fixed time τ entails that the transition probability matrix can be estimated only within the range $0 < s < t \leq \tau$. Next suppose that the covariate vector is partitioned into two blocks $Z = (Z_0, Z_1)$ and components of the vector Z_0 may be missing. To see that under the MAR condition, the matrix of transition probabilities cannot be in general recovered, let us first consider the term

$$\begin{aligned} &\Pr(J(t) = 0 | J(s) = 0, Z_0(R), Z_1, J_0 = 0, R) \\ &= \Pr(T_1 > t | T_1 > s, Z_0(R), Z_1, R) \end{aligned} \tag{3}$$

We have

$$\Pr(T_1 > t | T_1 > s, Z_0(r) = z_0(r), Z_1 = z_1, R = r) = \frac{\gamma(t, z_0(r), z_1, r)}{\gamma(s, z_0(r), z_1, r)} ,$$

where

$$\gamma(t, z_0(r), z_1, r) = \frac{\Pr(R = r, T_1 > t | Z_0(r) = z_0(r), Z_1 = z_1)}{\Pr(R = r | Z_0(r) = z_0(r), Z_1 = z_1)} .$$

Moreover, $\gamma(t, z_0(r), z_1, r) = \gamma'(t, z_0(r), z_1, r)/\gamma'(0, z_0(r), z_1, r)$, where

$$\begin{aligned} &\gamma'(t, z_0(r), z_1, r) = \\ &= \sum_{j=0}^{2} \int \gamma_j(t, z_0(r), z_0(\overline{r}), z_1, r) g_\theta(z_0(r), z_0(\overline{r}) | z_1, 0) \mu_{\overline{r}}(dz_0(\overline{r})) \end{aligned} \tag{4}$$

and for $z = (z_0, z_1)$,

$$\begin{aligned} \gamma_0(t, z, r) &= \Pr(R = r | N_.(\tau) = 0, Z_0(r) = z_0(r), Z_1 = z_1) \\ &\times \quad F(t \vee \tau | 0, 0, z) \\ \gamma_1(t, z, r) &= 1(t < \tau) \quad \times \\ &\sum_{i=1}^{2} \int_t^\tau \Big(\Pr(R = r | N_.(\tau) = 1, T_1 = u, J_1 = i, Z_0(r) = z_0(r), Z_1 = z_1) \\ &\times \quad f(u, i | 0, 0, z) F(\tau | u, i, z) \Big) du \\ &+ \quad \int_t^\tau \Pr(R = r | N_.(\tau) = 1, T_1 = u, J_1 = 3, Z_0(r) = z_0(r), Z_1 = z_1) \\ &\times \quad f(u, 3 | 0, 0, z) du \\ \gamma_2(t, z_0, z_1, r) &= 1(t < \tau) \quad \int_t^\tau \Big(\int_u^\tau \Pr(R = r | N_.(\tau) = 2, T_1 = u, J_1 = i, \\ &\qquad T_2 = v, J_2 = 3, Z_0(r) = z_0(r), Z_1 = z_1) \\ &\times \quad f(v, 3 | u, i, z) dv \Big) f(u, i | 0, 0, z) du . \end{aligned}$$

It is easy to see now that (3) depends in general on the distribution of the missing data indicator. However, under the partial MAR model (2), this distribution does not depend on the number of events observed in the interval $[0, \tau]$, or their types and times of the occurrence. In this case, the sum (4) reduces to the product

$$\sum_{j=0}^{2} \gamma_j(t, z_0, z_1, r) = \Pr(R = r | Z_0(r) = z_0(r), Z_1 = z_1) \quad \times$$
$$\Bigg(F(t \vee \tau | 0, 0, z) + 1(t < \tau) \Bigg[\sum_{i=1}^{2} \int_t^{\tau} f(u, i | 0, 0, z) F(\tau | u, i, z)$$
$$+ \sum_{i=1}^{2} \int_t^{\tau} f(u, i | 0, 0, z)[1 - F(\tau | u, i, z)] + \int_t^{\tau} f(u, 3 | 0, 0, z) \Bigg] \Bigg) =$$
$$= \Pr(R = r | Z_0(r) = z_0(r), Z_1 = z_1) \quad \times$$
$$\Big(F(t \vee \tau | 0, 0, z) + 1(t < \tau)[F(t|0, 0, z) - F(\tau | 0, 0, z)] \Big)$$
$$= \Pr(R = r | Z_0(r) = z_0(r), Z_1 = z_1) F(t | 0, 0, z)$$

Hence (3) does not depend on the conditional distribution of the missing data indicator. A similar algebra and Bayes theorem can be applied also to other entries of the matrix (1).

2.3 Estimation

The MAR assumption 2.2 implies that the conditional density of the sequence $(R, V, Z_0(R))$ given (J_0, Z_1) is proportional to

$$p(R, V, Z_0(R); \psi) = \tag{5}$$
$$\int \prod_{q=1}^{2} H_q(V, Z_0(R), z_0(\overline{r}); \psi) g_\theta(z_0(\overline{r}), Z_0(R) | Z_1, J_0) \mu_{\overline{r}}(dz_0(\overline{r})) \,,$$

where $\mu_{\overline{r}} = \otimes_{i: r_i = 0} \mu_i$,

$$H_1(V, Z_0(R), z_0(\overline{r}); \psi) = \left(\prod_{\ell=1}^{N_{.}(\tau)} f(T_\ell, J_\ell | T_{\ell-1}, J_{\ell-1}, Z_1, Z_0(R), z_0(\overline{r})) \right)^{1(N_{.}(\tau) > 0)}$$

and

$$H_2(V, Z_0(R), z_0(\overline{r}); \psi) = [F(\tau | T_{N_{.}(\tau)}, J_{N_{.}(\tau)}, Z_1, Z_0(R), z_0(\overline{r}))]^{1(J_{N_{.}(\tau)} \in T)} \,.$$

The right-hand side of (5) can also be represented as

$$p(R, V, Z_0(R); \psi) = \left(\prod_{\ell=1}^{N_.(\tau)} \alpha_{J_{\ell-1},J_\ell}(T_\ell) \right)^{1(N_.(\tau)>0)} \times \tag{6}$$
$$\int p(V, Z_0(R), z_0(\overline{r}); \psi) g_\theta(z_0(\overline{r}), Z_0(R)|Z_1, J_0)\mu_{\overline{r}}(dz_0(\overline{r})) ,$$

where

$$p(V, Z_0(R), z_0(\overline{r}); \psi) = \exp[1(N_.(\tau) > 0)H_3(V, Z_0(R), z_0(\overline{r}); \psi)]$$
$$\times \quad \exp[-1(J_{N_.(\tau)} \in \mathcal{T})H_4(V, Z_0(R), z_0(\overline{r}), V; \psi)] ,$$

and

$$H_3(V, Z_0(R), z_0(\overline{r}); \psi) =$$
$$\sum_{\ell=1}^{N_.(\tau)} \beta^T Z_{J_{\ell-1},J_\ell} - \sum_{h=(J_{\ell-1},\ell)\in E_0} e^{\beta^T Z_h}[A_h(T_\ell) - A_h(T_{\ell-1})] ,$$
$$H_4(V, Z_0(R), z_0(\overline{r}); \psi) = \sum_{h=(J_{N_.(\tau)},\ell)\in E_0} e^{\beta^T Z_h}[A_h(\tau) - A_h(T_{N_.(\tau)})]$$

In Appendix 2, we show that in analogy to the case of completely observable covariates, the integral $p(R, V, Z_0(R); \psi)$ can be written in the form of a product, evaluated over consecutive times of entrances into adjacent states of the model. However, as opposed to the case of completely observable covariates in Andersen et al. (1993), the factors share parameters in common. From the point of view of parameter estimation, it is easier to work with integrals (5)-(6).

By Bayes formula, for any measurable function Φ of the vector (V, Z_0), its conditional expectation given the data is

$$E_\psi[\Phi(V, Z_0)|V, Z_0(R), R] = \Phi(V, Z_0) \qquad \text{if} \quad R_j = 1 \quad \text{for all} \quad j = 1, \ldots, d$$

and

$$E_\psi[\Phi(V, Z_0)|V, Z_0(R), R] =$$
$$\frac{\int \Phi(V, Z_0(R), z_0(\overline{r}))p(V, Z_0(R), z_0(\overline{r}); \psi)g_\theta(z_0(\overline{r}), Z_0(r)|Z_1, J_0)\mu_{\overline{r}}(dz_0(\overline{r}))}{\int p(V, Z_0(R), z_0(\overline{r}); \psi)g_\theta(z_0(\overline{r}), Z_0(r)|Z_1, J_0)\mu_{\overline{r}}(dz_0(\overline{r}))}$$
$$\text{if} \quad R_j = 0 \quad \text{for some} \quad j = 1, \ldots, d .$$

In the following, we denote this conditional expectation by $\hat{E}_\psi \Phi(V, Z_0)$ for short.

We assume now that $(R_k, V_k, Z_{0,k}(R_k)), k = 1, \ldots, n$ is an iid sample of the missing data indicators and vectors $V_k = [N_{.,k}(\tau), (T_{j,k}, J_{j,k})_{j=1}^{N_{.,k}(\tau)}, Z_{1,k}]$, then the log-likelihood function is given by

$$L(\psi) = \sum_{k=1}^{n} \log p(R_k, V_k, Z_{0,k}(R_k); \psi) ,$$

plus a term depends only on the conditional distribution of the missing data indicators, but not on the parameter ψ. To estimate the unknown parameters, we shall approximate the hazard rates α_h using histogram estimates. For this purpose let $0 = \tau_1 < \tau_2 < \ldots < \tau_{\ell(n)} = \tau$ be a partition of the interval $[0, \tau]$, and define $I_p = [\tau_{p-1}, \tau_p)$ for $p = 2, \ldots, \ell(n)$ and $I_{\ell(n)} = [\tau_{\ell(n-1)}, \tau_{\ell(n)}]$. We assume that the number of partitioning points is either finite and independent of n $(l(n) = l)$, or else it grows to infinity with n, that is $\ell(n) \to \infty$ as $n \to \infty$. Set

$$\hat{\alpha}_h(t) = \sum_{p=1}^{\ell(n)} I(t \in I_p) a_{hp}$$

and let

$$\hat{\Lambda}_{h,k}(t) = \int_0^t Y_{h,k}(u) \hat{\alpha}_h(u) du = \sum_{p=1}^{\ell(n)} Y_{h,k}(I_p) a_{hp} ,$$

where for any pair of adjacent states, $h = (i, j) \in E_0$, we have

$$\begin{aligned} Y_{h,k}(I_p) &= \int_{I_p} Y_{h,k}(u) du \\ &= \sum_{m \geq 1} 1(J_{m-1,k} = i) \max\{0, \min(T_{m,k}, \tau_p) - \max(T_{m-1,k}, \tau_{p-1})\} . \end{aligned}$$

Substitution of $\hat{\Lambda}_{hk}$ into the likelihood function gives then an approximate likelihood $L_n(\psi_n), \psi_n = (\theta, \beta, \hat{\alpha} = [\hat{\alpha}_h : h \in E_0])$. The estimate is obtained by maximizing this function with respect to ψ_n.

In practice the function $L_n(\psi_n)$ may be too difficult to handle directly, so for purposes of estimation we can use EM algorithm. Define

$$\begin{aligned} Q_n(\psi|\psi') &= Q_{1n}(\psi|\psi') + Q_{2n}(\psi|\psi') , \\ Q_{1n}(\psi|\psi') &= \sum_{k=1}^{n} \hat{E}_{\psi'} \ell_1(V_k, Z_{0,k}; \beta, \alpha) , \\ Q_{2n}(\psi|\psi) &= \sum_{k=1}^{n} \hat{E}_{\psi'} \ell_2(V_k, Z_{0,k}; \theta) , \end{aligned}$$

where

$$\ell_1(V_k, Z_{0,k}; \alpha, \beta) = \sum_{h \in E_0} \int_0^\tau \alpha_h(t) N_{h,k}(dt) + \sum_{h \in E_0} \beta^T Z_{h,k} N_{h,k}(\tau)$$
$$- \sum_{h \in E_0} e^{\beta^T Z_{h,k}} \int_0^\tau Y_{h,k}(u) \alpha_h(u) du ,$$
$$\ell_2(V_k, Z_{0,k}; \theta) = \log g_\theta(Z_{0;k} | Z_{1k}, J_{0,k}) .$$

Here $Y_{h,k}$ is the risk process corresponding to subject k, and $N_{h,k}$ is the corresponding process counting transitions of type h, $h \in E_0$. The functions ℓ_1 and ℓ_2 represent the complete data log-likelihood functions for the parameters (α, β) and θ respectively. If $\hat{\psi}_q$ is the estimate of the parameter ψ obtained at the q-th step of the algorithm, then the $(q+1)$ step consists of the E-step in which we calculate the conditional expected

$$Q_n(\psi|\hat{\psi}_q) = Q_{1n}(\psi|\hat{\psi}_q) + Q_{2n}(\psi|\hat{\psi}_q) ,$$

In the M-step we maximize $Q_n(\psi|\hat{\psi}_q)$ with respect to $\psi = \psi_n = (\theta, \beta, \hat{\alpha} = [\hat{\alpha}_h : h \in E_0])$.

Let p_1 be the dimension of the vector θ, and let p_0 be the dimension of the vector of regression coefficients. Denote by S_n an $(p_1 + p_0 + |E_0| l(n)) \times (p_1 + p_0 + |E_0| l(n))$ the diagonal matrix with entries

$$S_n^{ii} = \frac{1}{n} \quad \text{for} \quad i = 1, \ldots, p_1 + p_2$$
$$= \frac{k(n)}{n} \quad \text{for} \quad i = p_1 + p_2 + 1, \ldots, p_1 + p_2 + |E_0|\ell(n)$$

where $|E_0|$ denotes the number of adjacent states in the model, and $k(n) = 1$, if $\ell(n) = \ell$ does not depend on n, and $k(n) = \ell(n)$, otherwise. The normalized score equation for estimation of the parameter $\psi = (\theta, \beta, \alpha)$ is given by $S_n \nabla Q_n(\psi|\hat{\psi}_q) = 0$. The vector $\nabla Q_n(\psi|\hat{\psi}_m)$ has components

$$\nabla_\theta Q_n(\psi|\hat{\psi}_q) = \sum_{k=1}^n \hat{E}_{\hat{\psi}_q} [\frac{\dot{g}_\theta}{g_\theta}(Z_k)] ,$$
$$\nabla_\beta Q_n(\psi|\hat{\psi}_q) =$$
$$\sum_{k=1}^n \sum_{h \in E_0} \left[\hat{E}_{\hat{\psi}_q}(Z_{hk}) N_{hk}(\tau) - \hat{E}_{\hat{\psi}_q}(Z_{hk} e^{\beta^T Z_{hk}}) \hat{\Lambda}_{h,k}(\tau) \right] .$$
$$\nabla_\alpha Q_n(\psi|\hat{\psi}_q) =$$
$$\left(\sum_{k=1}^n \left[\frac{N_{h,k}(I_p)}{a_{hp}} - \hat{E}_{\psi_q}(e^{\beta^T Z_{h,k}}) Y_{h,k}(I_p) \right] : p = 1, \ldots, \ell(n), h \in E_0 \right) .$$

Here for any pair of adjacent states, $h = (i,j) \in E_0$, we have

$$N_{h,k}(I_p) = \sum_{m\geq 1} 1(J_{m-1,k} = i, J_{m,k} = j)1(T_{m,k} \in I_p) .$$

With $\eta = (\theta, \beta)$ parameters fixed, the equation $\nabla_\alpha Q_n(\psi_n|\hat{\psi}_q) = 0$ can be solved for a_{hp}. The solution is given by

$$\hat{a}_{hp,q}(\eta) = \frac{\sum_{k=1}^{n} N_{h,k}(I_p)}{\sum_{k=1}^{n} Y_{h,k}(I_p)\hat{E}_{\hat{\psi}_q}(e^{\beta^T Z_{h,k}})} 1(\sum_{k=1}^{n} Y_{h,k}(I_p) > 0)$$

Thus setting

$$\hat{\alpha}_{h,q+1}(t,\eta) = \sum_{p=1}^{\ell(n)} I(t \in I_p)\hat{a}_{hp,q}(\eta) ,$$

at step $(q+1)$ of the EM algorithm, we obtain a pseudo- estimate of the hazard rate α_h. Set

$$\hat{\Lambda}_{hk,q+1}(\tau,\eta) = \int_0^\tau Y_{h,k}(u)\hat{\alpha}_{h,q+1}(u,\eta)du = \sum_{p=1}^{l(n)} \hat{a}_{hp,q}(\eta)Y_{h,k}(I_p) .$$

The profile likelihood score equation for the regression coefficients is

$$[\nabla_\beta Q_n](\theta,\beta,\hat{\alpha}_{q+1}(\eta)|\psi_q) =$$
$$\sum_{k=1}^{n}\sum_{h\in E_0} \left[\hat{E}_{\hat{\psi}_q} Z_{hk} N_h(\tau) - \hat{E}_{\hat{\psi}_q}[Z_{hk}e^{\beta^T Z_{hk}}]\hat{\Lambda}_{hk,q+1}(\tau,\beta)\right] .$$

The score equation for the unknown θ parameter is

$$[\nabla_\theta Q](\theta,\beta,\hat{\alpha}_{q+1}(\beta)|\psi_q) = \sum_{k=1}^{n} \hat{E}_{\hat{\psi}_q}(\frac{\dot{g}_\theta}{g_\theta})(Z_k) .$$

Assuming that the density g_θ is twice differentiable with respect to θ, the parameters $\eta = (\theta,\beta)$ can be updated using e.g. Newton-Raphson algorithm, by setting

$$\eta_{q+1} = \eta_q + H_n(\eta_q, \alpha_{q+1}(\eta_q))^-[\nabla_\eta Q_n](\eta_q, \alpha_{q+1}(\eta_q)|\psi_q) ,$$

where $H_n(\psi)^-$ is a generalized inverse of an estimate of the information matrix for parameter η. We give its form in Appendix 1.

In practice the partitioning points must be taken to depend on the sample size n. The rate of convergence of the histogram estimate depends in this case on the recurrence properties of the chain. Using results of Freedman (1982) for the piecewise constant proportional hazard model, we can show that if the total number of jumps of the process N is bounded by a fixed constant, and the dimension of the covariate space is fixed, then the asymptotically optimal choice of the binwidth corresponds to the choice $\sqrt{n}b(n)^2 \to 0, nb(n) \to \infty$. With this choice the regression coefficients can be estimated at $\sqrt{n}$ rate.

In Section 3 we assume that the covariate vector Z_0 is discrete, so that the conditional expected can be evaluated as sums taken over possible missing covariate values $z_0(\overline{\tau})$. The analysis of continuous or mixed discrete-continuous covariates is more difficult since the integrals with respect to the conditional distribution of the missing covariates given the observable variables must evaluated numerically using e.g. MCEM (Wei and Tanner, 1990, Sinha, Tanner and Hall, 1994).

2.4 Random censoring

So far we have assumed that data are subject to fixed censoring occurring at the termination of the study at time τ. Here we consider a censoring model in which the main finite state Markov chain model of interest has state space $E = \{1, \ldots, k\}$, whereas the observed marked point process has state space enlarged by one extra absorbing state "c" representing a withdrawal due to causes unrelated to the study. In particular, suppose that the censored data analogue of the four state illness model of section 2.2 can be represented by means of the following transition diagram.

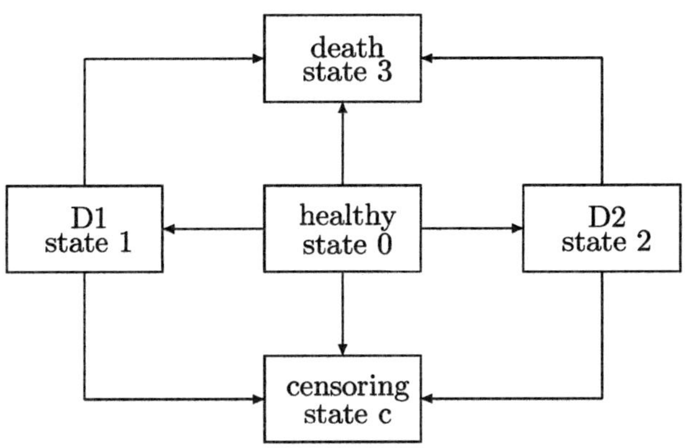

Figure 2.2. The censored four state illness model

Thus a healthy person may be observed either (i) to die without developing either of the two disease types, or (ii) to develop one of the two disease types and subsequently die, or (iii) to develop one of the two disease types and subsequently be censored or (iv) to be censored without developing either of the two disease types.

More generally, we assume that observed marked point process registers events $J_1, \dots, J_m, \dots$ at times $T_1 < T_2 < \dots < T_m$, according to the following assumptions.

Condition 2.3

(i) Given that at time $T_{m-1} = t_{m-1}$ the process enters a transient state $J_{m-1} = j_{m-1}$, the waiting time in state j_{m-1} has survival function

$$\begin{aligned}&\Pr(T_m > t | Z = z, W_{m-1}) = \\ &\quad = G_m(t, j_{m-1} | W_{m-1}, Z = z) F(t | t_{m-1}, j_{j-1}, z) \quad m \geq 1\end{aligned}$$

Moreover, the subdensities of the progression to an absorbing or transient state of the model are given by

$$\begin{aligned}&\Pr(T_m \leq t, J_m = j | Z = z, W_{m-1}) = \\ &= \int_{T_{m-1}}^{t} G_m(u-, j_{m-1} | W_{m-1}, z) f(u, j_m | t_{m-1}, j_{m-1}, z) dt \, .\end{aligned}$$

The probability of moving to the censoring state is given by

$$\begin{aligned}&\Pr(T_m \leq t, J_m = c | Z = z, W_{m-1}) = \\ &= \int_{T_{m-1}}^{t} \overline{G}_m(du, j_{m-1} | W_{m-1}, z) F(u, | t_{m-1}, j_{m-1}, z) \, ,\end{aligned}$$

where $\overline{G}_m = 1 - G_m$.

(ii) The parameters of the censoring survival functions G_m are noninformative on $\psi = (\theta, \alpha, \beta)$.

Denote by $E_0^c = E_0 \cup \{(i, c) : i \in \mathcal{T}\}$. As in section 2.1, let $N_h(t)$ and $Y_h(t), h \in E_0^c$ be given by

$$N_h(t) = \sum_{m \geq 1} N_{hm}(t) \, , \quad Y_h(t) = \sum_{m \geq 1} Y_{hm}(t) \, ,$$

where

$$\begin{aligned}N_{hm}(t) &= 1(T_{m-1} < T_m \leq t, J_{m-1} = i, J_m = j) \, , \\ Y_{hm}(t) &= 1(T_m \geq t > T_{m-1}, J_{m-1} = i) \, .\end{aligned}$$

Then $N(t) = \{N_h(t) : t \in [0, \tau], h = (i, j) \in E_0^c\}$ is a multivariate counting process whose compensator relative to the self–exciting filtration is given by

$$\Lambda_h(t) = \int_0^t Y_h(u) e^{\beta^T Z_h} \alpha_h(u) du \qquad \text{for} \quad h = (i,j) \in E_0$$
$$= \sum_{m \geq 0} \int Y_{hm}(u) \frac{\overline{G}_m(du, i|W_{m-1}, Z_1, Z_0)}{G_m(u-, i|W_{m-1}, Z_1, Z_0)} \quad \text{for} \quad h = (i,c), i \in \mathcal{T} .$$

In the case of the MAR model, we assume that the functions G_m do not depend on the covariate Z_0. Let

$$N_.(\tau) = \sum_{h \in E_0} N_h(\tau) \quad \widetilde{N}_.(\tau) = \sum_{h \in E_0^c} N_h(\tau) = N_.(\tau) + \sum_{i \in \mathcal{T}} N_{ic}(\tau)$$

Thus $N_.(\tau)$ counts the the total number of events observed in $[0, \tau]$, excluding withdrawals due to censoring, while $\widetilde{N}_.(\tau)$ counts the total number of withdrawals, including withdrawals due to censoring.

Condition 2.4.

(i) The conditional survival function G_m do not depend on the covariate Z_0.

(ii) The conditional distribution of the missing data indicator satisfies

$$Pr(R = r|V, Z_0) = \nu(V, Z_0(r)) \ ,$$

where $V = [\widetilde{N}_.(\tau), (J_\ell, T_\ell)_{\ell=0}^{\widetilde{N}_.(\tau)}, Z_1]$ and ν is a proper conditional probability measure not dependent on missing covariates.

(iii) The parameters of the conditional distribution ν of the missing data indicators are non-informative on the parameter $\psi = (\theta, \alpha, \beta)$ and on the family $\{G_m : m \geq 1\}$.

With this choice, the conditional density of the sequence $(R, V, Z_0(R))$ given (J_0, Z_1) is proportional to (5) or (6), with functions H_2 and H_4 replaced by

$$H_2(V, Z_0(R), z_0(\overline{r}); \psi) =$$
$$[F(T_{N_.(\tau)+1} \wedge \tau | T_{N_.(\tau)}, J_{N_.(\tau)}, Z_1, Z_0(R), z_0(\overline{r}))]^{1(J_{N_.(\tau)} \in \mathcal{T})}$$
$$H_4(V, Z_0(R), z_0(\overline{r}), V; \psi) =$$
$$\sum_{h=(J_{N_.(\tau)}, \ell) \in E_0} e^{\beta^T Z_h} [A_h(T_{N_.(\tau)+1} \wedge \tau) - A_h(T_{N_.(\tau)})]$$

Estimation of the parameters can be carried out much in the same way as in section 2.2 since the likelihood function is similar to the case of fixed censoring.

The assumption that the censoring distributions G_m do not depend on the missing covariates is in general quite restrictive. One method to alleviate this problem is to follow the approach of Fix and Neyman (1951) and Hoem (1969) analysis of Markov chain models, that is include among the possible states also states corresponding to possibly different forms of censoring. The hypothetical model corresponding to removal of "censoring" from the model, amounts then to evaluation of taboo probabilities of passage among states of interest without entering into censoring states.

3 A data example

For illustrative purposes we consider now data on 4144 patients treated for malignant melanoma cancer at the John Wayne Cancer Institute (JWCI) in Santa Monica. The data were collected during the time period 1980-1996.

The malignant melanoma neoplasm arises in the skin (state 0) from which it may spread to the regional lymphnodes (state 1) or to distant sites (stage 3) directly or indirectly progress first to nodal metastasis and then to distant metastatic sites (state 2). Distant metastasis is followed by death during a relatively short period of time. The survival characteristics of each of these stages are well known from natural history data in prospective databases (Barth et al. 1995, Morton et al. 1997).

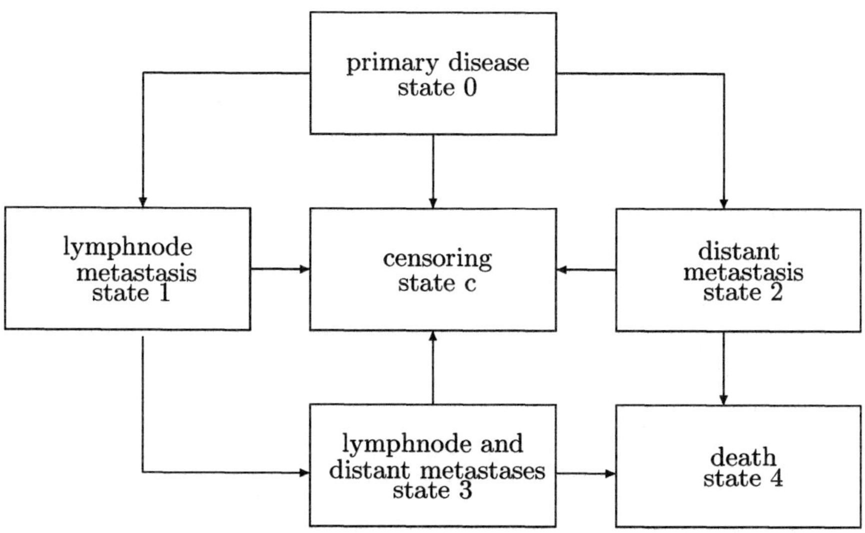

Figure 3.1. Censored melanoma progression process.

For purposes of analysis we use 5 covariates representing age, gender, site of primary tumor, its thickness and level of invasion. These covariates have been shown to form important prognostic factors for survival and metastasis in many clinical trials and database analyses. In the present study measurements of depth were missing for approximately 30% of patients whereas Clark's level was missing for 16% of patients

Since the remaining covariates were observed for all patients, their marginal distribution was taken as unspecified, whereas the conditional distribution of tumor depth and level of invasion was adjusted using logistic regression assuming that the cell probabilities satisfy

$$
\begin{aligned}
Pr(Z_{01} = \ell, Z_{02} = m|Z_1) &= g(\ell, m|Z_1, \theta) \\
&= \frac{e^{\theta_{l,m}^T Z_1}}{1 + \sum_{i,j \neq 0,0} e^{\theta_{ij}^T Z_1}} \\
&\qquad \text{for} \quad \ell, m = 0, 1, (\ell, m) \neq (0,0) \\
&= \frac{1}{1 + \sum_{i,j \neq 0,0} e^{\theta_{ij}^T Z_1}} \\
&\qquad \text{for} \quad \ell, m = 0, 0 ,
\end{aligned}
$$

The vector Z_1 was chosen to consist of the covariates age, gender and site of the primary tumor. Here $g(\ell, m|Z_1, \theta)$ are the conditional joint probabilities of Breslow's depth ($\ell = 0/1$ if depth is larger/smaller than 1.5 mm) and Clark's level ($m = 0/1$ for level $\geq / \leq$ III).

In the regression analysis we assumed that patients were randomly right censored and that the missing data mechanism satisfies MAR conditions 2.4. The corresponding baseline intensity matrix is given by

$$
\alpha(t) = \begin{pmatrix}
-\sum_{j=1,2} \alpha_{0j}(t) & \alpha_{01}(t) & \alpha_{02}(t) & 0 & 0 \\
0 & -\alpha_{13}(t) & 0 & \alpha_{13}(t) & 0 \\
0 & 0 & -\alpha_{24}(t) & 0 & \alpha_{24}(t) \\
0 & 0 & 0 & -\alpha_{34}(t) & \alpha_{34}(t) \\
0 & 0 & 0 & 0 & 0
\end{pmatrix}
$$

The intensities of the underlying Markov chain model of interest are of the form

$$
\begin{aligned}
\lambda_{0j}(t) &= Y_{0j}(t)\alpha_{0j}(t)e^{\beta_{0j}^T Z_{0j}} \quad \text{for} \quad j = 1, 2 \\
\lambda_{12}(t) &= Y_{12}(t)\alpha_{12}(t)e^{\beta_{12}^T Z_{12}} \\
\lambda_{24}(t) &= Y_{24}(t)\alpha_{24}(t)e^{\beta_{24}^T Z_{24}} \\
\lambda_{34}(t) &= Y_{34}(t)\alpha_{24}(t)e^{\beta_{34}^T Z_{34} + \rho}
\end{aligned}
$$

The risk processes are defined by $Y_{0j}(t) = I(T_1 \geq t, J_0 = 0)$, $Y_{13}(t) = I(T_2 \geq t > T_1, J_1 = 1)$, $Y_{24}(t) = I(T_2 \geq t > T_1, J_1 = 2)$ and $Y_{34}(t) = I(T_3 \geq t > T_2, J_1 = 1, J_2 = 3)$. In the regression analysis, the baseline hazard function for transitions into the death state (4) originating from state 2 and 3 are were taken to be the same, whereas the exponential part of the regression model depends on whether or not distant metastasis was preceded by lymphnode metastasis. The transition rates differ also in the risk processes Y_{24} and Y_{34}.

The diagram implies that each subject may contribute to the sample $N_.(\tau) = 0, \ldots, 3$ events. For a completely observable covariate vector

$Z = (Z_1, Z_0)$ we have

$$p(V, Z_0; \psi) = 1(N_\cdot(\tau) = 0) F(T_1 \wedge \tau | T_0, J_0, Z)$$
$$+ 1(N_\cdot(\tau) = 1) \sum_{j=1}^{2} 1(J_1 = j) F(T_2 \wedge \tau | T_1, J_1, Z) f(T_1, J_1 | T_0, J_0, Z)$$
$$+ 1(N_\cdot(\tau) = 2) 1(J_1 = 1, J_2 = 3) F(T_3 \wedge \tau | T_2, 3, Z) \prod_{q=1,2} f(T_q, J_q | T_{q-1}, J_{q-1}, Z)$$
$$+ 1(N_\cdot(\tau) = 2) 1(J_1 = 2, J_2 = 4) \prod_{q=1,2} f(T_q, J_q | T_{q-1}, J_{q-1}, Z)$$
$$+ 1(N_\cdot(\tau) = 3) 1(J_1 = 1, J_2 = 3, J_3 = 4) \prod_{q=1}^{3} f(T_q, J_q | T_{q-1}, J_{q-1}, Z_1, Z) ,$$

where

$$F(t|0, 0, z) = exp[-\sum_{q=1}^{2} e^{\beta_{0q}^T z} A_{0q}(t)] ,$$
$$F(t|t_1, 1, z) = exp[-e^{\beta_{13}^T z}[A_{13}(t) - A_{13}(t_1)] ,$$
$$F(t|t_1, 2, z) = exp[-e^{\beta_{24}^T z}[A_{24}(t) - A_{24}(t_1)] ,$$
$$F(t|t_2, 3, z) = exp[-e^{\beta_{34}^T z + \rho}[A_{34}(t) - A_{34}(t_2)]$$
$$f(t, j|t_0, j_0, z) = \alpha_{0j}(t) e^{\beta_{0j}^T z} F(t|0, 0, z), j = 1, 3 ,$$
$$f(t, j|t_1, j_1, z) = \alpha_{j_1,j}(t) e^{\beta_{j_1,j}^T z} F(t|t_1, j_1, z), \quad (j_1, j) = (1, 3), (2, 4) ,$$
$$f(t, j|t_2, j_2, z) = \alpha_{j_2,j}(t) e^{\beta_{j_2,j}^T z + \rho} F(t|t_1, j_2, z), \quad (j_2, j) = (3, 4) .$$

The first three terms of the density $p(V, Z_0, \psi)$ represent likelihood contributions corresponding subjects who are censored, whereas the last two terms are likelihood contributions for subjects who died of melanoma.

For any measurable function $\Phi(V, Z)$, its conditional expected given the data is

$$\hat{E}_\psi \Phi(V, Z) = \Phi(V, Z_0) \qquad \text{if} \quad R = (1, 1)$$
$$= \frac{\sum_{m=0,1} \Phi(V, \ell, m) e^{\theta_{lm} Z_1} p(V, \ell, m, \psi)}{\sum_{m=0,1} e^{\theta_{\ell m} Z_1} p(V, \ell, m, \psi)} \qquad \text{if} \quad R = (1, 0), Z_{01} = l$$
$$= \frac{\sum_{\ell=0,1} \Phi(V, \ell, m) e^{\theta_{\ell m}^T Z_1} p(V, \ell, m, \psi)}{\sum_{\ell=0,1} e^{\theta_{\ell m}^T Z_1} p(V, \ell, m; \psi)} \qquad \text{if} \quad R = (0, 1), Z_{02} = m$$
$$= \frac{\sum_{\ell,m=0,1} \Phi(V, \ell, m) e^{\theta_{\ell m}^T Z_1} p(V, \ell, m; \psi)}{\sum_{l,m=0,1} e^{\theta_{\ell m}^T Z_1} p(V, \ell, m, \psi)} \qquad \text{if} \quad R = (0, 0) ,$$

where $\theta_{00} = 0$.

In the EM algorithm, we replace the density $p(V, Z_0, \psi)$ by the function $p(V, Z_0, \hat{\psi}_q)$, where $\hat{\psi}_q$ is the estimate of the ψ parameter at the q-th step of the algorithm. In particular, in the case of the completely observable covariates, the score function for estimation of the θ parameter is given by

$$\nabla_{\theta_{\ell m}} \ell_2(Z_1, \ell, m) = Z_1[1(Z_{01} = \ell, Z_{02} = m) - \frac{e^{\theta_{\ell m}^T Z_1}}{1 + \sum_{(\ell', m') \neq (0,0)} e^{\theta_{\ell', m'}^T Z_1}}]$$

for $(\ell, m) \neq 0$. In the case of the missing covariates, the corresponding score function is

$$\nabla_{\theta_{\ell m}} Q_2(\psi | \hat{\psi}) = \sum_{k=1}^{n} Z_{1,k} \left[\hat{g}(\ell, m | Z_{1k}, \theta) - g(\ell, m | Z_{1k}, \theta) \right]$$

where

$$\begin{aligned} \hat{g}(\ell, m | Z_1, \theta) &= 1(R = (0,0)) 1(Z_{01} = \ell, Z_{0,2} = m) \\ &+ 1(R = (1,0)) 1(Z_{01} = \ell) \frac{e^{\theta_{\ell m} Z} p(V, \ell, m, \hat{\psi})}{\sum_{m'=0,1} e^{\theta_{\ell m'} Z} p(V, \ell, m', \hat{\psi})} \\ &+ 1(R = (0,1)) 1(Z_{02} = m) \frac{e^{\theta_{\ell m}^T Z} p(V, \ell, m, \hat{\psi})}{\sum_{\ell'=0,1} e^{\theta_{\ell' m}^T Z} p(V, \ell', m, \hat{\psi})} \\ &+ 1(R = (0,0)) \frac{e^{\theta_{\ell m}^T Z} p(V, \ell, m, \hat{\psi})}{\sum_{\ell', m'=0,1} e^{\theta_{\ell' m'}^T Z} p(V, \ell', m', \hat{\psi})} \end{aligned}$$

The conditional expected entering into the score function for the regression coefficients β are also simple to evaluate.

Numerous studies have shown that females have better survival rates than males. The primary reason for better performance of women is that their melanomas tend to occur more frequently on extremities, which is a more favorable location. Patients with melanomas located on extremities have in general better survival rate than patients whose primary lesion is located on trunk or head and neck. This is also shown by our results in Table 3.1. We found that primary tumor located on the extremities decreased the risk of all transitions. Positive site × gender interaction in the case of all transitions, except $0 \to 1$, indicates that primary site (extremities) decreases the rate of transitions among various states of the model but this effect is less marked among men.

Further, age associated with increased risk of lymphnode and distant metastases. Males experienced an increased risk of transition from state 0 to state 2, however, the negative (age) × (gender) interaction suggests that this increased risk is less pronounced among older men. Among pathological factors, Clark's level of invasion > III and Breslow's depth > 1.5 mm is associated with increased risk of the transition from state 0 to both state 1 and 2.

Table 3.1 compares results obtained from two regression analyses corresponding to the MAR model and the "case deletion" model (parenthesized regression coefficients and standard errors). In both cases we used partition of the observed range of the transition times into 10 intervals corresponding to a equidistant partition of the observed range of the transition times. The range of the observed transition times was $(0, 6.8)$ years for transition $0 \to 1$, $(0.93, 9.75)$ years for transition $0 \to 2$, $(0.69, 8.34)$ years for transition $2 \to 3$, and $(1.39, 9.08)$ years from transition from state $3 \to 4$. In the case of transitions between states $0 \to 2$, $2 \to 3$ and $3 \to 4$ the results from both analyses are quite similar, though the standard errors of the estimates obtained based on the MAR model are uniformly smaller as a result of the increased sample size. On the other hand, in the case of the transition from state 0 to state 1, the results differ. The MAR model suggests that location of the primary tumor and site $\times$ gender interaction are important risk factors for progression from state 0 into state 1, whereas the "case deletion" model does not identify these factors as significant.

Appendix 1

Let $\dot{\ell}_\theta$ and $\ddot{\ell}_\theta$ denote the first and second derivatives of the density g_θ with respect to θ, and for $k = 1, \ldots, n$, let $M_{h,k}(t, \psi) = N_{h,k}(t) - e^{\beta^T Z_{h,k}} \Lambda_{h,k}(t)$. We use Louis (1982) formula to get observed information,

$$\hat{\Sigma}_n(\psi) = S_n[\hat{\Sigma}_{1n}(\psi) - \hat{\Sigma}_{2n}(\psi)] \,,$$

where the first term in an estimate of the complete information and the second is an estimate of the expected conditional covariance of the score function given the data. The matrix $\hat{\Sigma}_{1n}(\psi)$ is the negative Hessian of the $Q_n(\psi|\psi)$ function with respect to the first argument. Similarly, $\hat{\Sigma}_{2n}(\psi)$ is the negative derivative of $\nabla Q_n(\psi|\psi)$ with respect to the second argument. For $q = 1, 2$, we have

$$\hat{\Sigma}_{qn}(\psi) = \begin{pmatrix} \hat{\Sigma}_{qn}^{12;12}(\psi) & \hat{\Sigma}_{qn}^{12;3}(\psi) \\ \hat{\Sigma}_{qn}^{3;12}(\psi) & \hat{\Sigma}_{qn}^{3;3}(\psi) \end{pmatrix} .$$

For $q = 1$, we have

$$\hat{\Sigma}_{1n}^{12;12}(\psi) = \sum_{k=1}^{n} \hat{E}_\psi \begin{pmatrix} -\ddot{\ell}_\theta(Z_k) & 0 \\ 0 & \sum_{h=1}^{m} Z_{h,k}^{\otimes 2} e^{\beta^T Z_{h,k}} \Lambda_{h,k}(\tau) \end{pmatrix} ,$$

$$\hat{\Sigma}_{1n}^{3;3}(\psi) = \mathrm{diag}[\frac{\sum_{k=1}^{n} N_{h,k}(I_p)}{a_{ph}^2} : p = 1, \ldots, \ell(n), h \in E_0] \,,$$

$$\hat{\Sigma}_{1n}^{3;12}(\psi) = \begin{pmatrix} 0 \\ \left[\sum_{k=1}^{n} Y_{hk}(I_p) \hat{E}_\psi[Z_{h,k} e^{\beta Z_{h,k}}]\right]_{\substack{p \le \ell(n) \\ h \le m}} \end{pmatrix}$$

Table 1. Regression estimates in melanoma example

factor	β	se	β	se
	state 0 → state 1		state 0 → state 2	
Age	1.46	0.38	1.39	0.34
	(1.31	0.46)	(1.34	.41)
Gender	0.31	0.14		
(male vs female)	(0.42	0.18)		
Clark	0.52	0.11	0.49	0.12
(> III vs ≤ III)	(0.42	0.13)	(0.42	0.15)
Depth	1.50	0.41	1.63	0.53
(> 1.5 mm vs ≤ 1.5 mm)	(1.60	0.54)	(1.67	0.60)
Site	-0.47	0.14	-1.20	0.17
(extremities vs other)	(-0.25	0.19)	(-1.08	0.19)
age × gender	-0.56	0.16		
interaction	(-.52	0.18)		
site × gender	0.42	0.20	0.87	0.20
	(0.41	0.24)	(0.74	0.25)
	state 2 → state 3		state 3 → state 4	
Age	1.53	0.50	-0.43	0.34
	(1.70	0.59)	(-0.31	0.42)
Clark	0.58	0.96	0.35	0.43
(> III vs ≤ III)	(0.66	1.09)	(0.39	0.50)
Depth	0.6	0.73	0.8	0.91
(> 1.5 mm vs ≤ 1.5 mm)	(0.69	0.82)	(0.89	1.12)
Site	-0.76	0.20	-0.17	0.10
(extremities vs other)	(-0.82	0.25)	(-0.27	0.13)
Age × gender	-0.34	0.18	0.29	0.10
	(-0.45	0.23)	(0.32	0.14)
Site × gender	1.08	0.25		
	(1.25	0.30)		
prior lymphnode	NA	NA	0.25	0.09
metastasis	NA	NA	(0.30	0.20)

and $\hat{\Sigma}_{1n}^{12;3}(\psi) = \hat{\Sigma}_{1n}^{3;12}(\psi)^T$. For $q = 2$,

$$\hat{\Sigma}_{2n}^{12;12}(\psi) = \sum_{k=1}^{n} \hat{\text{cov}}_\psi \left(\frac{\dot{\ell}_\theta(Z_k)}{\sum_{h \in E_0} \int_{[0,\tau]} Z_{h,k} M_{h,k}(du, \psi)} \right),$$

$$\hat{\Sigma}_{2n}^{3;12}(\psi) = -\sum_{k=1}^{n} \hat{\text{cov}}_\psi \left(\left(\frac{\dot{\ell}_\theta(Z_k)}{\sum_{h \in E_0} \int_{[0,\tau]} Z_{h,k} M_{h,k}(du, \psi)} \right), \left([Y_{h,k}(I_p) e^{\beta^T Z_{h,k}}]_{\substack{p \le \ell(n), \\ h \in E_0}} \right) \right),$$

$$\hat{\Sigma}_{2n}^{33}(\psi) = \sum_{k=1}^{n} \left[Y_{h,k}(I_p) Y_{h',k}(I_{p'}) \hat{\text{cov}}_\psi [e^{\beta^T Z_{h,k}}, e^{\beta^T Z_{h',k}}] \right]_{\substack{p,p' \le \ell(n) \\ h,h' \in E_0}}$$

and $\hat{\Sigma}_{2n}^{12;3}(\psi) = \hat{\Sigma}_{2n}^{3;12}(\psi)$. To update the $\eta = (\theta, \beta)$ coefficients, we use

$$\begin{pmatrix} \theta_{q+1} \\ \beta_{q+1} \end{pmatrix} = \begin{pmatrix} \theta_q \\ \beta_q \end{pmatrix} + H_n(\eta_q, \hat{\alpha}_{q+1}(\eta_q))^- [\nabla_\eta] Q(\eta_q, \hat{\alpha}_{q+1}(\eta_q) | \hat{\psi}_q) .$$

Here at the q-th step, we set

$$H_n(\psi) = [\hat{\Sigma}_n^{12;12}(\psi) - \hat{\Sigma}_n^{12;3}(\psi)[\hat{\Sigma}_n^{3;3}(\psi)]^- \hat{\Sigma}_n^{3;12}(\psi)] ,$$

where $\hat{\Sigma}_n^{p;r}(\psi) = \Sigma_{1n}^{p;r}(\psi) - \Sigma_{2n}^{p;r}(\psi)$, with conditional expectations and covariances evaluated at point $\hat{\psi}_q$.

Appendix 2

The following recurrent formulas can be easily verified using Bayes theorem.

We first consider the case of completely observable covariates and assume that Z is partitioned into two disjoint blocks $Z = (Z_0, Z_1)$. For $m \ge 1$, let $(j_0, j_1, \ldots, j_m)$ be a possible path in the model connecting the initial state j_0 with a transient or an absorbing state j_m and such that $(j_0, j_1, \ldots, j_{m-1}) \subset \mathcal{T}$. Let $(t_0 = 0, t_1, \ldots, t_m)$ be an ordered sequence $0 = t_0 < t_1 \ldots < t_m$ of potential times of entrances into states given by the sequence $\{j_\ell, \ell = 0, \ldots, m\}$. For $m \ge 0$, put $W_m = (T_\ell, J_\ell)_{\ell=0}^m$, $w_m = (t_\ell, j_\ell)_{\ell=0}^m$. Under assumptions of condition 2.1, the posterior distribution of the covariate Z_0 is of the form

$$\begin{aligned} &\Pr(Z_0 \in B | N_\cdot(\tau) = m, W_m = w_m, Z_1 = z_1) = \\ &= \int_B \mu_m(dz_0 | w_m, z_1) \quad \text{if} \quad j_m \in \mathcal{A}, t_m \le \tau, m \ge 1 \\ &= \int_B \widetilde{\mu}_m(dz_0, \tau | w_m, z_1) \quad \text{if} \quad j_m \in \mathcal{T}, t_m \le \tau < t_{m+1}, m \ge 0 , \end{aligned}$$

where $\mu_0(dz_0 | j_0, z_1)$ is the conditional distribution of the covariate Z_0 given the initial state $J_0 = j_0$ and the covariate $Z_1 = z_1$. For $m \ge 1$

$$\mu_m(dz_0|w_m,,z_1) = \frac{f(j_m,t_m|j_{m-1},t_{m-1},(z_1,z_0))\mu_{m-1}(dz_0|w_{m-1},z_1)}{E_{m-1}f(j_m,t_m|j_{m-1},t_{m-1},(z_1,z_0))}, \tag{7}$$

where E_{m-1} denotes conditional expectation with respect to $\mu_{m-1}(dz_0|w_{m-1},z_1)$. In addition, for $m \geq 0$

$$\widetilde{\mu}_m(dz_0,s|w_m,z_1) = \frac{F(s|(j_m,t_m),(z_1,z_0))\mu_m(dz_0|w_m,z_1)}{E_m F(s|(j_m,t_m),(z_1,z_0))}. \tag{8}$$

Next we collect parameters of the marginal model obtained by integrating out the covariate Z_0 from the model. If the covariate vector $Z = (Z_0, Z_1)$ is taken to assume values in the Cartesian product $\mathcal{Z}_0 \times \mathcal{Z}_1$ of a q and d-q dimensional Euclidean space, then the marginal model represents transformation X of the original probability space space $(\mathcal{Z}_0 \times \mathcal{Z}_1 \times \Omega, (\mathcal{G} \otimes \mathcal{F}_t), \Pr)$ into the space $(\mathcal{Z}_1, (\mathcal{G}_1 \otimes \mathcal{F}_t), \Pr_X)$ corresponding to the assignment $X(z_0, z_1, \omega) = (z_1, \omega)$. Thus the marginal model is adapted to the marginal self-exciting filtration, generated by $\mathcal{G}_1 \otimes \mathcal{F}_t = \sigma(Z_1) \otimes \sigma(J_0, N_h(s), Y_h(s+) : s \leq \tau)$. The probability $\Pr_X$ is the induced marginal probability, obtained by integrating out the covariate Z_0 from the model. In the following we write "Pr" for the induced probability $\Pr_X$, to simplify the notation.

For $m \geq 1$, let $(j_0, \ldots, j_{m-1})$ be a sequence of transient states. Set

$$\begin{aligned}
\widetilde{F}_m(t, j_{m-1}|w_{m-1}, z_1) &= E_{m-1}F(t, |t_{m-1}, j_{m-1}, z_1, Z_0) , \\
\widetilde{f}_m(t, j_m|w_{m-1}, z_1) &= E_{m-1}f(t, j_m|t_{m-1}, j_{m-1}, z_1, Z_0) , \\
\widetilde{\alpha}_m(t, j_m|w_{m-1}, z_1) &= \frac{\widetilde{f}_m(t, j_m|w_{m-1}, z_1)}{\widetilde{F}_m(t, |w_{m-1}, z_1)} , \\
q_m(j_m|t, w_{m-1}, z_1) &= \frac{\widetilde{\alpha}_m(t, j_m|w_{m-1}, z_1)}{\sum_{h=(j_{m-1},l)\in E_0} \widetilde{\alpha}_m(t, j_m|w_{m-1}, z_1)} ,
\end{aligned}$$

where E_{m-1} is the conditional expectation of Z_0 with respect to $\mu_{m-1}(dz_0|w_{m-1}, z_1)$. Under the assumption of the proportional hazard model, we have

$$\begin{aligned}
&\Pr(T_m > t|Z_1, W_{m-1}) = \widetilde{F}_m(t, J_{m-1}|W_{m-1}, Z_1) \\
&= E_{m-1}\left[\exp - \sum_{h:h=(J_{m-1},l)\in E_0} e^{\beta^T Z_h} \int_{(T_{m-1},t]} \alpha_h(u)du\right] .
\end{aligned}$$

In addition

$$\begin{aligned}
&\Pr[J_m = j|T_m = t, W_{m-1}, Z_1] = q_m(j|T_m = t, W_{m-1}, Z_1) \\
&= \frac{\alpha_{J_{m-1},j}(t)E_{m-1}[e^{\beta^T Z_{J_{m-1},j}}|T_m \geq t, W_{m-1}, Z_1]}{\sum_l \alpha_{J_{m-1},l}(t)E_{m-1}[e^{\beta^T Z_{J_{m-1},l}}|T_m \geq t, W_{m-1}, Z_1]}
\end{aligned}$$

and

$$\widetilde{\alpha}_m(t, j_m|W_{m-1}, Z_1) = \lim_{s \downarrow 0} \frac{1}{s} \Pr(T_m \in [t, t+s], J_m = j|T_m = t, W_{m-1}, Z_1]$$
$$= 1(T_m \geq t > T_{m-1})\alpha_{J_{m-1},j}(t)E_{m-1}[e^{\beta^T Z_{J_{m-1},j}}|T_m \geq t, W_{m-1}, Z_1]$$

The cumulative intensity of the process $[N_h(t) : h \in E_0]$ with respect to the marginal filtration $\sigma(Z_1) \otimes \mathcal{F}_t$ is given by

$$\Lambda_h(t) = \sum_{m \geq 1} \int_{[0,t]} Y_{hm}(u)\widetilde{\alpha}_m(u, j|W_{m-1}, Z_1)du \qquad \text{for} \quad h = (i, j) \in E_0$$

Thus marginal process has a compensator of a different form than the original process.

Next we consider the MAR condition 2.2. It is equivalent to the following two conditions.

(i) For $m \geq 0$, the conditional distribution of the missing data indicator satisfies

$$\begin{aligned} Pr(R = r|N_.(\tau) = m, W_m = w_m, Z = z) & \\ = \nu_m(r|w_m, z(r)) & \quad \text{if} \quad j_m \in \mathcal{A}, t_m \leq \tau, m \geq 1 \\ = \widetilde{\nu}_m(r, \tau|w_m, z(r)) & \quad \text{if} \quad j_m \in \mathcal{T}, t_m \leq \tau < \tau_{m+1}, m \geq 0 \end{aligned}$$

for some functions $(\nu_m, \widetilde{\nu}_m)$ depending only on the sequence w_m and the observed covariate $z(r) = (z_0(r), z_1)$, but not the missing covariates. In addition,

$$\sum_r \nu_m(r|w_m z(r)) = 1 \ , \quad \sum_r \widetilde{\nu}_m(r, \tau|w_m z(r)) = 1 \ , \tag{9}$$

where the sums extend over possible values of the missing data indicators.

(ii) The parameters of the conditional distribution of the missing data indicators, $(\nu_m, \widetilde{\nu}_m \geq 0)$ are non-informative on the parameters of the underlying model of interest.

The joint density of the vector $(V, Z_0(R), R)$, $V = [N_.(\tau), (J_\ell, T_\ell)_{\ell=0}^{N_.(\tau)}, Z_1]$ is given by

$$\begin{aligned} &\nu_m(r|w_m, z(r))p_m(w_m|z(r))\overline{\mu}_0(dz_0(r)|z_1, j_0) \\ &\qquad \text{if} \quad j_m \in \mathcal{A}, t_m \leq \tau, N_.(\tau) = m \geq 1 \\ &\widetilde{\nu}_m(r, \tau|w_m, z(r))\widetilde{p}_m(\tau, w_m|z(r))\overline{\mu}_0(dz_0(r)|z_1, j_0) \\ &\qquad \text{if} \quad j_m \in \mathcal{T}, t_m \leq \tau < t_{m+1}, N_.(\tau) = m \geq 0 \ . \end{aligned}$$

Here $\overline{\mu}_0(dz_0(r)|z_1, j_0)$ is the marginal conditional distributions of the covariate $Z_0(r)$ given (Z_1, J_0). In addition, $z(r) = (z_0(r), z_1)$ and for any sequence $0 = t_0 < t_1 < t_2 \ldots < t_m$, we have

$$p_m(w_m|z(r)) = \prod_{l=1}^{m} \widetilde{f}_l(t_l, j_l|w_{l-1}, z(r)) \qquad \text{if} \quad j_m \in \mathcal{A}, t_m \leq \tau, m \geq 1$$

$$\widetilde{p}_m(\tau, w_m|z(r)) = \widetilde{F}_{m+1}(\tau, j_m|w_m, z(r)) \prod_{l=1}^{m} \widetilde{f}_l(t_l, j_l|w_{l-1}, z(r))$$
$$\text{if} \quad j_m \in \mathcal{T}, t_m \leq \tau < t_{m+1}, m \geq 1$$
$$= \widetilde{F}_1(\tau, j_0|w_0, z(r)) \quad \text{if} \quad \tau < t_1, m = 0 \,.$$

The function $p_m(w_m|z(r))$ is the joint conditional subdensity density of a sequence $W_m = ((T_\ell, J_\ell) : \ell = 0, \ldots, m)$ terminating in an absorbing state, and evaluated conditionally on the initial state and the covariate $z(r)$ in the marginal model obtained by integrating out the covariate $z(\overline{r})$. Similarly, the function $\widetilde{p}_m(\tau, w_m|z(r))$ is the joint marginal conditional subdensity of survival in a transient state.

If parameters of the functions $\nu_m, \widetilde{\nu}_m$ do not depend on the Euclidean parameter φ of the Markov chain model, then in Section 2.3, the complete data likelihood is of the form

$$\mathrm{Lik}(\nu, \psi) = \prod_{k=1}^{n} \nu(V_k, Z_{0,k}(R_k)) p(R_k, V_k, Z_{0,k}(R_k); \psi) \,. \tag{10}$$

For each subject, $Z(R) = (Z_0(R), Z_1)$,

$$\nu(V, Z_0(R)) = \left(\nu_{N.(\tau)}(R|W_{N.(\tau)}, Z(R))\right)^{1(J_{N.(\tau)} \in \mathcal{A})}$$
$$\times \left(\widetilde{\nu}_{N.(\tau)}(R, \tau|W_{N.(\tau)}, Z(R))\right)^{1(J_{N.(\tau)} \in \mathcal{T})}$$

and

$$p(R, V, Z_0(R); \psi) = \left(\prod_{\ell=1}^{N.(\tau)} \widetilde{f}_\ell(T_\ell, J_\ell; \psi|W_\ell, Z(R)) \right)^{1(N.(\tau)>0)}$$
$$\times \left(\widetilde{F}_{N.(\tau)+1}(\tau, J_{N.(\tau)}; \psi|W_{N.(\tau)}, Z(R))\right)^{1(J_{N.(\tau)} \in \mathcal{T})}$$
$$\times \widetilde{g}_\theta(Z_0(R)|Z_1, J_0) \,,$$

where $\widetilde{g}_\theta(\cdot|Z_1, J_0)$ is the marginal conditional density of $Z_0(R)$ given (Z_1, J_0).

In the case of randomly censored data, the likelihood factorization (10) can be derived in an analogous fashion. We omit the details.

Acknowledgement. Research supported by the National Cancer Institute grant 1-R01-96-CA65595-02.

References

[VKVN02] Aalen O.O. and Johansen S. An empirical transition matrix for nonhomogeneous Markov chains based on censored observations. Scand. J. Statist., **5**, 141-150 (1978)

[VKVN02] Andersen, P. K. Hansen, L. S. and Keiding, N. Non- and semi–parametric estimation of transition probabilities from censored observations of a non-homogeneous Markov process. Scand. J. Statist., **18**, 153-167 (1991)

[BN02] Andersen P. K., Borgan O., Gill R. D. and Keiding N. Statistical Models Based on Counting Processes. Springer Verlag, New York (1993)

[VKVN02] Barth, A., Wanek, L. A., Morton, D. L. (1995). Prognostic factors in 1,521 melanoma patients with distant metastases. J. Amer. College Surg., **181**, 193-201 (1995)

[VKVN02] Chen, M. H. and Little, R. Proportional hazards regression with missing covariates. J. Amer. Statist. Assoc., **94**, 896-908 (1999)

[VKVN02] Chen, M. H. and Ibrahim, J. G. Maximum likelihood Moethods for cure rate models with missing Covariates. Biometrics, **57** 43-52 (2001)

[BN02] Chiang, C.L. Introduction to Stochastic Processes. Wiley, New York (1968)

[VKVN02] Friedman, M. Piecewise exponential models for survival with covariates. Ann. Statist., **10**, 101-113 (1982)

[VKVN02] Fix, E. and Neyman, J. A simple stochastic model for recovery, relapse, death and loss of patients. Human Biol., **23**, 25-241 (1951)

[VKVN02] Hoem. J. M. Purged and and partial Markov chains. Skand. Aktuarietidskr., **52**, 147-155 (1969)

[VKVN02] Lipsitz, S. R. and Ibrahim, J. G. Using the EM algorithm for survival data with incomplete categorical covariates. Lifetime Data Analysis, **2**, 5-14 (1999)

[VKVN02] Lipsitz, S. R. and Ibrahim, J. G. Estimating equations with incomplete categorical covariates in the Cox model. Biometrics **54** 1002-1013 (1998)

[BN02] Little, R. and Rubin, D. Statistical Analysis of Missing Data. Wiley, New York (1987)

[VKVN02] Lin, D. and Ying, Z. Cox regression with incomplete covariates measurements. J. Amer. Statist. Assoc., **88**, 1341-1349 (1993)

[VKVN02] Louis, T. Finding the observed information matrix when using the EM algorithm. J. Roy. Statist. Soc.-B, **44**, 226-233 (1982)

[VKVN02] Martinussen, T. Cox regression with incomplete covariates using EM algorithm. Scand. J. Statist., **26**, 479-491 (1999)

[17] Morton D. L., Essner, R., Kirkwood J. M. and Parker R. G. Malignant melanoma. In: Holland J, F. , Frei E., Bast R., Kuffe D, Morton D. L and Weichselbaum R., (eds.) Cancer Medicine. Williams & Wilkins, Baltimore (1997)

[VKVN02] Sinha, D., Tanner, M.A. and Hall, W.J. Maximization of the marginal likelihood of grouped survival data. Biometrika, **81**, 53-60 (1994)

[VKVN02] Wei, G.C.G. and Tanner, M.A. A Monte Carlo implementation of the EM algorithm and the poor man's data augmentation algorithm. J. Amer. Statist. Assoc., **85**,600-704 (1990)

[VKVN02] Zhou, Y. and Pepe, M. Auxiliary covariate data in failure time regression. Biometrika **82**, 139-149 (1995)

Tests of Fit based on Products of Spacings

Paul Deheuvels[1] and Gérard Derzko[2]

[1] L.S.T.A., Université Paris VI, 7 avenue du Château, F 92340 Bourg-la-Reine, France pd@ccr.jussieu.fr

[2] Sanofi-Synthélabo Recherche, 371 rue du Professeur Joseph Blayac, 34184 Montpellier Cedex 04, France Gerard.Derzko@sanofi-aventis.com

Summary. Let $Z = Z_1, \ldots, Z_n$ be an i.i.d. sample from the distribution $F(z) = P(Z \leq z)$ and density $f(z) = \frac{d}{dz}F(z)$. Let $Z_{1,n} < \ldots < Z_{n,n}$ be the order statistics generated by $Z_1, \ldots, Z_n$. Let $Z_{0,n} = a = \inf\{z : F(z) > 0\}$ and $Z_{n+1,n} = b = \sup\{z : F(z) < 1\}$ denote the end-points of the common distribution of these observations, and assume that f is continuous and positive on (a, b). We establish the asymptotic normality of the sum of logarithms of the spacings $Z_{i,n} - Z_{i-1,n}$, for $i = 1, \ldots, n+1$, under minimal additional conditions on f. Our results largely extend previous results in the literature due to Blumenthal [Blu68] and other authors.

1 Introduction and Main Results.

1.1 Introduction.

Let $Z = Z_1, Z_2, \ldots$ be independent and identically distributed [i.i.d.] random variables with distribution function $F(z) = \mathbb{P}(Z \leq z)$ and density $f(z)$, assumed throughout to be continuous and positive on (a, b) for $-\infty \leq a < b \leq \infty$, and equal to 0 otherwise. Here, $a = \inf\{z : F(z) > 0\}$ and $b = \sup\{z : F(z) < 1\}$ denote the distribution end-points. For each $n \geq 1$, denote by $a < Z_{1,n} < \ldots < Z_{n,n} < b$ the order statistics of $Z_1, \ldots, Z_n$, and set, for convenience, $Z_{0,n} = a$ and $Z_{n+1,n} = b$, with $F(Z_{0,n}) = F(a) = 0$ and $F(Z_{n+1,n}) = F(b) = 1$, for $n \geq 0$. Denote by

$$D_{i,n} = Z_{i,n-1} - Z_{i-1,n-1} \quad \text{for} \quad i = 1, \ldots, n, \tag{1}$$

the spacings of order $n \geq 1$ based upon $\{Z_{i,n-1} : 1 \leq i \leq n\}$. Darling [Dar53] introduced the class of statistics

$$T_n = T_n(p, q) = \sum_{i=p}^{n-q+1} \Big\{ -\log(nD_{i,n}) \Big\} = \log \Big(n^{n-p-q+2} \prod_{i=p}^{n-q+1} D_{i,n} \Big), \tag{2}$$

to test the null hypothesis (when $-\infty < a < b < \infty$)

$(H.0)\quad f(z) = (b-a)^{-1}\mathbb{I}_{(a,b)}(z)$ for $a < z < b$,

against the alternative $(H.1)$ that f is arbitrary on (a, b). In (2), p and q are fixed integers such that $1 \leq p \leq n - q + 1 \leq n$. When the distribution endpoints a and b are finite and known, a standard choice for p and q is given by $p = q = 1$. On the other hand, when a (resp. b) is unknown (or possibly infinite), $D_{1,n}$ (resp. $D_{n,n}$) is unknown (or possibly infinite), so it is more appropriate to choose $p \geq 2$ (resp. $q \geq 2$), otherwise $T_n(p, q)$ becomes meaningless. The aim of the present paper is to investigate the limiting behavior of $T_n = T_n(p, q)$ as $n \to \infty$. It will become obvious later on that the results we shall obtain are essentially independent of the choices of p, q, subject to the restrictions that

$$p \geq p_0 = \begin{cases} 1 & \text{when} \quad a > -\infty, \\ 2 & \text{when} \quad a = -\infty, \end{cases} \quad \text{and} \quad q \geq q_0 = \begin{cases} 1 & \text{when} \quad b < \infty, \\ 2 & \text{when} \quad b = \infty. \end{cases} \tag{3}$$

Because of this, we will use throughout the notation $T_n = T_n(p, q)$, and specify the values of p, q only in case of need.

Under rather strenuous regularity assumptions on f (assuming, in particular that f is twice differentiable on (a, b), see, e.g., (2.3a) in [Blu68]), implying finiteness of $\mathrm{Var}(\log f(Z))$, Blumenthal [Blu68] (see also Cressie [Cre76]) showed that, as $n \to \infty$,

$$\begin{aligned} n^{-1/2}&\Big\{T_n - n\gamma - n\mathbb{E}(\log f(Z))\Big\} \\ &\stackrel{d}{\to} N\Big(0, \zeta(2) - 1 + \mathrm{Var}(\log f(Z))\Big), \end{aligned} \tag{4}$$

where "$\stackrel{d}{\to}$" denotes weak convergence. In (4), $\zeta(\cdot)$ and γ denote, respectively, the Riemann zeta function and Euler's constant, conveniently defined by

$$\zeta(r) = \frac{1}{\Gamma(r)} \int_0^\infty \frac{t^{r-1}dt}{e^t - 1} = \sum_{j=1}^\infty \frac{1}{j^r} \quad \text{for} \quad r > 1, \tag{5}$$

$$\begin{aligned} \zeta(2) = \frac{\pi^2}{6}, \quad \gamma &= \int_0^\infty (-\log t)e^{-t}dt = \lim_{r\downarrow 1}\Big\{\zeta(r) - \frac{1}{r-1}\Big\} \\ &= \lim_{n\to\infty}\Big\{\sum_{j=1}^{n-1}\frac{1}{j} - \log n\Big\} = 0.577215\ldots, \end{aligned}$$

(see, e.g., Spanier and Oldham [SO87]). Here, $\Gamma(\cdot)$ stands for Euler's Gamma function, namely

$$\Gamma(r) = \int_0^\infty t^{r-1}e^{-t} \quad \text{for} \quad r > 0. \tag{6}$$

One of the purposes on the present paper is to give simple conditions implying the validity of (4). Our main result concerning this problem is stated in the following theorem.

Theorem 1.1 *Assume that*

$$\mathbb{E}\Big((\log f(Z))^2\Big) < \infty, \tag{7}$$

and either

(i) f *is continuous and bounded away from* 0 *on* $[a,b]$*; or*

(ii) f *is monotone in a right neighborhood of* a*, and monotone in a left neighborhood of* b*.*

Then, for each $p \geq p_0$ *and* $q \geq p_0$*, we have*

$$n^{-1/2}\Big\{T_n(p,q) - n\gamma - n\mathbb{E}(\log f(Z))\Big\}$$
$$\overset{d}{\to} N\Big(0, \zeta(2) - 1 + \mathrm{Var}(\log f(Z))\Big). \tag{8}$$

1.2 Some Relations with the Kullback-Leibler Information .

The limiting result in (8) is related to the Kullback-Leibler information in the following way. In general, for any two random variables Y_0 and Y_1 with densities g_0 and g_1 on $\mathbb{R}$, with respect to the Lebesgue measure, the Kullback-Leibler information $K(g_1, g_0)$ of g_1 with respect to g_0 is defined by (with the convention $0/0 = 1$)

$$K(g_1, g_0) = \mathbb{E}\Big(\log\Big\{\frac{g_1(Y_1)}{g_0(Y_1)}\Big\}\Big) = \int_{\mathbb{R}} \log\Big\{\frac{g_1(y)}{g_0(y)}\Big\} g_1(y)dy, \tag{9}$$

when $g_1(y)dy \ll g_0(y)dy$ (which we denote by $g_1 \ll g_0$), and

$$K(g_1, g_0) = \infty \quad \text{otherwise}. \tag{10}$$

The well-known property that

$$K(g_1, g_0) \geq 0, \tag{11}$$

with equality if and only if $g_1 = g_0$ a.e., follows from the fact that the function

$$\mathbf{h}(x) = \begin{cases} x\log x - x + 1 & \text{for } x > 0, \\ 1 & \text{for } x = 0, \\ \infty & \text{for } x < 0, \end{cases} \tag{12}$$

fulfills $\mathbf{h}(x) \geq 0$ with equality if and only if $x = 1$. This, in turn, implies that

$$K(g_1, g_0) = \int_{\mathbb{R}} \mathbf{h}\Big\{\frac{g_1(y)}{g_0(y)}\Big\} g_0(y)dy \geq 0, \tag{13}$$

with equality if and only if $g_1 = g_0$ a.e. (with $g_1 \ll g_0$). The inequality (13) also holds when $g_1 \not\ll g_0$, since then, by definition, $K(g_1, g_0) = \infty$. By applying (13) to $g_1 = f$ and $g_0 = (b-a)^{-1}\mathbb{I}_{(a,b)}$ when $-\infty < a < b < \infty$, we see that

$$K\big(f,(b-a)^{-1}\mathbb{I}_{(a,b)}\big) = \int_a^b f(z)\log f(z)dz + \log(b-a) \tag{14}$$

$$= \mathbb{E}(\log f(Z)) + \log(b-a) \geq 0, \tag{15}$$

with equality if and only $(H.0)$ holds, namely, when $f(t) = (b-a)^{-1}$ a.e. on (a,b). When the constants a and b are unknown (but finite), we may estimate these quantities by $Z_{1,n}$ and $Z_{n,n}$, respectively. Under $(H.0)$, it is straightforward that, as $n \to \infty$,

$$Z_{1,n} = a + O_{\mathbb{P}}(1/n) > a \quad \text{and} \quad Z_{n,n} = b + O_{\mathbb{P}}(1/n) < b. \tag{16}$$

By (16), the test rejecting $(H.0)$ when either (for a and b specified)

$$T_n \geq c^*_{n,\alpha} := n\gamma - n\log(b-a) + n^{1/2}\nu_\alpha\Big\{\frac{\pi^2}{6} - 1\Big\}^{1/2}, \tag{17}$$

or (for a and b unspecified)

$$T_n \geq c^{**}_{n,\alpha} := n\gamma - n\log(Z_{n,n} - Z_{1,n}) + n^{1/2}\nu_\alpha\Big\{\frac{\pi^2}{6} - 1\Big\}^{1/2}, \tag{18}$$

where ν_α denotes the upper quantile of order $\alpha \in (0,1)$ of the normal $N(0,1)$ law, is asymptotically consistent, with size tending to α as $n \to \infty$, against all alternatives for which (4) is satisfied. Moreover, the obvious inequality $Z_{n,n} - Z_{n,n} < b - a$ implies that $c^{**}_{n,\alpha} > c^*_{n,\alpha}$, so that we have always

$$T_n \geq c^{**}_{n,\alpha} \quad \Rightarrow \quad T_n \geq c^*_{n,\alpha}. \tag{19}$$

The exact critical value $c_{n,\alpha} = c^*_{n,\alpha} + o\big(n^{1/2}\big) = c^{**}_{n,\alpha} + o\big(n^{1/2}\big)$ defined by

$$\mathbb{P}\Big(T_n \geq c_{n,\alpha} \big| (H.0)\Big) = \alpha, \tag{20}$$

can be computed, making use of the methods of Deheuvels and Derzko [DD03], who described several methods to evaluate numerically the distribution of T_n under $(H.0)$. In particular, they gave a simple proof of the fact that, under $(H.0)$, with $a = 0$, $b = 1$ and $p = q = 1$,

$$\mathbb{E}\big(\exp(sT_n)\big) = \Gamma(1-s)^n\Big\{\frac{n^{-ns}\Gamma(n)}{\Gamma\big(n(1-s)\big)}\Big\} \quad \text{for} \quad s < 1. \tag{21}$$

We note that a version of (21) was obtained originally by Darling [Dar53] by different methods.

Unfortunately, the consistency of tests of the form (17)–(18), rejecting $(H.0)$ for values of T_n exceeding $c^*_{n,\alpha}$ or $c^{**}_{n,\alpha}$, is known to hold only for the rather narrow alternative class of density functions $f(\cdot)$ described in [Blu68] as sufficient to imply (4). One of the purposes of the present paper is to overcome this drawback by extending the validity of (4) to a more wider class of distributions. The just-given Theorem 1.1 provides this result by givien a new proof of (4), under much weaker conditions that that imposed by Blumenthal [Blu68], and Cressie [Cre76]. In the sequel, we will limit ourselves, unless otherwise specified, to gives details of the proof in the case where $-\infty < a < b < \infty$, and we will then set $a = 0$ and $b = 1$ without loss of generality. The following proposition, which will turn out to be an easy consequence of Theorem 1.1, gives an example of how these results apply in the present framework.

Proposition 1.1 *Let f be continuous and positive on (a, b), and either:*

(i) continuous and bounded away from 0 on $[a, b]$;

(ii) monotone in a right neighborhood of a, monotone in a left neighborhood of b, and such that, for some $\varepsilon > 0$,

$$\log f(x) = O\Big[\frac{1}{(F(x))^{\frac{1}{2}+\varepsilon}}\Big] \quad as \quad x \downarrow 0, \tag{22}$$

and

$$\log f(x) = O\Big[\frac{1}{(1-F(x))^{\frac{1}{2}+\varepsilon}}\Big] \quad as \quad x \uparrow 1. \tag{23}$$

Then, as $n \to \infty$,

$$\begin{aligned} n^{-1/2}&\Big\{T_n - n\gamma - ne(\log f(X))\Big\} \\ &\stackrel{d}{\to} N\Big(0, \zeta(2) - 1 + \mathrm{Var}(\log f(X))\Big). \end{aligned} \tag{24}$$

Proof. We observe that the conditions (23), (22) and (24) readily imply that $\mathbb{E}((\log f(Z))^2) < \infty$. Therefore, the proposition is a direct consequence of Theorem 1.1.□

Example 1.1 Let $F(x) = 1/(\log(e/x))^r$ for $0 < x \le 1$ and $r > 0$. Obviously, $f(x) = r/(x(\log(e/x))^{r+1})$ and $\log f(x) = (1 + o(1)) \log(e/x)$ as $x \downarrow 0$. Thus,

$$\begin{aligned} &\mathbb{E}((\log f(X))^2) < \infty \;\Leftrightarrow\; r > 2 \\ &\Leftrightarrow\; |\log f(x)| = O\Big[\frac{1}{(F(x))^{\frac{1}{2}+\varepsilon}}\Big] \quad \text{for some} \quad \varepsilon > 0. \end{aligned}$$

This show the sharpness of the conditions in Proposition 1.1, since the finiteness of $\mathbb{E}((\log f(X))^2) < \infty$ is a minimal requirement for (24) to hold.
The arguments used in our proofs, given in the next section, mix the methods of Deheuvels and Derzko [DD03], with classical empirical process arguments.

2 Proofs.

2.1 A useful Theorem.

We start by proving the following useful theorem, of independent interest.

Theorem 2.1 *Assume that* $\mathbb{E}\big((\log f(Z))^2\big) < \infty$. *Then, for each* $p \geq p_0$ *and* $q \geq p_0$, *we have, as* $n \to \infty$,

$$n^{-1/2} \sum_{i=p}^{n-q+1} \Big[-\log\Big\{\frac{n(F(Z_{i,n}) - F(Z_{i-1,n}))}{f(Z_{i,n})}\Big\} - \gamma - \mathbb{E}\big(\log f(Z)\big)\Big]$$

$$\overset{d}{\to} \quad N\Big(0, \zeta(2) - 1 + \mathrm{Var}\big(\log f(Z)\big)\Big). \quad (25)$$

Remark 2.1 *It will become obvious from the arguments given later on that the conclusion* (25) *of Theorem 1 remains valid when we replace formally in* (25), $F(Z_{i,n}) - F(Z_{i-1,n})$ *by* $F(Z_{i+1,n}) - F(Z_{i,n})$.

The remainder of the present sub-section is devoted to proving Theorem 1 in the case $p = 2$ and $q = 1$. The proof for arbitrary $p \geq p_0$ and $q \geq q_0$ is very similar, and left to the reader. We will show later on how Theorem 1 may be applied to prove Theorem 1.1. Below, the following notation will be in force. We will set $U_{0,n} = F(Z_{0,n}) = 0$ and $U_{n+1,n} = F(Z_{n+1,n}) = 1$, for each $n \geq 0$, and let $0 < U_{1,n} = F(Z_{1,n}) < \ldots < U_{n,n} = F(Z_{n,n}) < 1$ denote the order statistics of the first $n \geq 1$ observations from the i.i.d. sequence $U_1 = F(Z_1), U_2 = F(Z_2), \ldots$, of uniform $(0,1)$ random variables, defined on the probability space $(\Omega, \mathcal{A}, \mathbb{P})$ on which sit $Z_1, Z_2, \ldots$, as given in §1.

We set $Y_n = -\log(1 - U_n) = -\log(1 - F(Z_n))$ for $n = 1, 2, \ldots$, and observe that these random variables form an i.i.d. sequence of exponentially distributed random variables. Moreover, setting $Y_{0,n} = -\log(1 - F(Z_{0,n})) = 0$ for $n \geq 0$, the order statistics

$$Y_{0,n} = 0 < Y_{1,n} = -\log(1 - F(Z_{1,n})) < \ldots < Y_{n,n} = -\log(1 - F(Z_{n,n})), \quad (26)$$

of $Y_1, \ldots, Y_n$ fulfill, for $n \geq 1$, the equalities

$$Y_{i,n} = -\log(1 - U_{i,n}) = -\log(1 - F(Z_{i,n})) \quad \text{for} \quad 0 \leq i \leq n. \quad (27)$$

Set now $\omega_{i,n} = (n - i + 1)(Y_{i,n} - Y_{i-1,n})$ for $1 \leq i \leq n$, so that

$$Y_{i,n} = \sum_{j=1}^{i} \frac{\omega_{j,n}}{n - j + 1} \quad i = 0, \ldots, n. \quad (28)$$

In (28) and elsewhere, we use the convention that $\sum_\emptyset(\cdot) := 0$. It is noteworthy (refer to Sukhatme [Suk37], see, e.g., Malmquist [Mal50] and pp. 20-21 in David [Dav81]) that, for each $n \geq 1$, $\{\omega_{i,n} : 1 \leq i \leq n\}$ is a sequence of independent and exponentially distributed random variables. For convenience, we denote below by $\omega \stackrel{d}{=} \omega_{i,n}$, $i = 1, \ldots, n$, a standard exponential random variable, fulfilling $\mathbb{P}(\omega > y) = e^{-y}$ for $y \geq 0$.

Let $g(\cdot)$ be a measurable function on $\mathbb{R}_+ = [0, \infty)$. Below, we will assume that $g \in \mathcal{G}$, where $\mathcal{G} = L^2(\mathbb{R}_+, e^{-u}du)$ denotes the Banach space, with respect to the norm $\|\cdot\|_2$ of all such functions for which

$$\|g\|_2^2 := \mathbb{E}(g^2(\omega)) = \int_0^\infty g^2(u)e^{-u}du < \infty. \tag{29}$$

For each $g \in \mathcal{G}$, we will set

$$\mu_g = \mathbb{E}(g(\omega)) = \int_0^\infty g(u)e^{-u}du, \tag{30}$$

$$\sigma_g^2 = \mathrm{Var}(g(\omega)) = \int_0^\infty g^2(u)e^{-u}du - \mu_g^2. \tag{31}$$

Moreover, for each $0 \leq i \leq n$, we set

$$\begin{aligned} Y_{i,n}(g) &= \sum_{j=1}^{i} \Big\{ -\log \omega_{j,n} - \gamma + (\omega_{j,n} - 1) \Big\} + \sum_{j=1}^{i} \Big\{ g(Y_{j,n}) - \mu_g \Big\} \\ &=: \xi_{i,n} + \zeta_{i,n}(g), \end{aligned} \tag{32}$$

where, for $i = 0, \ldots, n$,

$$\xi_{i,n} := \sum_{j=1}^{i} \Big\{ -\log \omega_{j,n} - \gamma + \big(\omega_{j,n} - 1\big) \Big\}, \tag{33}$$

and

$$\zeta_{i,n}(g) := \sum_{j=1}^{i} \Big\{ g(Y_{j,n}) - \mu_g \Big\}. \tag{34}$$

We note for further use that the following inequalities hold for all $g_1, g_2 \in \mathcal{G}$ such that $\mu_{g_1} = \mu_{g_2}$. We have,

$$\begin{aligned} &\mathbb{E}\Big(|Y_{n,n}(g_1) - Y_{n,n}(g_2)|^2\Big) = \mathbb{E}\Big(|\zeta_{n,n}(g_1) - \zeta_{n,n}(g_2)|^2\Big) \\ &= \mathbb{E}\Big\{\Big|\sum_{j=1}^{n} \big(g_1(Y_j) - g_2(Y_j)\big)\Big|^2\Big\} = n\mathrm{Var}\big(g_1(Y_1) - g_2(Y_1)\big) \\ &= n\|g_1 - g_2\|_2^2. \end{aligned} \tag{35}$$

Lemma 2.1 *We have, as $n \to \infty$,*

$$\xi_{n,n} = \sum_{i=1}^{n}\Big\{ - \log \omega_{i,n} - \gamma + (\omega_{i,n} - 1)\Big\},$$
$$= \sum_{i=1}^{n}\Big\{ - \log\big(n(U_{i,n} - U_{i-1,n})\big) - \gamma\Big\} + O_{\mathbb{P}}(\log n). \qquad (36)$$

Proof. We will make use of the following inequalities (see, e.g., 8.368, p.947 in Gradshteyn and Ryzhik [GR65]). We have, for each $n \geq 1$,

$$-\frac{1}{2n} - \frac{1}{12n^2} \leq \sum_{k=1}^{n-1}\frac{1}{k} - \log n - \gamma \leq -\frac{1}{2n} - \frac{1}{12n^2} + \frac{1}{120n^4}, \qquad (37)$$
$$\frac{1}{2n} - \frac{1}{12n^2} \leq \sum_{k=1}^{n}\frac{1}{k} - \log n - \gamma \leq \frac{1}{2n} - \frac{1}{12n^2} + \frac{1}{120n^4}, \qquad (38)$$

which readily yield the following rough inequality. For each $1 \leq i \leq n$,

$$\Big|\sum_{j=1}^{i}\frac{1}{n-j+1} - \log\Big\{\frac{n}{n-i+1}\Big\}\Big| \leq \frac{1}{n} + \frac{1}{n-i+1}. \qquad (39)$$

Next, we write, via (27) and Taylor's formula, for $1 \leq i \leq n$,

$$U_{i,n} - U_{i-1,n} = \exp\big(- Y_{i-1,n}\big) - \exp(-Y_{i,n})$$
$$= \big(Y_{i,n} - Y_{i-1,n}\big)\exp\big(- \{Y_{i,n} - \rho_{i,n}(Y_{i,n} - Y_{i-1,n})\}\big), \qquad (40)$$

where $\rho_{i,n}$ fulfills $0 < \rho_{i,n} < 1$. By combining (28) with (40), we obtain readily that, as $n \to \infty$,

$$\sum_{i=1}^{n}\Big\{ - \log\big(n(U_{i,n} - U_{i-1,n})\big)\Big\} = -\sum_{i=1}^{n}\log\big(Y_{i,n} - Y_{i-1,n}\big) - n\log n$$
$$+\sum_{i=1}^{n}\Big\{Y_{i,n} - \rho_{i,n}(Y_{i,n} - Y_{i-1,n})\Big\} = -\sum_{i=1}^{n}\Big\{\log\omega_{i,n} + \log\Big\{\frac{n}{n-i+1}\Big\}\Big\}$$
$$+\sum_{i=1}^{n}\Big\{\sum_{j=1}^{i}\frac{\omega_{j,n}-1}{n-j+1}\Big\} + \sum_{i=1}^{n}\Big\{\sum_{j=1}^{i}\frac{1}{n-j+1}\Big\} - \sum_{i=1}^{n}\Big\{\rho_{i,n}(Y_{i,n} - Y_{i-1,n})\Big\}$$
$$= \sum_{i=1}^{n}\Big\{\omega_{i,n} - 1 - \log\omega_{i,n}\Big\} + \sum_{i=1}^{n}\Big\{\sum_{j=1}^{i}\frac{1}{n-j+1} - \log\Big\{\frac{n}{n-i+1}\Big\}\Big\}$$
$$-\sum_{i=1}^{n}\Big\{\rho_{i,n}(Y_{i,n} - Y_{i-1,n})\Big\}.$$

Observe that

$$0 \leq \sum_{i=1}^{n}\Big\{\rho_{i,n}(Y_{i,n} - Y_{i-1,n})\Big\} \leq \sum_{i=1}^{n}\Big\{Y_{i,n} - Y_{i-1,n}\Big\} = Y_{n,n}.$$

Now, it is easily checked that, for each choice of $c > 1$, as $n \to \infty$,

$$\mathbb{P}\big(Y_{n,n} \geq c \log n\big) \leq \sum_{i=1}^{n} \mathbb{P}\big(Y_1 \geq c \log n\big) = ne^{-c \log n} = n^{1-c} \to 0,$$

so that $Y_{n,n} = O_{\mathbb{P}}(\log n)$ as $n \to \infty$. Moreover, by (2), we have, as $n \to \infty$,

$$\sum_{i=1}^{n} \Big| \sum_{j=1}^{i} \frac{1}{n-j+1} - \log\Big\{\frac{n}{n-i+1}\Big\}\Big| \leq \sum_{i=1}^{n} \Big\{\frac{1}{n} + \frac{1}{n-i+1}\Big\}$$

$$= 1 + \sum_{i=1}^{n} \frac{1}{i} = O(\log n). \tag{41}$$

Therefore, as $n \to \infty$,

$$\sum_{i=1}^{n} \Big\{ - \log\big(n(U_{i,n} - U_{i-1,n})\big)\Big\} = \sum_{i=1}^{n} \Big\{\omega_{i,n} - 1 - \log \omega_{i,n}\Big\} + O_{\mathbb{P}}(\log n).$$

By substracting $n\gamma$ to the left- and right-hand side of the above equality, we obtain (36), as sought.□

Introduce now the following notation and facts. For each $n \geq 1$ and $0 \leq t \leq 1$, denote, respectively by

$$\mathbb{U}_n(t) = n^{-1}\#\{U_i \leq t : 1 \leq i \leq n\},$$

and

$$\mathbb{V}_n(t) = \inf\big\{s \geq 0 : \mathbb{U}_n(s) \geq t\big\}, \tag{42}$$

the uniform empirical and quantile functions based upon $U_1, \ldots, U_n$. Here and elsewhere, $\#A$ stands for the cardinality of A. The corresponding empirical and quantile processes are given, respectively, by

$$\alpha_n(t) = n^{1/2}\{\mathbb{U}_n(t) - t\} \quad \text{and} \quad \beta_n(t) = n^{1/2}\{\mathbb{V}_n(t) - t\}. \tag{43}$$

Fact 1 below is due to Kiefer [Kie67] (see, e.g., Deheuvels and Mason [DM90]).

Fact 1 *For each specified $0 \leq t_0 \leq 1$, we have, almost surely,*

$$\limsup_{n\to\infty} n^{1/4}(\log\log n)^{-3/4}|\alpha_n(t_0) + \beta_n(t_0)| = 2^{5/4}3^{-3/4}\{t_0(1-t_0)\}^{1/4}. \tag{44}$$

The next fact, stated below, is Lemma 3.1 of Deheuvels and Derzko [DD03] (see also Pyke [Pyk65], and Proposition 8.2.1 in Shorack and Wellner [SW86]).

Fact 2 *Let S_{n+1} denote a random variable, independent of $U_1, \ldots, U_n$, and following a $\Gamma(n+1)$ distribution, with density*

$$h(s) = \frac{s^n}{n!}\, e^{-s} \quad \textit{for} \quad s > 0, \quad h(s) = 0 \quad \textit{for} \quad s \leq 0.$$

Then, the random variables

$$\theta_{i,n} = S_{n+1}\{U_{i,n} - U_{i-1,n}\}, \quad i = 1, \ldots, n+1, \tag{45}$$

are independent, exponentially distributed with unit mean, and such that

$$S_{n+1} = \sum_{i=1}^{n+1} \theta_{i,n}. \tag{46}$$

In view of the above notation, letting $\{\theta_{i,n} : 1 \leq i \leq n+1\}$ be as in (46), Fact 3 below follows from Theorems 3.1-3.2 of Deheuvels and Derzko [DD03].

Fact 3 *We have, as $n \to \infty$,*

$$\sup_{0 \leq j \leq n+1} \Big| \sum_{i=1}^{j} \Big\{ -\log\big(n(U_{i,n} - U_{i-1,n})\big) - \gamma \Big\} - \sum_{i=1}^{j} \Big\{ -\log \theta_{i,n} - \gamma + (\theta_{i,n} - 1) \Big\} \Big| = O_{\mathbb{P}}(1). \tag{47}$$

We have now all the tools in hand to prove the following intermediary result of independent interest.

Theorem 2.2 *For each $g \in \mathcal{G}$, we have, as $n \to \infty$*

$$n^{-1/2} Y_{n,n}(g) \xrightarrow{d} N\Big(0, \zeta(2) - 1 + \sigma_g^2\Big). \tag{48}$$

Proof. Step 1. Recall the notation (32)–(33)–(34). To illustrate the arguments of our proof, we start by considering the simple case where $g(u) = a + bu$ is an affine function. Under this assumption, we infer from (30)–(31) that $\mu_g = a + b$ and $\sigma_g^2 = b^2$. Thus, by (28) and (34), we obtain that

$$\begin{aligned} \zeta_{n,n}(g) &= \sum_{i=1}^{n} \Big\{ g(Y_{j,n}) - \mu_g \Big\} = b \sum_{i=1}^{n} \Big\{ \sum_{j=1}^{i} \frac{\omega_{j,n} - 1}{n - j + 1} \Big\} \\ &= b \sum_{i=1}^{n} \Big\{ \omega_{i,n} - 1 \Big\}. \end{aligned} \tag{49}$$

This, in turn, shows that

$$\begin{aligned} Y_{n,n}(g) &= \xi_{n,n} + \zeta_{n,n}(g) \\ &= \sum_{i=1}^{n} \Big\{ -\log \omega_{i,n} - \gamma + (\omega_{i,n} - 1) \Big\} + b \sum_{i=1}^{n} \Big\{ \omega_{i,n} - 1 \Big\}, \end{aligned} \tag{50}$$

is the partial sum of order n of an i.i.d. sequence of random variables. Setting $\omega = \omega_{1,n}$, an easy calculus (see the Appendix in the sequel) shows that

$$\mathbb{E}\Big(-\log\omega - \gamma + (\omega - 1)\Big) = \mathrm{e}(\omega - 1) = 0, \tag{51}$$

$$\mathbb{E}\Big(\Big\{ -\log\omega - \gamma + (\omega - 1)\Big\}^2\Big) = \zeta(2) - 1, \tag{52}$$

$$\mathbb{E}\Big(\Big\{ -\log\omega - \gamma + (\omega - 1)\Big\}\Big\{\omega - 1\Big\}\Big) = 0, \tag{53}$$

$$\mathbb{E}\Big(\Big\{\omega - 1\Big\}^2\Big) = 1. \tag{54}$$

We readily infer from the above equalities that

$$\mathrm{Var}\Big\{(-\log\omega - \gamma + (\omega - 1)) + b(\omega - 1)\Big\} = \zeta(2) - 1 + b^2 = \zeta(2) - 1 + \sigma_g^2. \tag{55}$$

Given (50) and (55), the proof of (48) for the particular choice of $g(u) = a + bu$ is a simple consequence of the central limit theorem.

Step 2. In this step, we consider the setup where g is proportional to the indicator function of an interval, namely of the form

$$g(u) = \lambda \mathbb{I}_{(c,d]}(u) = \begin{cases} 0 & \text{for } 0 \le u \le c, \\ \lambda & \text{for } c < u \le d, \\ 0 & \text{for } u > d, \end{cases} \tag{56}$$

where $\lambda \in \mathbb{R}$, c and d are specified constants fulfilling $0 < c < d < \infty$. We will set, for convenience, $C = 1 - e^{-c}$ and $D = 1 - e^{-d}$, and observe that the constants C, D are such that $0 < C < D < 1$. An easy calculus based upon (30)–(31) shows that, under (56),

$$\mu_g = \int_c^d \lambda e^{-u} du = \lambda(D - C) \quad \text{and} \quad \sigma_g^2 = \lambda^2(D - C)(1 - D + C). \tag{57}$$

Letting $g \in \mathcal{G}$ be as in (56), we infer from (34), (42) and (43), in combination with Fact 1, that, almost surely as $n \to \infty$,

$$\begin{aligned} \zeta_{n,n}(g) &= \sum_{i=1}^{n} \Big\{ g(Y_i) - \lambda(D - C)\Big\} \\ &= \sum_{i=1}^{n} \Big\{ \mathbb{I}_{(c,d]}\big(-\log(1 - U_i)\big) - \lambda(D - C)\Big\} \\ &= \lambda \# \Big\{ -\log(1 - U_i) \in (c, d] : 1 \le i \le n\Big\} - n\lambda(D - C) \\ &= n^{1/2}\lambda\Big\{\alpha_n(D) - \alpha_n(C)\Big\} \\ &= -n^{1/2}\lambda\Big\{\beta_n(D) - \beta_n(C)\Big\} + O_{\mathbb{P}}\big(n^{1/4}(\log n)^{3/4}\big). \end{aligned} \tag{58}$$

Denote by $\lceil u \rceil \ge u > \lceil u \rceil - 1$ the upper integer part of u. Then, by (42),

$$\beta_n(t) = n^{1/2}(\mathbb{V}_n(t) - t) = n^{1/2}(U_{\lceil nt\rceil,n} - t) \quad \text{for} \quad 0 < t \leq 1. \tag{59}$$

We readily infer from (58) and (59) that, almost surely as $n \to \infty$,

$$\begin{aligned}\zeta_{n,n}(g) = -\lambda\Big\{nU_{\lceil nD\rceil,n} - nU_{\lceil nC\rceil,n} - (\lceil nD\rceil - \lceil nC\rceil)\Big\}\\ +O_{\mathbb{P}}\big(n^{1/4}(\log n)^{3/4}\big).\end{aligned} \tag{60}$$

Now, making use of (46), we may write

$$\begin{aligned}&nU_{\lceil nD\rceil,n} - nU_{\lceil nC\rceil,n} - (\lceil nD\rceil - \lceil nC\rceil)\\ &= \Big\{1 + \Big[\frac{n}{S_{n+1}} - 1\Big]\Big\} \sum_{i=\lceil nC\rceil+1}^{\lceil nD\rceil} \Big\{\theta_{i,n} - 1\Big\} + \Big[\frac{n}{S_{n+1}} - 1\Big](\lceil nD\rceil - \lceil nC\rceil).\end{aligned}$$

An application of the central limit theorem shows, in turn, that, as $n \to \infty$,

$$\frac{S_{n+1}}{n} = 1 + \frac{1}{n} + \frac{1}{n}\sum_{i=1}^{n+1}\Big\{\theta_{i,n} - 1\Big\} = 1 + O_{\mathbb{P}}(n^{-1/2}).$$

Likewise, we have, as $n \to \infty$,

$$\sum_{i=\lceil nC\rceil+1}^{\lceil nD\rceil} \Big\{\theta_{i,n} - 1\Big\} = O_{\mathbb{P}}(n^{1/2}).$$

Therefore, we obtain that, as $n \to \infty$,

$$\frac{n}{S_{n+1}} - 1 = \frac{1}{n}\sum_{i=1}^{n+1}\Big\{\theta_{i,n} - 1\Big\} + O_{\mathbb{P}}(n^{-1}) = O_{\mathbb{P}}(n^{-1/2}).$$

By all this, it follows that

$$\begin{aligned}&nU_{\lceil nD\rceil,n} - nU_{\lceil nC\rceil,n} - (\lceil nD\rceil - \lceil nC\rceil)\\ &= \sum_{i=\lceil nC\rceil+1}^{\lceil nD\rceil} \Big\{\theta_{i,n} - 1\Big\} - (D - C) \sum_{i=\lceil nC\rceil+1}^{\lceil nD\rceil} \Big\{\theta_{i,n} - 1\Big\} + O_{\mathbb{P}}(1),\end{aligned}$$

This, when combined with (60), entails that

$$\begin{aligned}\zeta_{n,n}(g) = -\lambda \sum_{i=\lceil nC\rceil+1}^{\lceil nD\rceil} \Big\{\theta_{i,n} - 1\Big\}\\ +\lambda(D - C)\sum_{i=1}^{n+1}\Big\{\theta_{i,n} - 1\Big\} + O_{\mathbb{P}}\big(n^{1/4}(\log n)^{3/4}\big).\end{aligned} \tag{61}$$

Set now, in view of (61),

$$\zeta_{n,n}^*(g) = -\lambda \sum_{i=\lceil nC\rceil+1}^{\lceil nD\rceil} \Big\{\theta_{i,n} - 1\Big\} + a(D-C)\sum_{i=1}^{n+1}\Big\{\theta_{i,n} - 1\Big\},$$

and, in view of (47),

$$\xi_{n,n}^* = -\sum_{i=1}^{j}\Big\{ - \log\theta_{i,n} - \gamma + (\theta_{i,n} - 1)\Big\}$$

A direct application of (53) and (57) shows that, as $n \to \infty$,

$$\begin{aligned}
&\mathrm{Var}\Big(\xi_{n,n}^* + \zeta_{n,n}^*(g)\Big) = \mathrm{Var}\Big(\xi_{n,n}^*\Big) + \mathrm{Var}\Big(\zeta_{n,n}^*(g)\Big)\\
&= n\Big(\zeta(2) - 1\Big)\Big) + \lambda^2\Big\{(D-C)^2(\lceil nC\rceil + (n+1-\lceil nD\rceil))\\
&\qquad +(D-C-1)^2(\lceil nD\rceil - \lceil nC\rceil)\Big\}\\
&= (1+o(1))n\lambda^2(D-C)(1-D+C) = (1+o(1))\sigma_g^2. \qquad (62)
\end{aligned}$$

Now, in view of (61), given that $\xi_{n,n}^* + \zeta_{n,n}^*(g)$ is a linear combination of three partial sums of i.i.d. centered random variables, an application of the central limit theorem shows that, as $n \to \infty$

$$\begin{aligned}
n^{-1/2}\Big\{\xi_{n,n} + \zeta_{n,n}(g)\Big\} &= n^{-1/2}\Big\{\xi_{n,n}^* + \zeta_{n,n}^*(g)\Big\} + O_{\mathbb{P}}(n^{-1/4}(\log n)^{3/4})\\
&\xrightarrow{d} N(0,\sigma_g^2).
\end{aligned}$$

In view of (32), (33) and (34), we thus obtain (48) in this particular case.

Step 3. We are now ready to establish the most general version of our theorem. In the first place, we consider the case where $g_L \in \mathcal{G}$ is a step-function, of the form

$$g_L(u) = \sum_{\ell=1}^{L} \lambda_\ell \mathbb{I}_{(c_\ell,d_\ell]}(u), \tag{63}$$

where $a_1,\ldots,a_L$ and $0 < c_1 < d_1 < \ldots < c_L < d_L < 1$ are specified constants. By repeating the arguments of Step 2 in this case, we readily obtain that the weak convergence

$$n^{-1/2}Y_{n,n}(g_L) \xrightarrow{d} N\Big(0, \zeta(2) - 1 + \sigma_{g_L}^2\Big), \tag{64}$$

holds. Now, if $g \in \mathcal{G} = L^2(\mathbb{R}_+, e^{-u}du)$ is arbitrary, for each $\varepsilon > 0$, we may select $L \geq 1$, together with $\lambda_1,\ldots,\lambda_L$ and $0 < c_1 < d_1 < \ldots < c_L < d_L < 1$, such that $\mu_g = \mu_{g_L}$ and

$$\|g - g_L\|_2^2 = \mathbb{E}\Big(|g(\omega) - g_L(\omega)|^2\Big) \leq \varepsilon. \tag{65}$$

By combining (35) with (64)–(65), we see that

$$\mathbb{E}\Big(\Big|n^{-1/2}Y_{n,n}(g) - n^{-1/2}Y_{n,n}(g_L)\Big|^2\Big) \le \varepsilon. \tag{66}$$

Since $\varepsilon > 0$ in (65)–(66) may be chosen arbitrarily small, we conclude to the validity of (48) by routine arguments.□

Proof of Theorem 1. Our assumptions imply that $F(Z_{i,n}) = U_{i,n}$ for $i = 1, \ldots, n$. Therefore, we infer readily (25) from (48), by setting in this last relation

$$g(y) = f(Q(1 - e^{-y})) \quad \text{for} \quad y > 0,$$

and making use of Lemma 2.1.□

Proof of Theorem 1.1. Denote by $Q(t) = \inf\{z : F(z) \ge t\}$, for $0 < t < 1$, the quantile function pertaining to F. Our assumptions imply that Q has a continuous derivative on $(0,1)$ given by

$$Q'(t) = 1/f(Q(t)).$$

By Taylor's formula, we obtain readily that

$$\begin{aligned} &-\log\Big\{n(Z_{i,n} - Z_{i-1,n})\Big\} \\ &= -\log\Big\{n(Q(F(Z_{i,n})) - Q(F(Z_{i-1,n})))\Big\} \\ &= -\log\Big\{\frac{n(F(Z_{i,n}) - F(Z_{i-1,n}))}{f(Z^*_{i,n})}\Big\}, \end{aligned} \tag{67}$$

where $Z^*_{i,n}$ lies within the interval $(Z_{i-1,n}, Z_{i,n})$. We now make use of the following well-known fact (see, e.g. Csörgő, Haeusler and Mason [CsHM88], and the references therein). Let $V_1, \ldots, V_n$ be an i.i.d. sequence of replicæof a random variable V with finite expectation $\mu = \mathbb{E}(V)$ and variance $0 < \sigma^2 = \mathrm{Var}(V) < \infty$. Denote by $V_{1,n} \le \ldots \le V_{n,n}$ the order statistics of $V_1, \ldots, V_n$. Then, independently of the fixed integers $p \ge 1$ and $q \ge 1$, we have

$$n^{-1/2} \sum_{i=p}^{n-q+1} \{V_{i,n} - \mu\} \xrightarrow{d} N(0, \sigma^2). \tag{68}$$

Moreover, we have, for each $\varepsilon > 0$

$$\begin{aligned} \lim_{r \downarrow 0} \Big\{ \limsup_{n\to\infty} \mathbb{P}\Big(n^{-1/2}\Big| &\sum_{i=\lceil nr \rceil+1}^{n-\lceil nr \rceil+1} \{V_{i,n} - \mu\} \\ &- \sum_{i=1}^{n} \{V_i - \mu\}\Big| \ge \varepsilon\Big)\Big\} = 0. \end{aligned} \tag{69}$$

Let us now assume, for the sake of simplicity, that f is nonincreasing on (a, b). In this case, we see that

$$S_n' := -\sum_{i=2}^{n} \log f(Z_{i-1,n}) \le R_n := -\sum_{i=2}^{n} \log f(Z_{i,n}^*)$$
$$\le S_n'' := -\sum_{i=2}^{n} \log f(Z_{i,n}).$$

Set $\mu = \mathbb{E}(\log f(Z))$ and $\sigma^2 = \mathrm{Var}(\log f(Z))$. In view of (68)–(69), we have, for each $t \in \mathbb{R}$,

$$\lim_{n\to\infty} \mathbb{P}\Big(n^{-1/2}(S_n' - n\mu) \le t\Big) \le \mathbb{P}\Big(n^{-1/2}(R_n - n\mu) \le t\Big)$$
$$\le \mathbb{P}\Big(n^{-1/2}(S_n' - n\mu) \le t\Big) = \Phi(t),$$

where $\Phi(t)$ is the $N(0, 1)$ distribution function. Thus, we may apply Theorem 1 in combination with Remark 2.1 to obtain (8). The proof of this result when f is monotone only in the neighborhood of the end-points is very similar, and obtained by splitting the range of Z into three component intervals, in combination with an application of (69). We omit details.□

2.2 Appendix.

Let ω denote a unit exponential random variable. The present sub-section is devoted to the computation of some moments of interest of slected functions of ω. We first observe that, for each $s \in \mathbb{R}$ such that $s < 1$,

$$\mathbb{E}\Big(\exp\big(-s\log\omega\big)\Big) = \int_0^\infty \omega^{-s} e^{-s} ds = \Gamma(1-s). \tag{70}$$

Recalling the expansion (see, e.g. Abramowitz and Stegun [AS70])

$$\log\Gamma(1-s) = \gamma s + \sum_{k=2}^{\infty} \frac{\zeta(k)}{k} s^k \quad \text{for} \quad |s| < 1, \tag{71}$$

we obtain readily that the k-th cumulant κ_k of $-\log\omega$ is given by $\kappa_1 = \gamma$ and $\kappa_k = (k-1)!\zeta(k)$ for $k \ge 2$, whence

$$\mathbb{E}(-\log\omega) = \int_0^\infty (-\log s)e^{-s} ds = \kappa_1 = \gamma, \tag{72}$$

$$\mathrm{Var}(\log\omega) = \int_0^\infty (\log s)^2 e^{-s} ds - \gamma^2 = \kappa_2 = \zeta(2) = \frac{\pi^2}{6}. \tag{73}$$

Let now ω_1 and ω_2 denote two independent unit exponential random variables. The following result, of independent interest, has potential applications in two-sample tests. This problem will be considered elsewhere.

Lemma 2.2 *The random variable $R = \log(\omega_1/\omega_2)$ follows a logistic law, with distribution and moments given by*

$$\mathbb{P}\big(R \leq t\big) = \frac{1}{1+e^{-t}} \quad for \quad t \in \mathbb{R}, \tag{74}$$

the moment-generating function of R is given by

$$\mathbb{E}\big(\exp(sR)\big) = \Gamma(1-s)\Gamma(1+s) \quad for \quad |s| < 1, \tag{75}$$

and the k-th cumulant of R is given by

$$\kappa_k = \begin{cases} 0 & fork = 1, 3, \ldots, 2p+1, \ldots, \\ 2(k-1)!\zeta(k) & fork = 2, 4, \ldots, 2p, \ldots. \end{cases} \tag{76}$$

In particular, we get

$$\mathbb{E}(R) = 0 \quad and \quad \mathrm{Var}(R) = 2\zeta(2) = \frac{\pi^2}{3}. \tag{77}$$

An easy calculus yields

$$\begin{aligned} \mathbb{E}(\omega \log \omega) &= \int_0^\infty (s \log s - s)e^{-s} ds + \int_0^\infty s e^{-s} ds \\ &= \big[-(x\log x - x)e^{-x}\big]_{x=0}^{x=\infty} + \int_0^\infty (\log s) e^{-s} ds + 1 = -\gamma + 1. \end{aligned} \tag{78}$$

Likewise, we get

$$\begin{aligned} \mathbb{E}\big(\omega(\log\omega)^2\big) &= \int_0^\infty s(\log s)^2 e^{-s} ds \\ &= \big[-x(\log x)^2 e^{-x}\big]_{x=0}^{x=\infty} + \int_0^\infty (\log s)^2 e^{-s} ds + 2\int_0^\infty (\log s) e^{-s} ds \\ &= \zeta(2) + \gamma^2 - 2\gamma + 2, \end{aligned} \tag{79}$$

and

$$\begin{aligned} \mathbb{E}\big(\omega^2(\log\omega)^2\big) &= \int_0^\infty s^2(\log s)^2 e^{-s} ds \\ &= \big[-x^2(\log x)^2 e^{-x}\big]_{x=0}^{x=\infty} + 2\int_0^\infty s(\log s)^2 e^{-s} ds + 2\int_0^\infty s(\log s) e^{-s} ds \\ &= 2\Big(\zeta(2) + \gamma^2 - 2\gamma + 2\Big) + 2\Big(-\gamma + 1\Big) \qquad (80) \\ &= 2\Big(\zeta(2) + \gamma^2 - 3\gamma + 3\Big). \qquad (81) \end{aligned}$$

References

[AS70] Abramowitz, M, Stegun, I. A.: Handbook of Mathematical Functions, Dover, New York (1970)

[Blu68] Blumenthal, S.: Logarithms of sample spacings. S.I.A.M. J. Appl. Math., **16**, 1184-1191 (1968)

[Cre76] Cressie, N.: On the logarithms of high-order spacings. Biometrika, **2**, 343-355 (1976)

[CsHM88] Csörgő, S., Haeusler, E., Mason, D. M.: A probabilistic approach to the asymptotic distribution of sums of independent identically distributed random variables. Advances in Appl. Probab. **9**, 259-333 (1988)

[Dar53] Darling, D. A.: On a class of problems related to the random division of an interval. Ann. Math. Statist., **24**, 239-253 (1953).

[Dav81] David, H. A.: Order Statistics. 2nd Ed., Wiley, New York (1981)

[DM90] Deheuvels, P. and Mason, D. M.: Bahadur-Kiefer-type processes. Ann. Probab., **18**, 669-697 (1990)

[Kie67] Kiefer, J.: On Bahadur's representation of sample quantiles. Ann. Math. Statist., **38**, 1323-1342 (1967)

[DD03] Deheuvels, P. and Derzko, G.: Exact laws for products of uniform spacings. Austrian J. Statist. **1-2**, 29-47 (2003)

[GR65] Gradshteyn, I. S., Ryzhik, I. M.: Tables of Integrals, Series and Products. Academic Press, New York (1965)

[Mal50] Malmquist, S.: On a property of order statistics from a rectangular distribution. Skand. Aktuarietidskr. **33**, 214-222 (1950)

[Pyk65] Pyke, R.: Spacings. J. Royal Statist. Soc. B. **27**, 395-436, and Discussion 437-449 (1965)

[SW86] Shorack, G. R. and Wellner, J. A.: Empirical Processes with Applications to Statistics. Wiley, New York (1986)

[SO87] Spanier, J. and Oldham, K. B.: An Atlas of Functions. Hemisphere Publ. Co., Washington (1987)

[Suk37] Sukhatme, P. V.: Tests of significance for samples of the χ^2 population with two degrees of freedom. Ann. Eugen., **8**, 52-56 (1937)

A Survival Model With Change-Point in Both Hazard and Regression Parameters

Dupuy Jean-François

Laboratoire de Statistique et Probabilités, Université Paul Sabatier,
118, route de Narbonne, 31062 Toulouse cedex 4, France
dupuy@math.ups-tlse.fr

1 Introduction

In this paper, we consider a parametric survival regression model with a change-point in both hazard and regression parameters. Change-point occurs at an unknown time point. Estimators of the change-point, hazard and regression parameters are proposed and shown to be consistent.

Let T be a random failure time variable. The distribution of T is usually specified by the hazard function $\lambda = f/1 - F$ where f and F are the density and distribution functions of T respectively. Change in hazard at an unknown time point has been extensively studied. Such a change may occur in medical studies after a major operation (e.g. bone marrow transplant) or in reliability (e.g. change of failure rate following temperature increase). Several authors have considered the following change-point hazard model:

$$\lambda(t) = \alpha + \theta 1_{\{t>\tau\}}, \tag{1}$$

with $\alpha > 0$, $\alpha + \theta > 0$, and where $\tau > 0$ is an unknown change-point time assumed to lie in a known interval $[\tau_1, \tau_2]$ such that $0 < \tau_1 < \tau_2 < \infty$ (see [CCH94], [MFP85], [MW94], [NRW84], [PN90]).
Wu et al. (2003) [WZW03] extend (1) to the hazard model

$$\lambda(t) = (\alpha + \theta 1_{\{t>\tau\}})\lambda_0(t;\gamma), \tag{2}$$

where $\lambda_0(\cdot;\gamma)$ is a baseline hazard function depending on an unknown parameter γ. This model allows hazard to be nonconstant anterior or posterior to change-point. In this paper, we extend (1) in a different way: we allow $\lambda(\cdot)$ to vary among individuals by incorporating covariates in the change-point model (1). Moreover, since the effect of covariates (e.g. age of a patient) may also change at the unknown time point τ, we allow for change in the regression parameter at τ (e.g. the risk of death of elderlies compared to young patients may increase after a major surgical operation). We specify the following hazard model:

$$\lambda(t|Z) = (\alpha + \theta 1_{\{t>\tau\}}) \exp\{(\beta + \gamma 1_{\{t>\tau\}})^T Z\}, \tag{3}$$

where $\alpha > 0$, $\alpha + \theta > 0$, τ is an unknown change-point time, β and γ are unknown regression coefficients. Previous work on change-point regression models has mainly focused on linear models (we refer to [CK94], [Hor95], [HHS97], [Jar03], [KQS03]). A detailed treatment and numerous references can be found in [CH97]. Gurevich and Vexler [GV05] consider change-point problem in the logistic regression model. Luo et al. [LTC97] and Pons ([Pon02], [Pon03]) consider a Cox model involving a change-point in the regression parameter.

In Section 2, we give a brief review of recent results on model (2), and we construct estimators for model (3). In Section 3, we prove that these estimators are consistent. Technical details are given in appendix.

2 Notations and construction of the estimators

2.1 Preliminaries

We consider a sample of n subjects observed in the time interval $[0, \zeta]$. Let T_i^0 be the survival time of the ith individual and Z_i be the related covariate. Z_i is assumed to be a q-dimensional random variable. Suppose that Z_i is bounded and $\mathrm{var}(Z_i) > 0$. We assume that T_i^0 may be right censored at a noninformative censoring time C_i such that C_i and T_i^0 are independent conditionally on Z_i. For individual i, let $T_i = T_i^0 \wedge C_i$ be the observed time and $\Delta_i = 1_{\{T_i^0 \leq C_i\}}$ be the censoring indicator.
The data consist of n independent triplets $\mathbf{X}_i = (T_i, \Delta_i, Z_i)$, $i = 1, \ldots, n$.

For model (2), which has no covariates, [WZW03] follow [CCH94] and define $Y_n(t)$ as

$$Y_n(t) = \left[\frac{\hat{\Lambda}_{NA}(\xi) - \hat{\Lambda}_{NA}(t)}{\hat{\Lambda}_0(\xi) - \hat{\Lambda}_0(t)} - \frac{\hat{\Lambda}_{NA}(t)}{\hat{\Lambda}_0(t)}\right] \left[\hat{\Lambda}_0(t)(\hat{\Lambda}_0(\xi) - \hat{\Lambda}_0(t))\right]^p,$$

where $0 \leq p \leq 1$, and $\hat{\Lambda}_{NA}(t)$ is the Nelson-Aalen estimator of $\Lambda(t) = \int_0^t \lambda(s)\,ds$, $\hat{\Lambda}_0(t) = \int_0^t \lambda_0(s; \hat{\gamma}_n)\,ds$ estimates $\Lambda_0(t) = \int_0^t \lambda_0(s; \gamma_0)\,ds$, $\hat{\gamma}_n$ is a consistent estimator of the true γ_0, and $\xi > \tau_2$ is a finite time point such that $P(T > \xi) > 0$. The asymptotic version of Y_n is

$$Y(t) = \left[\frac{\Lambda(\xi) - \Lambda(t)}{\Lambda_0(\xi) - \Lambda_0(t)} - \frac{\Lambda(t)}{\Lambda_0(t)}\right] [\Lambda_0(t)(\Lambda_0(\xi) - \Lambda_0(t))]^p.$$

Wu et al. [WZW03] remark that if the true θ_0 is strictly positive, then $Y(t)$ is increasing on $[0, \tau]$ and decreasing on $[\tau, \xi]$, hence they define an estimator $\hat{\tau}_n$ of τ by

$$\hat{\tau}_n = \inf\left\{t \in]\tau_1, \tau_2[: Y_n(t\pm) = \sup_{\tau_1 < u < \tau_2} Y_n(u)\right\}, \tag{4}$$

where $Y_n(t\pm)$ is the right or left-hand limit of $Y_n(\cdot)$ at t. If $\theta_0 < 0$, $Y(t)$ is decreasing on $[0, \tau]$ and increasing on $[\tau, \xi]$. In this case, Wu et al. (2003) define

$$\hat{\tau}_n = \inf \left\{ t \in]\tau_1, \tau_2[: Y_n(t\pm) = \inf_{\tau_1 < u < \tau_2} Y_n(u) \right\}. \tag{5}$$

Wu et al. [WZW03] show the following theorems under some regularity conditions:

Theorem 1. *The estimator $\hat{\tau}_n$ of τ defined in (4) or (5) is consistent.*

Let $l_n(\alpha, \theta, \tau, \gamma)$ denote the loglikelihood based on (T_i, Δ_i) $(i = 1, \ldots, n)$, and $\hat{\alpha}_n(\tau, \gamma)$ and $\hat{\theta}_n(\tau, \gamma)$ be respectively the solutions of $\partial l_n(\alpha, \theta, \tau, \gamma)/\partial \alpha = 0$ and $\partial l_n(\alpha, \theta, \tau, \gamma)/\partial \theta = 0$ for given (τ, γ).

Theorem 2. *$\hat{\alpha}_n(\hat{\tau}_n, \hat{\gamma}_n)$ and $\hat{\theta}_n(\hat{\tau}_n, \hat{\gamma}_n)$ are consistent estimators of α and θ respectively.*

In this paper, a different approach is taken to prove consistency of estimators in the model (3). It relies on modern empirical process theory as exposed in [Van98] and [VW96].

2.2 The estimators

We consider the statistical model defined by the family of densities

$$\begin{aligned} p_\varphi(\mathbf{X}) = &\left\{ \alpha e^{\beta^T Z} \right\}^{\Delta} \exp\left(-\alpha e^{\beta^T Z} T \right) 1_{\{T \le \tau\}} + \left\{ (\alpha + \theta) e^{(\beta+\gamma)^T Z} \right\}^{\Delta} \\ &\times \exp\left(-\alpha e^{\beta^T Z} \tau - (\alpha + \theta) e^{(\beta+\gamma)^T Z} (T - \tau) \right) 1_{\{T > \tau\}}, \end{aligned}$$

where $\varphi = (\tau, \xi^T)^T$, with $\xi = (\alpha, \theta, \beta^T, \gamma^T)^T$. Here α, θ, and the regression parameters β and γ belong respectively to bounded subsets $A \subset \mathbb{R}^+ \backslash \{0\}$, $B \subset \mathbb{R} \backslash \{0\}$, $C \subset \mathbb{R}^q$, and $D \subset \mathbb{R}^q \backslash \{0\}$. The change-point τ is a parameter lying in the open interval $]0, \zeta[$. Let $\varphi_0 = (\tau_0, \xi_0^T)^T$ be the true parameter value, lying in $\Phi =]0, \zeta[\times A \times B \times C \times D$. We suppose that φ_0 is such that $\theta_0 \neq 0$ and $\gamma_0 \neq 0$, so that a change-point actually occurs. We suppose also that $\alpha_0 + \theta_0 > 0$.
Under the true parameter values, we denote $P_0 \equiv P_{\varphi_0}$ the probability distribution of the variables (T_i^0, C_i, Z_i) and $\mathbb{E}_0$ the expectation of the random variables.

The log-likelihood function based on the observations $\mathbf{X}_i$ $(i = 1, \ldots, n)$ is

$$\begin{aligned} l_n(\varphi) = \sum_{i \le n} \Big\{ &N_i(\tau)[\ln \alpha + \beta^T Z_i] - 1_{\{T_i \le \tau\}} \alpha e^{\beta^T Z_i} T_i + [N_i(\infty) - N_i(\tau)] \\ &\times [\ln(\alpha + \theta) + (\beta + \gamma)^T Z_i] - 1_{\{T_i > \tau\}} \Big[\alpha e^{\beta^T Z_i} \tau \\ &+ (\alpha + \theta) e^{(\beta+\gamma)^T Z_i} (T_i - \tau) \Big] \Big\}, \end{aligned}$$

where $N_i(t) = \Delta_i 1_{\{T_i \le t\}}$ is the counting process for death of individual i. The estimator $\hat{\varphi}_n$ is obtained as follows: for a fixed τ, we let $\hat{\xi}_n(\tau)$ be the value of ξ which maximizes the log-likelihood $l_n(\varphi)$. Then τ_0 is estimated by $\hat{\tau}_n$ which satisfies the relationship

$$\hat{\tau}_n = \inf\left\{\tau \in]0,\zeta[: \max(l_n(\tau,\hat{\xi}_n(\tau)), l_n(\tau^+,\hat{\xi}_n(\tau^+))) = \sup_{\tau\in]0,\zeta[} l_n(\tau,\hat{\xi}_n(\tau))\right\},$$

where $l_n(\tau^+,\hat{\xi}_n(\tau^+))$ is the right-hand limit of l_n at τ. Then the maximum likelihood estimator of ξ is obtained as $\hat{\xi}_n = \hat{\xi}_n(\hat{\tau}_n)$.

For a given τ, we estimate α, θ, β and γ by considering the following score functions:

$$\frac{\partial l_n(\varphi)}{\partial \alpha} = \sum_{i\le n}\left\{\frac{N_i(\tau)}{\alpha} - 1_{\{T_i\le\tau\}} e^{\beta^T Z_i} T_i + \frac{N_i(\infty) - N_i(\tau)}{\alpha+\theta}\right.$$
$$\left. -1_{\{T_i>\tau\}}\left[e^{\beta^T Z_i}\tau + e^{(\beta+\gamma)^T Z_i}(T_i - \tau)\right]\right\},$$

$$\frac{\partial l_n(\varphi)}{\partial \theta} = \sum_{i\le n}\left\{\frac{N_i(\infty) - N_i(\tau)}{\alpha+\theta} - 1_{\{T_i>\tau\}} e^{(\beta+\gamma)^T Z_i}(T_i - \tau)\right\},$$

$$\frac{\partial l_n(\varphi)}{\partial \beta} = \sum_{i\le n}\left\{N_i(\infty) Z_i - 1_{\{T_i\le\tau\}}\alpha Z_i e^{\beta^T Z_i} T_i\right.$$
$$\left. -1_{\{T_i>\tau\}}\left[\alpha Z_i e^{\beta^T Z_i}\tau + (\alpha+\theta) Z_i e^{(\beta+\gamma)^T Z_i}(T_i - \tau)\right]\right\},$$

$$\frac{\partial l_n(\varphi)}{\partial \gamma} =$$
$$\sum_{i\le n}\left\{\left[N_i(\infty) - N_i(\tau)\right] Z_i - 1_{\{T_i>\tau\}}(\alpha+\theta) Z_i e^{(\beta+\gamma)^T Z_i}(T_i - \tau)\right\}.$$

3 Convergence of the estimators

Our main result is

Theorem 3. *The estimators $\hat{\tau}_n$ and $\hat{\xi}_n$ converge in probability to τ_0 and ξ_0.*

Proof. The proof of consistency is based on the uniform convergence of $X_n(\varphi) = n^{-1}(l_n(\varphi) - l_n(\varphi_0))$ to a function having a unique maximum at φ_0. Two lemmas will be needed, their proofs are given in appendix. By some rearranging, the process $X_n = n^{-1}(l_n - l_n(\varphi_0))$ can be written as a sum $X_n = X_{1,n} + X_{2,n} + X_{3,n} + X_{4,n}$ according to the sign of $\tau - \tau_0$, with

$$X_{1,n}(\varphi) = n^{-1} \sum_{i \leq n} \Delta_i 1_{\{T_i \leq \tau \wedge \tau_0\}} \left\{ \ln \frac{\alpha}{\alpha_0} + (\beta - \beta_0)^T Z_i \right\},$$

$$X_{2,n}(\varphi) = n^{-1} \sum_{i \leq n} \Delta_i 1_{\{T_i > \tau \vee \tau_0\}} \left\{ \ln \frac{\alpha + \theta}{\alpha_0 + \theta_0} + (\beta + \gamma - \beta_0 - \gamma_0)^T Z_i \right\},$$

$$X_{3,n}(\varphi) = n^{-1} \sum_{i \leq n} \Delta_i 1_{\{\tau < T_i \leq \tau_0\}} \left\{ \ln \frac{\alpha + \theta}{\alpha_0} + (\beta + \gamma - \beta_0)^T Z_i \right\}$$
$$+ n^{-1} \sum_{i \leq n} \Delta_i 1_{\{\tau_0 < T_i \leq \tau\}} \left\{ \ln \frac{\alpha}{\alpha_0 + \theta_0} + (\beta - \beta_0 - \gamma_0)^T Z_i \right\},$$

$$X_{4,n}(\varphi) = n^{-1} \sum_{i \leq n} \Big\{ 1_{\{T_i \leq \tau_0\}} \alpha_0 e^{\beta_0^T Z_i} T_i - 1_{\{T_i \leq \tau\}} \alpha e^{\beta^T Z_i} T_i$$
$$- 1_{\{T_i > \tau\}} \left[(\alpha + \theta) e^{(\beta + \gamma)^T Z_i} (T_i - \tau) + \alpha e^{\beta^T Z_i} \tau \right]$$
$$+ 1_{\{T_i > \tau_0\}} \left[(\alpha_0 + \theta_0) e^{(\beta_0 + \gamma_0)^T Z_i} (T_i - \tau_0) + \alpha_0 e^{\beta_0^T Z_i} \tau_0 \right] \Big\}.$$

Let X_∞ be the function defined as $X_\infty(\varphi) = X_{1,\infty}(\varphi) + X_{2,\infty}(\varphi) + X_{3,\infty}(\varphi) + X_{4,\infty}(\varphi)$, where

$$X_{1,\infty}(\varphi) = \mathbb{E}_0 \left[\Delta 1_{\{T \leq \tau \wedge \tau_0\}} \left\{ \ln \frac{\alpha}{\alpha_0} + (\beta - \beta_0)^T Z \right\} \right],$$

$$X_{2,\infty}(\varphi) = \mathbb{E}_0 \left[\Delta 1_{\{T > \tau \vee \tau_0\}} \left\{ \ln \frac{\alpha + \theta}{\alpha_0 + \theta_0} + (\beta + \gamma - \beta_0 - \gamma_0)^T Z \right\} \right],$$

$$X_{3,\infty}(\varphi) = \mathbb{E}_0 \Big[\Delta 1_{\{\tau < T \leq \tau_0\}} \left\{ \ln \frac{\alpha + \theta}{\alpha_0} + (\beta + \gamma - \beta_0)^T Z \right\}$$
$$+ \Delta 1_{\{\tau_0 < T \leq \tau\}} \left\{ \ln \frac{\alpha}{\alpha_0 + \theta_0} + (\beta - \beta_0 - \gamma_0)^T Z \right\} \Big],$$

$$X_{4,\infty}(\varphi) = \mathbb{E}_0 \Big[1_{\{T \leq \tau_0\}} \alpha_0 e^{\beta_0^T Z} T - 1_{\{T \leq \tau\}} \alpha e^{\beta^T Z} T$$
$$- 1_{\{T > \tau\}} \left[(\alpha + \theta) e^{(\beta + \gamma)^T Z} (T - \tau) + \alpha e^{\beta^T Z} \tau \right]$$
$$+ 1_{\{T > \tau_0\}} \left[(\alpha_0 + \theta_0) e^{(\beta_0 + \gamma_0)^T Z} (T - \tau_0) + \alpha_0 e^{\beta_0^T Z} \tau_0 \right] \Big].$$

The first lemma asserts uniform convergence of X_n to X_∞.

Lemma 1. $\sup_{\varphi \in \Phi} |X_n(\varphi) - X_\infty(\varphi)|$ *converges in probability to 0 as* $n \to \infty$.

Uniqueness of φ_0 as a maximizer of X_∞ comes from the following lemma, which asserts that the model is identifiable.

Lemma 2. $p_\varphi(\mathbf{x}) = p_{\varphi_0}(\mathbf{x})$ *a.s. implies* $\varphi = \varphi_0$.

Note that $X_\infty(\varphi)$ is minus the Kullback-Leibler divergence of p_φ and p_{φ_0}. Since $X_\infty(\varphi_0) = 0$, φ_0 is a point of maximum of X_∞. Moreover, it is a unique point of maximum since φ_0 is identifiable (Lemma 2). As X_n converges uniformly to X_∞ (Lemma 1), it follows that $\hat{\varphi}_n$ converges in probability to φ_0.

Remark The hazard function (3) specifies a multiplicative hazard model. An alternative formulation for the association between covariates and time-to-event is the additive hazard model, where $\lambda(t|Z) = \alpha + \beta^T Z$ (a semiparametric form may be specified by letting α be an unknown function of time). The results obtained for model (3) can be shown to hold for the additive change-point hazard regression model

$$\lambda(t|Z) = (\alpha + \theta 1_{\{t>\tau\}}) + (\beta + \gamma 1_{\{t>\tau\}})^T Z.$$

Proofs proceed along the same line as described above.

Appendix

Proof of Lemma 1. Writing $X_n(\varphi)$ as $X_n(\varphi) = n^{-1}\sum_{i\leq n} f_\varphi(\mathbf{X}_i)$, the uniform convergence stated by this lemma is equivalent to the class of functions $\mathcal{F} = \{f_\varphi : \varphi \in \Phi\}$ being Glivenko-Cantelli (we refer the reader to [Van98] and [VW96] for definition of Glivenko-Cantelli and Donsker classes, and for many useful results on these classes).
Since every Donsker class is also Glivenko-Cantelli, we show that $\mathcal{F}$ is Donsker by using results from empirical process theory [VW96].
To demonstrate how it works, we shall show that

$$\left\{g_\varphi(T, \Delta, Z) = 1_{\{T>\tau\}}(\alpha + \theta)e^{(\beta+\gamma)^T Z}T : \varphi \in \Phi\right\}$$

is Donsker. The set of all indicators functions $1_{\{(\tau,\infty)\}}$ is Donsker. From Theorem 2.10.1 of [VW96], $\{1_{\{(\tau,\infty)\}} : \tau \in]0, \zeta[\}$ is Donsker. The function $h_\varphi : (T, \Delta, Z) \mapsto T$ is bounded, which implies that $\{1_{\{(\tau,\infty)\}}T : \tau \in]0, \zeta[\}$ is Donsker (see Exemple 2.10.10 of [VW96]). The class $\{\alpha + \theta : \alpha \in A, \theta \in B\}$ is Donsker. By multiplying two Donsker classes, we get that $\{1_{\{(\tau,\infty)\}}(\alpha + \theta)T : \tau \in]0, \zeta[, \alpha \in A, \theta \in B\}$ is Donsker. Similarly, boundedness of Z implies that $\{(\beta + \gamma)^T Z : \beta \in C, \gamma \in D\}$ is Donsker. The exponential function is Lipschitz on compact sets of the real line, then we get from [VW96] (Theorem 2.10.6) that the class $\{e^{(\beta+\gamma)^T Z} : \beta \in C, \gamma \in D\}$ is Donsker. Again, by multiplication of two Donsker classes, we get that $\left\{g_\varphi(T, \Delta, Z) = 1_{\{T>\tau\}}(\alpha + \theta)e^{(\beta+\gamma)^T Z}T : \varphi \in \Phi\right\}$ is Donsker.
Using similar arguments and the fact that the sum of two Donsker classes is Donsker, we finally get that $\mathcal{F}$ is Donsker, and hence Glivenko-Cantelli.

Proof of Lemma 2. Considering the densities $p_\varphi(\mathbf{x}) = p_{\varphi_0}(\mathbf{x})$ on $\delta = 0$, we see that

$$\exp(-\alpha e^{\beta^T z} t) 1_{\{t \leq \tau\}} + \exp(-\alpha e^{\beta^T z} \tau - (\alpha + \theta) e^{(\beta+\gamma)^T z} (t - \tau)) 1_{\{t > \tau\}} = \exp(-\alpha_0 e^{\beta_0^T z} t) 1_{\{t \leq \tau_0\}} + \exp(-\alpha_0 e^{\beta_0^T z} \tau_0 - (\alpha_0 + \theta_0) e^{(\beta_0+\gamma_0)^T z} (t - \tau_0)) 1_{\{t > \tau_0\}} \tag{6}$$

for almost all (t, z). Suppose that $\tau \neq \tau_0$ (we suppose that $\tau < \tau_0$, the symetric case $\tau > \tau_0$ can be treated similarly).
Let Ω_z be the set of $t \in (\tau, \tau_0]$ such that (t, z) does not satisfy (6). Then, for almost all z (i.e. outside an exceptional zero-measure set $\mathcal{Z}$), $P_0(\Omega_z) = 0$. Given that $z_1 \neq z_2$ and that neither is in $\mathcal{Z}$, (t, z_1) and (t, z_2) satisfy the above relation for almost all $t \in (\tau, \tau_0]$. In particular, let $t_1 \neq t_2$ be two such values.
Evaluated at (t_1, z_1), (6) becomes

$$\exp(-\alpha e^{\beta^T z_1} \tau - (\alpha + \theta) e^{(\beta+\gamma)^T z_1} (t_1 - \tau)) = \exp(-\alpha_0 e^{\beta_0^T z_1} t_1),$$

which is equivalent to $\alpha e^{\beta^T z_1} \tau + (\alpha + \theta) e^{(\beta+\gamma)^T z_1} (t_1 - \tau) = \alpha_0 e^{\beta_0^T z_1} t_1$. Similarly, for (t_2, z_1) we obtain $\alpha e^{\beta^T z_1} \tau + (\alpha + \theta) e^{(\beta+\gamma)^T z_1} (t_2 - \tau) = \alpha_0 e^{\beta_0^T z_1} t_2$. By substracting these last two equalites, we obtain $(\alpha + \theta) e^{(\beta+\gamma)^T z_1} (t_2 - t_1) = \alpha_0 e^{\beta_0^T z_1} (t_2 - t_1)$, which implies

$$(\alpha + \theta) e^{(\beta+\gamma)^T z_1} = \alpha_0 e^{\beta_0^T z_1} \tag{7}$$

since $t_1 \neq t_2$. The same reasonment for the couples (t_1, z_2), (t_2, z_2) yields

$$(\alpha + \theta) e^{(\beta+\gamma)^T z_2} = \alpha_0 e^{\beta_0^T z_2}. \tag{8}$$

Since $\alpha + \theta \neq 0$ and $\alpha_0 \neq 0$, we can calculate the ratio (8)/(7): $e^{(\beta+\gamma)^T (z_2 - z_1)} = e^{\beta_0^T (z_2 - z_1)}$, which implies $(\beta + \gamma - \beta_0)^T (z_2 - z_1) = 0$. Note that the assumption $\text{var}(Z) > 0$ is necessary to achieve identifiability. If $\text{var}(Z) = 0$, then for any q-dimensional vector $a \neq 0$, $\text{var}(a^T Z) = 0$. Therefore $a^T (z_1 - z_2) = 0$ does not imply that $a = 0$. Consider all q-vectors that are orthogonal to $z_1 - z_2$, and select a pair (z_3, z_4) such that neither is in the zero-measure set $\mathcal{Z}$. Then we get $(\beta + \gamma - \beta_0)^T (z_4 - z_3) = 0$. In this way, we can select q pairs of z such that none of the pairs is in $\mathcal{Z}$, and the differences of the pairs are linearly independent. Thus $\beta + \gamma - \beta_0 = 0$ and finally $\beta + \gamma = \beta_0$. It follows from (8) that $\alpha + \theta = \alpha_0$.
Now, from (6), $\alpha e^{\beta^T z_1} \tau + \alpha_0 e^{\beta_0^T z_1} (t_1 - \tau) = \alpha_0 e^{\beta_0^T z_1} t_1$, which implies $\alpha e^{\beta^T z_1} \tau = \alpha_0 e^{\beta_0^T z_1} \tau$. Similarly, for the couple (t_1, z_2), $\alpha e^{\beta^T z_2} \tau = \alpha_0 e^{\beta_0^T z_2} \tau$. By taking the ratio of these two equalities, we obtain $e^{\beta^T (z_1 - z_2)} = e^{\beta_0^T (z_1 - z_2)}$, which implies $(\beta - \beta_0)^T (z_1 - z_2) = 0$. By the same reasonment as above, $\beta = \beta_0$, and hence $\gamma = 0$. This is a contradiction, hence $\tau \not< \tau_0$. Similarly, we can show that $\tau \not> \tau_0$. Hence if (6) holds, then $\tau = \tau_0$.
Using similar arguments, it is now easy to complete the proof and to show that $\xi = \xi_0$.

References

[AG82] Andersen, P.K., Gill, R.D.: Cox's regression model for counting processes: a large sample study. Ann. Statist., **10**, 1100–1120 (1982).

[CCH94] Chang, I.-S., Chen, C.-H., Hsiung, C.A.: Estimation in change-point hazard rate models with random censorship. In: Carlstein, E., Müller, H.-G., Siegmund, D. (eds.) Change-point Problems. IMS Lecture Notes-Monograph Ser. 23 (1994).

[CK94] Cohen, A., Kushary, D.: Adaptive and unbiased predictors in a change point regression model. Statist. Prob. Letters, **20**, 131–138 (1994).

[CH97] Csörgő, M., Horváth, L.: Limit Theorems in Change-Point Analysis. Wiley, New York (1997).

[GV05] Gurevich, G., Vexler, A.: Change point problems in the model of logistic regression. J. Statist. Plann. Inf., in press (2005).

[Hor95] Horváth, L.: Detecting changes in linear regressions. Statistics, **26**, 189–208 (1995).

[HHS97] Horváth, L., Hušková. M, Serbinowska, M.: Estimators for the time of change in linear models. Statistics, **29**, 109–130 (1997).

[Jar03] Jarušková, D.: Asymptotic distribution of a statistic testing a change in simple linear regression with equidistant design. Statist. Prob. Letters, **64**, 89–95 (2003).

[KQS03] Koul, H.L., Qian, L., Surgailis, D.: Asymptotics of M-estimators in two-phase linear regression models. Stoch. Proc. Appl., **103**, 123–153 (2003).

[LTC97] Luo, X., Turnbull, B.W., Clark, L.C.: Likelihood ratio tests for a changepoint with survival data. Biometrika, **84**, 555–565 (1997).

[MFP85] Matthews, D.E., Farewell, V.T., Pyke, R.: Asymptotic score-statistic processes and tests for constant hazard against a change-point alternative. Ann. Statist., **13**, 583–591 (1985).

[MW94] Müller, H.G., Wang, J.-L.: Change-point models for hazard functions. In: Carlstein, E., Müller, H.-G., Siegmund, D. (eds.) Change-point Problems. IMS Lecture Notes-Monograph Ser. 23 (1994).

[NRW84] Nguyen, H.T., Rogers, G.S., Walker, E.A.: Estimation in change-point hazard rate models. Biometrika, **71**, 299–304 (1984).

[PN90] Pham, T.D, Nguyen, H.T.: Strong consistency of the maximum likelihood estimators in the change-point hazard rate model. Statistics, **21**, 203–216 (1990).

[Pon02] Pons, O.: Estimation in a Cox regression model with a change-point at an unknown time. Statistics, **36**, 101–124 (2002).

[Pon03] Pons, O.: Estimation in a Cox regression model with a change-point according to a threshold in a covariate. Ann. Statist., **31**, 442–463. (2003)

[Van98] Van Der Vaart, A.W.: Asymptotic Statistics. Cambridge University Press, New York (1998).

[VW96] Van Der Vaart, A.W., Wellner, J.A.: Weak Convergence and Empirical Processes. Springer, New York (1996).

[WZW03] Wu, C.Q., Zhao, L.C., Wu, Y.H.: Estimation in change-point hazard function models. Statist. Prob. Letters, **63**, 41–48 (2003).

Mortality in Varying Environment

M.S. Finkelstein

Department of Mathematical Statistics
University of the Free State
PO Box 339, 9300 Bloemfontein,
Republic of South Africa
and Max Planck Institute for Demographic Research
Konrad-Zuse-Strasse 1
18057 Rostock, Germany
FinkelM.SCI@mail.uovs.ac.za

Summary. An impact of environment on mortality, similar to survival analysis, is often modelled by the proportional hazards model, which assumes the corresponding comparison with a baseline environment. This model describes the memory-less property, when the mortality rate at a given instant of time depends only on the environment at this instant of time and does not depend on the history. In the presence of degradation the assumption of this kind is usually unrealistic and history-dependent models should be considered. The simplest stochastic degradation model is the accelerated life model. We discuss these models for the cohort setting and apply the developed approach to the period setting for the case when environment (stress) is modelled by the functions with switching points (jumps in the level of the stress).

1 Introduction

The process of human aging is a process of accumulation of damage of some kind (e.g., accumulation of deleterious mutations). It is natural to model it via some stochastic process. Death of an organism uniquely defines the corresponding lifetime random variable in a cohort setting. We are interested in an impact of varying environment on the mortality rate, which is defined for a cohort via the lifetime distribution function in a standard way. There are two major possibilities. The first one is plasticity: a memory-less property, which says that mortality rate does not depend on the past trajectory of an environment and depends only on its current value. This is the unique property in some sense and a widely used proportional hazards (PH) model is a conventional tool for modelling plasticity. On the other hand, dependence on history is more natural for the hazard (mortality) rate of degrading objects, as it seems reasonable that the chance to fail in some small interval of time is higher for objects with higher level of accumulated degradation. There are various ways of modelling this dependence. The simplest one is via the accelerated life model (ALM), which performs the scale transformation in the lifetime distribution function. The ALM can be equivalently defined via the mortality rates as well (see Section 1).

These two models and their generalizations were thoroughly investigated in reliability and survival analysis studies (Bagdonavičius and Nikulin [BN02]), where the cohort setting is a natural one for defining the corresponding lifetime random variables. In Section 2 we discuss some traditional and new results for a cohort setting. In demography, however, period mortality rates play a crucial role, whereas defining 'proper' lifetime random variables is not straightforward and needs additional assumptions on a population structure. We mostly focus on the case when environment has switching points: jumps in severity from one level to another but the situation without switching points is also discussed. Generalization of the PH model to the period case is quite natural, whereas the corresponding generalization of the ALM needs careful reasoning. We perform this operation explicitly for the case of the linear ALM and discuss the idea how it can be generalized to the time-dependent scale transformation.

2 Damage accumulation and plasticity

2.1 Proportional hazards

Denote by X a cohort lifetime random variable (age at death) and by $\mu(x)$ and $l(x)$ the corresponding mortality rate and the survival probability, respectively. Then:

$$\overline{F}(x) \equiv l(x) = \exp\left\{-\int_0^x \mu(u)du\right\}, \tag{1}$$

where $F(x)$ is the cumulative lifetime distribution function (Cdf) and $\overline{F}(x) = 1 - F(x)$.

Let $z(x), x \geq 0$ be an explanatory variable, which for simplicity is assumed to be a scalar one. The function $z(x)$ describes environment or stress. We want to model an impact of a stress (environment) on X. Consider two stress functions: $z_0(x)$ and $z(x)$ - the baseline and the current, respectively. The stress $z_0(x)$ is an arbitrary function from a family of all admissible stresses A. The stress $z_0(x) \in A$ is usually a fixed function. Denote the mortality rate and the Cdf under the baseline stress by $\mu_0(x)$ and $F_0(x)$, respectively, and under the current stress, as in equation (1), by $\mu(x)$ and $F(x)$, respectively.

The most popular way to model a stress impact is via the PH model:

$$\mu(x) = w_P(z(x))\mu_0(x), \tag{2}$$

where $w_P(z(x))$ is a positive, strictly monotone, function (usually unknown), the sub-script "P" stands for "proportional" and $w_P(z_0(x)) \equiv 1$.

Consider now a step stress with switching from the baseline to the current stress at some $x_s > 0$. Several switching points can be considered similarly. This step stress models the abrupt change in environment (e.g., the development of a new critical for the healthcare drug, or the dramatic change in the lifestyle):

$$z_s(x) = \begin{cases} z_0(x), & 0 \le x < x_s \\ z(x), & x_s \le x < \infty \end{cases} \tag{3}$$

In accordance with definition (2), the mortality rate $\mu_s(t)$ for the stress $z_s(x)$ is:

$$\mu_s(x) = \begin{cases} \mu_0(x), & 0 \le x < x_s \\ \mu(x), & x_s \le x < \infty \end{cases} \tag{4}$$

Therefore, the change point in a stress results in the corresponding change point in $\mu_s(t)$: instantaneous jump to the level $\mu(x)$.

Definition (2) and properties (3)-(4) show that a *plastic, memory-less* reaction of the mortality rate on the changes in the stress function takes place. Denote by $F_s(x)$ the Cdf, which corresponds to the mortality rate $\mu_s(t)$. The remaining lifetime also does not depend on the mortality rate history in $[0, x_s)$, as clearly follows from the equation for the remaining lifetime Cdf $F_{rs}(x|x_s)$:

$$\overline{F}_{rs}(x|x_s) \equiv \frac{\overline{F}_s(x + x_s)}{\overline{F}_s(x_s)} = \exp\left\{ -\int_{x_s}^{x+x_s} \mu(u)du \right\}. \tag{5}$$

The PH model is usually not suitable for modelling an impact of stress on degrading (aging) objects, as it means that the stress in $[0, x)$ does not influence the degradation process in (x, ∞). This assumption usually does not hold, as the past changes in stress affect the history of the degradation process, changing its current value. These considerations, of course, are valid for any memory-less model (see the next section).

Mortality rates of humans are increasing in age x (for adults) as the consequence of biological degradation processes. However, there is at least one but a very important for the topic of our paper case, which shows that the PH model can be used for the human cohort mortality rate modelling as well. In this case the notion of stress has a more general meaning.

Example 1. Lifesaving. Describe the mortality environment for a population via the quality of a healthcare. Let $\mu_0(x)$, as previously, denote the mortality rate for some baseline, standard level of healthcare. Suppose that the better level of health care had been achieved, which usually results in lifesaving (Vaupel and Yashin [VY87]): each life, characterized by the initial mortality rate $\mu_0(x)$ is saved (cured) at each event of death with probability $1 - \theta(x)$, $0 < \theta(x) \le 1$ (or, equivalently, this proportion of individuals who would have died are now resuscitated and given another chance). Those who are saved, experience the *minimal repair*. The minimal repair is defined (Finkelstein [Fink00]), as the repair that brings an object back to the state it had just prior to the failure (death). It is clear that the new healthcare environment defined in such a way does not change the process of individual aging. If $\theta(x) = 0$, the lifetime is infinite and 'virtual deaths' form a memory-less nonhomogenous Poisson process. It can be proved (Vaupel and Yashin [VY87], Finkelstein [Fink99]) that under given assumptions the new mortality rate is given by:

$$\mu(x) = \theta(x)\mu_0(x), \tag{6}$$

which is the specific form of the PH model (2). The case, when there is no cure ($\theta(x) = 1$), corresponds to the baseline mortality rate $\mu_0(x)$ and switching from the "stress" $\theta(x) = 1$ to the stress $0 < \theta(x) < 1$ at age x_s results in the *plasticity property* given by equation (4).

Note, that the baseline mortality rate $\mu_0(x)$ can also model a possibility of lifesaving. In this case $\mu(x)$ defines the larger probability of lifesaving. Formally, the hypothetical mortality rate without lifesaving $\mu_h(x)$ should be then defined:

$$\mu_0(x) = \theta_h(x)\mu_h(x), \quad 1 - \theta_h(x) < 1 - \theta(x),\ x > 0.$$

The switching point in lifesaving, in fact, means that at a certain age x_s a switch from one probability of lifesaving to another is performed.

2.2 Accelerated life model

Another popular model describing an impact of a stress on X is the accelerated life model (ALM) (Cox and Oakes [CO84], Finkelstein [Fink99]). It performs the stress-dependent scale transformation of the baseline Cdf

$$F_0(x) = 1 - \exp\left\{-\int_0^x \mu_0(u)du\right\}$$

in the following way:

$$F(x) = F_0\left(\int_0^x w_A(z(u))du\right) \equiv F_0(W_A(x)), \tag{7}$$

where the subscript "A" stands for "accelerated", and notation

$$\int_0^x w_A(z(u))du = W_A(x)$$

is used for convenience. As previously, we assume that $w_A(z_0(x)) \equiv 1$. Note that $w_A(z(x))$ is unknown but can be estimated from the data.

This model is usually more appropriate for modelling additive degradation (accumulation of damage), as the effect of higher stress with $w_A(z(x)) > 1$, for instance, results in facilitation of degradation processes. The function $w_A(z(x))$ can be interpreted as a rate of degradation, whereas $W_A(x)$ is the accumulated damage in this case. We shall also assume in this model that mortality rates are increasing, as monotone degradation usually can be described by IFR (increasing failure rate) lifetime distributions. The mortality rate is obtained from equation (7) as (compare with equation (2)):

$$\mu(x) = w_A(z(x))\mu_0(W_A(x)). \tag{8}$$

Similar to equation (5) the survival function for the remaining lifetime is:

$$\begin{aligned} \overline{F}_r(x|a) &\equiv \frac{\overline{F}(x+a)}{\overline{F}(a)} = \frac{\overline{F}_0(W_A(x+a))}{\overline{F}_0(W_A(a))} \\ &= \frac{\overline{F}_0\left(W_A(a) + \int_0^x w_A(z(u+a))du\right)}{\overline{F}_0(W_A(a))}, \quad a > 0, \end{aligned} \tag{9}$$

where an important for the model additivity property is used:

$$\int_0^{x+a} w_A(z(u)du = W_A(a) + \int_a^{x+a} w_A(z(u))du.$$

Unlike equation (5)), the remaining lifetime already depends on the mortality rate history in $[0, a)$, but this dependence is only on the simple aggregated history characteristic $W_A(x)$.

Let the true biological age x be defined for the baseline stress $z_0(x)$, then the virtual age in the baseline environment of an organism that had survived time x under the current stress $z(x)$, in accordance with ALM, is defined as (Finkelstein [Fink92], Kijima [Kij89]):

$$x_V = W_A(x), \tag{10}$$

and the corresponding difference between these two ages is:

$$\triangle_V \equiv x_V - x.$$

Therefore, the ALM gives a simple and effective way for *age correspondence* under different stresses. If an organism had survived time x under the baseline stress, his virtual age under the current stress is $W_A^{-1}(x)$. Note that for the PH model the virtual age is equal to the calendar one.

If $w_A(z(x)) > x$, $\forall x > 0$, then $W_A(x) > x$ and the stress $z(x)$ is more severe than the baseline one, which in accordance with equation (10) means that $x_V > x$. Additionally, the corresponding mortality rates are ordered in this case as:

$$\mu_0(x) < \mu(x), \quad \forall x > 0, \tag{11}$$

which for increasing $\mu_0(x)$ immediately follows from equation (8).

Definition (7) reads:

$$\exp\left\{-\int_0^x \mu(u)du\right\} = \exp\left\{-\int_0^{W_A(x)} \mu_0(u)du\right\}$$

and

$$\int_0^x \mu(u)du = \int_0^{W_A(x)} \mu_0(u)du. \tag{12}$$

Therefore, given the mortality rates under two stresses in $[0, \infty)$, the function $W_A(x)$ can be obtained.

Similar to the previous subsection, consider now the stress $z_s(x)$ defined by equation (3) and assume for the definiteness that $z(x)$ is more severe than $z_0(x)$. The corresponding Cdf $F_s(x)$ for this stress is:

$$F_s(x) = \begin{cases} F_0(x), & 0 \le x < x_s \\ F_0\left(x_s + \int_{x_s}^x w_A(z(u))du\right), & x_s \le x < \infty. \end{cases} \tag{13}$$

Transforming the second row in equation (13):

$$\begin{aligned} F_0\left(x_s + \int_{x_s}^x w_A(z(u))du\right) &= F_0\left(\int_{x_s-\tau}^x w_A(z(u))du\right) \\ &= F_0(W_A(x) - W_A(x_s - \tau)), \end{aligned} \tag{14}$$

where τ is uniquely defined from the equation:

$$x_s = \int_{x_s-\tau}^{x_s} w_A(z(u))du. \tag{15}$$

Thus, the virtual age under the stress $z(x)$ (in other words, the re-calculated for the more severe stress the baseline age x_s) just after the switching is $x_s - \tau$. Equation (15) defines an interval $[x_s - \tau, x_s)$ in which the accumulated degradation under the stress $z(x)$ is equal to the accumulated degradation x_s under the stress $z_0(x)$ in the interval $[0, x_s)$.

A jump in the stress at x_s leads to a jump in mortality rate, which can be clearly seen by comparing equation (8) with

$$\mu_s(x) = \begin{cases} \mu_0(x), & 0 \le x < x_s \\ w_A(z(x))\mu_0\left(x_s + \int_{x_s}^x w_A(z(u))du\right), & x_s \le x < \infty \end{cases}$$

as for increasing $\mu_0(x)$ and for $w_A(z(x)) > 1$, $x \in [x_s, \infty)$:

$$\begin{aligned} \mu_0(x) &< w_A(z(x))\mu_0\left(x_s + \int_{x_s}^x w_A(z(u))du\right) \\ &< w_A(z(x))\mu_0\left(\int_0^x w_A(z(u))du\right) \\ &= w_A(z(x))\mu_0(W_A(x)) = \mu(x). \end{aligned} \tag{16}$$

Inequality (16) is a special case of inequality (11), obtained for a more severe stress $z_s(x)$.

It is important to note that, as follows from relations (7) and (14), for the general case $F_0\left(x_s + \int_{x_s}^{x} w_A(z(u))du\right)$ is not a segment of $F(x)$ for $x \geq x_s$ (and the corresponding mortality rate is not a segment of $\mu(x)$), but for the specific linear case $W_A(x) = w_A x$ it can be transformed to a segment:

$$F_0(w_A \cdot (x - x_s + \tau)) = F(x - x_s + \tau),$$

where τ is obtained from a simplified equation:

$$x_s = \int_{x_s-\tau}^{x_s} w_A du = \int_0^{\tau} w_A du \quad \Rightarrow \quad \tau = \frac{x_s}{w_A},\ w_A > 1 \tag{17}$$

and, finally, *only for this specific linear case* the Cdf (13) can be defined in the way usually referred to in the literature (Nelson, 1993):

$$F_s(x) = \begin{cases} F_0(x), & 0 \leq x < x_s \\ F(x - x_s + \tau), & x_s \leq x < \infty \end{cases}$$

Sometimes this equation written in terms of mortality rates:

$$\mu_s(x) = \begin{cases} \mu_0(x), & 0 \leq x < x_s \\ \mu(x - x_s + \tau), & x_s \leq x < \infty \end{cases} \tag{18}$$

is called the 'Sedjakin principle', although Sedjakin [Sed66] defined it in a more general way as the dependence on history only via the accumulated mortality rate. As $w_A = const$, $\mu(x)$ is also an increasing function. Taking into account that $\tau < x_s$:

$$\mu_s(x) = \mu(x - x_s + \tau) < \mu(x), \quad x_s \leq x < \infty, \tag{19}$$

which is a specific case of inequality (16).

2.3 Other models

There are not so many other candidates for memory-less models, the additive hazard (AH) model being probably the only one, which is widely used in applied statistical analysis:

$$\mu(x) = \mu_0(x) + w_{AD}(z(x)), \tag{20}$$

where $w_{AD}(x)$ is a positive function ($w_{AD}(z_0(x)) \equiv 1$) and the subscript "AD" stands for "additive". It is clear that the plasticity property (4), defined for the stress given by equation (3), holds also for this case. Similar to the PH model the stress in $[0, x)$ does not influence the degradation process in (x, ∞), but, probably, the AH model is more suitable when, for instance, the baseline

$\mu_0(x)$ describes some 'inherent' degradation process which is not influenced by the environment.

The memory-less property is a rather unique feature, whereas the dependence on a history can be modelled in numerous ways. Most of these generalizations are based on different extensions of the ALM or of the PH model (Bagdonavičius and Nikulin [BN02]). For instance, equation (8) can be generalized to:

$$\mu(x) = G\left(z(x), w_A(z(x), W_A(x))\right),$$

where $G(\cdot)$ is a positive function. The advanced statistical methods of analyzing the data via the chosen model also can be found in Bagdonavičius and Nikulin [BN02]. Our goal in this paper is, however, to discuss plasticity versus accumulated damage modelling for mortality rates in the cohort and period settings. The ALM is just a tractable example, which can be used for degradation modelling.

Let, as previously, $\mu_0(x)$ and $\mu(x)$ be two mortality rates for populations at baseline and current stresses, respectively. Assume that the rates are given or observed and this is the only information at hand. It is clear that without additional information on the degradation process or on the possible memory-lees property the 'proper' model for the stress influence is non-identifiable, as different models can result in the same. Indeed, by letting $w_P(z(x)) = \mu(x)/\mu_0(x)$ we arrive at the PH model (2), and by obtaining $W_A(x)$ from equation (12), which is always possible, results in the ALM (7). The following simple illustrative example will be also helpful for the reasoning of the next section.

Example 2. The Gompertz curve

Let

$$\mu_0(x) = a\exp\{bx\}, \quad a, b > 0 \tag{21}$$

$$\mu(x) = w_P \mu_0(x), \quad w_P > 0 \tag{22}$$

Therefore, equations (21) and (22) formally describe the PH model with a constant in age factor w_P. On the other hand, assuming the ALM defined by equation (7), the function $W_A(x)$ can be obtained from equation (12):

$$\int_0^x \mu(u)du = \int_0^{W_A(x)} \mu_0(x)du \Rightarrow w_P(\exp\{bx\} - 1) = \exp\{bW_A(x)\} - 1.$$

In accordance with the contemporary mortality data for the developed countries (Boongaarts and Feeney [BF02]) parameter b is approximately estimated as 0.1. Equation (22) can be simply approximately solved with a sufficient accuracy for $x > 30$ (when aging starts and the Gompertz curve is suitable for modelling):

$$W_A(x) \approx \frac{\ln w_P}{b} + x. \tag{23}$$

If $w_P < 1$, condition: $x > 30$ in combination with real values of parameters guarantees that $W_A(x) > 0$. Therefore, the ALM defined by relation (23) can formally explain equations (21) and (22), although it is not clear how to explain that the difference between the virtual and baseline ages $\triangle_V$, defined by equation (11), is approximately constant for this model. An explanation via the PH model seems much more natural.

If there is no sufficient information on the 'physical' processes of degradation in our objects, the simplest way to distinguish between the memory-less and accumulation of degradation models is to conduct an experiment and to apply the stress $z_s(x)$, defined by equation (3), to our cohort. If the resulting mortality rate $\mu_s(x)$ is obtained in the form, defined by equation (4), then we arrive at a memory-less property, which means that our object is 'degradation free'. The other option is that there is no dependence on the history of this degradation like in the lifesaving model or the degradation described by the baseline $\mu_0(x)$ does not depend on the environment. The latter possibility was already mentioned while discussing the AH model. On the other hand, if there is a dependence on the degradation history, then the resulting mortality rate should be

$$\mu_s(x) = \begin{cases} \mu_0(x_s), & 0 \leq x < x_s \\ \widetilde{\mu}(x_s), & x_s \leq x < \infty \end{cases} \tag{24}$$

where the mortality rate $\widetilde{\mu}(x)$, e.g., for the ALM, as follows from inequality (16), is contained between baseline and 'current' mortality rates:

$$\mu_0(x) < \widetilde{\mu}(x) < \mu(x), \quad x_s \leq x < \infty. \tag{25}$$

For a general case, if accumulated degradation in $[0, x_s)$ under the stress $z_0(x)$ is smaller than under the stress $z(x)$, inequality (25) should be considered as a *reasonable assumption.*

Inequality (16) defines a jump in mortality rate, which corresponds to a jump in the stress. For a general case the reaction in mortality rate should not be necessarily in the form of the jump: it can be some smooth function, showing some 'inertia' in the degradation process.

In simple electronic devices without degradation the failure rate pattern usually follows the stress pattern. In the lifesaving PH model, however, it is not often the case, as environmental changes are usually rather smooth which results in the smooth change in the probability of lifesaving. An important feature is that after some delay the mortality rate $\mu_s(x)$, $x > x_s$ reaches the level of $\mu(x)$. (Alternatively this delay can be modelled in the degradation framework with a short-term memory of the history of the degradation process).

The relevant example is the convergence of mortality rates of 'old cohorts' after unification of east and West Germany at $x_s = 1990$. ((Vaupel et al [VCC03]). This, of course is the consequence of a direct (better healthcare) and of an indirect (better environment eliminates some causes of death) lifesaving.

Another memory-less example, which is more likely to be modelled by the AH model, is the dietary restriction in Drosophila (Mair et al [MGPP03]). The results of this paper show practically absolute plasticity: the age-specific mortality of the flies with dietary restriction depends only on their age and their current nutritional status, with past nutrition having no detectable effect.

2.4 Damage accumulation and plasticity. Period Setting

The detailed modelling of previous subsections is helpful for considering the PH model and the ALM for the period setting. As far as we know, this topic was not considered in the literature. Denote by $N(x,t)$ a population density (age-specific population size) at time t - a number of persons of age x. See Keding [Kei90] and Arthur and Vaupel [AV84] for discussion of this quantity. We shall call $N(x,t)$, $x \geq 0$ a *population age structure at time t*. Let $\mu(x,t)$ denote the mortality rate as a function of age x and time t for a population with the age structure $N(x,t)$, $x \geq 0$:

$$\mu(x,t) = \lim_{\delta \to 0} \frac{(N(x+\delta, t+\delta) - N(x,t))/d\delta}{N(x,t)}. \tag{26}$$

On the other hand, it is clear that, as $\mu(x,t)d\delta$ is a local risk of death, it, in fact, does not depend on $N(x,t)$, $x \geq 0$. This means that for defining the PH model we do not need to define the corresponding lifetime variable. The stress now is a function of time: $z(t)$, and the cohort PH model (2) is generalized to:

$$\mu(x,t) = w_P(z(t),x)\mu_0(x,t). \tag{27}$$

If the stress (environment) is constant, the mortality rate does not depend on time and the population is stationary with additional assumptions that it is closed to migration and experience a constant birth rate. Consider now a step function *in time t*, which is a special case of the stress (3):

$$\tilde{z}(t) = \begin{cases} z_0, & 0 \leq t < t_s \\ z, & t_s \leq t < \infty \end{cases}.$$

Assuming the constant in age x PH model, the mortality rate for this stress is given by (compare with equations (2), (3) and (4)):

$$\mu_s(x,t) = \begin{cases} \mu_0(x), & 0 \leq t < t_s;\ x \geq 0 \\ w_P(z)\mu_0(x), & t_s \leq t < \infty;\ x \geq 0 \end{cases}. \tag{28}$$

Therefore, the baseline mortality rate after the change point is multiplied by $w_P(z)$ *for all ages* and not for the interval of ages as in equation (4). *This is an important distinction from the step stress modelling in the cohort setting.* The other important 'negative' feature of the period setting is that now the experiment with a step stress without analyzing concrete cohorts (see later) cannot indicate the memory-less property (if any) in a way it did for the purely cohort setting. (Note, that similar to (28), the AH model results for $t \geq t_s$ in the mortality rate $\mu_0(x) + w_{AD}(z)$).

A period ALM at time t should be *applied to each cohort* with varying age x ($0 \leq x < \infty$), and we must assume, as previously, that our population is closed to migration and experience a constant birth rate. In this case the corresponding lifetime random variable for each cohort is properly defined and the population is stationary before the change point and after it as well. We shall illustrate the construction of the period ALM for the step stress (3), where, as previously, the stress z is a more severe than the stress z_0. The opposite ordering of stresses is considered in the same way. Due to piecewise constant stress, the linear ALM with a constant rate w_A can be used and the mortality rate is defined via equation similar to equation (18) but with the age and time-dependent shift $\tau(x,t)$:

$$\mu_s(x,t) = \begin{cases} \mu_0(x), & 0 \leq t < t_s; \quad x \geq 0 \\ \mu(x + \tau(x,t)), & t_s \leq t < \infty; \quad x \geq 0 \end{cases}, \tag{29}$$

where $\tau(x,t)$ is obtained from equation similar to equation (17):

$$x - I(t - t_s) = \int_{x-\tau(x,t)}^{x} w_A du, \quad w_A > 1, \tag{30}$$

where $I(t - t_s)$ is an indicator:

$$I(t - t_s) = \begin{cases} 0, & 0 \leq t < t_s \\ 1, & t_s \leq t < \infty \end{cases}.$$

Equation (30) has the following solution:

$$\tau(x,t) = \begin{cases} \frac{x-(t-t_s)+(t-t_s)w_A}{w_A}, & x > (t - t_s) \\ 0, & x \leq (t - t_s). \end{cases} \tag{31}$$

Specifically, when $t = t_s$ similar to equation (17): $\tau(x, t_s) = x/w_A$, but now this solution is valid for all ages x. Therefore, for each $t > t_s$ the recalculation of initial age $\tau(x, t_s)$ is performed for each cohort. Specifically, if the age of the cohort is less than $t - t_s$ and therefore this cohort was born after the change point t_S and 'does not re-member' the previous stress z_0. All possibilities are incorporated by equation (29). An importance of the switching strategy is again in the fact that, if we look at concrete cohorts in the period framework, we are still able to detect the memory-less property or the absence of it.

A cumbersome generalization of this approach to the general time-dependent stress can be also performed using the similar considerations: at each time t the initial age $\tau(x,t)$ is obtained using the following expression for the age structure of a closed to migration population:

$$N(x,t) = B(t-x)\exp\left\{-\int_0^x \mu(u, t-x+u)du\right\}. \tag{32}$$

where $B(t-x)$ is the birth rate at time $(t-x)$.

Considering the time-dependent stress for the PH model, however, is much simpler. For the case with a switching point equations (27) and (28) is generalized to:

$$\mu_s(x,t) = \begin{cases} \mu_0(x,t), & 0 \le t < t_s; \quad x \ge 0 \\ w_P(z(t),x)\mu_0(x,t), & t_s \le t < \infty; \quad x \ge 0 \end{cases}, \tag{33}$$

where the multiplier already depends on the stress at time t and on the age x.

Different environments can be defined not necessarily by the switching point or by considering changing in time stresses. Let the stress $z_0(t_0)$ be a baseline stress at a baseline (fixed) time instant t_0. Denote the corresponding mortality rate, as previously, by $\mu_0(x, t_0)$. Then the stress $z(t)$ and the mortality rate $\mu(x,t)$ characterize the current instant of time t. Note that in this approach populations can be different and $t - t_0$ can be reasonably large (e.g., 10 or 20 years). The PH model for this case is naturally defined as (compare with (27))

$$\mu(x,t) = w_P(z(t),x)\mu_0(x,t_0), \quad x \ge 0,\, t > t_0, \tag{34}$$

where $\mu_0(x,t_0)$ plays the role of a baseline mortality rate. The analogue of the ALM, however, is not straightforward, as there should be a pair wise comparison between the corresponding cohorts of the same age x, using expression (32) for both instants of time. This topic needs further study.

Example 3. Gompertz shift model. As stated in Bongaarts and Feeney [BF02], the mortality rate in contemporary populations with high level of life expectancy tends to improve over time by a similar factor *at all adult ages* which results in our notation in the following Gompertz shift model (similar to equations (21) and (22)):

$$\mu_0(x,t_0) = a\exp\{bx\}, \quad a, b > 0, \tag{35}$$

$$\mu(x,t) = w_P(z(t))\mu_0(x,t_0), \quad w_P > 0, \tag{36}$$

This model was verified using contemporary data for different developed countries. Equations (35) and (36) define formally the age independent PH model. We do not have a switching in stress here which could help in verifying plasticity.

Most researchers agree that the process of human aging is the process of accumulation of damage of some kind (e.g., accumulation of deleterious mutations). Given the reasoning of the previous sections it means that the PH model (35), (36) is not suitable for this case unless it describes the lifesaving model. On the other hand, as it was stated in Example 2, it is really unnatural trying to explain (35)-(36) via some degradation model. Therefore, if the linear trend takes place (or, equivalently, the logarithms of mortality rates at different time instants are practically parallel), this can be explained by lifesaving in a general sense and not by slowing down the degradation processes, for instance. In other words: lifesaving is likely to be the main source of lifespan extension at present time. In fact, it is not a strict statement, but just a reasoning that seems to be true.

3 Concluding remarks

The most popular models that account for an impact of environment on a lifetime are the PH model and the ALM. The first one is the simplest way to describe the memory-less property, whereas the second describes the simplest dependence on a history in a form of accumulated damage. Various generalizations of these models are considered in Bagdonavičius and Nikulin [BN02]. In survival analysis these models were traditionally defined for the cohort setting.

The conventional demographic definition of the observed in a period (from t to $t + \triangle t$) age-specific and time-dependent mortality rate is given by equation(26). The generalization of the cohort PH model to this case is given by equations (27) and (28). The corresponding generalization of the ALM is explicitly performed for a specific case of the step stress $\widetilde{z}_s$. Therefore, the cohort ALM is applied to each cohort with varying age x ($0 \leq x < \infty$) at time t, which results in equations (29)-(31) defining the age specific mortality rate.

Although human aging is definitely a process of damage accumulation, the contemporary demographic data supports the Gompertz shift model (33)-(34), which is, at least formally, the PH model. In line with our reasoning of the previous section this means that lifesaving (versus the decrease in the rate of degradation) can explain the decrease in mortality rates with time.

References

[AV84] Arthur, W.B., and Vaupel, J.W. Some general relationships in population dynamics. Population Index, **50**, 214-226(1984)

[BN02] Bagdonavičius, V., and Nikulin, M. Accelerated Life Models. Modeling and Statistical Analysis, Chapman and Hall (2002)

[BF02] Bongaarts, J., and Feeney G. How long do we live? Population and Development Review, **28**, 13-29 (2002)

[CO84] Cox, D.R., and Oakes D. Analyses of Survival Data, **24** , Chapman and Hall, New York (1984)

[Fink92] Finkelstein M.S.A restoration process with dependent cycles, Automat. Remote Control, **53**, 1115-1120 (1992)

[Fink99] Finkelstein, M.S. Wearing-out components in variable environment, Reliability Engineering and System Safety, **66**, N3, 235-242 (1999)

[Fink00] Finkelstein, M.S. Modeling a process of non-ideal repair. In: Limnios N., Nikulin M. (eds) Recent Advances in Reliability Theory. Birkhauser, 41-53 (2000)

[Kei90] Keiding, N. Statistical inference in the Lexis diagram. Philosophical Transactions of the Royal Society of London, **A 332**, 487-509 (1990)

[Kij89] Kijima M. Some results for repairable systems with general repair, J. Appl. Prob., **26**, 89-102 (1989)

[MGPP03] Mair, W., Goymer, P., Pletcher, S., and Patridge, L. Demography of dietary restriction and death in Drosophila. Science, **301**, 1731-1733 (2003)

[Nel90] Nelson, W. Accelerated Testing. John Wiley and Sons, New York (1990)

[Sed66] Sedjakin N. M. On one physical principle of reliability theory. Techn. Kibernetika (in Russian), **3**, 80-82 (1966)

[VY87] Vaupel, J.W., and Yashin, A.I. Repeated resuscitation: how life saving alters life tables. Demography, **4**, 123-135 (1987)

[VCC03] Vaupel, J.W., Carey, J.R., and Christensen, K. It's never too late. Science, **301**, 1679-1681 (2003)

Goodness of Fit of a joint model for event time and nonignorable missing Longitudinal Quality of Life data

Sneh Gulati[1] and Mounir Mesbah[2]

[1] Department of Statistics, The Honors College, Florida International University, Miami, FL 33199,USA gulati@fiu.edu

[2] Laboratoire de Statistique Théorique et Appliquée (LSTA), Université Pierre et Marie Curie - Paris VI, Boîte 158, - Bureau 8A25 - Plateau A. 175 rue du Chevaleret, 75013 Paris, France mesbah@ccr.jussieu.fr

Abstract: In many survival studies one is interested not only in the duration time to some terminal event, but also in repeated measurements made on a time-dependent covariate. In these studies, subjects often drop out of the study before the occurrence of the terminal event and the problem of interest then becomes modelling the relationship between the time to dropout and the internal covariate. Dupuy and Mesbah (2002) (DM) proposed a model that described this relationship when the value of the covariate at the dropout time is unobserved. This model combined a first-order Markov model for the longitudinally measured covariate with a time-dependent Cox model for the dropout process. Parameters were estimated using the EM algorithm and shown to be consistent and asymptotically normal. In this paper, we propose a test statistic to test the validity of Dupuy and Mesbah's model. Using the techniques developed by Lin (1991), we develop a class of estimators of the regression parameters using weight functions. The test statistic is a function of the standard maximum likelihood estimators and the estimators based on the weight function. Its asymptotic distribution and some related results are presented.

1 Introduction and Preliminaries

In survival studies; for each individual under study; one often makes repeated observations on covariates (possibly time dependent) until the occurrence of some terminal event. The survival time T in such situations is often modeled by the Cox Regression Model (Cox, 1972) which assumes that its hazard function has the proportional form:

$$\lambda(t|Z) = \lambda_0(t)exp\{\beta^T Z(t)\} \tag{1}$$

In the above, t denotes the time to an event, $\lambda_0(t)$ denotes the baseline hazard function and $\boldsymbol{Z}$ denotes the vector of covariates. If the covariates are time dependent, we distinguish between two main types: *external* and *internal.* An external covariate is one that is directly related to the failure mechanism. An internal covariate is generated by the individual under study and therefore can be observed only as long as the individual in the study.

The survival times themselves may be censored on the right by a censoring variable C, so that what one observes is $X = \min\ (T, C)$ and the censoring indicator $\Delta = \mathrm{I}\{T \leq C\}$, where I is the indicator function and conditional on Z, T and C are assumed to be independent.

The first step in making any inferences about the survival time is the estimation of the baseline hazard λ_0 and the vector β_0. When all the covariates are observed and are external (if they are time dependent), one estimates the parameter vector β_0 by maximizing the following partial likelihood function (see Cox, 1972 for details):

$$L(\beta) = \prod_{i=1}^{n} \left[\frac{\exp\{\beta^T Z_i(X_i)\}}{\sum_{j=1}^{n} Y_j(X_i) \exp\{\beta^T Z_j(X_i)\}} \right] \tag{2}$$

where (X_i, Δ_i, Z_i), $1 \leq i \leq n$, is a random sample of the data and the variable $Y_j(t) = 1$ if $Xj \geq t$ and 0 otherwise. Given the MLE $\hat{\beta}$ of β, the estimator of the cumulative hazard function Breslow (1972, 1974) is the one obtained by linear interpolation between failure times of the following function:

$$\hat{\Lambda}(\tau) = \sum\nolimits_{X_i \leq \tau} \frac{\Delta_\iota}{\sum_{j=1}^{n} Y_j(X_i)exp\{\beta^T Z_j(X_i)\}} \tag{3}$$

In case of internal covariates, however, as noted before, the observed value of the covariate carries information about the survival time of the corresponding individual and thus such covariates must be handled a little differently. For internal covariates, Kalbfleisch and Prentice (1980) define the hazard of the survival time t as:

$$\lambda(t, \overline{Z}(t))dt = P\{T \in [t, t + dt)|\overline{Z}(t), \quad T \geq t\} \tag{4}$$

where $\overline{Z}(t)$ denotes the history of the covariate up to time t. Thus the hazard rate conditions on the covariate only up to time t, but no further. As pointed out by Dupuy and Mesbah (2002) fitting the Cox model with internal covariates can lead to several problems. The inclusion of a covariate whose path is directly affected by the individual can mask treatment effects when comparing two treatments. Similarly inferences about the survival time T will require integration over the distribution of $Z(t)$ or a model for failure time in which Z is suppressed. Several authors have dealt with the problem of fitting a Cox model involving internal covariates (see for example Tsiatis et. al. (1995), Wulfsohn and Tsiatis (1997), Dafni and Tsiatis (1998) etc.)

We focus here on the model developed by Dupuy and Mesbah (2002) who considered experiments where it was assumed that each subject leaves the study at a random time $T \geq 0$ called the dropout time and objective of the paper was to model the relationship between time to dropout and the longitudinally measured covariate. The work of Dupuy and Mesbah was motivated by a data set concerning quality of life (Qol) of subjects involved in a cancer clinical trial. The Qol values formed the covariate of interest. However, patients were likely to drop out of the study before disease progression and for such patients then, the value of the covariate was unobserved at the time of disease progression. Following the approach of Diggle and Kenward (1994), Dupuy and Mesbah (2002) fit a joint model to describe the relationship between the covariate and the time to dropout. Their model combined a first-order Markov model for the longitudinally measured covariate with a time-dependent Cox model for the dropout process, while Diggle and Kenward (1994) specified a logistic regression model for the dropout process. In this paper, we propose a test statistic to validate the model proposed by Dupuy and Mesbah (2002).

In Section II, we describe the various types of dropout processes which can be observed and the methods of dealing with them. The work of Dupuy and Mesbah (2002, 2004) is described in Section III. In Section IV, we develop the test statistic and study its properties.

2 The Dropout Process

Little and Rubin (1987) identify three main classifications of the drop-out process in longitudinal studies:

i) Completely Random (CRD): A drop-out process is said to be completely random when the drop out is independent of both the observed and the unobserved measurements.

ii) Random Drop-Out (RD): Here the drop-out process is dependent on the observed measurements, but is independent of the unobserved ones.

iii) Informative (or Non Ignorable) Dropout (ID): A drop-out process is noningnorable when it depends on the unobserved measurements, that is those that would have been observed if the unit had not dropped out.

Under the completely random drop-out process, drop-outs are equivalent to randomly missing values and so the data can be analyzed without requiring any special methods. In the random drop-out case, provided there are no parameters in common between the measurements and the drop-outs or any functional relationship between them, the longitudinal process can be completely ignored for the purpose of making likelihood based inference about the time to drop-out model. However, special methods are needed for the nonignorable case.

A number of authors have considered analysis under the ID model. Wu and Carroll (1988) considered informative drop-out in a random effects model where the experimental units follow a linear time trend whose slope and intercept vary according to a bivariate Gaussian distribution. Wang et. al. (1995) report on a simulation study to compare different methods of estimation under different assumptions about the drop-out process. Diggle and Kenward (1994) combined a multivariate linear model for the drop-out process with a logistic regression model for the drop-out. Molenberghs et. al. (1997) adopted a similar approach for repeated categorical data to handle informative drop-out, amongst others (see for example, Little, 1995, Hogan and Laird, 1997, Troxel et. al., 1998, Verbeke and Molenberghs, 2000 etc.)

In the setting of Dupuy and Mesbah (2002), the value of the covariate at drop-out is unobserved and since drop-out is assumed to depend on quality of life of the individual, the drop-out may be treated as nonignorable. The model suggested by them is a generalization of the model by Diggle and Kenward (1994), since as opposed to Diggle and Kenward, Dupuy and Mesbah (2002) allow censoring in their model. Next, we describe their model.

3 The Model of Dupuy and Mesbah (2002)

Assume that n subjects are to be observed at a set of fixed times t_j, j $=$ 0, 1, 2, . . ., such that $t_0 = 0 < \ldots < t_{j-1} < t_j < \ldots < \infty$ and $0 < \varepsilon_0 \leq \Delta t = t_j - t_{j-1} < \varepsilon_1 < \infty$ (times of measurement are fixed by study design). Let Z denote the internal covariate and $Z_i(t)$ denote the value of Z at time t for the ith individual under study (i $=$ 1, 2, . . ., n). Repeated measurements of Z are taken on each subject at common fixed times t_j, $j = 1, 2, \ldots$ $(t_0 = 0)$. Let $Z_{i,j}$ denote the response for the ith subject in $(t_j, t_{j+1}]$. The time to dropout model proposed by Dupuy and Mesbah (2002) assumes that the hazard of dropout is related to the covariate by the time dependent Cox model:

$$\lambda(t \mid \overline{z}(t)) \quad = \quad \lambda(t) \exp(r(\overline{z}(t), \beta)), \quad t \geq 0 \tag{5}$$

where $\lambda(.)$ is the unspecified baseline hazard function and $r(\overline{z}(t), \beta)$ is a functional of the covariate history up to time t. The functionals $r(\overline{z}(t), \beta)$ as considered by Dupuy and Mesbah (2004) are of the form $\beta_1(z(t - \Delta t)) + \beta_2 z(t)$ and $\beta_3(z(t) - z(t - \Delta t))$. The reason for using these functionals is the intuitive meaning behind them; in the first functional, we assume that the probability that T belongs to $(t_{j-1}, t_j]$ depends on the current unobserved covariate at the time of the dropout and on the last observed covariate value before t, $Z(t - \Delta t)$. In this setting, the covariate $Z(t)$ is referred to as nonignorable missing data. The second form for r is used when the interest is in studying whether the increase or decrease in the value of Z between the times $t - \Delta t$ and t influences dropout. As a result, equation (1.5) is reformulated as

$$\lambda(t \mid \overline{z}(t)) \quad = \quad \lambda(t) \exp(\beta^T w(t)), \quad t \geq 0 \qquad (6)$$

where $w(t) = (z(t - \Delta t), z(t))^T$ and $\beta = (\beta_1, \beta_2)^T$ or $\beta = (-\beta_3, \beta_3)^T$.

Once again, the dropout times are assumed to be censored on the right by the random variable C. In addition, the following conditions are assumed:

- i- The covariate vector $\boldsymbol{Z}$ is assumed to have the uniformly bounded continuous density $f(\boldsymbol{z}, \alpha)$, $\boldsymbol{z} = (z_0, z_1, z_2, \ldots) \in R^\infty$ depending on an unknown parameter α.
- ii- The censoring time C is assumed to have the continuous distribution function $G_C(u)$ on the $R_+ = (0, \infty)$.
- iii- The censoring distribution is assumed to be independent of the unobserved covariate, and of the parameters α, β and Λ.

Now, let τ denote a finite time point at which any individual who has not dropped out is censored, assume that $P(X \geq \tau) > 0$ and let $a_{t-1} = j$ *if* $t \in (t_j, t_{j+1}]$. With this notation, w(t) can be rewritten as $(z_{a_{t-1}}, z(t))^T$. We observe n independent replicates of $X = min\ (T, C)$, Δ and $\boldsymbol{Z}$ represented by the vector $\boldsymbol{y} = (x, \delta, z_0, \ldots, z_{a_{x-1}})$. The problem of interest is to estimate via maximum likelihood, the true regression parameters, denoted by α_0, β_0 and the baseline hazard function $\Lambda_0 = \int_0^t \lambda_0(u)du$ The probability measure induced by the observed $\mathbf{y}$ will be denoted by $P_\theta(dy) = f_Y(y;\theta)dy$, where $\theta = (\alpha, \beta, \Lambda)$.

The first step in the problem of maximum likelihood estimation is the development of the likelihood function. The likelihood $f_Y(y;\theta)$ for the vector of observations $\boldsymbol{y} = (x, \delta, z_0, \ldots, z_{a_{x-1}})$ was obtained by Dupuy, and Mesbah (2002) by first writing the density of ($\mathbf{y}$, z) for some value of z on $(t_{a_x-1}, t_{a_x}]$ and then by integrating over z. This gives the partial likelihood function as:

$$L(\theta) =$$

$$\int_{\Re} [\lambda(x)]^\delta \exp\left[\delta\beta^T w(x) - \int_0^x \lambda(u)\, e^{\beta^T w(u)} du\right] f(z_0, ..., z_{a_{x-1}}, z; \alpha) dz \qquad (7)$$

where $w(t) = (z_{a_{x-1}}, z)^T$ if $t \in (t_{a_{x-1}}, t_{a_x})$. To estimate the parameters, especially the hazard function, Dupuy and Mesbah (2002) use the well-known method of sieves. The method consists of replacing the original parameter space Θ of the parameters (α, β, Λ) by an approximating space Θ_n, called the sieve. More precisely, instead of considering the hazard function, $\Lambda = \Lambda(t)$, one considers increasing step wise versions $\Lambda_{n,i} = \Lambda_n(T_{(i)})$, at the points $T_{(i)}$, $i = 1, 2, \ldots, p(n)$, where $T_{(1)} \leq T_{(2)} \leq \ldots \leq T_{p(n)}$ are the order statistics corresponding to the distinct dropout times $T_1 \leq T_2 \leq \ldots \leq T_{p(n)}$ Hence the approximating sieve is $\Theta_n = \{\theta = (\alpha, \beta, \Lambda_n)\colon \alpha \in R^p, \beta \in R^2, \Lambda_{n,1} \leq \Lambda_{n,2} \leq \ldots \leq \Lambda_{n,p(n)}\}$. The estimates of the parameters α and

β and the values $\Lambda_{n,i}$ are obtained by maximizing the likelihood in (1.7) over the space Θ_n, in other words, one maximizes the pseudo-likelihood

$$L_n(\theta) = \prod_{i=1}^{n} L^{(i)}(\theta) \tag{8}$$

The above is obtained by multiplying over the subjects i (i = 1, . . ., n), the following individual contributions:

$$L^{(i)}(\theta) = \int_{\Re} \left[\prod_{k=1}^{p(n)} \Delta\Lambda_{n,k}^{\delta_i 1'_{T(k)=x_i\}}} \right] \exp \left[\delta_i \, \beta^T w_i(x_i) \; - \sum_{k:T_{(k)} \leq x_i} \Delta\Lambda_{n,k} e^{\beta^T w_i(T_{(k)})} \right]$$

$$\times f \left(z_{i,0}, \ldots, z_{i,a_{x_i}-1}, z; \; \alpha \right) dz \tag{9}$$

The semiparametric maximum likelihood estimator $\hat{\theta}_n = (\hat{\alpha}_n, \hat{\beta}_n, \hat{\Lambda}_n)$ of the parameter space is obtained by using the EM algorithm. The method is an iterative method, which iterates between an E-step where the expected loglikelihood of the complete data conditional on the observed data and the current estimate of the parameters is computed and an M-step where the parameter estimates are updated by maximizing the expected loglikelihood. Dupuy, Grama and Mesbah (2003) and Dupuy and Mesbah, 2004) have shown that the estimator $\hat{\theta}_n$ is identifiable and converges to a Gaussian process with a covariance that can be consistently estimated.

The purpose of this project is to validate the model in (1.7). In order to do so, we propose a method developed by Lin (1991) for the general Cox model. The method involves a class of weighted parameter estimates and works on the following idea: Parameter estimates for the standard Cox model are obtained by maximizing the score function, which assigns an equal weight to all the failures. The weighted parameter estimates are calculated by maximizing a weighted score function where different observations get different weights depending on their times of occurrence. Since both the weighted and the unweighted estimators are consistent, the rationale is that if there is no misspecification in the model, then they should be close to each other. However, in case of model misspecification, the two estimators will tend to be different. We propose the use of this method to validate the model of Dupuy et al (2002) to test model validity.

The proposed test statistic is studied and developed in the next section.

4 The Test of Goodness of Fit

As mentioned earlier, the results in this section are based on the methods of Lin (1991). To verify the model in (1.7), first define a class of weighted pseudo-likelihood functions (for a random weight function, WG(.)) as follows:

$$WL_n(\theta) = \prod_{i=1}^{n} WL^{(i)}(\theta) \tag{10}$$

$$\text{where } \mathrm{WL}^{(i)}(\theta) = \int\limits_{\Re} \left[\prod_{k=1}^{p(n)} \Delta\Lambda_{n,k}^{\;WG(x_i)\delta_i 1'_{T(k)=x_i\}}} \right]$$

$$\times \exp\left[WG(x_i)\delta_i\, \beta^T w_i(x_i) - \sum_{k:T_{(k)}\leq x_i} WG(x_i)\Delta\Lambda_{n,k} e^{\beta^T w_i(T_{(k)})} \right]$$

$$\times f\,(\,z_{i,0},\,.\,.,\,z_{i,a_{x_i-1}},\,z;\,\alpha)\,dz \tag{11}$$

The random weight function, $WG(t)$ is a predictable stochastic process, which converges uniformly in probability to a nonnegative bounded function on $(0, \infty)$. Note that when $WG(t) = 1$, we have the likelihood function in (1.7). Typically, for right censored data, $WG(t) = \hat{F}(t)$, the left continuous version of the Kaplan Meier estimator of the survival function.

Now define $\hat{\theta}_{W,\,n} = (\hat{\alpha}_{W,n},\, \hat{\beta}_{W,n},\, \hat{\Lambda}_{W,n})$ to be the maximizer of $WL_n(\theta)$ over $\theta \in \quad \Theta_n$. This estimator is obtained by following the steps of the EM algorithm of Dupuy and Mesbah. (2002) for the non-weighted likelihood function. Note that the weights themselves do not depend on the parameters, thus ensuring that all the steps for the EM algorithm for the weighted likelihood function go through as for they do for the non-weighted likelihood function.

Since the test statistic is a function of $\hat{\beta}_{W,n}$, we first present the following asymptotic results for the weighted parameter $\hat{\beta}_{W,n}$:

Theorem 1. *Under the model (1.7), the vector* $\sqrt{n}\ (\hat{\beta}_{W,n} - \beta_0)$ *converges in distribution to a bivariate normal distribution with zero mean and a covariance matrix* $\Sigma_{W,0}^{-1}$ *where* $\Sigma_{W,0} = E_{\theta_0}\left[\int_0^X WG(X)W(u)W(u)^T\ e^{\beta_0^T\, W(u)} d\Lambda_0(u) \right]$.

The proof of this theorem is similar to the proof of normality for the unweighted parameter $\hat{\beta}_n$ as presented in Dupuy, Grama and Mesbah (2003) and therefore will be omitted.

Hence once again, if model (1.7) is correct, $\hat{\beta}_{W,n}$ and $\hat{\beta}_n$ will be close to each other since the estimators are consistent. However, if model (1.7) is not correct, then the two estimators will differ from each other. This idea is reiterated in the following conjecture:

Theorem 2. *Under the model (1.7), the vector* $\sqrt{n}\ (\hat{\beta}_{W,n} - \hat{\beta}_n)$ *converges in distribution to a bivariate normal distribution with zero mean and a covariance matrix* $D_W = \Sigma_{W,0}^{-1} - \Sigma_0^{-1}$ *where* $\Sigma_0 = E_{\theta_0}\left[\int_0^X W(u)W(u)^T\ e^{\beta_0^T\, W(u)} d\Lambda_0(u) \right]$ *and* Σ_0^{-1} *is the variance-covariance matrix for the random vector* $\sqrt{n}\ (\hat{\beta}_n - \beta_0)$ *as presented in Dupuy and Mesbah (2004).*

The proof of the above theorem is straightforward, even if technically cumbersome. The steps in proving it are similar to the proof of Theorem 1. It involve showing that the score functions associated with the likelihoods defined in (1.7) and (4.1) are asymptotically jointly normal. This can be achieved by extending the steps of the proof of the asymptotic normality for the score function for the likelihood in (1.7).

Now, let $\hat{D}_W$be the consistent estimator of the matrix D_W. Note that this estimator exists for Σ_0^{-1} as shown by Dupuy and Mesbah (2002) and therefore naturally exists for $\Sigma_{W,0}^{-1}$ and D_W. Hence, we propose a test statistic to test the model (1.7) is as:

$$Q_W = n\,(\hat{\beta}_{W,n} \; - \; \hat{\beta}_n)^T \; \hat{D}_W^{-1} \; (\hat{\beta}_{W,n} \; - \; \hat{\beta}_n). \tag{12}$$

>From theorem 2, under the null hypothesis of a correct model, Q_W will have an asymptotic chi squared distribution with 2 degrees of freedom. Hence we can reject the model (1.7) for large values of the test statistic.

5 Conclusion

Testing the goodness of fit of a model involving nonignorable missing data is a first step, that cannot be overlooked. Of course, it is impossible to get a goodness of fit test for the process of missingness, as it is only partially observed. So applying a goodness of fit test to the marginal model (which include the process of missingness) is a necessary preliminary step. If the null hypothesis is rejected, then we can suspect an incorrect specification of i) the missingness process or of ii) the main statistical model for the data. But, if the null hypothesis is not rejected, we can go ahead, as usual, in similar situations.

If the null hypothesis is rejected, and if we have a strong belief that our main statistical model is correct, then we can suspect the missingness process. Building a test for the missingness process itself is not mathematically possible with similar data.

6 References

Breslow, N. (1972). Contribution to the discussion of paper by D.R. Cox. *Journal of the Royal Statistical Society, B,* **34,** 187-220.

Breslow, N. (1974). Covariance Analysis of Censored Survival Data. *Biometrics,* **30,** 89-99.

Chen, H.Y. and Little, R.J.A (1999). Proportional Hazards Regression with Missing Covariates. *Journal of the American Statistical Association,* **94,** 896-908.

Cox, D.R. (1972). Regression Models and Life Tables, with Discussion. *Journal of the Royal Statistical Society, B,* **34,** 187-220.

Dafni, U.G. and Tsiatis, A.A. (1998). Evaluating Surrogate Markers of Clinical Outcome when measured with Error. *Biometrics*, **54,** 1445-1462.

Diggle, P.J. and Kenward, M.G. (1994). Informative Dropout in Longitudinal Data Analysis (with discussion). *Applied Statistics*, **43**, 49-93.

Dupuy, J.F and Mesbah, M. (2002). Joint Modeling of Event Time Data and Nonignorable Missing Longitudinal Data. *Lifetime Data Analysis,* **8**,99-115

Dupuy, J.-F.; Mesbah, M.(2004) Estimation of the asymptotic variance of SPML estimators in the Cox model with a missing time-dependent covariate. *Communications in Statistics - Theory and Methods 33,* **.6**, 1385-1401 (2004).

Dupuy, J.-F.; Grama, I. and Mesbah, M. (2003) Normalité asymptotique des estimateurs semi paramétriques dans le modèle de Cox avec covariable manquante non-ignorable (In french) *C. R. Acad. Sci. Paris Sér. I Math.* **336, No.1**, 81-84.

Hogan, J.W., and Laird, N.M. (1997). Model Based Approaches to Analysing Incomplete Longitudinal and Failure Time Data. *Statistics in Medicine,* **16,** 239-257.

Kalbfleisch, J.D. and Prentice, R.L. (1980). *The Statistical Analysis of Failure Time Data.* Wiley: New-York.

Lin, D.Y. (1991). Goodness-of-Fit Analysis for the Cox Regression Model Based on a Class of Parameter Estimators. *Journal of the American Statistical Association,* **86,** 725-728.

Lin, D.Y. and Ying, Z. (1993). Cox Regression with Incomplete Covariate Measurements. *Journal of the American Statistical Association,* **88,** 1341-1349.

Little, R.J.A. and Rubin, D.B. (1987). *Statistical Analysis with Missing Data.* Wiley: New-York.

Little, R.J.A. (1995). Modeling the Dropout Mechanism in Repeated Measure Studies. *Journal of the American Statistical Association,* **90**, 1112-1121.

Martinussen, T. (1999). Cox Regression with Incomplete Covariate Measurements using the EM-algorithm. *Scandinavian Journal of Statistics,* **26,** 479-491.

Molenberghs, G., Kenward, M.G. and Lesaffre, E. (1997). The Analysis of Longitudinal Ordinal Data with Nonrandom Dropout. *Biometrika,* **84,** 33-44.

Paik, M.C. and Tsai, W.Y. (1997). On Using the Cox Proportional Hazards Model with Missing Covariates. *Biometrika,* **84,** 579-593.

journal Tsiatis, A.A., DeGruttola, V. and Wulfsohn, M.S. (1995). Modeling the Relationship of Survival to Longitudinal Data Measured with Error. Applications to Survival and CD4 Counts in Patients with AIDS. *Journal of the American Statistical Association,* **90,** 27-37.

Verbeke, G. and Molenberghs, G. (2000). *Linear Mixed Models for Longitudinal Data.* Springer-Verlag: New-York.

Troxel, A.B., Lipsitz, S.R. and Harrington, D.P. (1998). Marginal Models for the Analysis of Longitudinal Measurements with Nonignorable Nonmonotone Missing Data. *Biometrika,* **85**, 661-672.

Wang-Clow, F. Lange, M., Laird, N.M. and Ware, J.H. (1995). A Simulation Study of Estimators for Rate of Change in Longitudinal Studies with Attrition. *Statistics in Medicine,* **14,** 283-297.

Wu, M.C. and Carroll (1988). Estimation and Comparison of Changes in the Presence of Informative Right Censoring by Modeling the Censoring Process. *Biometrics,* **44,** 175-188.

Wulfsohn, M.S. and Tsiatis, A.A. (1997). A Joint Model for Survival and Longitudinal Data Measured with error. *Biometrics,* **53,** 330-339.

Three approaches for estimating prevalence of cancer with reversibility. Application to colorectal cancer

C.Gras[1], J.P.Daurès[1,2] and B.Tretarre[2]

[1] Laboratoire de Biostatistique, Institut Universitaire de Recherche Clinique, 641 avenue de Doyen Gaston Giraud, 34093 Montpellier, France.
claudine.gras@iurc.montp.inserm.fr

[2] Registre des Tumeurs de l'Hérault, bâtiment recherche, rue des Apothicaires B.P. 4111, 34091 Montpellier Cedex 5.

1 Introduction

Estimates of disease-specific incidence, prevalence and mortality specified by age are important information factors for estimating the burden of disease. Publications on prevalence estimates from the population-based registry generally consider all people with a past diagnosis of cancer as prevalent cases without taking into account the possibility of getting better. But today, many cancer patients are actually cured of the disease. It is therefore important to take recovery into account in the estimates of prevalence.

In this work three methods of estimating age-specific prevalence are presented, two of which allow us to estimate age-specific prevalence of non recovery. On the one hand, thanks to a four-state stochastic model and Poisson process, expressions of age-specific prevalence and non recovery prevalence can be built. From these expressions and using an actuarial method of estimation, this leads to the Transition Rate Method (TRM) (the first approach). Moreover, assuming the rare disease and using a parametric method of estimation, this leads to the Parametric Model (PM) (the second approach). On the other hand, the Counting Method (CM) is presented. Contrary to the other method, transition rate estimates are not required. The Counting Method counts all subjects who are known to have survived for a certain calendar time t and adds an estimate of the number of survivors among those who were alive or lost from follow-up before t.

In section 1, the concept of recovery is defined. The definitions of age-specific prevalence and non recovery prevalence are also outlined. One approach is to use data from disease registries to estimate various intensity functions. A second approach [CD97] is to use data from disease registries to estimate age-specific incidence and relative survival. In both approaches, prevalence and non recovery prevalence are thus determined. A third approach the counting method [GKMS99] [FKMN86], estimates the number of disease survivors in the population. These methods are described in section 2. In section 3, the application of these models is illustrated using data from the **S**urveillance, **E**pidemiology, and **E**nd **R**esults [SEERD03] colorectal Cancer Registry in Connecticut. Colorectal cancer is the second most common cancer in developed countries.

2 Definitions

The notion of recovery has to be defined. The purest definition of recovery would be based on complete eradication of the disease in the individual. Unfortunately, it is not possible to determine it for disease such as cancer, indeed people who appear to be cured by clinical criteria often have recurrences. That's why the concept of recovery requires careful definition. The statistical definition of recovery rely on the excess mortality risk becoming zero [VDCSMGB98] that can be revealed by the fact that the relative survival function tends to a value greater than zero

$$\lim_{d \to \infty} S_r(d) = p, p > 0. \tag{1}$$

p is called the probability of recovery. This kind of relative survival function can be modelled by a mixture survival model [G84], [G90], [VDCSMGB98], [DCHSV99] and [MZ96].

$$S_r(d) = p + (1-p) S_r^*(d), \tag{2}$$

in which $S_r^*(t)$ is the non cure relative survival.

The population of cancer patients may be considered as being composed of two different groups that have different risk of dying from the disease. The first one represents the "non recovered" patients: they should have an excess of mortality with respect to the general population. The second one represents the "recovery" patients: they should have an excess of risk parallel to the one of general population.

The probability of being cured given the individual has survived for a time $d > 0$ is defined as follows

$$p(d) = \frac{p}{p + (1-p) S_r^*(d)}. \tag{3}$$

Moreover, it is also interesting to specify a time prior to which disease is present with certainty (the active period of treatment) and after which disease is present with some probability such as equation 3. This time is called the time to cure, noted by Tc.

Then following [MZ96], a probability of recovering from cancer given currently elapsed survival time can be attributed to each individual of the cancer registry database.

In this paper, time- and age-specific total prevalence refers to all persons in a given population diagnosed in the past with cancer and alive on a determined date. The time- and age-specific L-year partial prevalence at time t, age z, including all persons who have developed the disease in the age interval $[z, z-L)$, $\pi(z, L)$, is therefore formulated as follows

$$\pi(z, L) = \frac{P(\text{a subject at age } z \text{ is diseased, diagnosed in } [z, z-L))}{P(\text{a subject at age } z \text{ is alive at time } t)}. \quad (4)$$

The notion of time-, age-specific non recovery L-year partial prevalence, $\pi_{NR}(t, z, L)$, that is defined as, on a determined date, the proportion of people alive who have been diagnosed with the disease, have not been cured and have developed the disease in the age interval $[z, z-L)$. This prevalence at time t and age z is therefore formulated as follows

$$\pi_{NR}(z, L) = \quad (5)$$
$$\frac{P(\text{a subject at age } z \text{ is diseased and non cured, diagnosed in } [z, z-L))}{P(\text{a subject at age } z \text{ is alive at time } t)}.$$

3 Three approaches for estimating prevalences

3.1 Transition Rate Method

The Transition Rates Method allows us to estimate not only age-specific prevalence but also age-specific non recovery prevalence.

Method

Let us assume that life history of the individual can be modelled by a stochastic process with four states (Alive and disease free, Alive with the disease and non recovery, Alive considered as cured, Dead). Let us consider the compartment model of Figure 1 with two life states. Denote the healthy state by H, the disease by I, the cure by C and death by D. Assume that the disease is reversible, i.e. that each person who has cancer can recover from the disease.

A subject, at calendar time s, in the healthy state H, may transit to the disease I with intensity $\alpha(x)$ which depends age x. Alternatively, the individual may die directly from state H with intensity $\mu(x)$. A subject in the disease I may transit to state C with the intensity $\lambda(x, d)$ which depends on age x and duration of the disease d. A person in state I is at risk of death with intensity $\nu(x,d)$ which depends on duration of the disease d as well as calendar age x (Figure 1). These intensities allow us to establish the age-specific non recovery prevalence of a chronic disease. In order to obtain the expression of the age-specific prevalence, the probabilities of being in the various states of the process are required. Following [B86] and [K91], these numbers are obtained.

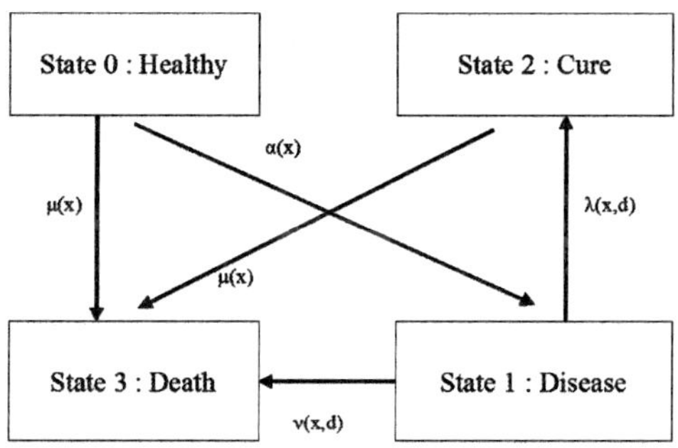

Fig. 1. Four-state stochastic model with state 0 : Alive and disease free, state 1: Alive with the disease and non recovery, state 2 : Alive considered as cured and state 3 : Dead

The probability of being alive with disease (i.e. in state I) at age z is expected by

$$P_I(z,L) = P(\text{a subject at age } z \text{ is diseased and non cured, diagnosed in } [z, z-L)),$$
$$= \int_{z-L}^{z} \underbrace{\exp\left\{-\int_0^y (\mu+\alpha)(u)\,du\right\}}_{(i)} \underbrace{\alpha(y)}_{(ii)} \underbrace{\exp\left\{-\int_y^z (\nu+\lambda)(u, u-y)\,du\right\}}_{(iii)} dy. \tag{6}$$

The justification for equation 6 is as follows.

(i) represents the probability of surviving disease-free up to age y,
(ii) represents the conditional "probability" of disease onset at age y,
(iii) represents is the conditional probability of surviving and not being cured to age z given that the individual is diagnosed with disease at age y.

The probability of being alive without disease (i.e. in state H) at age z is expected by

$$P_H(z) = P(\text{a subject at age } z \text{ is alive}), \\ = \exp\left\{-\int_0^y \mu(u)\,du\right\}\exp\left\{-\int_0^y \alpha(u)\,du\right\}. \tag{7}$$

and the probability of being alive in the cure state at age z is expected by

$$\begin{aligned} P_C(z) &= P(\text{a subject at age } z \text{ is cured, in state } C), \\ &= \int_0^z \int_0^w \left[\exp\left\{-\int_0^y (\mu(u)+\alpha(u))\,du\right\}\alpha(y)\right. \\ &\times \exp\left\{-\int_y^w (\nu+\lambda)(u,u-y)\,du\right\}\lambda(w,w-y)\exp\left\{-\int_w^z \mu(u)\,du\right\}dydw. \end{aligned} \tag{8}$$

Then, at time t, the probability that an individual is alive at age z is given by

$$P(\text{subject alive at age } z) = P_I(z) + P_C(z) + P_H(z). \tag{9}$$

So, thanks to the definition in Section 2, the non recovery L-year partial prevalence of the disease, $\pi(z,L)$, can be formulated as

$$\pi_{NR}(z,L) = \tag{10}$$

$$\frac{\int_{z-L}^z \exp\left\{-\int_0^y (\mu+\alpha)(u)\,du\right\}\alpha(y)\exp\left\{-\int_y^z (\nu+\lambda)(u,u-y)\,du\right\}dy}{P_I(z)+P_C(z)+P_H(z)}.$$

It should be noted that assuming

$$\lambda(x,d) = 0 \tag{11}$$

.i.e there is no transition from state I to state C, we therefore find the illness-death model [K91]. This model does not admit the possibility of recovery but allows us to estimate age-specific L-year partial prevalence as follows

$$\pi(z,L) = \frac{\int_{z-L}^z \exp\left\{-\int_0^y (\mu+\alpha)(u)\,du\right\}\alpha(y)\exp\left\{-\int_y^z \nu(u,u-y)\,du\right\}dy}{P_I(z)+P_C(z)+P_H(z)}. \tag{12}$$

At this point, it is necessary to note that equations 3.1 and 12 give the more general expression prevalences.

In the following, the probability that an individual is alive at age z is assumed to be approximated by $S^*(z)$, the overall survival of the population at age z provided by vital statistics.

$$P(\text{subject alive at age } z) = P_I(z) + P_C(z) + P_H(z), \\ = S^*(z). \tag{13}$$

Model specifications

In order to use equations provided by the section 3.1.1, a number of parameters and quantities must be specified. As regards the model specification, following [GKMS99], a semi-parametric model is used.

Mortality rates

Age-specific mortality rates for all causes of death, written as $\mu^*(x)$, is used to estimate $S^*(x)$. They are provided by vital statistics and assumed to be without error. $S^*(x)$ is computed as

$$S^*(x) = \exp\left\{-\mu_j^*(x - g_{j-1}) - \sum_{k=1}^{j-1} \mu_g^*(g_k - g_{k-1})\right\}. \tag{14}$$

The probability of not dying of other causes than cancer to age x is computed exactly as for $S^*(x)$; however, instead of using the overall mortality rates $\mu^*(x)$, we set $\mu(x)$ equal to the mortality rate from all causes of death except the cause which interests us.

Incidence rates

A finite partition of the age axis is constructed, $0 < g_1 < ... < g_J$ with $g_J > y_i$ for all $i = 1, 2, ..., n$. Thus, we obtain the J intervals $(0, g_1]$, $(g_1, g_2]$,..., $(g_{J-1}, g_J]$. We thus assume that the hazard is equal to α_j for the jth interval, $j = 1, 2, ..., J$, leading to

$$\widehat{\alpha}_j = \frac{\sum_{s=s_o}^{t} I_{is}}{\sum_{s=s_o}^{t} N(i,s)}. \tag{15}$$

I_{is} cases are diagnosed in the disease state at the age interval i the year s and $N(i,s)$ is the corresponding number at risk of an incident cancer in the population.

Transition rates from the disease

For the survival function for the "illness" population, we construct a finite partition of the age incidence axis $0 < g_1 < ... < g_J$ with $g_J > y_i$ for all $i = 1, 2, ..., n$ and a finite partition of the duration in the disease axis $0 < r_1 < ... < r_K$ with $r_K > d_i$ for all $i = 1, 2, ..., n$. $J \times K$ intervals $(0, g_1] \times (0, r_1]$, $(g_1, g_2] \times (0, r_1]$,..., $(g_{J-1}, g_J] \times (r_{K-1}, r_K]$ are therefore obtained. We thus assume that the hazards are equal to $\widehat{\lambda}_j^i$ leading to

$$\widehat{\lambda}_j^i = -\ln\left(1 - \frac{C_{ij}}{R_{ij} - 0.5L_{ij}}\right), \tag{16}$$

in which C_{ij} cases transit from the disease to cure in the jth year following cancer diagnosis among those who were in age interval i at time of diagnosis and in which R_{ij} and L_{ij} are respectively the corresponding number at risk of transiting to death at the beginning of interval j and the number of those who were lost from follow-up in this interval.

The number of cases that transit from the disease to cure is determined using the definition of recovery described in section 2. The cure proportion is estimated according to the age at diagnosis. Then following [MZ96] a probability of recovering from cancer given currently elapsed survival time is attributed to each individual of the cancer registry database. A binary variable of recovery is built using a Bernoulli framework, a time of recovery is therefore generated for each recovered individual. This technic allows us to simulate the event cure, because it is impossible to diagnose recovery by clinical exams or by the information available in registries.

Likewise $\nu(.,.)$ is assumed to be piecewise constant in age at diagnosis and in duration of the disease. ν_j^i is the hazard of death in year j following diagnosis of cancer for individuals diagnosed at any age in the age interval i.

Age-specific non recovery prevalence estimates

The estimated age-specific non recovery partial prevalence $\widehat{\pi}_{NR}(z,L)$ is obtained from equation 3.1 using $\widehat{\alpha}(x)$, $\widehat{\lambda}(x,d)$ and $\widehat{\nu}(x,d)$ described in section 4.1 and using $\mu(x)$ and $\mu^*(x)$ assumed as known without error.

$$\widehat{\pi}_{NR}(z,L) = \sum_{i=Q_1}^{Q_2} \widehat{\pi}_{NR}^i(z), \tag{17}$$

in which $z = g_{Q_2+1}$, $z - L = gQ_1$ and $\widehat{\pi}^i$ is the prevalence of people who were diagnosed in state I in the interval age $[g_i, g_{i+1})$ and who have not been cured,

$$\widehat{\pi}_{NR}^i(z) = \tag{18}$$
$$\frac{\int_{[g_i,g_{i+1})} \exp\left\{-\int_0^y (\widehat{\mu}+\widehat{\alpha})(u)\,du\right\} \widehat{\alpha}(y) \exp\left\{-\int_y^z \left(\widehat{\nu}+\widehat{\lambda}\right)(u,u-y)\,du\right\} dy}{\widehat{S}^*(z)}.$$

Analytical expressions of the integral over $[g_i, g_{i+1})$ are provided in the appendix A.

3.2 A parametric model [CD97]

The model developed by [CD97] allows us to estimate age-specific prevalence and age-specific non recovery prevalence using a parametric model.

Method

Let $\mu^*(x)$ represent the general mortality rates at age x. Let $\alpha(x)$ be the incidence rate at age x. Let $\nu(x, x-y)$also be the death rates at age x for people who had a cancer diagnosed at age y. Let $S_r(x, x-y)$ be the relative survival

$$S_r(x, x-y) = \exp\left\{-\int_y^x \nu(u, u-y)\, du - \int_0^x \mu^*(u)\, du\right\}. \tag{19}$$

Let $1 - k(d)$ be the probability of being cured given that an individual has survived for a time d in the disease. The age-specific prevalence provided by [CD97] is therefore expressed as

$$\pi(z, L) = \int_{z-L}^{z} \alpha(x)\, k(d)\, S_r(x, d)\, dx \tag{20}$$

in which $k(d)$ specifies the hypotheses made on disease reversibility. If $k(d) = 0$, $\pi(z, L)$ is the partial prevalence of the disease, if $k(d) > 0$, $\pi(z, L)$ is the partial non recovery prevalence of the disease.

The equation 20 has to be compared to expressions built by the Transition Rate Method 3.1 and 12. Indeed, assuming that

- the disease is rare i.e. the incidence rate is low

$$\alpha \ll 1 \Longrightarrow e^{-\int_0^u \alpha(y) dy} \simeq 1, \tag{21}$$

- the mortality rate of non diseased people $\mu(x)$ is approximated by the mortality rate of the general population $\mu^*(x)$ i.e. $\mu(x) \equiv \mu^*(x)$,
- the probability that an individual is alive at age $z$is approximated by $S^*(z)$

$$\begin{aligned} P(\text{subject alive at age } z) &= P_I(z) + P_C(z) + P_H(z), \\ &= S^*(z). \end{aligned} \tag{22}$$

age-specific L-year partial prevalence (cf equation 3.1) can be reformulated as follows

$$\pi_{NR}(z, L) = \frac{\int_{z-L}^{z} \exp\left\{-\int_0^y \mu^*(y)\, dy\right\} \alpha(y) \exp\left\{-\int_y^z (\nu + \lambda)(u, u-y)\, du\right\} dy}{S^*(z)}. \tag{23}$$

Leading to

$$\pi_{NR}(z, L) = \int_{z-L}^{z} \alpha(y) \exp\left\{-\int_y^x \lambda(u, u-y)\, du\right\} S_r(x-y)\, dy, \tag{24}$$

in which $\exp\left\{-\int_y^x \lambda(u, u-y)\, du\right\}$ is the probability of surviving from the event "recovery", then it corresponds to the probability of not being cured given that an individual has survived for a time d in the disease $k(d)$.

Model specifications

In order to use equations provided by the previous section, incidence rate and relative survival must be specified.

The incidence rate is modelled by

$$\alpha(x) = ax^b, \tag{25}$$

this exponential shape has been validated for a quite general class of cancers.

The relative survival function is parametrized by a mixture model [DCHSV99] as follows

$$S_r(x, x-y) = p + (1-p)\exp(-\lambda(x-y)), \tag{26}$$

p represents the cure proportion and an exponential model is used for the non cure relative survival S_r^*.

If $k(d) = 1$, both cured and non cured cases contribute to estimate prevalence

$$\pi(z, L) = \int_{z-L}^{z} ax^b \{p + (1-p)\exp(-\lambda d)\}\, dx. \tag{27}$$

If $k(d) = \begin{cases} 1 & for\ d \le Tc \\ \frac{(1-p)\exp(-\lambda d)}{S(x,d)} & for\ d > Tc \end{cases}$, prior to the survival time Tc, disease is present with certainty so that the probability of being prevalent case is one, and after Tc the probability of being prevalent cases depends on the cure proportion and on the non cure survival. The non recovery prevalence could be expressed as follows

$$\pi_{NR}(z, L) = p\int_{\min(z-L, z-Tc)}^{z} ax^b dx + (1-p)\int_{z-L}^{z} ax^b \exp(-\lambda d)\, dx. \tag{28}$$

These prevalences can be computed numerically and variance estimates are obtained by the Delta method [DFSW88].

3.3 Counting Method estimates

The Counting method was developed by [GKMS99]. The notations used are the following

- let X_i be the exact age at cancer incidence for the ith case of a cancer registry,
- let T_i be the exact calendar time of cancer incidence for that member,
- let Y_i be the exact time of death,
- let U_i be the exact time of loss from follow-up.
- let $S(d, x, t)$ be the probability that a person who develops cancer at age x and date t will survive beyond duration d after cancer incidence. $\widehat{S}(d, x, t)$ is an estimates of $S(d, x, t)$ obtained by actuarial methods

The probability that an individual who is alive at calendar time t and is in the age group $[z, z+1)$ and has disease incidence in the age interval $[c_1, c_2)$ with $c_2 \leq z$ is estimated by

$$\begin{aligned}&\widehat{\pi}(z, z-L, t)\\&= \frac{1}{N(z,t)} \left[\sum I(z - L \leq X_i < z, Y_i \geq t, U_i \geq t, z \leq X_i + s - T_i < z+1)\right.\\&\left.+ \sum I(z - L \leq X_i < z, Y_i > U_i, U_i < t, z \leq X_i + t - T_i < z+1) \frac{\widehat{S}(t-T_i, X_i, T_i)}{\widehat{S}(U_i - T_i, X_i, T_i)}\right]\end{aligned} \quad (29)$$

the summations are overall disease cases in the registry and $I(.)$ is an indicator function equaling one when the argument is true and zero otherwise.

The justification for equation 29 is as follows.

(i) the first summation represents cancer cases known to have survived up to age z,

(ii) the second summation represents cancer cases who were lost from follow-up before age z.

This method was implemented by the SEER program [SEERD03], then in order to obtain the estimates of partial prevalence, the SEER*Stat Software is used.

For the estimate of the variance, they used a method based on a Poisson approximation proposed by [CGF02].

4 Results

Previous methods were applied to colorectal cancer. Data were collected by the SEER Program [SEERD03] during the 1990-1999 period. The material consists of 23334 cases that were followed until the end of 1999. The follow-up cut-off date for this analysis was 31 December 1999. Survival time was defined as the time from diagnosis to death if prior to 1999 or to the end of 1999 . Overall status was defined as unity if the subject was dead before the end of 1999, and zero otherwise. The population used for calculating prevalence rates was provided by SEER program [SEERD03], as well as the U.S mortality rates for the population [SEERM03] and the size of population [SEERP03]. Mortality rates are published in 5-year age intervals.

Table 1 shows the age-specific incidence of colorectal cancer in Connecticut between 1990 and 1999 used for estimating age-specific prevalences by the Transition Rate Method.

For the parametric model, estimates of the parameters of incidence are b equals 5 and a equal 1.51^{-12}. The estimates of the proportion of cured cases and the parameter of survival of uncured cases were performed using S-plus

Table 1. Table 1 : Colorectal cancer observed incidence rates by age in Connecticut between 1990 and 1999. Rates are per 100,000.

Age range	Rate (SE)
40-44	13.7 (0.7)
45-49	28.9 (1.1)
50-54	53.6 (1.6)
55-59	97.1 (2.5)
60-64	163.2 (3.5)
65-69	247.8 (4.4)
70-74	327.9 (5.3)

function (nlminb in order to minimize the negative of the log of the likelihood function). The estimate of proportion of cure is about 63% corresponding to the standard error of 0.028 and the parameter of the exponential survival is about 0.56 year^{-1} corresponding to the standard error of 0.026.

The age-specific prevalence for people diagnosed between 1990 and 1999 are presented in table 2. The Transition Rates Method estimate of age-specific 10-year partial prevalence of colorectal cancer for people in the 55-59 age range is 466.6 per 100,000. The estimated standard error, 11.1, corresponds to a coefficient of variation of 2.3% [D93]. The parametric method and the counting method yield prevalence estimates in people aged 55-59 of 427.8 and 419.2, respectively, with corresponding standard errors of 11.8 and 15.5 and corresponding coefficients of variation of 2,7% and 3.7%.

Table 2. Colorectal cancer estimates 10-year partial prevalence by age in Connecticut in 1999 using the Transition Rate Method, the Parametric Method and the Counting Method. Rates are per 100,000.

Age range	TRM (SE)	PM (SE)	CM (SE)
40-44	61.1 (3.2)	82.9 (1.9)	52.7 (4.3)
45-49	127.7 (4.8)	152.3 (3.8)	121.1 (6.7)
50-54	253.2 (7.4)	262.1 (6.9)	212.7 (9.7)
55-59	466.6 (11.1)	427.8 (11.8)	419.2 (15.5)
60-64	790.1 (15.1)	668.9 (19.1)	734.1 (23.6)
65-69	1223.6 (19.4)	1008.2 (29.9)	1192.9 (31.8)
70-74	1664.6 (22.5)	1473.6 (45.3)	1673.9 (38.24)

As regards the age-specific prevalence of non-recovery, the time required to be cured is assumed to be 5 years. Table 3 presents the estimates of non recovery 10-year partial prevalence of colorectal cancer with the corresponding standard error. The Transition Rate and parametric methods yield estimates of non recovery prevalence in the age range 55-59 of 225.9 and 188.1, respectively, with corresponding standard errors 6.7 and 5.9. The coefficient of variation are, respectively 2.9% and 3.1%.

Table 3. Colorectal cancer estimates non recovery 10-year prevalence by age in Connecticut in 1999 using the Transition Rate Method and the Parametric Method. Rates are per 100,000.

Age range	Estimated non cure prevalence	
	TRM	PM
40-44	29.4 (2)	45.9 (1.2)
45-49	63 (2.7)	77.7 (2.1)
50-54	125.2 (4.6)	123.9 (3.7)
55-59	225.9 (6.7)	188.1 (5.9)
60-64	375.9 (9.4)	274.7 (9.4)
65-69	571.7 (11.4)	388.4 (14.1)
70-74	761.7 (12.7)	534.3 (20.6)

Figure 2 shows the estimates of prevalence and non recovery prevalence using all previous methods. For the under sixties, the estimates of 10-year partial prevalence are similar and for the over sixties, the Parametric Method provides estimates lower than using the two other methods. As regards the non recovery 10-year partial prevalence estimates, the same trend is noted, the Transition rate method provides estimates higher than using the parametric method.

Using the TRM and the parametric method, the non recovery 10-year prevalence represents around 45% of the prevalence (Figure 3). This point is consistent with the estimate of the proportion of cure (63%).

5 Discussion

This paper presents three procedures for estimating age-specific prevalence : the Counting Method [GKMS99], the Transition Rate Method [GKMS99] [FKMN86] and a parametric method [CD97]. We have also developed a method to estimate age-specific non recovery prevalence using a transition rate model. The variances of these prevalences are estimated using the Delta-Method [DFSW88].

The Counting Method was developed by [GKMS99] [FKMN86] and is used to estimate prevalence based on tumor registry data. The estimates of the standard error for the Counting Method are based on the Poisson method [CGF02]. This method is implemented by the SEER*Stat software developed by the Statistical Research and Applications Branch.

Generally, publications about prevalence assume that the disease is irreversible, no return to the healthy state is allowed [MCFM00], [VC89]. But

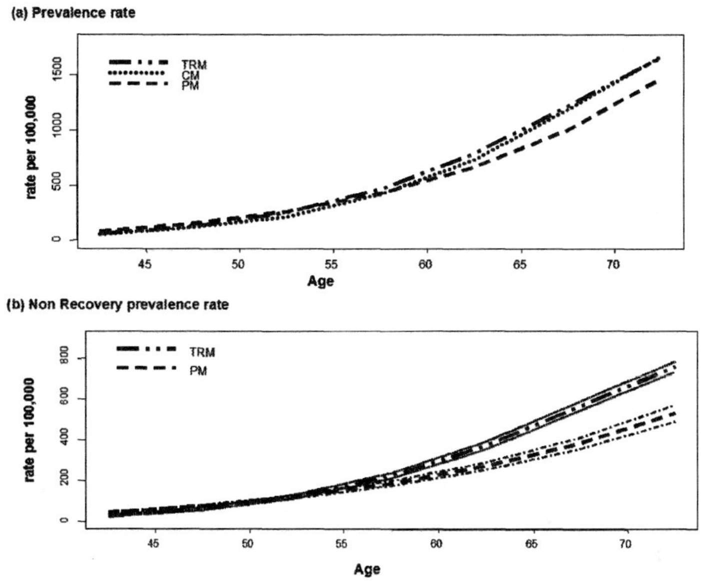

Fig. 2. (a) Age-specific prevalence of colorectal cancer in 1999 estimated by the Transition Rate Method (TRM), the Counting Method (CM) and the Parametric Method (PM).(b)Age-specific prevalence of non recovery of colorectal cancer in 1999 estimated by the Transition Rate Method (TRM) and the Parametric Method PM).

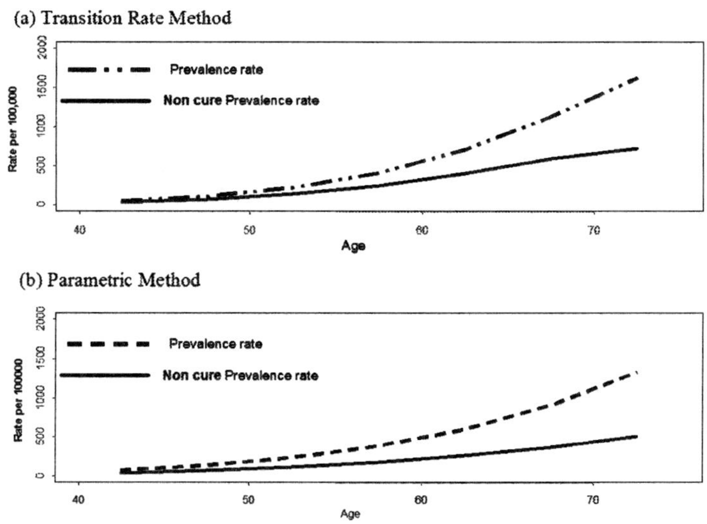

Fig. 3. (a) Estimates of age-specific prevalence and prevalence of non recovery using the Transition Rate Method (TRM). (b) Estimates of age-specific prevalence and prevalence of non recovery using the Parametric Method (PM).

today, thanks to improvements of treatment, the word "cure" can be used for certain cancers. So, in order to better understand the burden of cancer on the population, it is important to estimate non recovery prevalence. Indeed, a subject considered as "cured" requires fewer health resources than a subject who is not cured. [CD97] provides models of prevalence with the hypothesis of disease reversibility. Both reports estimate cancer prevalence using mixture models for cancer survival. They model the relative survival function by a mixture model without covariates and an exponential distribution for non recovery survival. It is the parametric method which is presented here. This method requires a choice of model and programming with statistical software.

In our report, in order to estimate prevalence and non recovery prevalence, we proposed a transition rate estimate method. The added plus represents estimates of variance using the Delta-Method [DFSW88]. As regards rates estimates, these are estimated according to actuarial intervals. In order to use this method, we have developed a software called SSPIR [GDT04] that implements the Transition Rate Method. The Counting Method and the Transition Rate Method are therefore easy to use.

Table 4. Comparison of the three approaches.

	TRM	PM	CM
Type of database	Cancer Registry + Vital statistics	Cancer Registry + Vital statistics	Cancer Registry
Cancer Registry	Exhaustive	Exhaustive	Exhaustive
Closed Population	Yes	Yes	No
Estimation Method	Actuarial	Parametric	Non Parametric
Non recovery prevalence	Yes	Yes	No
Estimation Model			
Incidence Rate	Observed incidence	Exponential shape	No
No diseased mortality rate	All other mortality	No	No
Diseased mortality rate	Vital tables method	Relative Survival	Survival of losts
Cure rate	Vital tables method	Mixture model	No
Software	SSPIR	No	SEER*Stat

As regards the estimates of age-specific prevalence, we note that the estimates of using the three methods are close. But the variances of the Transition Rate Method are smaller than the variance of the Counting Method and the Parametric Method estimates. Moreover, the estimates of the Transition Rate Method are slightly higher than the estimates of the Parametric Method. The estimates of age-specific non recovery prevalence are slightly higher using the Transition Rate Method compared to using the parametric method. We can also note that the coefficients of variation are equivalent to the one of the parametric method. This parametric model seems to be well adapted to colorectal cancer but it may not be the case for other diseases.

We described three methods of estimating prevalence, two of which are non parametric methods. These are both attractive since they are easy to use and robust. The parametric method requires, when it is possible, to find the model which is best adapted to data. This point directly raises the well known problem of choice between parametric and non parametric methods.

APPENDIX A : Expression of non cure prevalence

$$\widehat{\pi}_{NR}^{i}(z)=\frac{\int_{[g_i,g_{i+1})}\widehat{S}_H(y)\,\widehat{S}_o(y)\,\widehat{\alpha}(y)\,\widehat{S}_D(y,z-y)\,\widehat{S}_R(y,z-y)\,dy}{\widehat{S}^*(z)}. \tag{30}$$

Assuming that $z-y\in[e_h,e_{h+1}[$

1. If $\lambda_h^i+\nu_h^i-\alpha^i-\mu^i\neq 0$:

$$\begin{aligned}\widehat{\pi}_{NR}^{i}(z)&=\frac{1}{\widehat{S}^*(z)}\exp\left\{-\int_0^{g_i}\alpha(u)\,du\right\}\exp\left\{-\int_0^{g_i}\mu(u)\,du\right\}\\&\times\alpha^i\exp\left\{-\int_0^{e_h}\lambda^i(u)+\nu^i(u)\,du\right\}\\&\times\frac{\exp\left\{-(z-g_{i+1}-e_h)\left(\lambda_h^i+\nu_h^i\right)\right\}}{\lambda_h^i+\nu_h^i-\alpha^i-\mu^i}\\&\times\exp\left\{-(g_{i+1}-g_i)\left(\alpha^i+\mu^i\right)\right\}-\exp\left\{-(g_{i+1}-g_i)\left(\lambda_h^i+\nu_h^i\right)\right\}.\end{aligned}$$

2. If $\lambda_h^i+\nu_h^i-\alpha^i-\mu^i=0$:

$$\begin{aligned}\widehat{\pi}_{NR}^{i}(z)&=\frac{1}{\widehat{S}^*(z)}\exp\left\{-\int_0^{g_i}\alpha(u)\,du\right\}\exp\left\{-\int_0^{g_i}\mu(u)\,du\right\}\\&\times\alpha^i\exp\left\{-\int_0^{e_h}\lambda^i(u)+\nu^i(u)\,du\right\}\exp\left\{g_i\left(\alpha^i+\mu^i\right)\right\}\\&\times\exp\left\{-(z-e_h)\left(\lambda_h^i+\nu_h^i\right)\right\}(g_{i+1}-g_i).\end{aligned}$$

References

[B86] Brillinger, D.R.: The natural variability of vital rates and associated statistics (with discussion). Biometrics. **42**, 693–734 (1986)

[CD97] Capocaccia, R., De Angelis, R.: Estimating the completeness of prevalence based on cancer registry data. Statistics in Medicine. **16**, 425–440 (1997)

[CGF02] Clegg, L.X., Gail, M.H., Feuer, E.J.: Estimating the variance of disease-prevalence. Estimates from population-based registries. Biometrics. **58**, 684–688 (2002)

[DCHSV99] De Angelis, R., Capocaccia, R., Hakulinen, T., Soderman, B., Verdecchia, A.: Mixture models for cancer survival analysis : application population-based data with covariates. Statistics in Medicine. **18**, 441–454 (1999)

[DFSW88] DeGroot, M.H., Ferber, R., Frankel, M.R., Seneta, E., Watson, G.S.: Encyclopedia of statistical sciences. Wiley, New York (1988)

[D93] Dodge, Y.: Statistique Dictionnnaire encyclopédique. Dunod, Paris (1993)

[FKMN86] Feldman, A.R., Kessler, L., Myers, M.H., Naughton, M.D.: The prevalence of cancer. Estimates based on the Connecticut Tumor Registry. New England Journal of Medicine. **315**, 1394–1397 (1986)

[GKMS99] Gail, M.H., Kessler, L., Midthune, D., Scoppa, S.: Two approaches for estimating disease prevalence from population-based registries of incidence and total mortality. Biometrics. **55**, 1137–1144 (1999)

[G90] Gamel, J.W.: Proportion cured and mean long survival time as functions of tumor size. Statistics in Medicine. **9**, 999–1006 (1990)

[G84] Goldman, A.I.: Survivorship analysis when cure is a possibility : a Monte carlo study. Statistics in Medicine. **3**, 153–163 (1984)

[GDT04] Gras, C., Daures, J.P., Tretarre, B.: A computer software to estimate chronic disease age- and stage-specific incidence and prevalence using cancer registries data. In submission.

[K91] Keiding N.: Age-specific incidence and prevalence: A statistical perspective. Journal of the Royal Statistical Society, Series A. **154**, 371–412 (1991)

[MZ96] Maller R., Zhou, X.: Survival analysis with long-term survivors. Wiley, New York (1996)

[MCFM00] Merrill, R.M., Capocaccia, R., Feuer, E.J., Mariotto, A.: Cancer Prevalence Estimates Based on Tumor Registry Data in the SEER Program. International Journal of Epidemiology. **29**, 197-207 (2000)

[SEERD03] Surveillance, Epidemiology, and End Results (SEER) Program (www.seer.cancer.gov) SEER*Stat Database: Incidence - SEER 9 Regs, Nov 2002 Sub (1973-2000), National Cancer Institute, DC-CPS, Surveillance Research Program, Cancer Statistics Branch, released April 2003, based on the November 2002 submission.

[SEERM03] urveillance, Epidemiology, and End Results (SEER) Program (www.seer.cancer.gov) SEER*Stat Database: Mortality - All COD, Public-Use With State, Total U.S. (1969-2000), National Cancer Institute, DCCPS, Surveillance Research Program, Cancer Statistics Branch, released April 2003. Underlying mortality data provided by NCHS (www.cdc.gov/nchs).

[SEERP03] Surveillance, Epidemiology, and End Results (SEER) Program (www.seer.cancer.gov) SEER*Stat Database: Populations - Total U.S. (1969-2000), National Cancer Institute, DCCPS, Surveillance Research Program, Cancer Statistics Branch, released April 2003.

[VC89] Verdecchia, A., Capocaccia, R.: A method for the estimation of chronic disease morbidity and trends from mortality data. Statistics in Medicine. **8**, 201–216 (1989)

[VDCSMGB98] Verdecchia, A., De Angelis, R., Capocaccia, R., Sant, M., Micheli, A., Gatta, G., Berrino, F.: The cure for colon cancer: results from the EUROCARE study. International Journal of Cancer. **77**, 322–329 (1998)

On statistics of inverse gamma process as a model of wear

B.P. Harlamov[1]

Institute of Problems of Mechanical Engineering, Russian Academy of Sciences, Saint-Petersburg, harlamov@random.ipme.ru

Summary. Some aspects of statistics of inverse gamma process as a model of wear are considered. Formulae for finite-dimensional distribution densities of the process are given. Partial derivatives with respect to parameters of one-dimensional densities of both the direct, and inverse processes with independent positive increments are derived. Methods for estimation of parameters of the inverse gamma process are investigated.

1 Introduction

In the present work we continue to investigate the wear process which has been considered in papers [har02, har04a, har04b, har04d]. In these works there were proposed inverse processes with independent positive increments as models of wear processes. For brevity we will call such a process as I-process. A partial case of such a process is gamma process [har04b], which has some advantages comparatively with other I-processes. Varying a wear diagnostic parameter under law of I-process seams to be completely justified, because the sense of this assumption reduces to the condition for times of wearing of non-overlapping portions of material to be independent random values. Such a model combines two necessary properties of a practical model: "good" features of the process realizations (continuity, monotonicity), and sufficiently simple mathematical apparat. Advantages of the model explicitly appears for non-standard registration of wear data.

In this work two ways for gathering information about I-processes are investigated. The first one is the classical (direct) way, when an observer can measure (random) portions of material which had been worn during determinate time intervals. And the second one is the inverse way, when he finds (random) time intervals having been spent by determinate portions of worn material. Both ways have technical and organizational base and their sphere of application.

In the recent time there arises possibility to record practically the whole continuous wear curve, for example, the acoustical method (see [fad04]). In this case the statistical analysis option depends only on computation possibility of the observer.

For the first variant of data gathering in works [har04b, har04c] there were derived formulae for both one-dimensional, and multi-dimensional distribution densities, which can be used for determination of maximum likelihood estimates and criterions. In the present work we revise and supplement results of these works.

In the second variant we propose for the time increment distributions to be known to within finite number of parameters, and for the increments themselves to be mutually independent. Thus one can use methods of classical mathematical statistics.

Under recording a continuous trajectory of wear it seems to be reasonable to quantize date in the second manner. Under this choice the problem arises how one should split all the interval of wear for obtaining corresponding family of wearing time increments which possesses optimal statistical properties. In the work we discuss the optimal choice of partition fineness with regard to variance of estimate and computation expenditure.

2 Inverse process with independent positive increments

Initial definitions

Let Φ be set of all continuous non-decreasing functions $\xi : R_+ \mapsto R_+$, such that $\xi(t) \to \infty$ $(t \to \infty)$; (P_x) $(x \geq 0)$ be consistent semi-Markov family of probability measures on $(\Phi, \mathcal{F})$, where $\mathcal{F}$ is Borel sigma-algebra of subsets of the set Φ, generated by topology of homogeneous convergency on all bounded intervals. Let us denote $\tau_x(\xi)$ the first exit time of the process from the interval $[0, x)$ $(x \geq 0)$.

A random process ξ, determined by the family of measures (P_x), is said to be an inverse process with independent positive increments (I-process for brevity), if the random function $\tau_x(\xi)$ as a function of x is a (proper) process with independent positive increments. The natural characteristic of the I-process is its the first exit time distribution which is assumed to be absolutely continuous: $P_x(\tau_y \in dt) = f_y(t|\, x)\, dt$, where $x < y$ and for $x_0 < x_1 < x_2$ the following equation holds

$$f_{x_1+x_2}(t|\, x_0) = \int_0^t f_{x_1}(s|\, x_0) f_{x_2}(t-s|\, x_1)\, ds.$$

It is well known that for Laplace image of this density Lévy formula is true (see [sko64, har01])

$$P_x(\exp(-\lambda\, \tau_y)) = \exp(-b(\lambda, x, y)) \quad (y > x > 0),$$

where

$$b(\lambda, x, y) = \lambda\, a([x, y)) + \int_{0+}^{\infty} (1 - e^{-\lambda\, u})\, n(du \times [x, y)) - \sum_{x \le x_i < y} \log P_x(e^{-\lambda\, \tau_i});$$

$a(dx)$ is a locally finite measure on the half line R_+; $n(du \times dx)$ is a locally finite measure on the quadrant $(0, \infty) \times [0, \infty)$ (it is so called Lévy measure); (x_i, τ_i) is a sequence of pairs, where (x_i) $(x_i > 0)$ is determinate sequence, describing space points of temporary stops of the process, (τ_i) is a sequence of independent positive random values, determining intervals of constancy (durations of temporary stops) at the corresponding space points of the trajectory. In the neighborhood of the line $\{0\} \times [0, \infty)$ Lévy measure can be infinite, however it satisfies the conditions

$$\int_{0+}^{1} u\, n(du \times [0, x)) < \infty, \quad n([1, \infty) \times [0, x)) < \infty.$$

We will assume that the process (τ_x) $(x \in R_+)$ is stochastically continuous, thus the third member of exponent power in Lévy formula is absent. Besides we consider the case, when measures a and n are absolutely continuous in their domains with respect to Lebesgue measures. In this case for some positive functions α and ν the following representations hold

$$a([0, x)) = \int_0^x \alpha(s)\, ds, \quad n([u_1, u_2) \times [x, y)) = \int_x^y \int_{u_1}^{u_2} \nu(u,\, s)\, du\, ds \quad (u_1 > 0).$$

Because of positiveness of increments of the process τ_x (P_0-almost sure) the following property is true: if the function $\alpha \equiv 0$ then for any x $\nu(u, x) \longrightarrow \infty$ as $u \longrightarrow 0$.

A typical I-process is not Markov. According to our terminology (see [har01]) it is monotone continuous semi-Markov process. Violation of the Markov property is connected with intervals of constancy, which either do not have fixed position in the space, or are distributed with respect to an arbitrary law (not necessary exponential). For intervals of constancy one can easier find a reasonable physical interpretation than that for point of jumps. But the main merit of I-process is its analytical form permitting simple evaluation of reliability functionals for degradation problems (see [har04d]).

Moments of the first exit time distributions

Moments of distribution of the random value τ_x can be find from Lévy formula by means of differentiating it with respect to λ as $\lambda = 0$. So we have

$$A_1(x) \equiv E_0(\tau_x) \equiv \int_0^{\infty} t\, P_0(\tau_x \in dt) = -\frac{\partial}{\partial \lambda} E_0(e^{-\lambda \tau_x})\Big|_{\lambda=0} =$$

$$= a([0,x)) + \int_{0+}^{\infty} u\, n(du \times [0,x)), \tag{1}$$

$$M_2(x) \equiv \sigma^2(x) \equiv E_0(\tau_x - A_1(x))^2 = \int_{0+}^{\infty} u^2 n(du \times [0,x)), \tag{2}$$

$$M_3(x) \equiv E_0(\tau_x - A_1(x))^3 = \int_{0+}^{\infty} u^3 n(du \times [0,x)), \tag{3}$$

$$M_4(x) \equiv E_0(\tau_x - A_1(x))^4 = \int_{0+}^{\infty} u^4 n(du \times [0,x)) + 3M_2^2(x). \tag{4}$$

It is not difficult to evaluate other moments.

Inverse gamma process

In the works [har04a, har04d] and others we gave arguments justifying I-process to be used in reliability problems. We showed examples, where I-process analytical properties are useful in optimization problems of prophylaxis and reservation. For practical aim one should consider more narrow class of the processes with a finite number of parameters. In [har04b, har04c] such processes were shown. Now we consider such a process, namely gamma process, in more details.

Inverse (homogeneous) gamma process is an I-process where the function τ_x is distributed according to gamma distribution: $P_x(\tau_y \in dt) = f_y(t|\, x)\, dt$ $(x < y)$, where $f_y(t|\, x) = f_{y-x}(t|\, 0) \equiv f_{y-x}(t)$ and

$$f_x(t) = \frac{\gamma}{\Gamma(x\delta)}(\gamma\, t)^{x\delta - 1} e^{-\gamma t} \quad (x > 0),$$

$\Gamma(\cdot)$ is the gamma function $(\Gamma(x) = \int_0^\infty t^{x-1} e^{-t}\, dt)$, $\delta > 0$ is a form parameter, and $\gamma > 0$ is a scale parameter. Evidently,

$$f_x(t) \equiv f_x(t; \gamma, \delta) = \gamma f_{\delta\, x}(\gamma\, t; 1, 1) \equiv \gamma f^0_{\delta\, x}(\gamma\, t),$$

where

$$f^0_x(t) = \frac{1}{\Gamma(x)} t^{x-1} e^{-t}.$$

The Lévy exponent of the gamma process is of the form $x\delta \ln(\gamma + \lambda)/\gamma$. The view of its Lévy measure follows from the formula

$$\ln \frac{\gamma + \lambda}{\gamma} = \int_0^\infty (1 - e^{-\lambda u}) \frac{e^{-\gamma\, u}}{u}\, du \tag{5}$$

(see [har01]). Thus the density of Lévy measure has representation:

$$\nu(u, x) = \delta e^{-\gamma u}/u.$$

In applications they consider more general class of inverse gamma processes when parameters γ and δ depend on position x. This class of I-processes seems to be naturally called as class of generalized inverse gamma processes in spite of the distribution of the value τ_x in this case is not gamma distribution. The generating function of the first exit time of such a more general process has the form

$$E_0(e^{-\lambda \tau_x}) = \exp\left(-\int_0^x \delta(y)\,\ln\frac{\gamma(y)+\lambda}{\gamma(y)}\,dy\right),$$

which implies formulae for moments:

$$A_1(x) = \int_0^x \frac{\delta(y)}{\gamma(y)}\,dy, \qquad M_2(x) \equiv \sigma^2(x) = \int_0^x \frac{\delta(y)}{\gamma^2(y)}\,dy,$$

$$M_3(x) = \int_0^x \frac{2\delta(y)}{\gamma^3(y)}\,dy, \qquad M_4(x) = \int_0^x \frac{6\delta(y)}{\gamma^4(y)}\,dy + 3M_2^2(x).$$

The main advantage of the inverse gamma process comparatively with other I-processes are both its simplicity, and flexibility due to its two parameters. For a fixed wear level the gamma distribution is used for description of a random failure time in work [gk66]. In work [bn01] the direct (proper) gamma process had been considered as a model of wear.

In work [har04b] we show how a continuous strictly increasing function can be approximated by I-processes (partially, gamma processes). This property of I-processes permits to deny the determinate component in Lévy representation in any case when such a component does not have explicit interpretation. In turn the denial of the determinate component makes more simple using of absolute continuous property for I-processes (see [sko64]) and analysis of its finite-dimensional densities.

One-dimensional distribution

Let us consider I-process without determinate component. For such a process

$$b(\lambda, x, y) = \int_x^y \int_0^\infty (1 - e^{-\lambda u})\,\nu(u, s)\,du\,ds. \tag{6}$$

Lemma 1. *If condition (6) is fulfilled, and functions $f_y(t|\,x)$ and $\nu(t,x)$ are continuous in their arguments, and $\int_0^\infty u\,\nu(u,x)\,du < \infty$, then for any $t > 0$*

$$\frac{1}{y-x} f_y(t|\,x) \to \nu(t,x) \quad (y \downarrow x),$$

besides for any continuous bounded function φ

$$\frac{1}{y-x}\int_0^\infty \varphi(t)\,t\,f_y(t|\,x)\,dx \to \int_0^\infty \varphi(t)\,t\,\nu(t,x)\,dx \qquad (y \downarrow x)$$

Proof. We have

$$P_x(e^{-\lambda\tau_y}) = \exp(-\int_x^y \int_0^\infty (1-e^{-\lambda u})\,\nu(u,s)\,du\,ds).$$

From here it follows that for any $\lambda_0 > 0$ uniformly for all $\lambda < \lambda_0$

$$L_c(\lambda,x) \equiv \frac{1}{c}\int_0^\infty (1-e^{-\lambda u})\,f_{x+c}(u|\,x)\,du \equiv \frac{1}{c}\,(1-E_x(e^{-\lambda\tau_{x+c}})) =$$

$$= \frac{1}{c}\left(1-\exp(-\int_x^{x+c}\int_0^\infty (1-e^{-\lambda u})\,\nu(u,s)\,du\,ds)\right) \to$$

$$\to \int_0^\infty (1-e^{-\lambda u})\,\nu(u,x)\,du \equiv L_0(\lambda,x) \quad (c\to 0).$$

Derivatives with respect to λ of the functions $L_0(\lambda,x)$ and $L_c(\lambda,x)$ are continuous in λ and decrease. It follows $L_c'(\lambda,x) \to L_0'(\lambda,x)$. Actually,

$$\frac{1}{h}\,(L_c(\lambda,x)-L_c(\lambda-h,x)) > L_c'(\lambda,x) > \frac{1}{h}\,(L_c(\lambda+h,x)-L_c(\lambda,x)),$$

$$\frac{1}{h}\,(L_0(\lambda,x)-L_0(\lambda-h,x)) > L_0'(\lambda,x) > \frac{1}{h}\,(L_0(\lambda+h,x)-L_0(\lambda,x)).$$

Hence

$$L_c'(\lambda,x)-L_0'(\lambda,x) \le \frac{1}{h}\,(L_c(\lambda,x)-L_c(\lambda-h,x)) - \frac{1}{h}\,(L_0(\lambda+h,x)-L_0(\lambda,x)),$$

$$L_0'(\lambda,x)-L_c'(\lambda,x) \le \frac{1}{h}\,(L_0(\lambda,x)-L_0(\lambda-h,x)) - \frac{1}{h}\,(L_c(\lambda+h,x)-L_c(\lambda,x))$$

and consequently,

$$|L_c'(\lambda,x)-L_0'(\lambda,x)| \le \frac{1}{h}\,(L_0(\lambda,x)-L_0(\lambda-h,x)) - \frac{1}{h}\,(L_0(\lambda+h,x)-L_0(\lambda,x)) +$$

$$+\frac{1}{h}\,|L_c(\lambda,x)-L_0(\lambda,x)| + \frac{1}{h}\,|L_c(\lambda-h,x)-L_0(\lambda-h,x)| +$$

$$+\frac{1}{h}\,|L_c(\lambda+h,x)-L_0(\lambda+h,x)| = \varepsilon_h + \varepsilon_c(h),$$

wher $\varepsilon_h \to 0$ as $h\to 0$ and for any $h>0$ $\varepsilon_c(h)\to 0$ as $c\to 0$.

Consequently, uniformly in $\lambda < \lambda_0$

$$\int_0^\infty u\,e^{-\lambda u}\,\frac{f_y(u|\,x)}{y-x}\,du \to \int_0^\infty u\,e^{-\lambda u}\,\nu(u,x)\,du.$$

From here according to the theorem about continuous correspondence between image and preimage of Laplace transformation (see [fel67]) we obtain that the distribution $F_y(\cdot|\,x)$ $(dF_y(u|\,x) = u\,\frac{f_y(u|\,x)}{y-x}\,du)$ converges weakly to the distribution $F_0(\cdot|\,x)$ $(dF_0(u|\,x) = u\,\nu(u,x)\,du)$. Lemma is proved.

Example 1

For inverse gamma process with parameters γ and δ we have $\nu(t,x) = \nu(t,0) \equiv \nu(t)$:

$$\frac{1}{x}\, f_x(t) = \frac{\delta\gamma}{\delta x\, \Gamma(\delta x)}\, e^{-\gamma t}(\gamma t)^{\delta x-1} =$$

$$= \frac{\delta\gamma}{\Gamma(\delta x+1)}\, e^{-\gamma t}(\gamma t)^{\delta x-1} \to \frac{\delta\, e^{-\gamma t}}{t} = \nu(t) \quad (x \to 0).$$

Example 2

For the process of records of standard Wiener process (see [har01]) we have:

$$f_x(t) = \frac{x\, e^{-\frac{x^2}{2t}}}{\sqrt{2\pi t^3}}, \qquad \frac{1}{x}\, f_x(t) = \frac{e^{-\frac{x^2}{2t}}}{\sqrt{2\pi t^3}} \to \frac{1}{\sqrt{2\pi t^3}} = \nu(t).$$

Corollary 1. *If conditions of lemma 1 are fulfilled, and the function $f_y(t|\, x)$ is differentiable with respect to t, then*

$$\lim_{h\downarrow 0} \frac{1}{h}\, (f_{y+h}(t|\, x) - f_y(t|\, x)) =$$

$$= \int_0^t (f_y(t-s|\, x) - f_y(t|\, x))\, \nu(s,y)\, ds - f_y(t|\, x) \int_0^\infty \nu(s,y)\, ds. \qquad (7)$$

Proof. We have for $h > 0$

$$\frac{1}{h}\, (f_{y+h}(t|\, x) - f_y(t|\, x)) = \frac{1}{h} \left(\int_0^t f_y(t-s|\, x)\, f_{y+h}(s|\, y)\, ds - f_y(t|\, x) \right) =$$

$$= \frac{1}{h} \left(\int_0^t f_y(t-s|\, x) - f_y(t|\, x))\, f_{y+h}(s|\, y)\, ds - f_y(t|\, x) \int_t^\infty f_{y+h}(s|\, y)\, ds \right) \to$$

$$\to \int_0^t (f_y(t-s|\, x) - f_y(t|\, x))\, \nu(s,y)\, ds - f_y(t|\, x) \int_t^\infty \nu(s,y)\, ds \quad (h \to 0)$$

because both the function $(f_y(t-s|\, x) - f_y(t|\, x))/s$ in the first integral, and $1/s$ in the second integral are continuous and bounded. Corollary is proved.

Let us find the density $g_t(y|\, x)$ of the one-dimensional distribution of I-process, where $P_x(\xi(t) \in dy) = g_t(y|\, x)\, dy$. We have

$$g_t(y|\, x) = \lim_{h\to 0} \frac{1}{h}\, P_x(\tau_y < t, \tau_{y+h} \geq t),$$

and also

$$P_x(\tau_y < t, \tau_{y+h} \geq t) = \int_0^t f_y(s|\, x) \int_{t-s}^\infty f_{y+h}(u|\, y)\, du\, ds.$$

From here

$$\frac{1}{h}\int_0^t f_y(s|\,x)\int_{t-s}^{\infty} f_{y+h}(u|\,y)\,du\,ds =$$

$$= \frac{1}{h}\left(\int_0^t f_{y+h}(u|\,y)\int_{t-u}^{t} f_y(s|\,x)\,ds\,du + \int_t^{\infty} f_{y+h}(u|\,y)\int_0^t f_y(s|\,x)\,ds\,du\right) \to$$

$$\to \int_0^t \nu(u,y)\int_{t-u}^{t} f_y(s|\,x)\,ds\,du + \int_t^{\infty} \nu(u,y)\int_0^t f_y(s|\,x)\,ds\,du \quad (h\to 0)$$

because both the functions $\int_{t-u}^{t} f_y(s|\,x)\,ds/u$ in the first integral, and $\int_0^t f_y(s|\,x)\,ds/u$ in the second integral are continuous and bounded. Consequently, if the conditions of lemma 1 are fulfilled, then

$$g_t(y|\,x) = \int_0^t f_y(s|\,x)\int_{t-s}^{\infty} \nu(u,y)\,du\,ds. \tag{8}$$

Theorem 1. *If condition (6) is fulfilled, and the functions $f_y(t|\,x)$, $\nu(t,x)$ are continuous in their arguments, besides $f_y(t|\,x)$ is differentiable with respect to t, and $\int_0^\infty u\,\nu(u,x)\,du < \infty$, then*

$$\frac{\partial}{\partial y} f_y(t|\,x) = -\frac{\partial}{\partial t} g_t(y|\,x), \tag{9}$$

where

$$\frac{\partial}{\partial y} f_y(t|\,x) = \int_0^t (f_y(t-s|\,x)-f_y(t|\,x))\,\nu(s,y)\,ds - f_y(t|\,x)\int_0^\infty \nu(s,y)\,ds. \tag{10}$$

Proof. Let us note that for any point (t,x) $(t,x>0)$ we have

$$P_0(\tau_x \le t) + P_0(\xi(t) < x) = 1.$$

This identity is valid for any non-decreasing process, beginning from the point $(0,0)$. Moreover, for $0<t_1<t$ and $0<x_1<x$ we have

$$0 = P_0(\tau_x \le t) + P_0(\xi(t) < x) - (P_0(\tau_x \le t_1) + P_0(\xi(t_1) < x)) -$$

$$-(P_0(\tau_{x_1} \le t) + P_0(\xi(t) < x_1)) + 1 =$$

$$= (P_0(\tau_x \le t) - P_0(\tau_x \le t_1)) + (P_0(\xi(t) < x) - P_0(\xi(t) < x_1)) -$$

$$-P_0(\xi(t_1) < x) - P_0(\tau_{x_1} \le t) + 1 =$$

$$= \int_{t_1}^t f_x(s)\,ds + \int_{x_1}^x g_t(y)\,dy - \int_0^x g_{t_1}(y)\,dy - \int_0^t f_{x_1}(s)\,ds + 1,$$

where for brevity we write $f_x(t|\,0) \equiv f_x(t)$ and $g_t(x|\,0) \equiv g_t(x)$. If $(1/h)(f_x(s|0) - f_{x-h}(s|0))$ tends uniformly with respect to $s \in (t_1, t]$ to corresponding partial derivative as $h \to 0$, we have

$$0 = \int_{t_1}^{t} \frac{\partial}{\partial x} f_x(s)\, ds + g_t(x) - g_{t_1}(x).$$

Consequently, if $\partial f_x(s)/\partial x$ is continuous at the point t we obtain

$$0 = \frac{\partial}{\partial x} f_x(t) + \frac{\partial}{\partial t} g_t(x).$$

We continue our proof for non-uniform case. Let $f_x(t)$ satisfy Lipschitz condition in some neighborhood of the point t_1. We have for $t_1 < t_2$

$$g_{t_2}(x) - g_{t_1}(x) = \int_0^{t_2} f_x(s) \int_{t_2-s}^{\infty} \nu(u,x)\, du\, ds - \int_0^{t_1} f_x(s) \int_{t_1-s}^{\infty} \nu(u,x)\, du\, ds =$$

$$= \int_{t_1}^{t_2} f_x(s) \int_{t_2-s}^{\infty} \nu(u,x)\, du\, ds - \int_0^{t_1} f_x(s) \int_{t_1-s}^{t_2-s} \nu(u,x)\, du\, ds =$$

$$= \int_{t_1}^{t_2} f_x(s) \int_{t_1}^{\infty} \nu(u,x)\, du\, ds + \int_{t_1}^{t_2} f_x(s) \int_{t_2-t_1}^{t_1} \nu(u,x)\, du\, ds +$$

$$+ \int_{t_1}^{t_2} f_x(s) \int_{t_2-s}^{t_1} \nu(u,x)\, du\, ds - \int_0^{t_2-t_1} f_x(s) \int_{t_1}^{t_2-s} \nu(u,x)\, du\, ds -$$

$$- \int_{t_2-t_1}^{t_1} \nu(u,x) \int_{t_1-u}^{t_2-u} f_x(s)\, ds\, du - \int_0^{t_2-t_1} \nu(u,x) \int_{t_1-u}^{t_1} f_x(s)\, ds\, du.$$

the first member in this sum has an order

$$(t_2 - t_1) f_x(t_1) \int_{t_1}^{\infty} \nu(u,x) du + o(t_2 - t_1).$$

The sum of the second and fifth members can be represented as follows

$$\int_{t_2-t_1}^{t_1} \nu(u,x) \int_{t_1}^{t_2} (f_x(s) - f_x(s-u))\, ds\, du =$$

$$= (t_2 - t_1) \int_0^{t_1} (f_x(t_1) - f_x(t_1 - u))\, \nu(u,x)\, du + o(t_2 - t_1)$$

because the ratio $(f_x(s) - f_x(s-u))/u$ is uniformly bounded on the integration domain. The sum of the third and sixth members has the form

$$\int_0^{t_2-t_1} \nu(u,x) \int_{t_2-u}^{t_2} (f_x(s) - f_x(s - t_2 + t_1))\, ds\, du =$$

$$= (t_2 - t_1) \int_0^{t_2-t_1} \nu(u,x) \int_{t_2-u}^{t_2} (f_x(s) - f_x(s - t_2 + t_1))/(t_2 - t_1)\, ds\, du = o(t_2 - t_1),$$

due to the Lipschitz condition and integrability of the function $u\,\nu(u,x)$ in the neighborhood of zero. The fourth member is estimated as

$$\int_0^{t_2-t_1} f_x(s) \int_{t_1}^{t_2-s} \nu(u,x)\,du\,ds \le$$

$$\le \max(\nu(u,x) : u \in (t_1,t_2))\,(t_2-t_1) \int_0^{t_2-t_1} f_x(s)\,ds = o(t_2-t_1)$$

due to integrability of the density $f_x(s)$. So we obtain the formula

$$\frac{\partial}{\partial t} g_t(x) = \int_0^t (f_x(t) - f_x(t-s))\,\nu(s,x)\,ds + f_x(t) \int_t^\infty \nu(s,x)\,ds. \qquad (11)$$

Evidently, this formula is true for any other initial point x_0 ($0 \le x_0 < x$). Comparing this formula with (11), we obtain proof of the theorem.

Multi-dimensional distribution

Let us show that under conditions of lemma 1 there exists a multi-dimensional distribution density of I-process. Let $t_0 < t_1 < \ldots, t_n$, $x_0 < x_1 < \ldots, x_n$ and $0 < h_i < x_{i+1} - x_i$ ($x_{n+1} = \infty$). Then we have

$$g_{t_1,\ldots,t_n}(x_1,\ldots,x_n|\,x_0) =$$

$$= \lim_{h_1\to 0,\ldots,h_n\to 0} \frac{1}{h_1\ldots h_n} P_{x_0}((\tau_{x_1} < t_1, \tau_{x_1+h_1} \ge t_1),\ldots,(\tau_{x_n} < t_n, \tau_{x_n+h_n} \ge t_n)),$$

and also

$$P_{x_0}((\tau_{x_1} < t_1, \tau_{x_1+h_1} \ge t_1),\ldots,(\tau_{x_n} < t_n, \tau_{x_n+h_n} \ge t_n)) =$$

$$= \int_0^{t_1} f_{x_1}(s|\,x_0) \int_{t_1-s}^{t_2-s} f_{h_1}(u|\,x_1) \times$$

$$\times P_{x_1+h_1}((\tau_{x_2} < t'_2, \tau_{x_2+h_2} \ge t'_2),\ldots,(\tau_{x_n} < t'_n, \tau_{x_n+h_n} \ge t'_n))\,du\,ds,$$

where $t'_i = t_i - s - u$. From here the formula follows

$$g_{t_1,\ldots,t_n}(x_1,\ldots,x_n|\,x_0) =$$

$$= \int_0^{t_1} f_{x_1}(s|\,x_0) \int_{t_1-s}^{t_2-s} \nu(u,x_1)\, g_{t'_2,\ldots,t'_n}(x_2,\ldots,x_n|\,x_1)\,du\,ds.$$

So we obtain

$$g_{t_1,\ldots,t_n}(x_1,\ldots,x_n|\,x_0) = \int\limits_{\Delta_n} \prod_{k=1}^n A_{x_k}(t_k - u_{k-1}, u_k - u_{k-1}|\,x_{k-1})\,du_k, \qquad (12)$$

where $\Delta_n = (t_1,t_2) \times \cdots \times (t_{n-1},t_n) \times (t_n,\infty)$, $\quad u_0 = 0$,

$$A_y(t,u|\,x) = \int_0^t f_y(s|\,x)\,\nu(u-s,y)\,ds.$$

3 Estimation of parameters

Testing of hypophysis about independence of wear times for non-overlapping parts of an experimental specimen relates to the set of non-parametric statistics problems, which operate with infinitely many properties of the sample. That is why statistical verifying of this hypophysis is impossible for any large value of experimental data. The independence hypophysis is usually the object of believe until a case of its essential contradiction.

The direct way of data gathering

Traditional testing of wear at fixed time epochs sometimes is the uniquely possible under exploitation conditions of the technical product. Taking into account this fact, we investigate the problem of estimating the process parameters starting from the table of data, where random observation correspond to a priori fixed time epochs. In this connection we can meet two variants of data gathering: with restoration of the initial conditions (statical method), and without of such a restoration (dynamical method). Under the statical method one can consider values of wear in every circle of measurement as independent random values. For example, the testing device stops after a determinate time. The tested specimen is being taken out and weighted for determining of wear value, and then it is being established again for new testing and so on. In this case the one-dimensional distribution corresponding to a fixed time epoch contains all the information about the sample distribution. Some complications can arise for different time intervals between measurements because it gives not identical distributed members of the sample.

For the inverse gamma process, beginning at the point $(0,0)$, the one-dimensional density can be found from formula (8). Thus for a space homogeneous process we have

$$g_t(x;\gamma,\delta) = \gamma\delta \int_0^t \frac{e^{-\gamma s}(\gamma s)^{\delta x-1}}{\Gamma(\delta x)} \int_{t-s}^{\infty} \frac{e^{-\gamma u}}{u}\, du\, ds. \tag{13}$$

After not difficult transformations we obtain identity

$$g_t(x;\gamma,\delta) = \delta\, g_{\gamma t}(\delta x; 1, 1) \equiv \delta g_{\gamma t}(\delta x).$$

For independent testing the likelihood function is equal to product of these densities

$$L(x_1,\ldots,x_n;\gamma,\delta) = \delta^n \prod_{k=1}^{n} g_{\gamma t_k}(\delta x_k).$$

For to obtain maximum likelihood estimates one can search the maximum of this function by something suitable evaluating method. The analytical search of maximum is being reduced to search of roots of a system of two equations arising as a result of partial differentiating of the function with respect to its parameters. For evaluating the partial derivative of $g_{\gamma\, t}(\delta\, x)$ with respect to parameter δ one can use formula (7), which can be rewritten as

$$\frac{\partial}{\partial y} f_y(t|\, x) = \int_0^\infty (f_y(t-s|\, x) - f_y(t|\, x))\, \nu(s,y)\, ds,$$

taking into account that $f_y(t|\, x) = 0$ as $t < 0$. For evaluating the partial derivative of $g_{\gamma\, t}(\delta\, x)$ with respect to parameter γ one can use theorem 1, because analytical representation of $f_x(t)$ is considerably simpler than that of $g_t(x)$.

Approximate maximum likelihood estimates

Essential difficulties arise under dynamical method of registration of wear data. In this case increments of wear values are not independent. Hence the likelihood function represents the whole multi-dimensional joint density (12) for values of the process at the fixed time epochs. Operation time for evaluating this $2n$-dimension integral increases exponentially as n increases. Apparently it is impossible to use this formula for obtaining maximum likelihood estimates with reasonable precise. That is why some approximate methods for parameters estimation deserves attention, in partial, constructing an approximate likelihood function. For this aim one can use the property of decreasing of dependency for increments of the process on time intervals separated by sufficiently long time gaps.

It is well-known (see for example [har01]) that trajectories of inverse gamma process consists of intervals of constancy almost wholly. Therefore for any $t > 0$ (non-random) the trajectory ξ is constant on some interval containing t. It implies dependence of increments of the process ξ. The nearest regenerative point of the process (the process has Markov property with respect to the point) is the right edge of this interval. From ergodic theory it follows there exists the limit distribution of right parts of such intervals. Let P_{st} be the stationary distribution of an embedded regenerative process, and R_t^+ be length of the right part of the interval covering the point t. Then

$$P_{st}(R_t^+ > r) = \gamma \int_r^\infty (u-r) \frac{e^{-\gamma\, u}}{u}\, du$$

(see [har01, c.368]). One have to take into account this interval when evaluating stationary distribution of the increment of the process on given time interval. Hence $P_{st}(\xi(t) - \xi(0) \in dx) = \overline{g}_t(x)\, dx$, where

$$\overline{g}_t(x) = \int_0^t p_{st}(r) g_{t-r}(x)\, dr,$$

and

$$p_{st}(r) = \gamma \int_r^t \frac{e^{-v}}{v}\, dv.$$

If a statistician has observations on N "small" intervals with lengths t_i separated by "large" gaps with lengths T_i he can search approximate maximum likelihood estimates as a point of maximum of the product of stationary densities

$$\widetilde{g}_{t_1,\dots,t_N}(x_1,\dots,x_N) = \prod_{k=1}^{N} \overline{g}_{t_k}(x_k).$$

The obtained estimate the more precise, the more values T_i.

For I-processes the rate of convergence to the stationary distribution can be derived from general ergodic theorems. In our case this rate can be estimated more precisely using special properties of gamma-process. Let

$$E(t) = e^{-t} \int_0^\infty \frac{t^{x-1}}{\Gamma(x)}\, dx.$$

Existence of the limit

$$\lim_{t\to\infty} E(t) = \lim_{\lambda\to 0} \lambda/\ln(1+\lambda) = 1.$$

follows from Tauberian theorem (see [fel67, c.513]). In work [har04b] it has been shown that convergency rate of $E(t)$ to its limit determines the value of mistake when we substitute the product of one-dimensional stationary densities instead of the proper likelihood function.

Inverse way of data gathering

In this case we propose there exists a table of fixed wear levels and corresponding increments of hitting times for these levels. In frames of the theory of inverse gamma processes for to find reasonable estimates of two its parameters it is sufficient to know two first distribution moments. Consistent estimates can be obtain by the method of moments. More precise estimates can be obtain with the help of maximum likelihood method. Let, for example, levels of wear be $(x_1, x_2, \dots, x_n)$ fixed and their corresponding hitting times be measured. Taking into account independence along the time axis we construct likelihood function, i.e. the joint density with unknown parameters:

$$L(t_1,\dots,t_n;\gamma,\delta) \equiv \prod_{k=1}^{n} f_{y_i}(s_i;\gamma,\delta) = \gamma^n \prod_{k=1}^{n} \frac{e^{-\gamma\, s_k}(\gamma\, s_k)^{\delta\, y_k - 1}}{\Gamma(\delta\, y_k)},$$

where $s_k = t_k - t_{k-1}$, $y_k = x_k - x_{k-1}$ ($t_0 = 0$, $x_0 = 0$). The maximum likelihood estimates one can obtain either by standard analytical method, or with the help of computer.

Inverse way of data gathering when dealing with a continuous wear curve

In accordance with non-contact (non-stop) methods of wear registration, which recently have increasing expansion (see [fad04]), the problem arises how to estimate parameters when dealing with continuous trajectory of the process. Reducing the continuous record $\xi(t)$ $(0 \le t \le T)$ to a finite sample of n independent and identically distributed random values can be obtain by splitting the realized wear interval $(0, x)$ on n equal parts by points $(x/n, 2x/n, \ldots, x(n-1)/n)$ and determining n the first hitting times $(t_1, t_2, \ldots, t_{n-1}, t_n)$ of the boundaries of these intervals, where $t_k = \tau_{xk/n}(\xi)$. The first problem is to choose the number n in a reasonable way. In the case of inverse gamma process the random value $t_k - t_{k-1}$ has distribution with the density $f_x(t; \gamma, \delta)$. Thus $E_0\tau_x = x\,\delta/\gamma$, $D_0(\tau_x) \equiv E_0(\tau_x - E_0\tau_x)^2 = x\,\delta/\gamma^2$. It serves base for application of the method of moments for estimating both the ratio δ/γ, and δ/γ^2, and corresponding parameters of the process with accordance to formulae

$$\frac{\widehat{\delta}}{\widehat{\gamma}} = \frac{n}{x}\,\overline{\tau}_{x/n}, \qquad \frac{\widehat{\delta}}{\widehat{\gamma^2}} = \frac{n}{x}\,S^2,$$

where

$$\overline{\tau}_{x/n} = \frac{1}{n}\sum_{k=1}^{n}(t_k - t_{k-1}) = \frac{T}{n}, \qquad S^2 = \frac{1}{n}\sum_{k=1}^{n}(t_k - t_{k-1} - \overline{\tau}_{x/n})^2.$$

As follows from these formulae, the estimate of the ratio $\widehat{\delta}/\widehat{\gamma}$ does not depend on n. For reasonable choice of n we have to find variance of the estimate $\widehat{\delta}/\widehat{\gamma^2}$:

$$D_0(\widehat{\delta}/\widehat{\gamma^2}) = D_0(n\,S^2/x) = \frac{n^2}{x^2}\,D_0(S^2).$$

By the way of not-difficult but awkward transformations we obtain the formula

$$D_0(S^2) = M_4(x/n)\,\frac{(n-1)^2}{n^3} - \sigma^4(x/n)\,\frac{(n-1)(n-3)}{n^3},$$

which implies the formula for a homogeneous I-process:

$$D_0(S^2) = (x\,m_4/n + 3\sigma^4(x/n))\,\frac{(n-1)^2}{n^3} - \sigma^4(x/n)\,\frac{(n-1)(n-3)}{n^3} =$$

$$= x\,m_4\,\frac{(n-1)^2}{n^4} + 2\sigma^4(x/n)\,\frac{n-1}{n^2} =$$

$$= x\,m_4\,\frac{(n-1)^2}{n^4} + 2x^2 m_2^2\,\frac{n-1}{n^4},$$

where $m_k = \int_0^\infty u^k\,\nu(u)\,du$ $(k \ge 1)$. From here we obtain

$$D_0(\widehat{\delta}/\widehat{\gamma^2}) = \frac{n^2}{x^2}\left(x\,m_4\,\frac{(n-1)^2}{n^4} + 2x^2 m_2^2\,\frac{n-1}{n^4}\right) = \frac{1}{x}\,m_4\,\frac{(n-1)^2}{n^2} + 2m_2^2\,\frac{n-1}{n^2}.$$

Analysis of this function of n shows that it can have a local maximum in the neighborhood of the point $n = 2$ (for $x > m_4/m_2^2$), and after this point it decreases monotonically till the meaning m_4/x. Because the gain in precise is negligible when n rises, one should take into account another considerations, for example, rate of computing, which is the more, the less n.

Soft ware

In Laboratory of Reliability Analysis Methods of Institute of Problems of Mechanical Engineering of Russian Academy of Sciences there are programs permitting to find estimates of parameters of inverse gamma processes for different methods wear data gathering.

References

[bn01] Bagdonavicius, V., Nikulin, M.S.: Estimation in Degradation Models with Explanatory Variables. Lifetime Data Analysis, **7**, 85–103 (2001).

[fad04] Fadin, Yu.A., Kirienko, O.F.: Determining of wear of friction pairs during their exploitation. Vestnik mashinostroenija, **3**, 27–32 (2004) (in Russian).

[fel67] Feller, W.: An introduction to probability theory and its applications, 2. Mir, Moscow (1967) (in Russian).

[har01] Harlamov, B.P.: Continuous semi-Markov processes. Nauka, St.Petersburg (2001) (in Russian).

[har02] Harlamov, B.P.: Continuous monotone model of wear, Abstracts of Communications, Third Int. Conf. on Math. Methods in Reliability, Trondheim, 275–275 (2002).

[har04a] Harlamov, B.P.: Continuous semi-Markov processes and their applications. In: Limnios, N. (ed) Semi-Markov processes and their applications. Communications in Statistics, Theory and Methods. v.33, **3**, 569–589 (2004).

[har04b] Harlamov, B.P.: Inverse gamma process as a model of wear. In: V.Antonov, C.Huber, M.Nikulin and V.Polischook (eds) Longevity, Aging and Degradation Models, **v.2**, St.Petersburg, Politechnical University, 180–190 (2004).

[har04c] Harlamov, B.P.: Inverse process with independent positive increments: finite-dimensional distributions. Zapiski nauch. semin. POMI, v.311, 286–297 (2004) (in Russian).

[har04d] Harlamov, B.P.: Semi-Markov model of a wear process. In: Bulatov, V.P. (ed) Problems of Engineering: preciseness, friction and wear, reliability, perspective technologies. Nauka, St.Petersburg (2005) (to appear, in Russian).

[gk66] Herzbach, I.B., Kordonski, H.B.: Models of failures. Sov. radio, Moscow (1966) (in Russian).

[sko64] Skorokhod, A.V.: Random processes with independent increments. Nauka, Moscow (1964) (in Russian).

Operating Characteristics of Partial Least Squares in Right-Censored Data Analysis and Its Application in Predicting the Change of HIV-I RNA

Jie Huang[1] and David Harrington[2]

[1] Department of Preventive Medicine, Feinberg School of Medicine, Northwestern University, 680 N. Lake Shore Drive Suite 1102, Chicago, Illinois 60611, U.S.A. `jjhuang@northwestern.edu`

[2] Department of Biostatistics, Harvard School of Public Health, and Department of Biostatistical Science, Dana-Farber Cancer Institute, 44 Binney Street, Boston, Massachusetts 02115, U.S.A. `dph@jimmy.harvard.edu`

Abstract. It is often of interest to effectively use the information on a large number of covariates in predicting response or outcome. Various statistical tools have been developed to overcome the difficulties caused by the high-dimensionality of the covariate space in the setting of a linear regression model. This paper focuses on the situation where the outcomes of interest are subjected to right censoring. We implement the extended partial least squares method along with other commonly used approaches for analyzing the high dimensional covariates to a data set from AIDS clinical trials (ACTG333). Predictions were computed on the covariate effect and the response for a future subject with a set of covariates. Simulation studies were conducted to compare our proposed methods with other prediction procedures for different numbers of covariates, different correlations among the covariates and different failure time distributions. Mean squared prediction error and mean absolute distance were used to measure the accuracy of prediction on the covariate effect and the response, respectively. We also compared the prediction performance of different approaches using numerical studies. The results show that the Buckley-James based partial least squares, stepwise subset model selection and principal components regression have similar predictive power and the partial least squares method has several advantages in terms of interpretability and numerical computation.

Keywords: Dimension reduction, partial least squares, accelerative failure time model, cross-validation, prediction

1 Introduction

In the study of chronic diseases such as HIV infection and cancer, the rapid improvement in the technology of measuring disease characteristics at the molecular or genetic level makes it possible to collect large amounts of data on potential predictors of outcome. In cancer, these data often include measurements of gene or protein expression in the tumor; in HIV, the data may include characterizations of genetic mutations in the HIV-1 virus that lead to amino acid substitutions in the protease gene at specified codons. In either situation, the existing knowledge of the biology of the disease may not be sufficient to guide a study team to a definitive (or at least plausible) set of predictor variables for outcome. When used judiciously, data-dependent methods for variable selection or dimension reduction can be a useful part of exploratory analyses.

This paper examines methods for finding low-dimensional predictors of outcome when the response variable is a potentially censored measurement of change in HIV viral load (HIV-1 RNA response) and the predictors include both traditional prognostic measures, such as the history of prior treatment, and patient-specific biomarkers of mutations at specified locations in the HIV-1 virus. More specifically, we use the data to cluster subjects into groups with predicted good or poor response. We use the data from the AIDS Clinical Trials Group (ACTG) randomized trial 333 [PG00]. The primary outcome for the trial was the change in HIV-1 RNA level ($\log_{10}$ copies/mL) measured at randomization (considered baseline) compared to times during the course of therapy (weeks 2, 4, 6, 8, 16 and 24). The assay used to quantify levels of HIV-1 RNA was unable to detect virus present in blood plasma at lower than 500 (2.70 $\log_{10}$) copies/mL. For patients whose HIV-1 RNA level could be measured at baseline (all patients in this analysis), the change between baseline and later time points was right-censored when the RNA level was below the limit of quantification. This paper uses a particular method of dimension reduction (partial least squares) for a detailed analysis of this data set. The operation of the method used here is examined in more detail in [HH05].

Methods for right-censored data can be used to estimate the association of potential prognostic variables or treatment with RNA levels. Because the censored data are incomplete observations on laboratory parameters and not event times, linear models for censored data, rather than the more common proportional hazards model, can sometimes be easier to interpret. In [PG00], parametric linear regression models with normally distributed errors are used to model the dependence between changes in RNA levels and treatment or other patient level characteristics. The justification for using these models in studies of HIV is discussed in [Mar99]. Clinical response was defined in the study as the change in viral RNA between randomization and week 8, so the study report emphasizes linear models for this change, although changes at weeks 4 and 6 are also analyzed. In this paper, we focus on methods for predicting the changes from baseline to week 8, using semiparametric models for censored data in linear models, so-called accelerated failure time model (AFT). We then use the predicted changes to construct prognostic subgroups.

Some authors [Hug99, JT00] have investigated the use of linear mixed models [LW82] for the longitudinal measurements of RNA levels. The time-dependent RNA levels are left-censored when they fall below the limit of quantification for the assay, so linear mixed models must be extended to allow for partially observed measurements. We do not use the longitudinal model approach here.

Polymerase chain reaction (PCR) was used in ACTG 333 to amplify genes in the HIV-1 RNA extracted from patient plasma at baseline. The HIV-1 protease gene was fully sequenced, enabling the detection of mutations to the wild-type of this gene and amino acid substitutions at 99 protease residues. [PG00] describes in detail the association of substitutions at 12 selected residues with the change in viral RNA between baseline and week 8. These substitutions had been implicated in previous literature with resistance of the virus to the treatments used in this trial. The data for the trial present an opportunity to explore the value of the additional mutation data, along with clinical measurements, in predicting week 8 viral response. The data set analyzed here contained mutation data on 25 residues, or codon positions, and 10 clinical variables for 60 patients (details in Section 4). The large number of covariates compared to the number of subjects emphasizes the need for dimension reduction in the covariate space. We examine the behavior of stepwise regression (Step-AFT) in this context as well as extensions of principal component regression (PCR) and partial least squares (PLS).

This paper gives a more extended treatment of partial least squares with censored data than can be found in the companion paper [HH04] published in Lifetime Data Analysis. In Section 2, we have added the use of the conditional median of the estimated error distribution to predict the response for a future subject with a given set of covariates. Extensive simulation studies show the small and moderate sample size properties of partial least squares. These simulation results are discussed in a new Section 3 and sections 3 and 4 from [HH04] have been moved to sections 4 and 5 accordingly. In Section 5 the data analysis for the HIV data set, we have added analysis showing the prediction of the response for a future subject and the use of resampling to examine the leave-two-out cross validation method.

2 Analysis Methods

Let $\boldsymbol{Y}$ be an $n \times 1$ column vector of responses for n subjects and $\boldsymbol{Z}$ an $n \times p$ predictor matrix of $p-$dimensional covariate vectors. Some of the methods discussed in the paper are not scale-invariant, that is, they may yield different results when response and/or covariates are rescaled. In this paper, the columns of $\boldsymbol{Z}$ will always be centered to have mean zero and scaled to have variance one, even for binary covariates. The notation $\boldsymbol{Z}$ is used for the covariate matrix to emphasize that point. We denote row i of $\boldsymbol{Z}$ by $\boldsymbol{Z}_i$. We assume temporarily that the responses are not censored. When $\boldsymbol{Z}$ is singular or nearly so in the linear model $\boldsymbol{Y} = \boldsymbol{\beta}_0'\boldsymbol{Z} + \varepsilon$, ordinary least squares (OLS) estimates of the $p \times 1$ parameter vector $\boldsymbol{\beta}_0$ are not estimable or may be numerically unstable. Data analysts commonly use two classes of methods to mitigate the effect of collinearity in the predictor matrix. One set of methods selects a subset of the original predictors, and numerous subset selection methods are available [DS81, Hoc76, Mil90]. The other set of methods are based on biased (typically shrinkage) estimators of regression coefficients which may reduce mean-squared estimation or prediction error by reducing the variance of the estimators. These shrinkage methods include well known methods such as ridge regression, principal components regression, partial least squares, and some newer methods, such as the LASSO [Tib96]. Both sets of methods are sometimes used when $\boldsymbol{Z}$ is of full rank as well.

Stepwise regression methods are widely used, in part because software for stepwise model selection is available in nearly all standard statistical software packages. There is an extensive literature on efficient numerical algorithms for stepwise fitting of regression models, for incorporating penalty terms such as the AIC or Schwarz criterion (BIC) to reduce the likelihood of over-fitting, and to reduce the potential bias in estimates of coefficients for variables selected. For linear models, the recent monograph by [Mil90] contains an account of both the benefits and drawbacks of stepwise selection techniques for linear regression.

[Hot33] originally proposed principal component analysis to reduce the column dimension of a data matrix of highly correlated variables while retaining a large portion of the variation in the data. Let $\lambda_1 \geq \lambda_2 \geq \cdots \geq \lambda_p \geq 0$ be the eigenvalues of $\boldsymbol{Z}'\boldsymbol{Z}$, with corresponding orthogonal eigenvectors $\boldsymbol{v}_1, \boldsymbol{v}_2, \ldots \boldsymbol{v}_p$. The vectors $\boldsymbol{Z}\boldsymbol{v}_j$ are called the principal components of $\boldsymbol{Z}'\boldsymbol{Z}$. Let r be the rank of $\boldsymbol{Z}'\boldsymbol{Z}$. Principal component regression (PCR) replaces the columns in original predictor matrix by the $K \leq r$ vectors $\boldsymbol{Z}\boldsymbol{v}_1, \ldots \boldsymbol{Z}\boldsymbol{v}_K$ and fits a regression model using the new predictor matrix. When $K < r$, the new vectors do not span the column space of $\boldsymbol{Z}$, and the estimated parameters will not be unbiased estimates of $\boldsymbol{\beta}_0$. In addition, there is no theoretical basis for the new predictors satisfying any statistical optimality criteria when $K < r$. Nevertheless, the approach has some appeal, primarily because the new predictor matrix will have orthogonal columns and the fit will be numerically more stable. In addition, $\boldsymbol{v}_1$ has largest variance among the $\boldsymbol{v}_i$, $\boldsymbol{v}_2$ the second largest variance, etc, so that the first few principal components may account for a substantial proportion of the variation in the original covariates. There are a variety of suggestions in the literature for choosing K [Jol86], including minimizing a cross-validated error sums of squares or choosing K so that $\sum_1^K \lambda_j / \sum_1^r \lambda_j$ is large.

Unlike PCR, PLS uses both response and predictor values to construct transformations of the covariates to be used as new predictors. The method of PLS was first proposed by [Wol66, Wol76] for modeling information-scarce situations in social sciences. It has also been used in chemometrics, for instance, to predict a chemical composition from a near infrared reflectance spectrum [Gou96]. [WM03] provide a detailed comparison of the use of PCR and PLS in chemometrics. The original development of PLS was motivated by a heuristically appealing representation of both the vector of responses $\boldsymbol{Y}$ and the predictor matrix $\boldsymbol{Z}$ as linear combinations (with error) of a common set of latent variables, so that

$$\boldsymbol{Z} = \boldsymbol{t}_1\boldsymbol{p}_1' + \boldsymbol{t}_2\boldsymbol{p}_2' + \cdots + \boldsymbol{t}_K\boldsymbol{p}_K' + \boldsymbol{\varrho}_{K+1}$$

and

$$\boldsymbol{Y} = \boldsymbol{t}_1 q_1 + \boldsymbol{t}_2 q_2 + \cdots + \boldsymbol{t}_K q_K + \boldsymbol{\varepsilon}_{K+1}.$$

The $N \times 1$ vectors $\boldsymbol{t}_i$ are the latent variables, the $p \times 1$ vectors $\boldsymbol{p}_i$ are called the loading vectors, and the scalars q_i are called loading scores. Wold's original algorithm for computing the latent variables and their loadings has been discussed in [Hel88] and [SB90]. We have adopted Helland's notation here; interested readers should see that paper for a heuristic motivation of the algorithm.

[Wol84] gives the following algorithm for partial least squares on data $\{(Y_i, \boldsymbol{Z}_i)\}$ with a fixed number $K << \min(p, n)$ latent variables:

1. Initialize $\boldsymbol{\varrho}_0 = \boldsymbol{Z}$ and $\boldsymbol{\varepsilon}_0 = \boldsymbol{Y} - n^{-1}\mathbf{1}\mathbf{1}'\boldsymbol{Y}$.
2. For $k = 1$ to K, compute the kth
   (1) weight vector $\boldsymbol{w}_k = \boldsymbol{\varrho}_{k-1}'\boldsymbol{\varepsilon}_{k-1}$ and latent variable $\boldsymbol{t}_k = \boldsymbol{\varrho}_{k-1}\boldsymbol{w}_k$;
   (2) loading score $q_k = (\boldsymbol{t}_k'\boldsymbol{t}_k)^{-1}\boldsymbol{t}_k'\boldsymbol{\varepsilon}_{k-1} = (\boldsymbol{t}_k'\boldsymbol{t}_k)^{-1}\boldsymbol{t}_k'\boldsymbol{Y}$ and loading vector $\boldsymbol{p}_k = (\boldsymbol{t}_k'\boldsymbol{t}_k)^{-1}\boldsymbol{t}_k'\boldsymbol{\varrho}_{k-1} = (\boldsymbol{t}_k'\boldsymbol{t}_k)^{-1}\boldsymbol{t}_k'\boldsymbol{Z}$;
   (3) residuals $\boldsymbol{\varepsilon}_k = \boldsymbol{\varepsilon}_{k-1} - q_k\boldsymbol{t}_k$ and $\boldsymbol{\varrho}_k = \boldsymbol{\varrho}_{k-1} - \boldsymbol{t}_k\boldsymbol{p}_k'$.
3. The predicted value of the response is $\widehat{\boldsymbol{Y}} = n^{-1}\mathbf{1}\mathbf{1}'\boldsymbol{Y} + \sum_{k=1}^{K} q_k\boldsymbol{t}_k$.

The small data set in ACTG 333 makes model checking difficult, so in the analysis presented here we use extensions of the methods presented above to semiparametric linear models for right censored data, called the accelerated failure time (AFT) model in the time to event literature. In the AFT model, no assumption is made about the form of the error distribution. As usual, right-censored data is denoted by $\{(T_i \wedge C_i, \delta_i, \boldsymbol{Z}_i),\ i = 1, \ldots, n\}$, where $T_i \geq 0$ is the response variable, $C_i \geq 0$ is the censoring variable, $\delta_i = I_{\{T_i \leq C_i\}}$, $\boldsymbol{Z}_i$ is a $p \times 1$ covariate vector, $A \wedge B$ is the minimum of A and B. The indicator $I_{\{A\}}$ assumes value 1 if the A occurs and 0 otherwise. We take T_i and C_i to be conditionally independent given $\boldsymbol{Z}_i$. The $p \times 1$ regression coefficient $\boldsymbol{\beta}_0$ in the AFT model satisfies

$$g(T_i) = \boldsymbol{\beta}_0'\boldsymbol{Z}_i + \varepsilon_i,$$

where $\{\varepsilon_i\}$ are independent, identically distributed with finite variance and an unspecified distribution function $F_{\boldsymbol{\varepsilon}}$. Since the intercept is not specified in the model, $\boldsymbol{\varepsilon}$ may have non-zero mean. The known monotone transformation $g(\cdot)$

is usually chosen to be the identity function or a logarithm transformation. Because of the presence of censoring, we do leave the response variable T, equivalently $Y = g(T)$, in its original measurement scale.

To estimate the coefficients in the semiparametric AFT, Buckley and James, (1979) used the transformation $\varphi(\cdot)$ on the observed response $Y_i^o{=}g(T_i) \wedge g(C_i)$, where $\varphi(Y_i^o) = \delta_i Y_i^o + (1-\delta_i)\mathrm{E}\{Y_i|Y_i \geq Y_i^o, \boldsymbol{Z}_i\}$. If $\varphi(\cdot)$ were known, $\mathrm{E}\{\varphi(Y_i^o)|\boldsymbol{Z}_i\} = \mathrm{E}(\boldsymbol{\varepsilon}_i) + \boldsymbol{\beta}_0'\boldsymbol{Z}_i$, and ordinary least squares could be used with the transformed responses $\{\varphi(Y_i^o)\}$. The Buckley-James estimating algorithm simultaneously updates $\{\hat{\varphi}(Y_i^o)\}$ and $\hat{\boldsymbol{\beta}}$ at each step and proceeds iteratively:

1. Select an initial estimator $\boldsymbol{\beta}^{(0)}$, and let $\widetilde{\boldsymbol{Y}} = \boldsymbol{Z}\boldsymbol{\beta}^{(0)}$.
2. Compute the residuals $\boldsymbol{\varepsilon} = \boldsymbol{Y}^o - \widetilde{\boldsymbol{Y}}$ and the estimated transformation

$$
\begin{aligned}
\widehat{\varphi}(Y_i^o) &= \delta_i Y_i^o + (1-\delta_i)\hat{E}(Y_i|Y_i \geq Y_i^o, \boldsymbol{Z}_i) \\
&= \delta_i Y_i^o + (1-\delta_i)\left[\widetilde{Y_i^o} - \{\widehat{S}_{\boldsymbol{\varepsilon}}(\varepsilon_i)\}^{-1}\int_{\varepsilon_i}^{\infty} s d\widehat{S}_{\boldsymbol{\varepsilon}}(s)\right], \; i = 1, \cdots, n,
\end{aligned}
$$

   where $\widehat{S}_{\boldsymbol{\varepsilon}}(\cdot)$ is the Kaplan-Meier estimator of the survival function $1 - F_{\boldsymbol{\varepsilon}}$ using the censored residuals $\{\varepsilon_i, \delta_i\}$.
3. Apply ordinary least squares (OLS) to $\{(\widehat{\varphi}(Y_i^o), \boldsymbol{Z}_i)\}$. Update $\widetilde{\boldsymbol{Y}} = \widehat{\boldsymbol{\beta}}'\boldsymbol{Z}$.
4. Stop if $\widetilde{\boldsymbol{Y}}$ converges or oscillates. Otherwise, return to step 2.

Incorporating PCR into the Buckley-James algorithm is straightforward, since the calculation of the principal components uses only the matrix of covariates and is done before any regression models are estimated. A forward stepwise regression using Wald tests (Step-AFT) to enter new variables requires only parameter estimates and standard errors. As described below, we used a nonparametric bootstrap to estimate standard errors of regression parameters estimated using the Buckley-James algorithm. We did not incorporate a penalty term (e.g., AIC or BIC) in the stepwise regression since no theory has been worked out for these penalties in the AFT model.

Incorporating partial least squares into the AFT is more difficult. In [HH05], we have studied replacing step 3 in the algorithm with the partial least squares algorithm originally proposed by [Wol76], leading to an iterative partial least squares algorithm (BJ-PLS). The modified step 3 is:

3. Apply partial least squares with a fixed number K of latent variables on the transformed data $\{\widehat{\varphi}(Y_i^o), \boldsymbol{Z}_i\}$ with initial values $\boldsymbol{\varrho}_0 = \boldsymbol{Z}$ and $\boldsymbol{\varepsilon}_0 = \widehat{\varphi}(\boldsymbol{Y}^o) - n^{-1}\mathbf{1}\mathbf{1}'\widehat{\varphi}(\boldsymbol{Y}^o)$. For $k = 1$ to K, compute q_k and $\boldsymbol{t}_k$. Update $\widetilde{\boldsymbol{Y}} = \sum_{k=1}^{K} q_k \boldsymbol{t}_k$.

We used a method of cross validation to select the number of latent variables in the PLS. A detailed evaluation of the particular cross validation we used reported elsewhere (Huang and Harrington, 2004), and is summarized here. In the semiparametric AFT, the intercept is a nuisance parameter that is absorbed into the error distribution and not directly estimated. The estimated model provides information only on how subjects differ from the overall mean response. Our main objective for the analysis of the ACTG 333 HIV data is to rank the subjects according to their reduction of HIV-1 RNA levels from baseline to week 8, which can be achieved by estimating or predicting the difference $(Y_i - Y_j)$ between two subjects. Specifically, suppose $\hat{\boldsymbol{\beta}}$ is an estimate of the model coefficients and temporarily assume that the nuisance intercept α_0 is known. The error in estimating the response difference is

$$(Y_i - Y_j) - \{(\alpha_0 + \hat{\boldsymbol{\beta}}' \boldsymbol{Z}_i) - (\alpha_0 + \hat{\boldsymbol{\beta}}' \boldsymbol{Z}_j)\} = (Y_i - Y_j) - (\hat{\boldsymbol{\beta}}' \boldsymbol{Z}_i - \hat{\boldsymbol{\beta}}' \boldsymbol{Z}_j),$$

which does not involves the intercept. Therefore, there is no need to estimate the intercept in the regression model for our purpose. While the traditional leave-one-out cross validation evaluates model performance in predicting the mean response and involves the estimation of the intercept, our proposed leave-two-out cross validation procedure estimates the error in predicting response difference and leaves the intercept as a nuisance parameter.

The leave-two-out cross validation uses each pair of observations as a validation sample, with the remaining data serving as a training sample. In the training sample that excludes subjects i and j, let $\widehat{\varphi}^k_{-(i,j)}(\cdot)$, $\widehat{\boldsymbol{\beta}^k_{-(i,j)}{}'\boldsymbol{Z}_i}$ and $\widehat{\boldsymbol{\beta}^k_{-(i,j)}{}'\boldsymbol{Z}_j}$ be the partial least squares estimates of $\varphi(\cdot)$, $\boldsymbol{\beta}_0'\boldsymbol{Z}_i$ and $\boldsymbol{\beta}_0'\boldsymbol{Z}_j$ with k latent variables, respectively. If these estimates are recomputed for all $n(n-1)/2$ possible pairs (i,j), the mean-squared prediction error for the response difference between two cases with k latent variables can be estimated by $\mathcal{C}(k) =$

$$2\{n(n-1)\}^{-1} \sum_{1 \le i < j \le n} \left\{ \widehat{\varphi}^k_{-(i,j)}(Y_i^o) - \widehat{\varphi}^k_{-(i,j)}(Y_j^o) - (\widehat{\boldsymbol{\beta}^k_{-(i,j)}{}'\boldsymbol{Z}_i} - \widehat{\boldsymbol{\beta}^k_{-(i,j)}{}'\boldsymbol{Z}_j}) \right\}^2 .$$

For any k, $\mathcal{C}(k)$ requires $O(n^2)$ partial least squares model estimates, so we recommend using a stochastic estimate $\mathcal{C}^*(k)$ of $\mathcal{C}(k)$. The most natural estimate is the observed mean-squared prediction error over randomly selected training samples. The number of training samples should be chosen so that the estimated standard error for $\mathcal{C}^*(k)$ is no larger than 10% of $\mathcal{C}^*(k)$. The number of latent variables K is selected to minimize $\mathcal{C}^*(k)$.

In the linear regression model, the response of a future subject with a set of covariates is usually predicted by its conditional expectation given the covariates. However, in right-censored data, there is generally no unbiased estimator of the conditional mean of the response. Often the conditional median of the response can be well estimated if the censoring proportion is not too large. In those cases, the conditional median can be used to predict the response of a future subject in the accelerated failure time model, which corresponds to minimizing the mean absolute difference loss function. Another advantage of using the conditional median is that the median of T, the response variable of our primary interest, can be obtained easily from the monotone transformation function $T = g^{-1}(Y)$. This method is similar to that of [YWL92] for predicting the response of future subjects.

For a future subject with a covariate vector $\boldsymbol{Z}_f$, the predicted response based on the estimated conditional median of Y is given by $\hat{\boldsymbol{\beta}}'\boldsymbol{Z}_f + \hat{S}_{\boldsymbol{\varepsilon}}^{-1}(0.5)$, where $\hat{\boldsymbol{\beta}}$ is the partial least squares parameter estimate from the observed data set $\{(Y_i, \delta_i, \boldsymbol{Z}_i)\}$ and $\hat{S}_{\boldsymbol{\varepsilon}}(\cdot)$ is the Kaplan-Meier estimator of the survival function of the residuals using the empirical residuals $\{(Y_i - \hat{\boldsymbol{\beta}}'\boldsymbol{Z}_i, \delta_i)\}$. The corresponding prediction for T is $g^{-1}(\hat{\boldsymbol{\beta}}'\boldsymbol{Z}_f + \hat{S}_{\boldsymbol{\varepsilon}}^{-1}(0.5))$.

The next section shows the simulation studies exploring the predictive power of the accelerated failure time model using partial least squares and the Buckley-James fitting algorithm.

3 Simulation studies

We used simulation studies to explore the predictive power of the accelerated failure time model using partial least squares and the Buckley-James fitting algorithm. Mean squared prediction error was used to measure how well the covariate effect was predicted, and mean absolute prediction error was used to measure how well the response was predicted. Simulations were done using different numbers of explanatory variables ($p = 10$, 25, 40, 50, and 100), with different correlations among the covariates ($\rho = 0$ and 0.3), and for different underlying error distributions. The simulation design modeled a situation where many variables have moderate effects, a difficult situation for model fitting when sample sizes are not sufficiently large.

The simulations used the model

$$\log(T_i) = \boldsymbol{\beta}_0'\boldsymbol{Z}_i + \varepsilon_i, \; i = 1, \cdots, n,$$

where $\boldsymbol{Z}_i \sim N_p\left(\boldsymbol{0}, \sigma^2[(1-\rho)\boldsymbol{I} + \rho\boldsymbol{1}\boldsymbol{1}']\right)$, and $\{\varepsilon_i\}$ were independent and identically distributed. The initial parameter vector $\boldsymbol{\beta}_0$ was selected using independent draws from a uniform distribution on (-0.2, 0.2) to reflect a setting where all variables have moderate effect. We generated $\{\varepsilon_i\}$ from two different distributions. The first was an extreme value distribution (Table 1) with the survival function $S_{\boldsymbol{\varepsilon}}(x) = \exp(-\exp(x/\sigma))$, with $\sigma = 0.5$, corresponding to an increasing hazard over time. This resulted in a Weibull distribution for the response variable. The other error distribution was a normal distribution (Table 2) with variance $\sigma_0^2 = 0.4$, which produces a similar variance for $\log(T_i)$ as the chosen extreme value error distribution. The censoring times were generated from a uniform distribution $U(0, c)$ and c was chosen to produce an average censoring proportion of 20%.

Fixing the sample size n (= 50), design matrix $\boldsymbol{Z} = (\boldsymbol{Z}_1, \cdots, \boldsymbol{Z}_n)'$ and parameter vector $\boldsymbol{\beta}_0$, we generated a training sample

$\{(\min\{T_i, C_i\}, 1_{\{T_i<C_i\}}, \boldsymbol{Z}_i),\ i = 1, \cdots, n\}$ and a validation sample $\{(T_i^*, \boldsymbol{Z}_i^*), i = 1, \cdots, m\}$, where $m = 100$ and $\{(T_i^*, \boldsymbol{Z}_i^*)\}$ had the same distribution as $(T, \boldsymbol{Z})$. The true covariate effect for subject i in the validation sample was given by $\boldsymbol{\beta}_0' \boldsymbol{Z}_i^*$.

To obtain the mean squared prediction error of the covariate effects, we fit an accelerated failure time model on the training sample using the Buckley-James algorithm with all covariates in the model (when $p < n$) and with partial least squares, then used the resulting parameter estimates to predict the covariate effects for subjects in the validation sample. We computed mean squared prediction error $m^{-1} \sum_{i=1}^{m} (\hat{\boldsymbol{\beta}}' \boldsymbol{Z}_i^* - \boldsymbol{\beta}_0' \boldsymbol{Z}_i^*)^2$ for various numbers of latent variables and repeated this process for $B_1 = 50$ times. We calculated the average of the mean squared prediction error over the B_1 validation samples for different numbers of latent variables and compared the performance of partial least squares and the Buckley-James method (Table 1 and Table 2). Table 1 also appears in [HH05] and is listed here for convenience.

In linear regression with censored data, when the number of explanatory variables is close to the number of uncensored observations, some dimension reduction technique would very likely be used on the covariates before fitting a linear model with the Buckley-James algorithm. Because no such dimension reduction techniques have been widely studied for the accelerated failure time model, we chose to compare the performance of model estimates using partial least squares with models using all of the data.

The "optimal" mean squared prediction error and number of latent variables were computed, respectively, by averaging over the minimum mean squared prediction error and the corresponding number of latent variables over the validation samples.

The "dominant" number of latent variables was defined as the number of latent variables that provided the minimum average mean squared prediction error over all the validation samples, and the corresponding mean squared prediction error was called the "dominant" mean squared prediction error.

Leave-two-out cross-validation (CV) method was applied to each validation sample to select the number of the latent variables for the partial least squares method. The CV mean squared prediction error was the average of the mean squared prediction error given by the cross-validated number of latent variables, and the corresponding average of the number of latent variables gave the CV number of latent variables.

Across all the simulations, the mean squared prediction errors of the covariate effects from the partial least squares method using leave-two-out cross-validation to select the number of latent variables are close to that from the partial least squares method using the optimal number of latent variables. The mean squared prediction error of covariate effects from partial least squares was 50% or less of that from a model fit with the Buckley-James algorithm when $p < n$. The mean squared prediction error using the cross-validated number of latent variables is comparable to the optimal mean squared prediction error, even when the number of predictors p is twice the sample size n. This suggests that the leave-two-out cross validation method efficiently identifies the number of latent variables.

Number of covariates		Mean Squared Prediction Error of Covariate Effects: correlation $\rho = 0$				correlation $\rho = 0.3$			
		BJ[a]	**Optimal**[b]	**Dominant**[c]	**CV**[d]	**BJ**	**Optimal**	**Dominant**	**CV**
	p	10	1.2	1	1.2	10	1.9	2	1.7
10	**MSE**	0.103	0.060	0.060	0.063	0.135	0.069	0.082	0.093
	(SE)	(0.008)	(0.004)	(0.004)	(0.004)	(0.011)	(0.004)	(0.006)	(0.006)
	p	25	1.4	1	1.2	25	1.8	1	1.5
25	**MSE**	0.487	0.187	0.195	0.204	0.590	0.181	0.191	0.223
	(SE)	(0.030)	(0.008)	(0.007)	(0.010)	(0.057)	(0.004)	(0.001)	(0.009)
	p	40	1.5	1	1.2	40	1.9	2	1.6
40	**MSE**	2.298	0.222	0.235	0.243	3.242	0.288	0.307	0.326
	(SE)	(0.209)	(0.010)	(0.009)	(0.011)	(0.350)	(0.007)	(0.013)	(0.011)
	p		2.4	2	1.6		1.9	2	1.8
50	**MSE**	N/A	0.452	0.517	0.542	N/A	0.298	0.301	0.373
	(SE)		(0.014)	(0.021)	(0.022)		(0.009)	(0.009)	(0.018)
	p		3.6	3	2.1		3.6	3	5.4
100	**MSE**	N/A	1.136	1.176	1.287	N/A	0.738	0.749	0.912
	(SE)		(0.026)	(0.028)	(0.027)		(0.016)	(0.016)	(0.028)

[a] The Buckley-James algorithm.
[b] The optimal number of latent variables used at each run.
[c] The same number of latent variables used for all runs.
[d] The cross-validated number of latent variables used at each run.

Table 1. Comparison of mean squared prediction error of covariate effects from the Buckley-James algorithm and partial least squares given $n = 50$ and approximately 20% censoring, assuming an extreme value error distribution.

We used the conditional median to predict the response of a future subject and mean absolute prediction error to measure the accuracy of the response prediction:

$$MAE = m^{-1} \sum_{i=1}^{m} |W_i^* - \hat{W}_i^*|,$$

where $W_i^* = \log(T_i^*)$ and $\hat{W}_i^* = \hat{\boldsymbol{\beta}}' \boldsymbol{Z}_i^* + \hat{S}_{\boldsymbol{\varepsilon}}^{-1}(0.5)$, $i = 1, \cdots, m$ were the true and predicted responses, respectively. Note that $\hat{S}_{\boldsymbol{\varepsilon}}(0.5)$ was the median of the Kaplan-Meier estimate of the survival function for the error term and

$\{(T_i^*, \boldsymbol{Z}_i^*)\}$ gave a set of true responses and covariate vectors for m future subjects.

Number of covariates		Mean Squared Prediction Error of Covariate Effects							
		correlation $\rho = 0$				correlation $\rho = 0.3$			
		BJ[a]	Optimal[b]	Dominant[c]	CV[d]	BJ	Optimal	Dominant	CV
	p	10	1.1	1	1.1	10	1.7	2	1.6
10	**MSE**	0.110	0.062	0.062	0.065	0.132	0.067	0.074	0.086
	(SE)	(0.008)	(0.005)	(0.005)	(0.005)	(0.009)	(0.005)	(0.005)	(0.006)
	p	25	1.9	1	1.2	25	2.3	2	1.7
25	**MSE**	0.431	0.190	0.207	0.209	0.493	0.202	0.219	0.264
	(SE)	(0.029)	(0.007)	(0.006)	(0.008)	(0.026)	(0.009)	(0.010)	(0.015)
	p	40	1.8	2	1.4	40	2	2	1.3
40	**MSE**	3.088	0.344	0.360	0.378	2.897	0.306	0.324	0.363
	(SE)	(0.355)	(0.013)	(0.015)	(0.013)	(0.273)	(0.008)	(0.011)	(0.006)
	p		1.7	2	1.4		2.08	2	1.9
50	**MSE**	N/A	0.344	0.368	0.417	N/A	0.240	0.243	0.336
	(SE)		(0.009)	(0.020)	(0.033)		(0.008)	(0.016)	(0.020)
	p		2.9	1	3.4		2.2	1	2.4
100	**MSE**	N/A	0.914	0.987	1.053	N/A	0.599	0.628	0.706
	(SE)		(0.017)	(0.027)	(0.026)		(0.007)	(0.015)	(0.029)

[a] The Buckley-James algorithm.
[b] The optimal number of latent variables used at each run.
[c] The same number of latent variables used for all runs.
[d] The cross-validated number of latent variables used at each run.

Table 2. Comparison of mean squared prediction error of covariate effects from the Buckley-James algorithm and partial least squares given $n = 50$ and approximately 20% censoring, assuming a normal error distribution.

We constructed the mean absolute prediction error of responses over a sample of size $m = 100$ with different covariate numbers ($p = 25,\ 40,\ 50,\ 100$) and different correlations in the covariate space ($\rho = 0,\ 0.3$) for extreme value (Table 3) or normal (Table 4) error distribution of variance $\sigma^2 = 0.2,\ 0.4,\ 0.6$. The responses were predicted in two ways. The first method assumed that the true covariate effects $\boldsymbol{\beta}_0'\boldsymbol{Z}_i^*,\ i = 1, \cdots, m$ were known and the predicted responses were computed by $\hat{W}_i^* = \boldsymbol{\beta}_0'\boldsymbol{Z}_i^* + \hat{S}_{\boldsymbol{\varepsilon}}^{-1}(0.5),\ i = 1, \cdots, m$. The estimated prediction error was thus due to the estimation of the median of the error distribution and the variation of the future subjects and not errors in estimating the regression coefficients. The other method estimated covariate effects using the partial least squares with the cross-validated number of latent variables, and estimated the median of the error term from the empirical error distribution. The predicted responses from the PLS prediction were computed by $\hat{W}_i^* = \hat{\boldsymbol{\beta}}'\boldsymbol{Z}_i^* + \hat{S}_{\boldsymbol{\varepsilon}}^{-1}(0.5)$, $i = 1, \cdots, m$. The first method would of course not be available to a data analyst in practice, and was used simply for comparison purpose.

For both error distributions, when the percentage of censoring was small (20%), partial least squares appeared to give an accurate prediction of the response for a future subject with a set of covariates. The prediction was better when the covariates were moderately correlated and when the variance of the error distribution was moderately small. Increasing the number of parameters reduced the accuracy of the response prediction. The optimal mean absolute mean error stayed around 0.36.

Number of Covariates	Error Variance	Mean Absolute Predicted Error of Responses			
		correlation $\rho = 0$		correlation $\rho = 0.3$	
		Method I	Method II	Method I	Method II
	0.2	0.32	0.39	0.33	0.39
10	0.4	0.45	0.51	0.44	0.52
	0.6	0.56	0.60	0.55	0.62
	0.2	0.31	0.47	0.31	0.52
25	0.4	0.46	0.64	0.44	0.52
	0.6	0.53	0.75	0.53	0.75
	0.2	0.32	0.56	0.31	0.66
40	0.4	0.40	0.64	0.44	0.67
	0.6	0.55	0.81	0.53	0.76
	0.2	0.32	0.69	0.30	0.71
50	0.4	0.47	0.68	0.44	0.68
	0.6	0.51	0.86	0.51	0.74
	0.2	0.32	0.87	0.31	0.83
100	0.4	0.43	0.95	0.46	0.86
	0.6	0.53	1.19	0.55	0.96

Table 3. Comparison of the mean absolute prediction error of responses from methods I and II, assuming an extreme value error distribution.

The next section describes the data set from ACTG 333 in more detail and presents an analysis of that data.

4 A Description of the Data

ACTG 333 was a randomized trial with a primary objective of determining whether substituting hard capsule saquinavir (SQVhc) with indivinar (IDV) or soft gelatin capsule saquinvir (SQVsgc) would show a greater decrease plasma HIV-1 RNA levels for patients with a previous prolonged (more than

Number of Covariates	Error Variance	Mean Absolute Predicted Error of Responses			
		correlation $\rho = 0$		correlation $\rho = 0.3$	
		Method I	Method II	Method I	Method II
	0.2	0.38	0.43	0.37	0.42
10	0.4	0.50	0.59	0.52	0.59
	0.6	0.68	0.74	0.62	0.66
	0.2	0.37	0.47	0.37	0.53
25	0.4	0.54	0.68	0.52	0.62
	0.6	0.64	0.74	0.62	0.70
	0.2	0.35	0.64	0.37	0.54
40	0.4	0.53	0.76	0.49	0.65
	0.6	0.64	0.80	0.61	0.81
	0.2	0.37	0.69	0.38	0.64
50	0.4	0.52	0.83	0.51	0.68
	0.6	0.61	0.86	0.59	0.76
	0.2	0.36	1.00	0.36	0.73
100	0.4	0.53	1.12	0.52	0.83
	0.6	0.66	1.16	0.63	0.95

Table 4. Comparison of the mean absolute prediction error of responses from methods I and II, assuming a normal error distribution.

one year) use of SQVhc. A secondary objective was to assess the predictive power of mutations in the protease gene at baseline for the in vivo anti-viral response. The mutant strains of the virus were conjectured to have developed during the prior exposure to SQVhc and could confer drug resistance. Study participants were randomized to one of the three treatment arms: 8 weeks of SQVhc, followed by IDV; 8 weeks of SQVsgc, followed by crossover to IDV if no HIV-1 RNA response; or 8 weeks of IDV, followed by crossover to SQVsgc if no HIV-1 RNA response. There were two stratification factors, one is viral load at screening ($\geq 50,000$ or $< 50,000$ RNA copies/mL) and the other is the number of nucleoside reverse-transcriptase (RT) inhibitors in the anti-retroviral drug regimen (0-1 or ≥ 2) at study entry. The original enrollment goal of the trial called for 144 participants, but the trial was stopped by the ACTG and its review board after eighty-nine subjects had been enrolled when an interim analysis demonstrated the superiority of the IDV arm.

Increased drug resistance in HIV disease has been observed with mutations leading to amino acid substitutions in the protease gene at codons $10, 46, 48, 82$ and 84 [CC96, JHO96, VIS99] and with the accumulation of multiple mutations [CS95, CH96]. Although the HIV-1 protease gene was fully sequenced in this study, only amino acid substitutions at the 12 selected protease residues 10, 20, 24, 46, 48, 54, 71, 73, 82, 84, 88, and 90 were analyzed in the study report, because of their recognized association with resistance to SQV and/or IDV [PG00]. We explore here the use of as much as possible of baseline protease genotype, along with the treatment assignment and other baseline clinical measurements, in predicting the in vivo anti-viral response measured by the reductions in HIV−1 RNA level from baseline to week 8.

Sixty-five study subjects had measurements on HIV-1 RNA protease gene sequence. After the deletion of 5 subjects with missing CD4 measurements or detectable HIV-1 RNA level at baseline, the data set used here consisted of 60 patients who had information on protease sequence, treatment assignment, baseline clinical measurements (HIV-1 RNA viral load, the percentage of white cells that are CD4 positive (called CD4 percentile), CD4 cell counts (measured in cells/mm^3), CD8 percentile, CD8 counts, prior experience with SQVhc (measured in number of weeks of therapy), and the two stratification factors. Response was defined as the reduction of HIV-1 RNA level ($\log_{10}$ copies/mL) from baseline to week 8; a negative reduction indicated a rise in HIV-1 RNA from baseline. If the patient's RNA viral load at week 8 dropped below the quantification limit of 500 copies/mL, the corresponding observation would be right-censored. Out of the 60 observations, 12 (20%) were censored. The potential censoring value for each change in $\log_{10}$RNA is the difference between baseline $\log_{10}$RNA and $\log_{10}(500)$. Large potential censoring values correspond to subjects with high initial viral load who, because of disease burden, might respond poorly to treatment, leading to a potential dependence between censoring and response. We assume, as others have in similar situations, that including the baseline viral RNA load among the covariates mitigates this possible dependence. The analysis depends more heavily on the conditional independence of censoring and response, given the covariates, than is often the case in the analysis of censored event times from clinical trials.

Table 6 shows the distribution of the protease gene mutations among 60 subjects. One patient had no mutation, 3 had 1 mutation and the remainder had at least 2 mutations. Seventy-four codon positions had no more than 2 patients with mutations at those positions and thus were deleted as explanatory variables.

Table 5 gives the list of codon positions with at least 3 mutations. As a result, the analysis presented here used a data set of 60 subjects with 35 covariates (including 25 variables for 25 codon positions with mutations).

5 The Data Analysis

In chronic diseases such as cancer or HIV, statistical models are more often used to identify groups with predicted good or poor response to treatment than to predict individual outcomes. This approach is consistent with semi-parametric models such as the proportional hazards model [Cox72] and the AFT, where baseline failure rates and mean response values (the intercept in the linear model) are not included in the estimating equations. The goal of the analysis presented here was to find low dimensional predictors to rank subjects according to predicted outcome using stepwise regression, principal

Number of mutations	Codon positions
3	69, 73*, 82*
4	12, 48*, 74
5	16, 19, 20*, 72
6	13
7	14, 41
9	36
10	10*
11	62
12	15, 35
15	64, 77
19	71*
21	37
23	93
27	90*
50	63

*These are the 7 out of 12 codon positions at which mutations are known to have association with resistance to SQV and/or IDV. Codons 24, 46, 54, 84, and 88 are left out since no more than two subjects with mutation at these positions.

Table 5. Codons for which 3 or more patient samples indicated mutations

Number of mutations	0	1	2	3	4	5	6	7	8	9
Number of patients	1	3	7	8	4	16	9	4	6	2

Table 6. Distribution of mutations

component regression (PCR) and partial least squares, all adapted to the Buckley-James algorithm for fitting the AFT. The information for ranking is contained in values estimated for $\boldsymbol{\beta}'\boldsymbol{Z}_i$; in the proportional hazards model for right-censored event times, these values denote log relative risk, and are sometimes called risk scores. In the AFT, positive values of the covariate effect $\boldsymbol{\beta}'\boldsymbol{Z}_i$ denote changes in viral load that are larger than the mean change of the group, so in this setting, we call the value $\boldsymbol{\beta}'\boldsymbol{Z}_i$ a "beneficial score".

Fitting the AFT model with the Buckley-James algorithm presents several computational challenges. The algorithm sometimes fails to converge, and variances of parameter estimates can be difficult to estimate. With this data set, we did not encounter convergence problems, even though the Buckley-James algorithm had to be repeated many times for both the stepwise fitting and partial least squares. We used a non-parametric bootstrap to estimate the standard errors of parameter estimates, drawing 500 bootstrap samples with replacement from the 60 observations and re-estimating regression coefficients using Buckley-James algorithm. We then used the empirical variance of the estimated regression coefficients based on these bootstrap samples to approximate the variance of the estimated regression coefficients. Empirically, we have found that the inference results changed little with more than 500 bootstrap samples, which implies the sufficiency of 500 bootstrap samples. We did not use the [Tsi90] estimating equations based on the marginal likelihood of the ranks of the observed failure times because those equations are also difficult to solve numerically, and we did not use the recent iterative linear programming approach by [JLW03] because it was not easily adapted to the ideas of PLS.

Because of the large number of covariates, the forward stepwise regression with the AFT was preceded by bootstrap-based Wald tests of significance for association between change in $\log_{10}$ RNA and the covariates in univariate regression models. Nine covariates were statistically significant at 0.1 level (Table 7).

In the last column of the table, we reported the proportion of reaching convergence within 50 iterations in the 500 bootstrap samples for each univariate regression, where the convergence of Buckley-James algorithm is reached when the relative change of the consecutive estimated coefficients is smaller than 1%. We have found that the proportions of reaching convergence are fairly high and the algorithm is often settled with few closed stable points when the convergence is not reached. The proportion of reaching convergence of Buckley-James algorithm for multivariate case is similarly high.

The final model from the stepwise forward procedure includes mutation in codon 19 and 69, the number of RT inhibitors and the number of weeks of prior saquinavir with respective point estimates (standard errors) for the centered and scaled covariates -0.112(0.033), -0.113(0.018), -0.228(0.064) and -0.192(0.083). In this setting, positive values of the regression coefficient indicate reductions in RNA levels between baseline and week 8, so the mutation at codon 19, 69, more RT inhibitors and longer time of prior saquinavir predict for a poor response. Because of the need to conduct the Buckley-James algorithm on 500 bootstrap samples for every model encountered in the stepwise subset selection procedure, this approach was computationally intensive.

With principal component regression, we made the somewhat arbitrary choice to include the first seven principal components as predictors since they explained 52% of the variation in the covariate space. The low proportion of explained variation by the first 7 principal components indicates a lack of correlation structure in the covariate space. The point estimates (standard errors) for these 7 components were -0.080 (0.045), 0.102 (0.048), -0.018 (0.053), 0.088 (0.050), 0.096 (0.053), 0.029 (0.054), and -0.024 (0.052). Only the estimated regression coefficient for the second principal component is statistically significant at 0.05 level.

The BJ-PLS approach using the leave-two-out cross validation produces a model which has one latent variable with point estimate (standard error) 0.262 (0.058). The single latent variable selected in BJ-PLS had the largest

Covariates	coef.	se.	p-value	% of conv.*
codon10	-0.195	0.273	0.476	97%
codon12	0.218	0.224	0.330	82%
codon13	-0.126	0.334	0.707	70%
codon14	0.227	0.243	0.351	85%
codon15	0.140	0.206	0.495	85%
codon16	0.309	0.299	0.302	93%
codon19	-0.366	0.166	0.027	86%
codon20	-0.103	0.195	0.598	97%
codon35	-0.033	0.182	0.856	89%
codon36	-0.128	0.173	0.460	96%
codon37	0.162	0.213	0.446	91%
codon41	-0.062	0.189	0.741	89%
codon48	0.456	0.308	0.139	99%
codon62	-0.110	0.236	0.642	99%
codon63	-0.325	0.241	0.176	85%
codon64	0.014	0.180	0.937	87%
codon69	-0.528	0.134	0.000	100%
codon71	0.027	0.211	0.899	91%
codon72	0.377	0.472	0.425	89%
codon73	-0.178	0.289	0.538	89%
codon74	-0.086	0.257	0.739	99%
codon77	-0.005	0.194	0.978	90%
codon82	0.137	0.261	0.599	98%
codon90	-0.369	0.201	0.067	98%
codon93	-0.258	0.185	0.162	99%
wks. prior Saq.	-0.005	0.002	0.055	99%
CD4 percent	0.023	0.012	0.053	96%
CD4 count	0.001	0.001	0.032	90%
CD8 percent	-0.009	0.007	0.234	97%
CD8 count	0.000	0.000	0.593	86%
screening viral load	0.024	0.197	0.904	90%
Num. RT inhibitors	-0.479	0.148	0.001	100%
baseline log rna	-0.185	0.106	0.081	93%
SQVsqc	-0.050	0.145	0.730	91%
IDV	0.580	0.200	0.004	100%

* percentages of convergence reached in 500 bootstrap samples within 50 iterations to obtain the standard error estimates.

Table 7. Univariate analysis with AFT model

observed z-score among all original or transformed variables. The loadings for the original 35 covariates are given in Figure 1, with the covariates listed in the caption in the order in which they appear along the horizontal axis. The latent variable is linear combination of the original covariates, and the loadings are the weights for the covariates in the combination. The loadings can be viewed as a measure of the contribution from the individual covariate to the latent variable. A closer look at the loadings for these data reveals that BJ-PLS and stepwise regression provide similar and somewhat complementary information.

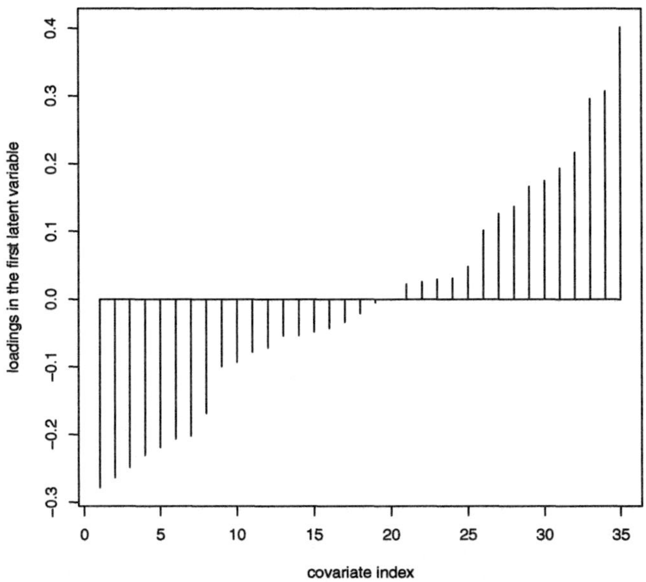

Fig. 1. Loadings for the standardized covariates. The covariates along the horizontal axis are, from right to left, (1) number of RT inhibitors, (2) number of weeks of prior saquinavir, (3) baseline log $_{10}$ viral RNA, (4) codon90, (5) CD8 percentile, (6) codon93, (7) codon63, (8) codon69, (9) codon10, (10) CD8 cell count, (11) randomization to SQVsgc (12) codon19, (13) codon20, (14) codon62, (15) codon36, (16) codon74, (17) codon77, (18) codon71, (19) codon73, (20) codon13, (21) codon41, (22) codon64, (23) codon82, (24) codon35, (25) screening viral load, (26) codon72, (27) codon15, (28) codon48, (29) codon16, (30) codon37, (31) codon14, (32) codon12, (33) CD4 cell count, (34) CD4 percentile, (35) randomization to IDV

The two of the four estimated coefficients of the covariates chosen by the stepwise model fitting (number of RT inhibitors and the number of weeks of prior saquinavir) lie at the extreme ends of the loading values and all have the same signs as the loadings in BJ-PLS. The loadings present a more detailed picture, however, suggesting a smooth transition in effect of variables from those that adversely effect response (number of RT inhibitors, number of weeks of prior saquinavir, baseline HIV-1 RNA level and mutations in codon 90) to those that strongly predict good response (mutations in codon 12, CD4 percentile and cell counts, and randomizations to IDV). The middle grouping of loadings (consisting largely of mutations in codons such as 13, 41, 71, 73, etc) seem to have little association with response. Seven of the twelve codons that have previously been associated with drug resistance are marked in Table 1; the remaining 5 of those 12 codons were dropped since fewer than three patients harbored virus with mutations in those positions. None of these 7 codons have loadings in the group with largest absolute value. Finally, the primary conclusion of ACTG 333 was that randomization to IDV significantly improved response, and that variable has the largest positive loading. Even in the presence of the information in the 35 covariates, treatment remains an important predictor in this data set.

The objective of our analysis was to use baseline clinical measurements and HIV virus mutation information to classify future subjects into potentially "good" or "poor" responders. We computed estimated beneficial scores $\boldsymbol{\beta}'\boldsymbol{Z}_i$ for all patients based on one latent variable estimated in BJ-PLS, the four variables selected in the stepwise approach to the AFT, and the first 7 principal components. All patients were divided into two groups according to whether or not their estimated beneficial scores were above or below the median score. The two groups were compared using the non-parametric Kaplan-Meier estimates of their distribution functions (Figure 2). For the groups constructed using PLS, the median reduction of HIV RNA from baseline was 0.64 $\log_{10}$ copies/mL (4.32 fold reduction) in the potentially good responders and -0.001 $\log_{10}$ copies (essentially, no reduction) in the other groups. Because of the data dependent way in which the groups were constructed, any p-value comparing these two groups would not be valid. We also divided the subjects into potentially good and poor responders using the fitted model from Step-AFT and PCR, with Kaplan-Meier estimates for the distributions of beneficial scores in two groups also shown in Figure 2. The pairs of survival curves are all similar; each method seems to adequately identify the cases whose response is in the right tail of the distribution of changes in RNA value. The principal components are least able to discriminate between individuals whose change in viral load is in the left tail (increases in viral load). The BJ-PLS and Step-AFT generated curves appear similar, but there are differences. There are 20 subjects who are placed in different groups by Step-AFT and by BJ-PLS. To examine the detailed ordering of the 20 subjects, the predictions from Step-AFT and BJ-PLS for all subjects were plotted in Figure 3. All the subjects that were inconsistently grouped were marked in the figure. Evidently, most subjects with inconsistent grouping by the two methods have relatively small estimated covariate effects by both methods and do not contradict the general trend of consistency between the two. The only exception is that the two methods made quite different prediction for one censored subject with change in viral load of at least 1.1 fold. One possible reason of the discrepancy is that the subject has high CD4 count and CD4 percentile whose beneficial effects are not reflected in the prediction of Step-AFT where the final model does not include CD4 count and percentile.

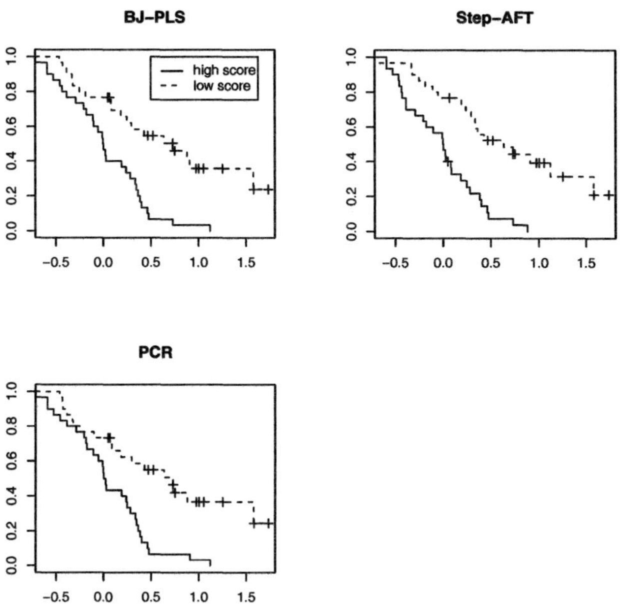

Fig. 2. Kaplan-Meier estimates for distributions of change in $\log_{10}$ RNA for good and poor risk groups estimated by BJ-PLS, Step-AFT and PCR

For a future subject with a covariate vector $\boldsymbol{Z}_f$, we can compute the predicted covariate effect by $\hat{\boldsymbol{\beta}}'\boldsymbol{Z}_f$ and the predicted response, the reduction in HIV−1 RNA level ($\log_{10}$ copies/mL) from baseline to week 8, by $\hat{\boldsymbol{\beta}}'\boldsymbol{Z}_f + 0.365$, where $\hat{\boldsymbol{\beta}}$ is the partial least squares estimate of the covariate coefficients with one latent variable and 0.365 is the estimated median of the error term, obtained by inverting the Kaplan-Meier estimate of the survival function of the empirical residuals.

We used a resampling experiment with these data to examine the ability of the leave-two-out cross validation method to prevent over-fitting. We ran-

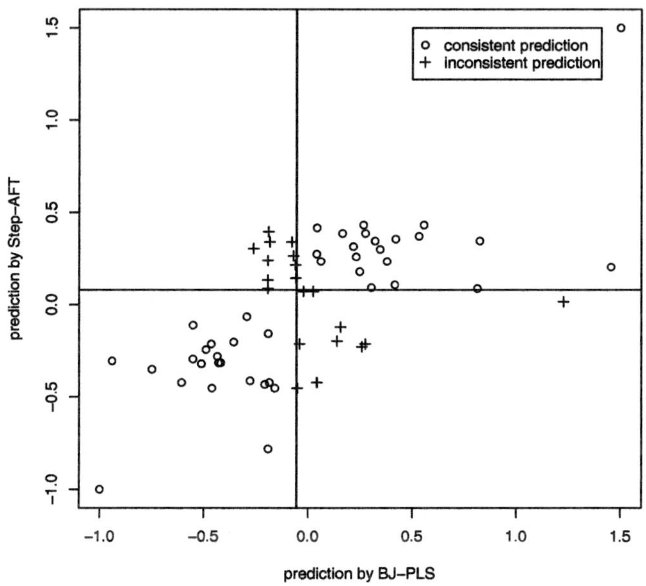

Fig. 3. Scatter plot of predictions for covariate effects made by Step-AFT and BJ-PLS (subjects with inconsistent grouping are marked with cross)

domly divided the original data set into a training sample and a validation sample with sample sizes of 32 and 31, respectively. We then fit an accelerated failure time model to the training sample using partial least squares with different numbers of latent variables ($k = 1, 3, 5$, and 7) and used the resulting parameter estimates to predict the beneficial scores on the subjects in the validation sample. The validation sample was then divided into two groups using the median of the predicted beneficial scores as the cutoff point (high versus low). The difference between the two groups in RNA reduction was compared using the log rank test and the p-values were computed. We repeated this process for $B_2 = 100$ times. Figure 4 gives the Q-Q plots of the p-values obtained from using different numbers of latent variables ($k = 1, 3, 5$, and 7) against the uniform distribution $U(0,1)$. If the partial least squares estimates do not have much predictive power, the p-values should be uniformly distributed between 0 and 1. The observed pattern of the p-values using one latent variable differed the most from a uniform distribution.

To examine the reproducibility of predictions from these methods, we conducted two resampling experiments. The first, labeled CV I (for cross-validation I) took the models arrived at in the whole data sets from the three methods as fixed, then examined how well coefficients for these models pre-

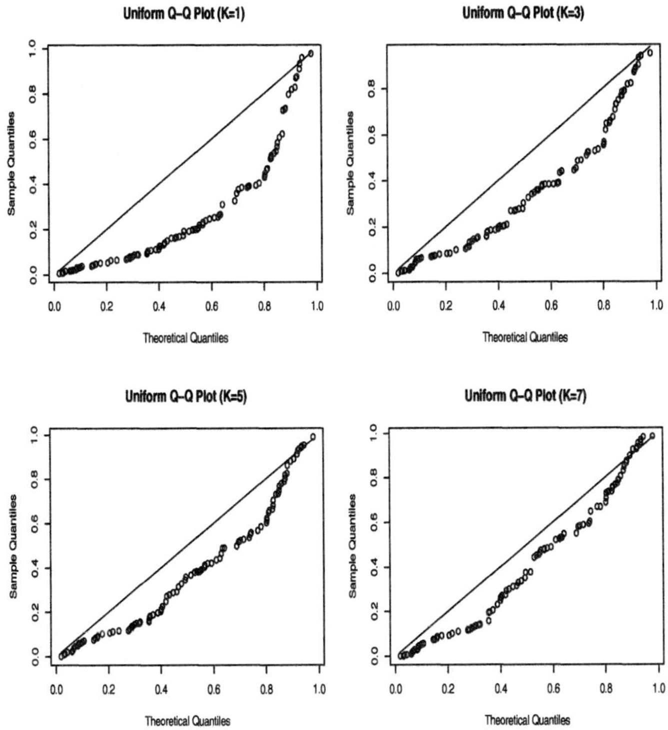

Fig. 4. Distribution of p-values

dicted outcome in validation samples when the coefficients were re-estimated in training samples. The second, CV II, re-estimated the model (including the model selection process) in each training sample then used a validation sample for predictions. In the CV II, we randomly split the data into training and validation samples with equal sizes. Then we estimated AFT, PCR and PLS model coefficients for the models arrived at earlier using the training sample; the parameter estimates were then used to split the validation sample into high and low beneficial score groups. We used a logrank test to assess the association between HIV-1 RNA change from baseline to week 8 with the two groups. A small p-value indicated a good separation between the groups. We repeated the cross-validation procedure $B_3 = 200$ times. Models with less predictive ability will produce p-values in these 200 replicates that are closer to the uniform distribution. Models with more predictive ability will have p-values clustered near 0. Table 8 summarizes the empirical distributions of the observed p-values.

Overall, the model selected from BJ-PLS gave the smallest p-values, while the model from PCR seemed to have almost no predictive power. The differ-

Methods	Min	1st Q.	Median CV I	3rd Q.	Max	Min	1st Q.	Median CV II	3rd Q.	Max
BJ-PLS	0.000	0.002	0.009	0.035	0.971	0.000	0.094	0.227	0.485	0.978
Step-AFT	0.000	0.005	0.016	0.059	0.975	0.000	0.085	0.251	0.585	0.988
PCR	0.000	0.200	0.419	0.720	0.993					

Table 8. Summary of the empirical distribution of p-values from cross-validation procedures for BJ-PLS, Step-AFT and PCR

ences between the modeling techniques were smaller in this experiment, most likely because the small training samples made the models chosen less reliable.

Since the PLS method led to a single latent variable, it is possible to use relatively simple methods for model checking. With one latent variable, the model reduces to a simple linear regression of the response variable on the latent variable. The Buckley-James algorithm replaces a censored response with an imputed value, an estimated conditional mean $\hat{\varphi}(Y^o) = \widehat{E}(Y|Y \geq Y^o)$. The appropriateness of the single latent variable in PLS can be checked in scatter plots of response by the latent variable, where censored responses are replaced by their imputed values, and by residual plots. These two plots are shown in Figure 5 and Figure 6. The least squares and lowess lines on the first of these plots show a strong linear relationship between the latent and response variables. Here, we used the lowess function in R (version 1.6.2) with the default smoother span of 2/3. The slope of the least squares line is 0.263, the same as the coefficient of the latent variable in PLS. The intercept (0.361) of the line can be used as an estimate of the mean of the error distribution, and is interpreted as the average RNA response across the subjects. The lowess line fit to the residual plot also suggests a strong linear relationship between the RNA response and the latent variable.

Figure 7 shows the estimated density of the values of $\hat{\boldsymbol{\beta}}' Z_i$ from the latent variable, estimated from the observed values using BJ-PLS. The beneficial scores appear approximately normally distributed, supporting breaking the group of patients into two groups using the median score. Smaller groups could be constructed using the quartiles of this distribution.

6 Summary and Discussion

Partial least squares algorithm has been extended to the proportional hazards model to analyze right censored data [NR02, PT02]. Though proportional hazards model is very popular to analyze right censored survival data, the accelerated failure time model is more interpretable under certain circumstances. The approach described in this paper extends principal component

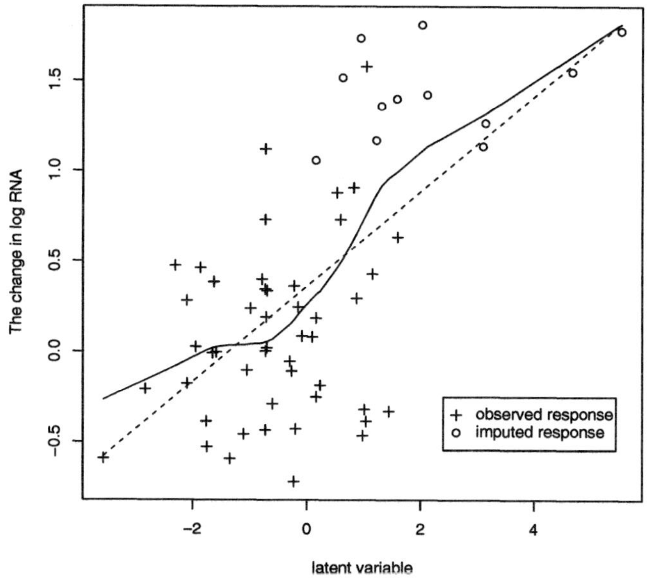

Fig. 5. Scatter plot of change in $\log_{10}$ RNA versus the BJ-PLS latent variable (censored responses are replaced by imputed values)

regression (PCR) and partial least squares (PLS) to the accelerated failure time model(AFT), and compares the exploratory analysis of an HIV data set using these methods to more traditional stepwise regression. Even in the simplest setting of linear regression, model selection and prediction can present difficulties, and those difficulties are amplified in the presence of censored data. Nonparametric estimates of a mean response with censored data are well-known to be biased and estimating the intercept in the AFT model presents the same issues. We have chosen to absorb the intercept as an unmodeled term in the error distribution, treating the covariate effect $\boldsymbol{\beta}'\boldsymbol{Z}_i$ as the main quantity of interest. Because of the unknown intercept and the incomplete observation of censored responses, we use the leave-two-out cross validation described in section 2, which relies only on predicted covariate effect for cases dropped from the training data set, instead of the usual prediction error sum of squares. The leave-two-out cross-validation suits the primary objective of the analysis, i.e., grouping subjects according to prognosis. In such an analysis, the error in minimizing the difference in response between two subjects should be minimized.

Principal component analysis performed poorly with this data set. The empirical experience with principal component analysis [WM03] suggests that

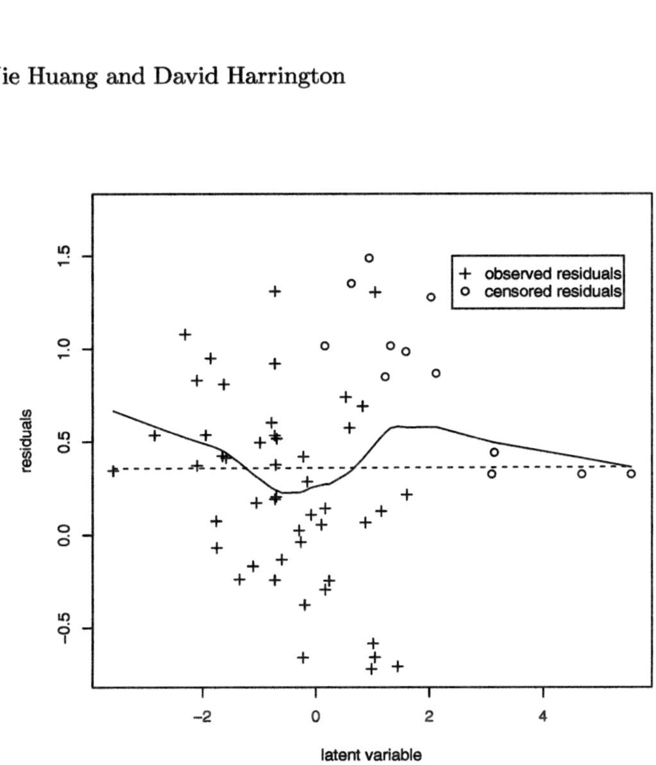

Fig. 6. Residuals from the least squares fit of the response variable on the BJ-PLS latent variable (censored residuals are replaced by imputed values)

it often leads to a larger number of latent variables than partial least squares to achieve the same prediction error, so it is possible that more than 7 principal components were necessary in this data set. We are reporting elsewhere the results of detailed simulations comparing PCR and PLS. Those simulations also show that PCR in the AFT with censored data also leads to more latent variables when the number of latent variables is chosen by cross-validation.

The proposed BJ-PLS method takes advantage of the fact that every iterative step of Buckley-James algorithm is an ordinary least squares fit and replaces the regular least squares fitting with the PLS fitting. Since the major computational burden of BJ-PLS is on the PLS algorithms performed at each iteration step, it is expected that the BJ-PLS shares similar scalability of PLS, which is known to be numerically adaptive to high-dimensional data sets.

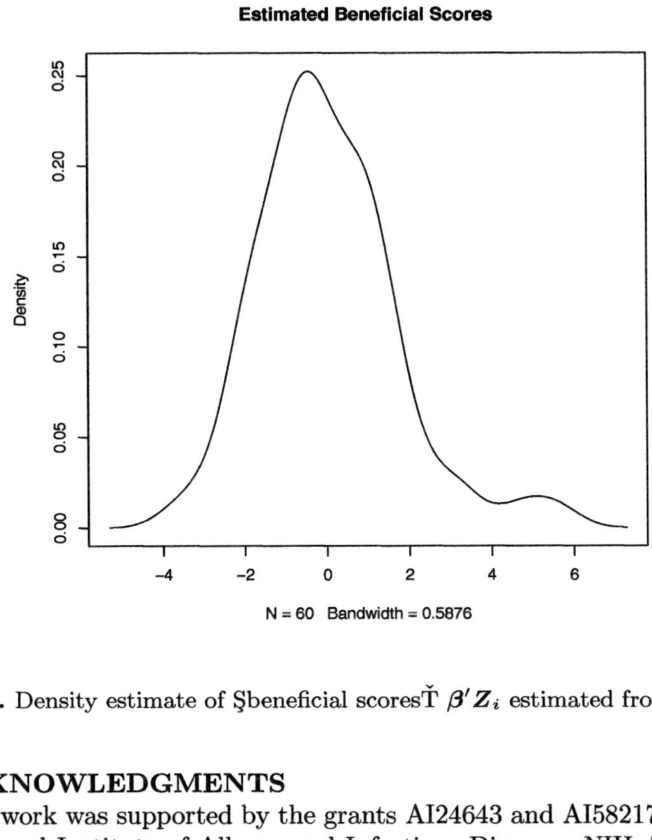

Fig. 7. Density estimate of Şbeneficial scoresŤ $\boldsymbol{\beta}'\boldsymbol{Z}_i$ estimated from BJ-PLS

ACKNOWLEDGMENTS

This work was supported by the grants AI24643 and AI58217 awarded by the National Institute of Allergy and Infectious Diseases, NIH. The data set from protocol 333 was graciously provided by the Statistics and Data Analysis Center (SDAC) of the AIDS Clinical Trials Group (ACTG).

References

[Bro93] Brown,P., Measurement, Regression, and Calibration. Clarendon: Oxford (1993)

[BJ79] J. Buckley and I. James, "Linear regression with censored data," *Biometrika* vol. 66, pp. 429–436, 1979.

[BD00] N. Butler and M. Denham, "The peculiar shrinkage properties of partial least squares regression," *J. Roy. Stat. Soc., Ser. B* vol. 62, pp. 585–593, 2000.

[CC96] Collier, A., Coombs, R., Schoenfeld, D., Bassett, R., Timpone, J., Baruch, A., Jones, M., Facey, K., Whitacre, C., McAuliffe, V., Friedman, H., Merigan, T., Reichman, R., Hopper, C., Corey L.: Treatment of human immunodeficiency virus infection with saquinavir, zidovudine, and zalcitabine: AIDS Clinical Trial Group. N. Engl. J. Med. **16**, 1011–1017 (1996)

[CS95] J. Condra, W. Schleif, O. Blahy, L. Gabryelski, D. Graham, J. Quintero, A. Rhodes, H. Robbins, E. Roth, M. Shivaprakash, D. Titus, T. Yang, H. Tepplert, K. Squires, P. Deutsch and E. Emini, "In vivo emergence of HIV-I variants resistant to multiple protease inhibitors," *Nature* vol. 374, pp. 569–571, 1995.

[CH96] J. Condra, D. Holder, W. Schleif, and et al., "Genetic correlates of in vivo viral resistance to indinavir, a human immunodeficiency virus type I protease inhibitor," *J. Virol.* vol. 70, 8270–8276, 1996.

[Cox72] D. Cox, "Regression models and life tables," *J. Roy. Stat. Soc., Ser. B* vol. 34, pp. 187–220, 1972.

[Cur96] I. Currie, "A note on Buckley-James estimators for censored data," *Biometrika* vol. 83, pp. 912–915, 1996.

[dej93] S. de Jong, "SIMPLS: an alternative approach to partial least squares regression," *Chem. Intell. Lab. Syst.* vol. 18, pp. 251–263, 1993.

[Den91] M. Denham, Calibration in infrared spectroscopy, Ph.D. Dissertaion, University of Liverpool, 1991.

[DS81] N. Draper and H. Smith, *Applied Regression Analysis,* John Wiley and Sons: New York, 1981.

[Fra93] I. Frank and J. Friedman, "A statistical view of some chemometrics regression tools," *Technometrics* vol. 35, pp. 109–134, 1993.

[Gou96] C. Goutis, "Partial least squares algorithm yields shrinkage estimators," *Ann. Stat.* vol. 24, pp. 816–824, 1996.

[Gun83] R. Gunst, "Regression analysis with multicollinear predictor variables: Definition, detection, and effects," *Commun. Stat. Theo. Meth.* vol. 12, pp. 2217–2260, 1983.

[Hel88] I. Helland, "On the structure of partial least squares regression," *Commun. Stat. Simu. Comp.* vol. 17, pp. 581–607, 1988.

[HS92] G. Heller and J. Simonoff, "Prediction in censored survival data: a comparison of the proportional hazards and linear regression models," *Biometrics* vol. 48, pp. 101–115, 1992.

[Hoc76] R. Hocking, "The analysis and selection of variables in linear regression," *Biometrics* vol. 32, pp. 1–49, 1976.

[Hot33] H. Hotelling, "Analysis of a complex of statistical variables into principal components," *J. Educ. Psychol.* vol. 24, pp. 417–441, 498-520, 1933.

[Hug99] J. Hughes, "Mixed effects models with censored data with applications to HIV RNA levels," *Biometrics* vol 55, pp. 625–629, 1999.

[HH05] J. Huang and D. Harrington, "Iterative partial least squares with right-censored data analysis: A comparision to other dimension reduction technique," *Biometrics*, in press, March 2005.

[HH04] J. Huang and D. Harrington, "Dimension reduction in the linear model for right-censored data: predicting the change of HIV-I RNA levels using clinical and protease gene mutation data," *Lifetime Data Analysis*, in press, December 2004.

[JHO96] H. Jacobsen, M. Hanggi, M. Ott, I. Duncan, S. Owen, M. Andreoni, S. Vella, and J. Mous, "In vivo resistance to a human immunodeficiency virus type I protease inhibitor: mutations, kinetics, and frequencies," *J. Inf. Dis.* vol. 173, pp. 1379–1387, 1996.

[JT00] H. Jacqmin-Gadda and R. Thiébaut, "Analysis of left censored longitudinal data with application to viral load in HIV infection," *Biostatistics* vol. 1, pp. 355–368, 2000.

[JLW03] Z. Jin, D. Lin, L. Wei, and Z. Ying, "Rank-based inference for the accelerated failure time model," *Biometrika* vol. 90, pp. 341–353, 2003.

[Jol86] I. Jolliffe, *Principal Component Analysis,* Springer-Verlag: New York, 1986.

[LW82] N. Laird and J. Ware, "Random effects models for longitudinal data," *Biometrics* vol. 38, pp. 963–974, 1982.

[Mar99] I. Marschner, R. Betensky, V. Degruttola, S. Hammer, and D. Kuritzkes, "Clinical trials using HIV-1 RNA-based primary endpoints: statistical analysis and potential biases,"Ť *J. Acq. Imm. Def. Syndr. Hum. Retr.* vol. 20, pp. 220Ů227, 1999.

[Mil90] A. Miller, *Subset Selection in Regression,* Chapman and Hall: London, 1990.

[MH82] R. Miller and J. Halpern, "Regression with censored data," *Biometrika* vol. 69, pp. 521–531, 1982.

[NR02] D. Nguyen and D. Rocke, "Partial least squares proportional hazard regression for application to DNA microarray survival data," *Bioinformatics* vol. 18, pp. 1625–1632, 2002.

[PG00] M. Para, D. Glidden, R. Coombs, A. Collier, J. Condra, C. Craig, R. Bassett, S. Leavitt, V. McAuliffe, and C. Roucher, "Baseline human immunodeficiency virus type I phenotype, genotype, and RNA response after switching from long-term hard-capsule saquinavir to indinavir or soft-gel-capsule in AIDS clinical trials group protocol 333," *J. Inf. Dis.* vol. 182, pp. 733–743, 2000.

[PT02] P. Park, L. Tian and I. Kohane, "Linking gene expression data with patient survival times using partial least squares," *Bioinformatics* vol. 18, pp. S120–S127, 2002.

[SB90] M. Stone and R. Brooks, "Continuum regression: cross-validation sequentially constructed prediction embracing ordinary least squares, partial least squares and principal components regression," *J. Roy. Stat. Soc., Ser. B* vol. 52, pp. 237–269, 1990.

[Tib96] R. Tibshirani, "Regression shrinkage and selection via the lasso," *J. Roy. Stat. Soc., Ser. B* vol. 58, pp. 267–288, 1996.

[Tsi90] A. Tsiatis, "Estimation regression parameters using linear rank tests for censored data model with censored data," *Ann. Stat.* vol. 18, pp. 354–372, 1990.

[VIS99] M. Vaillancourt, R. Irlbeck, T. Smith, R. Coombs, and R. Swanstrom, "The HIV type I protease inhibitor saquinavir can select for multiple mutations that confer increasing resistance," *AIDS Res. Hum. Retr.* vol. 15, pp. 355–363, 1999.

[WM03] P. Wentzell and L. Montoto, "Comparison of principal components regression and partial least squares through generic simulations of complex mixtures," *Chem. Intell. Lab. Syst.* vol. 65, pp. 257–279, 2003.

[Wol66] H. Wold, "Nonlinear estimation by iterative least squares procedures," *Research papers in Statistics: Festschrift for J. Neyman* John Wiley and Sons: New York, pp. 411–444, 1966.

[Wol76] H. Wold, "Soft modeling by latent variables: The non-linear iterative partial least squares (NIPALS) approach," *Perspectives in Probability and Statistics, In Honor of M. S. Bartlett* Academic: New York, pp. 117–144, 1976.

[Wol84] S. Wold, H. Wold, W. Dunn, and A. Ruhe, "The collinearity problem in linear regression: The partial least squares (PLS) approach to generalized inverse," *SIAM J. Sci. Stat. Comput.* vol. 5, pp. 735–743, 1984.

[YWL92] Z. Ying, L. Wei, and D. Lin, "Prediction of survival probability based on a linear regression model," *Biometrika* vol. 79, pp. 205–209, 1992.

Inference for a general semi-Markov model and a sub-model for independent competing risks

Catherine Huber-Carol[1], Odile Pons[2], and Natacha Heutte[3]

[1] University Paris 5, 45 rue des Saints-Pères, 75270 Paris Cedex 06, France and U 472 INSERM, 16bis avenue P-V Couturier, 94 800, Villejuif, France catherine.huber@univ-paris5.fr
[2] INRA Applied Mathematics and Informatics, 78352 Jouy-en-Josas Cedex, France odile.pons@jouy.inra.fr
[3] IUT de Caen, Antenne de Lisieux, Statistique et Traitement Informatique des Données. 11, boulevard Jules Ferry 14100 Lisieux, France N.Heutte@lisieux.iutcaen.unicaen.fr

1 Introduction

The motivation for this paper is the analysis of a cohort of patients where not only the survival time of the patients but also a finite number of life states are under study. The behavior of the process is assumed to be semi-Markov in order to weaken the very often used, and often too restrictive, Markov assumption. The behavior of such a process is defined through the initial probabilities on the set of possible states, and the transition functions defined as the probabilities, starting from any specified state, to reach another state within a certain amount of time. In order to define this behavior, the set of the transition functions may be replaced by two sets. The first one is the set of direct transition probabilities $p_{jj'}$ from any state j to any other state j'. The second one is the set of the sojourn times distributions $F_{|jj'}$ as functions of the actual state j and the state j' reached from there at the end of the sojourn (section 2).

The most usual model in this framework is the so-called competing risk model. This model may be viewed as one where, starting in a specific state j, all states that may be reached directly from j are in competition: the state j' with the smallest random time $W_{jj'}$ to reach it from j will be the one. It is well known that the joint distribution and the marginal distribution of the latent sojourn times $W_{jj'}$ is not identifiable in a general competing risk model [TSI75]. In a semi-Markov model as well as in a competing risk model, only the sub-distribution functions $F_{j'|j} = p_{jj'}F_{|jj'}$ are identifiable and it is always possible to define an independent competing risk (ICR) model by assuming that the variables $W_{jj'}$, $j' = 1, \dots, m$, are independent with distributions $F_{|jj'} = F_{j'|j}/F_{j'|j}(\infty)$. Without an assumption about their dependence, their joint distribution is not identifiable and a test of an ICR model against an alternative of a general competing risk model is not possible. Similarly, there is always a representation of any general semi-Markov model as a competing risk model with possibly dependent $W_{jj'}$ but it is not uniquely defined. When the random variables $W_{jj'}$, $j' \in J(j)$, are assumed to be independent, the semi-Markov model simplifies : the transition probabilities can be deduced from the laws of the sojourn times $W_{jj'}$ (section 3). As the term "competing risk" is also used in case of dependence of the $W_{jj'}$, we shall emphasize the independence we always assume in a competing risk model, by denoting it the ICR model (Independent Competing Risk model).

For a general right-censored semi-Markov process, Lagakos, Sommer and Zelen [LSZ78] proposed a maximum likelihood estimator for the direct transition probabilities and the distribution functions of the sojourn times, under the assumption of a discrete function with a finite number of jumps. In non-parametric models for censored counting processes, Gill [GILL80], Voelkel and Crowley [VC84] considered estimators of the sub-distribution functions $F_{j'|j} = p_{jj'}F_{|jj'}$ and they studied their asymptotic behavior. Here, we consider maximum likelihood estimation for the general semi-parametric model defined by the probabilities $p_{jj'}$ and the hazard functions related to the distribution functions $F_{|jj'}$ (section 4). If the mean number of transitions by an individual tends to infinity, then, the maximum likelihood estimators are asymptotically equivalent to those of the uncensored case. In section 5, we present new estimators defined for the case of a right-censored process with a bounded number of transitions [PONS04]. The difficulty comes from the fact that we do not observe the next state after a right-censored duration in a state.

Under the ICR assumption, specific estimators of the distribution functions $F_{|jj'}$ and of the direct transition probabilities $p_{jj'}$ are deduced from Gill's estimator of the transition functions $F_{j'|j}$. A comparison of those estimators to the estimators for a general semi-Markov process leads to tests for an ICR model against the semi-Markov alternative (section 6).

2 Framework

For each individual i, $i = 1, \cdots, n$, we observe, during a period of time t_i, the successive states $J(i) = (J_0(i), J_1(i), \cdots, J_{K(i)}(i))$, where $J_0(i)$ is the initial state, $J_{K(i)}(i)$ the final state after $K(i)$ transitions. The total number of possible states is assumed to be finite and equal to m. The successive observed sojourn times are denoted $X(i) = (X_1(i), X_2(i), \cdots, X_{K(i)}(i))$, where $X_k(i)$ is the sojourn time i spent in state $J_{k-1}(i)$ after $(k-1)$ transitions, and the cumulative sojourn times are $T_k = \Sigma_{\ell=1}^{k} X_\ell$.
One must notice that, if i changes state $K(i)$ times, the sojourn time i spent in the last state $J_{K(i)}$ is generally right censored by $t_i - T_{K(i)}(i)$, where t_i is

the total period of observation for subject i. We simplify the rather heavy notation for this last duration, and the last state $J_{K(i)}(i)$ as

$$X^*(i) \equiv t_i - T_{K(i)}(i), \quad J^*(i) \equiv J_{K(i)}(i).$$

The subjects are assumed independent and the probability distribution of the sojourn times absolutely continuous. The two models we propose for the process describing the states of the patient are renewal semi-Markov processes. Their behavior is defined through the following quantities:

1. The initial law $\rho = (\rho_1, \rho_2, \cdots, \rho_m)$:

$$\rho_j = P(J_0 = j), \quad j \in \{1, 2, \cdots, m\},$$
$$\sum_{j \in \{1,2,\cdots,m\}} \rho_j = 1. \tag{1}$$

2. The transition functions $F_{j'|j}(t)$:

$$F_{j'|j}(t) = P(J_k = j', X_k \leq t | J_{k-1} = j) \ , \ j, j' \in \{1, 2, \cdots, m\}. \tag{2}$$

Equivalent to the set of the transition functions $F_{j'|j}$, is the set of the transition probabilities, $p = \{p_{jj'} \ , \ j, j' \in \{1, 2, \cdots, m\}$, together with the set of the distribution functions $F_{|jj'}$ of the sojourn times in each state conditional on the final state as defined below

1. The direct transition probabilities from a state j to another state j' :

$$p_{jj'} = P(J_k = j' | J_{k-1} = j), \tag{3}$$

2. The law of the sojourn time between two states j and j' defined by its distribution function:

$$F_{|jj'}(t) = P(X_k \leq t | J_{k-1} = j, J_k = j'), \tag{4}$$

$$\text{where} \sum_{j'=1}^{m} p_{jj'} = 1 \ , \ p_{jj'} \geq 0 \ , \ j, j' \in \{1, 2, \cdots, m\}. \tag{5}$$

We notice that the distribution functions $F_{|jj'}$ conditional on states (j, j') do not depend on the value of k, the rank of the state reached by the patient along the process, which is a characteristic of a renewal process. We can define the hazard rate conditional on the present state and the next one,

$$\lambda_{|jj'}(t) \ = \ \lim_{dt \longrightarrow 0} \frac{P(t \leq X_k \leq t + dt | X_k \geq t, J_{k-1} = j, J_k = j')}{dt}, \tag{6}$$

as well as the cumulative conditional hazard

$$\Lambda_{|jj'}(t) = \int_0^t \lambda_{|jj'}(u)du. \tag{7}$$

Let W_j be a sojourn time in state j when no censoring is involved, F_j its distribution function, and $\overline{F}_j \equiv 1 - F_j$ its survival function, such that

$$\overline{F}_j(x) \equiv P(W_j > x) = \sum_{j'=1}^{m} p_{jj'}\overline{F}_{|jj'}(x). \tag{8}$$

The potential sojourn time in state j may be right censored by a random variable C_j having distribution function G_j, density g_j and survival function $\overline{G}_j$. The observed sojourn time in state j is $W_j \wedge C_j$.
A general notation will be $\overline{F}$ for the survival function corresponding to a distribution function F, so that, for example, $\overline{F}_{|jj'} = 1 - F_{|jj'}$ and similarly, for the transition functions, $\overline{F}_{j'|_j} = p_{jj'} - F_{j'|j}$.

3 Independent Competing Risks Model

We assume now that, starting from a state j, the potential sojourn times $W_{jj'}$ until reaching each of the states j' directly reachable from j are independent random variables having distribution functions defined through (4). The final state is the one for which the duration is the smallest. One can thus say that all other durations are right censored by this one. Without restriction of the generality, we assume that the subject is experiencing the kth transition. The competing risks model is defined by

$$\begin{aligned} X_k &= \min_{j'=1,\dots,m} W_{jj'}, \\ J_k &= j' \text{ such that } W_{jj'} < W_{jj"}, j" \neq j', \end{aligned} \tag{9}$$

where $W_{jj'}$ has the distribution function $F_{|jj'}$.

In this simple case, independence, both of the subjects and of the potential sojourn times in a given state, allows us to write down the likelihood as a product of factors dealing separately with the time elapsed between two specific states (j, j'). For the Independent Competing Risk model, one derives from (6), (8) and(9) that

$$\begin{aligned} F_{j'|j}(t) &= P(J_k = j', X_k \leq t | J_{k-1} = j) \\ &= \int_0^t \{ \prod_{j" \neq j'} \overline{F}_{|jj"}(u) \} \, dF_{|jj'}(u) \\ &= \int_0^t \lambda_{|jj'}(u) e^{-\sum_{j"} \Lambda_{|jj"}(u)} du. \end{aligned} \tag{10}$$

A consequence is that the direct transition probabilities $p_{jj'}$ defined in (3) may be derived from the probabilities defined in (4),

$$p_{jj'} = P(J_{k+1} = j'|J_k = j) = \int_0^\infty \lambda_{|jj'}(u) e^{-\sum_{j"} \Lambda_{|jj"}(u)} du. \tag{11}$$

In this special case, the likelihood is fully determined by the initial ρ_j and the functions $\lambda_{|jj'}$ defined in (6). The likelihood $L_{rc,n}$ for the independent competing risks is proportional to

$$L_{rc,n} = \prod_{i=1}^{n} \rho_{J_0(i)} \prod_{k=1}^{K(i)} \lambda_{|J_{k-1}(i),J_k(i)}(X_k(i)) \\ \times e^{-\sum_{j"} \Lambda_{J_{k-1}(i),j"}(X_k(i))} e^{-\sum_{j"} \Lambda_{|J^*(i),j"}(X^*(i))} . \tag{12}$$

It can be decomposed into the product of terms each of which is relative to an initial state j and a final state j'. When gathering the terms in $L_{rc,n}$ that are relative to a same hazard rate $\lambda_{|jj'}$ or else $\Lambda_{|jj'}$, one observes that the hazard rates appear separately in the likelihood for each pair (j, j')

$$L_{rc,n} = \{\prod_{i=1}^{n} \rho_{J_0(i)}\} \prod_{j} \prod_{j'=1}^{m} L_{rc,n}(j,j'),$$

$$L_{rc,n}(j,j') = \prod_{i=1}^{n} \prod_{k=1}^{K(i)} [\lambda_{|jj'}(X_k(i)) e^{-\Lambda_{|jj'}(X_k(i))}]^{1\{J_{k-1}(i)=j,\ J_k(i)=j'\}} \\ \times \left[e^{-\Lambda_{|jj'}(X_k(i))}\right]^{1\{J_{k-1}(i)=j,\ J_{k(i)}\neq j'\}} \\ \times \left[e^{-\Lambda_{|jj'}(X^*(i))}\right]^{1\{J^*(i)=j\}}. \tag{13}$$

This problem may be treated as m parallel and independent problems of right censored survival analysis. The only link between them is the derivation of the direct transition probabilities using (11).

4 General Model

The patients are assumed to be independent, while the potential times for a given subject are no longer assumed to be independent. We model separately the hazard rate and the transition functions ρ_j, $p_{jj'}$ and $\lambda_{|jj'}$ defined as in (1), (3) and (6). The direct transition probabilities $p_{jj'}$ can no longer be derived from the hazard rates.They are now free, except for the constraints (5). The distributions of the time elapsed between two successive states j and j' and those of the censoring are assumed to be absolutely continuous. The likelihood L_n is proportional to

$$
\begin{aligned}
L_n &= \prod_{i=1}^{n} \rho_{J_0(i)} \prod_{k=1}^{K(i)} \overline{G}_{J_{k-1}(i)}(X_k(i)) p_{J_{k-1}(i),J_k(i)} \lambda_{|J_{k-1}(i),J_k(i)}(X_k(i)) \\
&\quad \times e^{-\Lambda_{|J_{k-1}(i),J_k(i)}(X_k(i))} g_{J^*(i)}(X^*(i)) \\
&\quad \times \Big\{ \sum_{j'=1}^{m} p_{J^*(i),j'} e^{-\Lambda_{|J^*(i),j'}(X^*(i))} \Big\} \\
&= \prod_{i=1}^{n} \prod_{j=1}^{m} \rho_j^{1\{J_0(i)=j\}} \prod_{k=1}^{K(i)} \prod_{j'=1}^{m} [p_{jj'} \lambda_{|jj'}(X_k(i)) e^{-\Lambda_{|jj'}(X_k(i))} \\
&\quad \times \overline{G}_j(X_k(i))]^{1\{J_{k-1}(i)=j, J_k(i)=j'\}} \\
&\quad \times \Big\{ g_j(X^*(i)) \sum_{j'=1}^{m} p_{jj'} e^{-\Lambda_{|jj'}(X^*(i))} \Big\}^{1\{J^*(i)=j\}}.
\end{aligned}
$$

This likelihood may be written as a product of terms each of which implies sojourn times exclusively in one specific state j, $L_n = \prod_{j=1}^{m} L_n(j)$.

For each subject i, and for each $k \in \{1, 2, \cdots, K(i)\}$, we denote $1 - \delta_k(i)$ the censoring indicator of its sojourn time in the kth visited state, $J_{k-1}(i)$, with the convention that $\delta_0(i) \equiv 1$ for every i. If j' is an absorbing state, and if $J_k(i) = j'$, then j' is he last state observed for subject i, $k \equiv K(i)$, and we denote it $X^*(i) = 0$ and $\delta_{K(i)+1}(i) = 1$.
Another convention is that subject i is censored, when the last visited state $J^*(i)$ is not absorbing and the sojourn time in this state $X^*(i)$ is strictly positive and we denote $1 - \delta_i$ the censoring indicator. In all other cases, in particular if the last visited state is absorbing or if the sojourn time there is equal to 0, we say that the subject is not censored and we thus have $\delta_i = 1$. We can then write

$$
\delta_k(i) = \prod_{k'=1}^{k} \delta_{k'}(i), \quad \delta_i = 1\{X^*(i) = 0\}.
$$

For each state j of $\{1, 2, \cdots, m\}$, we define the following counts where k varies, for each subject i, between 1 and $K(i)$, $i \in \{1, 2, \cdots, n\}$, and $x \geq 0$,

$$
\begin{aligned}
N_{i,k}(x, j, j') &= 1\{J_{k-1}(i) = j, J_k(i) = j'\} 1\{X_k(i) \leq x\}, \\
Y_{i,k}(x, j, j') &= 1\{J_{k-1}(i) = j, J_k(i) = j'\} 1\{X_k(i) \geq x\}, \\
N_i^c(x, j) &= (1 - \delta_i) 1\{J^*(i) = j\} 1\{X^*(i) \leq x\}, \\
Y_i^c(x, j) &= (1 - \delta_i) 1\{J^*(i) = j\} 1\{X^*(i) \geq x\}.
\end{aligned}
$$

By summation of the counts thus defined on the indices j', i, or k, we get

$$N(x,j,j',n) = \sum_{i=1}^{n}\sum_{k=1}^{K(i)} N_{i,k}(x,j,j'), \tag{14}$$

$$N^{nc}(x,j) = \sum_{j'=1}^{m} N(x,j,j',n),$$

$$N(x,j,n) = \sum_{i=1}^{n} N_i^{nc}(x,j) + N^{nc}(x,j),$$

$$Y^{nc}(x,j,j',n) = \sum_{i=1}^{n}\sum_{k=1}^{K(i)} Y_{i,k}(x,j,j'),$$

$$Y(x,j,n) = \sum_{j'=1}^{m} Y^{nc}(x,j,j',n) + \sum_{i=1}^{n} Y_i^c(x,j).$$

By taking for x the limiting value ∞ we define $N_{i,k}(j,j') = N_{i,k}(\infty,j,j')$, $N_i^c(j) = N_i^c(\infty,j)$, $N(j,j',n) = N(\infty,j,j',n)$, $N^{nc}(j,n) = N^{nc}(\infty,j,n)$, so that $N(j,j',n)$ is the number of direct transitions from j to j' that are fully observed,$N(j,n)$ is the number of sojourn times in state j, whose $N^{nc}(j,n)$ (nc for not censored) are fully observed and $N^c(j,n)$ (c for censored) are censored. For $x = 0$, we denote $Y_i^c(j) = Y_i^c(0,j)$. The number of individuals initially in state j is $N^0(j,n) = \sum_{i=1}^{n} 1\{J_0(i) = j\}$.

The true parameter values are denoted ρ_j^0 and $p_{jj'}^0$, and the true functions of the model are $\overline{F}^0_{j'|j}$, $\overline{F}^0_{|jj'}$, $\overline{F}^0_j$, $\overline{G}^0_j$ and $\Lambda^0_{|jj'}$.

Let $l_n = \log(L_n)$ and $l_n(j) = \log(L_n(j))$. The log-likelihood relative to state j is proportional to

$$\begin{aligned} l_n(j) &= \log \rho_j N^0(j,n) + \sum_{j'=1}^{m} N(j,j',n)\log(p_{jj'}) \\ &\quad + \sum_{i=1}^{n}\sum_{k=1}^{K(i)}\sum_{j'=1}^{m} N_{i,k}(j,j')[\log(\lambda_{|jj'}(X_k(i))) - \Lambda_{|jj'}(X_k(i))] \\ &\quad + \sum_{i=1}^{n} N_i^c(j)[\log\{\sum_{j'=1}^{m} p_{jj'} e^{-\Lambda_{|jj'}(X^*(i))}\}] \qquad (15) \\ &= l_n^0(j) + l_n^{nc}(j) + l_n^c(j), \end{aligned}$$

Among the sum of four terms giving (15), let l_n^0 be the first term relative to the initial state, l_n^{nc} (nc for non censored) the sum of the second and third terms, which involve exclusively fully observed sojourn times in state j, and finally l_n^c (c for censored) the last term which deals with censored sojourn times in state j.

We denote $K_n = \max_{i=1,2\cdots,n} K(i)$ and $n\overline{K}_n = \sum_{i=1}^{n} K(i)$ respectively the maximum number of transitions and the total number of transitions for the n subjects. We consider two different designs of experiments, whether or not observations are stopped after a fixed amount K of direct transitions.

It is obvious that if the densities f_j of the sojourn times, without censoring, for every state j, are strictly positive on $]0;t_0[$ for some $t_0 > 0$, and if the distribution functions G_j of the censoring times are such that $G_j(t) < 1$ for all $t > 0$, the maximal number $K_n = \max_i K(i)$ of transitions experienced by a subject tends to infinity when n grows. If moreover the mean number of transitions $\overline{K}_n$ goes also to infinity, then the term relative to censored times $l_n^c(j)$ is the sum of terms of order n while the term $l_n^{nc}(j)$ is a sum of terms of order $n\overline{K}_n$. Therefore

Proposition 1. *If $\overline{K}_n \to \infty$ and if $N^{nc}(j,n)(n\overline{K}_n)^{-1}$ converges to a strictly positive number for every $j \in \{1, 2, \cdots, m\}$, then*

$$\lim_{n\longrightarrow\infty} \frac{l_n(j)}{n\overline{K}_n} = \lim_{n\longrightarrow\infty} \frac{l_n^{nc}(j)}{n\overline{K}_n}$$

and the maximum likelihood estimators of $p_{jj'}$, $\Lambda_{|jj'}$ and $\overline{F}_{|jj'}$ are asymptotically equivalent to

$$\begin{aligned}
\widehat{p}_{jj'} &= \frac{N(j,j',n)}{N^{nc}(j,n)}, \\
\widehat{\Lambda}_{|jj'}(x) &= \int_0^x \frac{dN(s,j,j',n)}{Y^{nc}(s,j,j',n)}, \\
\widehat{\overline{F}}_{|jj'}(x) &= \prod_{i=1}^{n}\prod_{k=1}^{K(i)}\Big\{1 - \frac{N_{i,k}(x,j,j')}{Y^{nc}(X_k(i),j,j',n)}\Big\}.
\end{aligned}$$

5 Case of a bounded number of transitions

We now assume that the number of transitions is bounded by a finite number K. For each subject $i = 1, \cdots, n$, the observation ends at time $t_i = \sum_{k=1}^{K(i)} X_k(i)$ if $K(i) = K$ or if $J_{K(i)}$ is an absorbing state, and at time t_i where there is a right censoring in the $K(i)^{\text{th}}$ visited state, $K(i) < K$.

Using notations in (14), the likelihood term relative to the initial state j may be written

$$l_n^0(j) = N^0(j,n)\log(\rho_j),$$

the terms relative to the fully observed sojourn times in state j is

$$\begin{aligned}
l_n^{nc}(j) = \sum_{j'=1}^{m} \Big\{ & N(j,j',n)\log(p_{jj'}) \\
& + \sum_{i=1}^{n}\sum_{k=1}^{K} N_{i,k}(j,j')[\log(\lambda_{|jj'}(X_k(i))) - \Lambda_{|jj'}(X_k(i))] \Big\},
\end{aligned}$$

and the term relative to the censored sojourn times in state j is

$$l_n^c(j) = \sum_{i=1}^n N_i^c(j)[\log\{\sum_{j'=1}^m p_{jj'} e^{-\Lambda_{|jj'}(X^*(i))}\}].$$

The score equations for $p_{jj'}$ and $\Lambda_{jj'}$ do not lead to explicit solutions because they involve the survival function $\overline{F}_j$ and the transition function $\overline{F}_{j'|j}$. We define estimators $\widehat{p}_{n,jj'}$ and $\widehat{\Lambda}_{n,|jj'}$ by plugging in the score equations the Kaplan-Meier estimator of $\overline{F}_j$ and the estimator of $F_{j'|j}$ given by Gill [GILL80],

$$\widehat{\overline{F}}_{n,j}(x) = \prod_{i=1}^n \prod_{k=1}^{K(i)} \left\{1 - \frac{N_{i,k}^{nc}(x,j)}{Y(X_k(i),j,n)}\right\} = \prod_{y \le x} \left\{1 - \frac{dN(y,j,n)}{Y(y,j,n)}\right\}, \quad (16)$$

$$\begin{aligned} \widehat{F}_{n,j'|j}(x) &= \sum_{i=1}^n \sum_{k=1}^{K(i)} \widehat{\overline{F}}_{n,j}(X_k^-(i)) \frac{N_{i,k}(x,j,j')}{Y(X_k(i),j,n)} \\ &= \int_0^x \widehat{\overline{F}}_{n,j}(y^-) \frac{dN(y,j,j',n)}{Y(y,j,n)}. \end{aligned} \quad (17)$$

We obtain the estimators

$$\begin{aligned} \widehat{\rho}_{n,j} &= \frac{N^0(j,n)}{n}, \\ \widehat{p}_{n,jj'} &= \frac{N(j,j',n) + \widehat{N}^c(j,j',n)}{N^{nc}(j,n) + N^c(j,n)}, \\ \widehat{\Lambda}_{n,|jj'}(x) &= \int_0^x \frac{dN(y,j,j',n)}{Y^{nc}(y,j,j',n) + \widehat{Y}^c(y,j,j',n)}, \end{aligned} \quad (18)$$

with

$$\widehat{Y}^c(y,j,j',n) = \sum_{i=1}^n Y_i^c(y,j) \frac{\widehat{\overline{F}}_{n,j'|j}(X^*(i))}{\widehat{\overline{F}}_{n,j}(X^*(i))},$$

$$\widehat{N}^c(j,j',n) = \sum_{i=1}^n N_i^c(j) \frac{\widehat{\overline{F}}_{n,j'|j}(X^*(i))}{\widehat{\overline{F}}_{n,j}(X^*(i))}.$$

The variable $(n^{1/2}(\widehat{p}_{n,jj'} - p_{jj'}^0))_{j'}$ and the process $(n^{1/2}(\widehat{\Lambda}_{n,|jj'} - \Lambda_{|jj'}^0))_{j'}$ are asymptotically Gaussian, on every interval $[0,\tau]$ such that $\int_0^\tau (\overline{F}_{j'|j}^0 \overline{G}_j^0)^{-1} \, d\Lambda_{j'|j}^0$ is finite [PONS04].

6 A Test of the Hypothesis of Independent Competing Risks.

In the ICR case, the initial probabilities jointly with the survival functions $\overline{F}_{|jj'}$ of the sojourn times conditional on states on both ends, are sufficient to determine completely the law of the process. In the general case, however, the two sets of parameters $p_{jj'}$ and $\overline{F}_{|jj'}$ are independent and may be modeled separately. Our aim is to derive a test of the hypothesis of Independent Competing Risks (ICR):

$$\begin{aligned} H_0 &: \text{The process is ICR} \\ H_1 &: \text{The process is not ICR} \end{aligned}$$

The Kaplan-Meier estimator $\widehat{\overline{F}}_{n,j}$ of $\overline{F}_j$, given in (16), and the estimator $\widehat{F}_{n,j'|j}$ of $F_{j'|j}$, given in (17), are consistent and asymptotically Gaussian both under H_0 and under H_1. It is also true for the straightforward estimator $\widehat{\rho}_{n,j}$ of the initial probabilities. From those estimators, one may derive general estimators of the transition probability $p_{jj'}$ and of the survival function $\overline{F}_{|jj'}$ of the time elapsed between two successive jumps in states j and j'. For these estimators, we shall use the same notations as the estimators of $p_{jj'}$ and $\overline{F}_{|jj'}$ defined in section 5, though they are now given by

$$\widehat{p}_{n,jj'} = \max_t \widehat{F}_{n,j'|j}(t) \tag{19}$$

$$\widehat{\overline{F}}_{n,|jj'}(t) = 1 - \frac{\widehat{F}_{n,j'|j}(t)}{\widehat{p}_{n,jj'}}. \tag{20}$$

In the independent competing risk model, the transition probability $F_{j'|j}$ satisfies (10) and thus may be estimated as

$$\begin{aligned} \widehat{F}^{RC}_{n,j'|j}(t) &= -\int_0^t \prod_{j"\neq j'} \widehat{\overline{F}}_{n,|jj"}(s)\ d\widehat{\overline{F}}_{n,|jj'}(s) \\ &= \frac{1}{\prod_{j"} \widehat{p}_{n,jj"}} \int_0^t \prod_{j"\neq j'} \widehat{\overline{F}}_{n,j"|j}(s)\ \widehat{\overline{F}}_{n,j}(s^-)\ d\widehat{\Lambda}_{n,j'|j}(s), \end{aligned} \tag{21}$$

where

$$\widehat{\Lambda}_{n,j'|j}(t) = \int_0^t 1\{Y(s,j,n) > 0\} \frac{dN(s,j,j',n)}{Y(s,j,n)} \tag{22}$$

is the estimator of the cumulative hazard function $\Lambda_{n,j'|j}$ in the general model. A competitor to $\widehat{p}_{n,jj'}$ is deduced as

$$\widehat{p}^{RC}_{n,jj'} = \max_t \widehat{F}^{RC}_{n,j'|j}(t). \tag{23}$$

Let π^0_j be the mean number of sojourn times in state j for subject i.

Proposition 2. *If* $p^0_{jj'} > 0$ *and* $\int_0^\infty \{\overline{G}^0_j(s)\overline{F}^0_j(s)\}^{-1} d\Lambda^0_j(s) < \infty$, *then* $\sqrt{n}(\widehat{p}_{n,jj'} - p^0_{jj'})$ *is asymptotically distributed as a normal random vector with mean 0, variances and covariances*

$$\sigma^2_{jj'} = \frac{1}{\pi^0_j}\int_0^\infty \frac{1}{\overline{G}^0_j(s)\overline{F}^0_j(s)}[(\overline{F}^0_{j'|j}(s) - p^0_{j'|j})^2 \frac{dF^0_j(s)}{\overline{F}^0_j(s)}$$
$$+\{\overline{F}^0_j(s) + 2(\overline{F}^0_{j'|j}(s) - p^0_{j'|j})\}\, d\overline{F}^0_{j'|j}(s)],$$

$$\sigma^2_{jj'j"} = \frac{1}{\pi^0_j}\int_0^\infty \frac{1}{\overline{G}^0_j(s)\overline{F}^0_j(s)}[(\overline{F}^0_{j'|j}(s) - p^0_{j'|j})(\overline{F}^0_{j"|j}(s) - p^0_{j"|j})\frac{dF^0_j(s)}{\overline{F}^0_j(s)}$$
$$+(\overline{F}^0_{j'|j}(s) - p^0_{j'|j}\}\, d\overline{F}^0_{j"|j}(s) + (\overline{F}^0_{j"|j}(s) - p^0_{j"|j})\}\, d\overline{F}^0_{j'|j}(s)].$$

Moreover, $\sqrt{n}(\widehat{p}^{RC}_{n,jj'} - p^0_{jj'})$ is asymptotically distributed as a centered Gaussian variable.

Estimators of the asymptotic variance and covariances of $(\widehat{p}_{n,jj'})_{j'\in J(j)}$ may be obtained by replacing the functions $\overline{F}^0_j$, $F^0_{j'|j}$ and $\Lambda^0_{j'|j}$ by their estimators in the general model, (16), (17) and (22). Due to their intricate formulas, it seems difficult to use an empirical estimator of the asymptotic variance of $\widehat{p}^{RC}_{n,jj'}$ and a bootstrap estimator should be preferred. Asymptotic confidence intervals for $p^0_{jj'}$ at the level α are deduced from the $(1-\alpha/2)$-quantile c_α of their boostrap distributions, $I_{n,jj'}(\alpha)$ in the general case and $I^{RC}_{n,jj'}(\alpha)$ under the null hypothesis of Independent Competing Risks.

A test of the Independent Competing Risks hypothesis may be defined by rejecting H_0 if $I_{n,jj'}(\alpha)$ and $I^{RC}_{n,jj'}(\alpha)$ are not overlapping for some j'. As the estimators of the parameters $p^0_{jj'}$ are not independent, the level α^* of this test with critical region

$$R_{nj}(\alpha) = \cap_{j'=1}^m R_{njj'}(\alpha), \text{ where } R_{njj'}(\alpha) = \{I_{n,jj'}(\alpha) \cap I^{RC}_{n,jj'}(\alpha) \neq \emptyset\},$$

satisfies $\alpha^* \geq 1-(1-\alpha)^m$.

7 Proofs

Proof of Proposition 2.
Let $\tau_{n,j} = \arg\max_t \widehat{F}_{n,j}(t)$. The asymptotic behavior of $\widehat{p}_{n,jj'}$ is derived from theorem 3 in Gill [GILL80] which states the weak convergence of the process

$$(\sqrt{n}(\widehat{F}_{n,j'|j}(t\wedge\tau_{n,j}) - F^0_{j'|j}(t\wedge\tau_{n,j}))_{j'\in J(j)}, \sqrt{n}(\widehat{\overline{F}}_{n,j}(t\wedge\tau_{n,j}) - \overline{F}^0_j(t\wedge\tau_{n,j}))_{t\geq 0}$$

to a Gaussian process defined, for continuous transition functions $F^0_{j'|j}$, as

$$\left\{\int_0^t \frac{\overline{F}^0_{j'|j}(s)\, dV_{jj'}(s)}{EY_i(s,j)} - \overline{F}^0_{j'|j}(t)\int_0^t \frac{dV_j(s)}{EY_i(s,j)} + \int_0^t \frac{\overline{F}^0_{j'|j}(s)\, dV_j(s)}{EY_i(s,j)},\right.$$
$$\left.\overline{F}^0_j(t)\int_0^t \frac{dV_j(s)}{EY_i(s,j)}\right\}$$

where $V_{jj'}, j, j' \in \{1, 2, \cdots, m\}$ is a multivariate Gaussian process with independent increments, having mean 0 and covariances

$$var(V_{jj'}(t)) = \int_0^t EY_i(s,j) \frac{d\overline{F}^0_{j'|j}(s)}{\overline{F}^0_j(s)},$$

$cov(V_{jj'}(t), V_{jj''}(t)) = 0$ if $j' \neq j''$ and $cov(V_{jj'}(t), V_{j_1 j_1'}(t_1)) = 0$ if $j_1 \neq j$ or $t_1 \neq t$, and $V_j = \sum_{j'} V_{jj'}$.

As $EY_i(s,j) = \pi^0_j \overline{G}^0_j(s) \overline{F}^0_j(s)$, it follows that $\sqrt{n}(\widehat{p}_{n,jj'} - p^0_{jj'})$ is asymptotically distributed as

$$\int_0^\infty \frac{dV_{jj'}(s)}{\pi^0_j \overline{G}^0_j(s)} - p^0_{jj'} \int_0^\infty \frac{dV_j(s)}{\pi^0_j \overline{G}^0_j(s) \overline{F}^0_j(s)} + \int_0^\infty \overline{F}^0_{j'|j}(s) \frac{dV_j(s)}{\pi^0_j \overline{G}^0_j(s) \overline{F}^0_j(s)}.$$

Denoting this limit as $A - B + C$, we have

$$var(A) = \frac{1}{\pi^0_j} \int_0^\infty \frac{1}{\overline{G}^0_j(s)} d\overline{F}^0_{j'|j}(s)$$

$$var(B) = \frac{p^2_{jj'}}{\pi^0_j} \int_0^\infty \frac{1}{\overline{G}^0_j(s) \overline{F}^0_j(s)^2} dF^0_j(s)$$

$$var(C) = \frac{1}{\pi^0_j} \int_0^\infty \frac{\overline{F}^0_{j'|j}(s)^2}{\overline{G}^0_j(s) \overline{F}^0_j(s)^2} dF^0_j(s)$$

$$cov(A, B) = \frac{p^0_{jj'}}{\pi^0_j} \int_0^\infty \frac{1}{\overline{G}^0_j(s) \overline{F}^0_j(s)} d\overline{F}^0_{j'|j}(s)$$

$$cov(A, C) = \frac{1}{\pi^0_j} \int_0^\infty \frac{\overline{F}^0_{j'|j}(s)}{\overline{G}^0_j(s) \overline{F}^0_j(s)} d\overline{F}^0_{j'|j}(s)$$

$$cov(B, C) = \frac{p^0_{jj'}}{\pi^0_j} \int_0^\infty \frac{\overline{F}^0_{j'|j}(s)}{\overline{G}^0_j(s) \overline{F}^0_j(s)^2} dF^0_j(s),$$

and $\sigma^2_{jj'}$ is the variance of $A - B + C$. The covariance $\sigma^2_{jj'j''}$ is obtained by similar calculations, but the covariance between the corresponding terms $A(jj')$ and $A(jj'')$ is zero.

From (21), the asymptotic Gaussian distribution of $\sqrt{n}(\widehat{p}^{RC}_{n,jj'} - p^0_{jj'})$ is a consequence of the asymptotic behavior of the estimators $\widehat{\overline{F}}_{n,j}$ and $\widehat{\overline{F}}_{n,j'|j}$ and of the estimator $\widehat{\Lambda}_{n,j'|j}$ given by (22), using again theorem 3 in Gill [GILL80].

Limiting covariances.
The limiting covariance of $\sqrt{n}(\widehat{p}^{RC}_{n,jj'} - p^0_{jj'})$ may be calculated using the following expressions

$$\sigma^2_{jj'}(t) = \frac{1}{\pi^0_j}\int_0^t \frac{1}{\overline{G}^0_j(s)\overline{F}^0_j(s)}\Big\{(\overline{F}^0_{j'|j}(s) - \overline{F}^0_{j'|j}(t))^2\frac{dF^0_j(s)}{\overline{F}^0_j(s)}$$
$$+\{\overline{F}^0_j(s) + 2(\overline{F}^0_{j'|j}(s) - \overline{F}^0_{j'|j}(t))\}\, d\overline{F}^0_{j'|j}(s)\Big\},$$

$$\sigma^2_{jj'j"}(t) = \frac{1}{\pi^0_j}\int_0^t \frac{1}{\overline{G}^0_j(s)\overline{F}^0_j(s)}\Big\{(\overline{F}^0_{j'|j}(s) - \overline{F}^0_{j'|j}(t))(\overline{F}^0_{j"|j}(s) - \overline{F}^0_{j"|j}(t))\frac{dF^0_j(s)}{\overline{F}^0_j(s)}$$
$$+(\overline{F}^0_{j'|j}(s) - \overline{F}^0_{j'|j}(t))\, d\overline{F}^0_{j"|j}(s) + (\overline{F}^0_{j"|j}(s) - \overline{F}^0_{j"|j}(t))\, d\overline{F}^0_{j'|j}(s)\Big\},$$

$$c^{(1)}_{jj'}(t) = \lim_n \ \mathrm{Cov}\{\sqrt{n}(\widehat{\overline{F}}_{n,j}(t) - \overline{F}^0_j(t)), \sqrt{n}(\widehat{\overline{F}}_{n,j'|j}(t) - \overline{F}^0_{j'|j}(t))\}$$
$$= \overline{F}^0_j(t) Bigl\{\int_0^t \frac{\overline{F}^0_{j'|j}}{\overline{G}^0_j(\overline{F}^0_j)^2}(dF^0_{j'|j} + dF^0_j) - \overline{F}^0_{j'|j}(t)\int_0^t \frac{dF^0_j}{\overline{G}^0_j(\overline{F}^0_j)^2}\Big\},$$

$$v^{(1)}_{jj'}(t) \equiv \lim_n \ \mathrm{Var}\sqrt{n}\{\widehat{\overline{F}}_{n,j}(t^-)\prod_{j_1\neq j'}\widehat{\overline{F}}_{n,j_1|j}(t) - \overline{F}^0_j(t^-)\prod_{j_1\neq j'}\overline{F}^0_{j_1|j}(t)\}$$
$$= \lim_n \ \{\prod_{j_1\neq j'}\widehat{\overline{F}}_{n,j_1|j}(t)\}^2\Big[\sum_{j_2\neq j'}\mathrm{Var}\sqrt{n}\{\widehat{\overline{F}}_{n,j_2|j}(t) - \overline{F}^0_{j_2|j}(t)\}\{\frac{\overline{F}^0_j(t)}{\overline{F}^0_{j_2|j}(t)}\}^2$$
$$+\mathrm{Var}\sqrt{n}\{\widehat{\overline{F}}_{n,j}(t^-) - \overline{F}^0_j(t)\} + \sum_{j_2\neq j'}\sum_{j_3\neq j',j_2}\frac{(\overline{F}^0_j(t))^2}{\overline{F}^0_{j_2|j}(t)\overline{F}^0_{j_3|j}(t)}$$
$$\times\mathrm{Cov}\{\sqrt{n}(\widehat{\overline{F}}_{n,j_2|j}(t) - \overline{F}^0_{j_2|j}(t)), \sqrt{n}(\widehat{\overline{F}}_{n,j_3|j}(t) - \overline{F}^0_{j_3|j}(t))\}$$
$$+\sum_{j_2\neq j'}\frac{\overline{F}^0_j(t)}{\overline{F}^0_{j_2|j}(t)}\mathrm{Cov}\{\sqrt{n}(\widehat{\overline{F}}_{n,j}(t^-) - \overline{F}^0_j(t)), \sqrt{n}(\widehat{\overline{F}}_{n,j_2|j}(t) - \overline{F}^0_{j_2|j}(t))\}\Big]$$
$$= \{\overline{F}^0_j(t)\prod_{j_1\neq j'}\widehat{\overline{F}}_{n,j_1|j}(t)\}^2\Big[\sum_{j_2\neq j'}\frac{\sigma^2_{jj_2}(t)}{(\overline{F}^0_{j_2|j}(t))^2} + \int_0^\infty \frac{dF^0_j(s)}{\pi^0_j\overline{G}^0_j(s)(\overline{F}^0_j)^2(s)}$$
$$+\sum_{j_2\neq j'}\sum_{j_3\neq j',j_2}\frac{\sigma^2_{jj_2j_3}}{\overline{F}^0_{j_2|j}(t)\overline{F}^0_{j_3|j}(t)} + \sum_{j_2\neq j'}\frac{c^{(1)}_{jj_2}(t)}{\overline{F}^0_{j_2|j}(t)}\Big].$$

and, for any sequence A_{nj} converging to A_j,

$$\lim_n \mathrm{Var}\sqrt{n}(\prod_j A_{nj} - \prod_j A_j) = \sum_j \prod_{j' \neq j} A_{j'}^2 \lim_n n\mathrm{Var}(A_{nj} - A_j)$$
$$+ \sum_j \sum_{j' \neq j} A_j A_{j'} \prod_{j', j" \neq j} A_{j"}^2 \lim_n n\mathrm{Cov}(A_{nj} - A_j, A_{nj'} - A_j).$$

Thus

$$(\sigma_{jj'}^{RC})^2 = \frac{1}{\prod_{j"} p_{jj"}^0 \}^2} \{v_{jj'}^{(2)} + v_{jj'}^{(3)} - 2c_{jj'}^{(2)}\}$$

with

$$v_{jj'}^{(2)} \equiv \lim_n \mathrm{Var}\sqrt{n}\{\int_0^\infty \widehat{\overline{F}}_{n,j}(s^-) \prod_{j" \neq j'} \widehat{\overline{F}}_{n,j"|j}(s)\, d\widehat{\Lambda}_{n,j'|j}(s) - p_{jj'}^0 \prod_{j"} p_{jj"}^0\}$$
$$= \int_0^\infty \lim_n \mathrm{Var}\sqrt{n}\{\widehat{\overline{F}}_{n,j}(s^-) \prod_{j" \neq j'} \widehat{\overline{F}}_{n,j"|j}(s) - \overline{F}_j^0(s) \prod_{j" \neq j'} \overline{F}_{j"|j}^0(s)\}\, d\Lambda_{j'|j}^0(s)$$
$$+ \int_0^\infty \{\overline{F}_j^0(s) \prod_{j" \neq j'} \overline{F}_{j"|j}^0(s)\}^2 \lim_n \mathrm{Var}\sqrt{n}(d\widehat{\Lambda}_{n,j'|j}(s) - d\Lambda_{j'|j}^0(s))$$
$$= \int_0^\infty v_{jj'}^{(1)}(s)\, d\Lambda_{j'|j}^0(s) + \int_0^\infty \{\overline{F}_j^0(s) \prod_{j" \neq j'} \overline{F}_{j"|j}^0(s)\}^2 \frac{dF_j^0(s)}{\pi_j^0 \overline{G}_j^0(s)(\overline{F}_{j'|j}^0)^2(s)},$$
$$v_{jj'}^{(3)} = \lim_n \mathrm{Var}\sqrt{n}\{\prod_{j'} \widehat{p}_{n,jj'} - \prod_{j'} p_{jj'}^0\}$$
$$= \sum_{j_1} \sigma_{jj_1}^2 \{\prod_{j_2 \neq j_1} p_{jj_2}^0\}^2 + \sum_{j_1} \sum_{j_2 \neq j_1} p_{jj_1}^0 p_{jj_2}^0 (\prod_{j_3 \neq j_1, j_2} p_{jj_3}^0)^2 \sigma_{jj_1j_2}^2,$$

and similar calculations give the expression of

$$c_{jj'}^{(2)} \equiv \lim_n \mathrm{Cov}\big[\sqrt{n}\{\int_0^\infty \widehat{\overline{F}}_{n,j}(s^-) \prod_{j" \neq j'} \widehat{\overline{F}}_{n,j"|j}(s)\, d\widehat{\Lambda}_{n,j'|j}(s) - p_{jj'}^0 \prod_{j"} p_{jj"}^0,$$
$$\sqrt{n}\{\prod_{j'"} \widehat{p}_{n,jj"} - \prod_{j"} p_{jj"}^0\}\}.$$

References

[ABGK93] Andersen,P.K., Borgan,O, Gill,R.D. and Keiding,N.: Statistical Models Based on Counting Processes. Springer, New York (1993)

[BN02] Bagdonavicius,V. and Nikulin,M.N. : Accelerated Life Models, Modeling and Statistical Analysis. Chapman and Hall, London (2002)

[GILL80] Gill,R. : Nonparametric estimation based on censored observations of a Markov renewal process. Z. Wahrsch. verw. Gebiete, **53**, 97–116 (1980)

[GILL83] Gill,R. : Large sample behaviour of the product-limit estimator on the whole line. Ann. Statist., **11**, 49–58 (1983)

[HEUT02] Heutte,N. and Huber-Carol,C. : Semi-Markov Models for Quality of Life Data with Censoring. In: Emery, M. (ed)? Statistical Methods for Quality of Life Studies, Kluwer Academic Publishers, Boston (2002)

[PONS04] O. Pons: Estimation of semi-Markov models with right-censored data. In: Balakrishnan, N. and Rao, C.R. (ed) Handbook of Statistics, **23**, Advances in Survival Analysis, Elsevier,? (2004)

[LSZ78] S.W. Lagakos, C. Sommer and M. Zelen: Semi-markov models for partially censored data. Biometrika, **65**, 311–317 (1978)

[PYKE61] R. Pyke: Markov renewal processes: definitions and preliminary properties, *Ann. Math. Statist.*, 32, 1231–1342 (1961)

[TSI75] A. Tsiatis: A nonidentifiability aspect of the problem of competing risks, Proc. Nat. Acad. Sci. USA , **37**, 20–22 (1975)

[VC84] J. Voelkel and J. Crowley: Nonparametric inference for a class of semi-markov processes with censored observations, Ann. Statist., **12**, 142–160 (1984)

Estimation Of Density For Arbitrarily Censored And Truncated Data

Catherine Huber[1], Valentin Solev[2], and Filia Vonta[3]

[1] Université René Descartes - Paris 5, 45 rue des Saints-P'eres, 75006 Paris, catherine.huber@univ-paris5.fr
[2] Steklov Institute of Mathematics at St. Petersburg, nab. Fontanki, 27 St.Petersburg 191023 Russia, solev@pdmi.ras.ru
[3] Department of Mathematics and Statistics, University of Cyprus P.O. Box 20537, CY-1678, Nicosia, Cyprus, vonta@ucy.ac.cy

Summary. We consider survival data that are both interval censored and truncated. Turnbull [Tur76] proposed in 1976 a nice method for nonparametric maximum likelihood estimation of the distribution function in this case, which has been used since by many authors. But, to our knowledge, the consistency of the resulting estimate was never proved. We prove here the consistency of Turnbull's NPMLE under appropriate conditions on the involved distributions: the censoring, truncation and survival distributions.

Key words: incomplete observations, censored and truncated data, nonparametric maximum likelihood estimation, consistency.

1 Introduction.

Very often in practice, survival data are both interval censored and truncated, as observation of the process is not continuous in time and is done through a window of time which could exclude totally some individuals from the sample. For example, the time of onset of a disease in a patient, like HIV infection or toxicity of a treatment, is not exactly known, but it is usually known to have taken place between two dates t_1 and t_2; this occurs in particular when the event of interest results in an irreversible change of state of the individual: at time t_1, the individual is in state one, while at time t_2, he is in state two. Moreover, some people can escape the sample if they are observed during a period of time not including some pair of dates t_1 , t_2 having the above property. Turnbull [Tur76] proposed a nice method for nonparametric maximum

[2] The research of the second author was supported by grants RFBR 02-01-00262, grant RFBR-DFG 04-01-04000

likelihood estimation of the distribution function in this case. His method, slightly corrected by Frydman [Fry94], has been used extensively since by several authors, and extended to semi-parametric cases (Alioum and Commenges [ACo96], Huber-Carol and Vonta, [HbV04]). But, to our knowledge, the consistency of the resulting estimates was never proved, even in the simple totally nonparametric case. We give here conditions on the involved distributions, the censoring, truncation and survival distributions, implying the consistency of Turnbull's estimate. The proofs use results of Sara Van de Geer [VdG93], Xiatong Shen [Sh97], Wing Hung Wong and Xiatong Shen [WSh95], Lucien Birgé and Pascal Massart [BiM98], Luc Devroye and Gabor Lugosi [DeL01], Nikulin and Solev [NiS02], [NiS04], on non-parametric estimation.
In section two, we give a representation of the censoring and truncation mechanisms. As it is due to a non continuous observation of the survival process, the censoring mechanism is represented as a denumerable partition of the total interval of observation time (a, b). Then a truncation is added to the censoring, conditioning the observations both of the survival and the censoring processes. The particular case of right truncation is considered.
In the next three sections, three distributions are successively studied, each being conditional on fixed values which become random in the next section.
In section three, the distribution associated with a random covering, which is a censoring set conditional on a fixed value x of the survival process. It is considered as the sum of a denumerable number of elementary probabilities, and it is proved to have a density with respect to a baseline probability.
In section four, we define the joint distribution of a pair of intervals, a censoring $L(x), R(x)$ and a truncating one $L(z), R(z)$, conditional on fixed values x and z respectively of the survival X and the right truncation Z.
Finally, in section five, we consider the distribution of the incomplete observation of X: $L(X), R(X), L(z), R(z)$, conditional on the truncating variable $Z = z$.
In section six, the non parametric maximum likelihood estimate of the density of the survival is defined in the presence of the nuisance infinite dimensional parameters introduced by the censoring and the truncation laws, using Kullback and Hellinger distances.
Finally, in the last section, conditions are found on the sets of probabilities that govern the survival process and the censoring and truncation processes that lead to consistency of the NPMLE of the density of the survival process.

2 Partitioning the total observation time

2.1 Random covering.

Let $\vartheta(x) = (L(x), R(x)]$, $x \in (a, b) \subset \mathbb{R}$ be a random covering of interval (a, b). That is $\vartheta(x)$ is a process, indexed by $x \in (a, b)$, which values are intervals, and such that with probability one

$$x \in (L(x), R(x)] \subset (a,b), \quad \bigcup_{x \in (a,b)} \vartheta(x) = (a,b).$$

When it is clear from the context, we shall identify process $\vartheta(x)$ with vector valued process $v(x) = (L(x), R(x))$, whose coordinates are the left and right ends of interval $\vartheta(x)$.

In the special case of a random covering $\vartheta(x)$ generated by a random partition, for any $x, y \in (a,b)$, with probability one

$$\vartheta(x) = \vartheta(y), \text{ or } \vartheta(x) \cap \vartheta(y) = \emptyset\,. \tag{1}$$

Conversely, let us assume that condition (1) is true. Then, with probability one, the random function $R(x)$ is a left continuous step function. Therefore, there exists a partition τ

$$\tau = \{(Y_j, Y_{j+1}],\ j = 0, \pm 1, \ldots\},$$

$$a < \ldots < Y_{-m} < \ldots < Y_0 < \ldots < Y_n \ldots < b, \ \bigcup_j (Y_j, Y_{j+1}] = (a,b), \tag{2}$$

such that

$$\vartheta(x) = \left(Y_{k(x)}, Y_{k(x)+1}\right], \ x \in (a,b). \tag{3}$$

Here

$$k = k(x) = \inf\{j : \ x \le Y_{j+1}\}\,. \tag{4}$$

From now on we assume that the random covering $\vartheta(x) = (L(x), R(x)]$, $x \in (a,b)$, satisfies condition (1) and hence may be generated by a partition τ defined in (2) – (4). Such a random covering will be called a simple random covering. For simplicity we suppose that $a = -\infty$, $b = \infty$

2.2 Short-cut covering.

Let $\vartheta(x) = (L(x), R(x)]$, $x \in \mathbb{R}$, be a simple random covering, τ be the partition associated with $\vartheta(x)$,

$$\tau = \{(Y_j, Y_{j+1}],\ j = 0, \pm 1, \ldots\},$$

$$\ldots < Y_{-m} < \ldots < Y_0 < \ldots < Y_n \ldots$$

and $\triangle = (z_1, z_2]$ be an interval, and $z = (z_1, z_2)$, $z_1 \le z_2$

For a fixed value of $\tau = t$,

$$t = \{(y_j, y_{j+1}],\ j = 0, \pm 1, \ldots\},$$

$$\ldots < y_{-m} < \ldots < y_0 < \ldots < y_n \ldots,$$

define functions

$$\kappa_1 = \kappa_1(t, z_1) = \inf\{k : y_k \geq z_1\} \quad \mathfrak{z}_1 = \mathfrak{z}_1(t, z_1) = Y_{\kappa_1}$$

$$\kappa_2 = \kappa_2(t, z_2) = \sup\{k : y_k \leq z_2\} \quad \mathfrak{z}_2 = \mathfrak{z}_2(t, z_2) = Y_{\kappa_2}$$

The short-cut covering $\vartheta_\triangle(x) = (L_\triangle(x), R_\triangle(x)]$, $x \in \triangle$, is defined below. The short-cut covering $\vartheta_\triangle(x)$ is trivial: $\vartheta_\triangle(x) = (z_1, z_2]$, if

$$\mathfrak{z}_1(t, z_1) > \mathfrak{z}_2(t, z_2),$$

else

$$(L_\triangle(x), R_\triangle(x)] = \begin{cases} (L(x), R(x)], & \text{if } x \in (\mathfrak{z}_1, \mathfrak{z}_2] \\ (z_1, \mathfrak{z}_1], & \text{if } x \in (z_1, \mathfrak{z}_1] \\ (\mathfrak{z}_2, z_2], & \text{if } x \in (\mathfrak{z}_2, z_2] \end{cases}$$

In the special case when

$$\triangle = (-\infty, z]$$

we shall use notations for corresponding short-cut covering $\vartheta_\triangle(x), x \in \triangle$, and connected objects

$$\begin{array}{c} \vartheta_z(x) = \vartheta_\triangle(x), \\ \kappa_z = \kappa(t, z) = \sup\{k : y_k \leq z\}, \ \mathfrak{z}_z = \mathfrak{z}(t, z) = Y_{\kappa_2} \\ L_z(x) = L_\triangle(x), \ R_z(x) = R_\triangle(x) \end{array} \tag{5}$$

2.3 The mechanism of truncation and censoring

The mechanism of censoring and truncating of a random variable X is defined as follows. Let X be a random variable, $\Delta = (Z_1, Z_2]$ be a random interval, $\vartheta(x) = (L(x), R(x)]$, $x \in \mathbb{R}$, be a random covering , generated by a partition τ

$$\tau = \{(Y_j, Y_{j+1}], \ j = 0, \pm 1, \ldots\},$$

$$\ldots < Y_{-m} < \ldots < Y_0 < \ldots < Y_n \ldots.$$

We denote

$$\Lambda = \Lambda(\tau) = \{Y_j,, \ j = 0, \pm 1, \ldots\}$$

We suppose that random covering $\vartheta(\cdot)$, random variable X and random interval Δ are independent, but we have not complete observations. More precisely, we suppose that random vector (X, Z_1, Z_2) is partly observable only in the case when $(L(X), R(X)] \subset \Delta$:

$$Z_1 \leq L(X) < R(X) \leq Z_2.$$

In that case the available observations are the interval $(L(X), R(X)]$ of the covering $\vartheta(\cdot)$, which contains X, and random interval $\Delta^* = (R(Z_1), L(Z_2)]$.

When $(L(X), R(X)] \not\subset \Delta$ we have not any observation.

We have to think that

1) Conditionally on a fixed value t of τ the random interval Δ is taken from the truncated distribution

$$\mathcal{P}_t\{A\} = P\{\Delta \in A \,|\, \text{ the interval } [Z_1, Z_2] \text{ contains at least two points of } \Lambda\}.$$

In other words, conditionally on fixed values of $\tau = t$ the random vector $Z = (Z_1, Z_2)$ is taken from the truncated distribution

$$P_t\{B\} = P\{Z \in B \,|\mathfrak{z}_1(t, Z_1) < \mathfrak{z}_2(t, Z_2)\};$$

2) Conditionally on a fixed value of $\tau = t$ and $\Delta = \triangle = (z_1, z_2]$, the random variable X is taken from truncated distribution

$$P_\triangle\{A\} = P\{X \in A \,|\, X \in (R(z_1), L(z_2)]\}.$$

In other words conditionally on fixed values of $\tau = t$ and $Z_1 = z_1, Z_2 = z_2$ the random variable X is taken from truncated distribution

$$P\{A \,|\, t, z_1, z_2\} = P\{X \in A \,|\, X \in (\mathfrak{z}_1(t, z_1), \mathfrak{z}_2(t, z_2)]\}.$$

We consider the simple case when for a random variable Z random interval $\Delta = (-\infty, Z]$, and use the notations that were given in (5). We denote $\mathfrak{Z}$ the random variable

$$\mathfrak{Z} = \mathfrak{z}(\tau, Z).$$

We have to think that

1) The random covering $\vartheta(\cdot)$ and the random variable Z are independent.

2) Conditionally on a fixed value of $\mathfrak{Z} = \mathfrak{z}$, the random variable X is taken from the truncated distribution

$$P_\mathfrak{z}\{A\} = P\{X \in A \,|\, X \le \mathfrak{z}\}.$$

In other words, conditionally on fixed values of $\tau = t$ and $Z = z$ the random variable X is taken from the truncated distribution

$$P\{A|t, z\} = P\{X \in A \,|\, X \le \mathfrak{z}(t, z)\}.$$

3 The distribution associated with random covering.

Let $\vartheta(x) = (L(x), R(x)]$, $x \in \mathbb{R}$, be a simple random covering. The distribution P_x of random vector $v(x) = (L(x), R(x))$ will be called the distribution, associated with random covering $\vartheta(x)$.

We assume that for all x the distribution P_x has density with respect to Lebesgue measure λ^2 on the plane $\mathbb{R}^2$,

$$r_x(u,v) = \frac{dP_x}{d\lambda^2},$$

and plan to prove in this case, that there exists a nonnegative function $r(u,v)$ such that for all x

$$r_x(u,v) = r(u,v)\mathbb{1}_{(u,\,v]}(x) \text{ (a.s.)}$$

Function $r(u,v)$ will be called the *baseline density of simple random covering* $\vartheta(x)$. It is clear that function $r(u,v)$ is the density of a σ-finite measure, but, for all x, function $r(u,v)\mathbb{1}_{(u,\,v]}(x)$ is the density of a probability measure.

It is clear that for all x

$$r_x(u,v) = r_x(u,v)\mathbb{1}_{(u,v]}(x).$$

For positive $x < y$ and nonnegative measurable function $\psi(u,v)$ such that

$$\psi(u,v) = 0, \text{ if } u < x \le v < y \text{ or } x \le u < y \le v. \tag{6}$$

Condition (6) is equivalent to the condition (on function ψ)

$$\psi(u,v)\mathbb{1}_{(u,v]}(x) = \psi(u,v)\mathbb{1}_{(u,v]}(y).$$

Therefore

$$\mathbf{E}\,\psi(L(x),R(x)) = \mathbf{E}\,\psi(L(x),R(x))\sum_k \mathbb{1}_{(Y_k,\,Y_{k+1}]}(x) =$$

$$= \sum_k \mathbf{E}\,\psi(Y_j,Y_{j+1})\,\mathbb{1}_{(Y_j,\,Y_{j+1}]}(x) = \sum_k \mathbf{E}\,\psi(Y_j,Y_{j+1})\,\mathbb{1}_{(Y_j,\,Y_{j+1}]}(y) =$$

$$= \mathbf{E}\,\psi(L(x),R(x))\sum_k \mathbb{1}_{(Y_k,\,Y_{k+1}]}(y) = \mathbf{E}\,\psi(L(y),R(y)).$$

Thus, under condition (6) on function ψ

$$\iint\limits_{u<v} \psi(u,v)r_x(u,v)\,dudv) = \iint\limits_{u<v} \psi(u,v)r_y(u,v)\,dudv,$$

and we obtain for all $u < x \le y \le v$

$$r_x(u,v) = r_y(u,v). \tag{7}$$

From (7) we conclude that there exists a nonnegative function $r(u,v)$, whose support is the set $\{(u,v):\ u < v\}$, and such that for x

$$r_x(u,v) = r(u,v)\mathbb{1}_{(u,\,v]}(x) \text{ (a.s.)}$$

It is easy to see that the baseline density $r(u,v)$ depends only on the joint distributions of vectors (Y_j, Y_{j+1}).

Now we prove that measure P_x is absolutely continuous with respect to the Lebesgue measure for all x if and only if

(i) for all j the distribution of vector $(Y_j,\, Y_{j+1})$ has density $r^j(u,v)$ with respect to the Lebesgue measure,

(ii) the series $\sum_j r^j(u,v)$ converges a.s. to a function $r(u,v)$,

(iii) the function r(u,v) satisfies the following condition:
for all x

$$r_x(u,v) = r(u,v)\mathbb{1}_{(u,\,v]}(x).$$

Indeed, suppose that for all x the distribution P_x has density $r_x(u,v)$. Let $\psi(u,v)$ be a nonnegative function, then for all j

$$\mathbf{E}\,\psi(Y_j,\, Y_{j+1})\mathbb{1}_{(Y_j,\, Y_{j+1}]}(x) \le \mathbf{E}\,\psi(L(x),\, R(x)) =$$

$$= \iint \psi(u,v)r(u,v)\mathbb{1}_{(u,\,v]}(x)\, dudv.$$

Therefore, for all x the distribution of vector $(Y_j,\, Y_{j+1})\mathbb{1}_{(Y_j,\, Y_{j+1}]}(x)$ has a density. Hence, the distribution of vector $(Y_j,\, Y_{j+1})$ also has a density $r^j(u,v)$.

We have

$$\mathbf{E}\,\psi(L(x),\, R(x)) = \sum_j \mathbf{E}\,\psi(Y_j,\, Y_{j+1})\mathbb{1}_{(Y_j,\, Y_{j+1}]}(x) =$$

$$= \sum_j \iint \psi(u,v)r^j(u,v)\mathbb{1}_{(u,\,v]}(x)\, dudv =$$

$$= \iint \psi(u,v)\left\{\sum_j r^j(u,v)\mathbb{1}_{(u,\,v]}(x)\right\} dudv.$$

So, we obtain

$$r(u,v) = \sum_j r^j(u,v) \text{ (a.s.)}.$$

Now suppose that (i), (ii) are fulfilled. Then we obtain for a nonnegative measurable function $\psi(u,v)$ (by the same way as above)

$$\mathbf{E}\,\psi(L(x),\, R(x)) =$$

$$= \iint \psi(u,v)\left\{\sum_j r^j(u,v)\right\}\mathbb{1}_{(u,\,v]}(x)\, dudv.$$

From this equality we conclude that series

$$r(u,v) = \sum_j r^j(u,v) < \infty \ \text{(a.s.)},$$

and

$$r_x(u,v) = r(u,v)\mathbb{1}_{(u,\,v]}(x).$$

4 The distribution of random vector $(L(x), R(x), L(z), R(z))$.

Now for $x < z$ we denote by $P_{x,z}$ the distribution of random vector $(L(x), R(x), L(z), R(z))$. Denote by λ^n the Lebesgue measure on $\mathbb{R}^n$. The distribution $P_{x,z}$ is not absolutely continuous with respect to the measure on λ^4. Denote by ν the measure, which is defined for continuous nonnegative functions $\psi(s) = \psi(s_1, s_2, s_3, s_4)$ by the relation

$$\iiiint \psi(s)\, d\nu = \iint \psi(s_1, s_2, s_1, s_2)\, ds_1 ds_2 +$$

$$+ \iiint \psi(s_1, s_2, s_2, s_4)\, ds_1 ds_2 ds_4 + \iiiint \psi(s_1, s_2, s_3, s_4)\, ds_1 ds_2 ds_3 ds_4.$$

We suppose that the distribution $P_{x,z}$ is absolutely continuous with respect to the measure ν and denote its density $q_{x,\,z}(s)$:

$$q_{x,\,z}(s) = q_{x,\,z}(s_1, s_2, s_3, s_4) = \frac{dP_{x,z}}{d\nu}$$

We suppose that for all $n, m > 0$ the random vector $(Y_{-m}, \ldots, Y_n)$ has a density with respect to the corresponding Lebesgue measure. For $i + 1 < j$, let function

$r_{i,j}(y_1, y_2, y_3, y_4)$ be the density of random vector $(Y_i, Y_{i+1}, Y_j, Y_{j+1})$,

$r_j(y_1, y_2, y_3)$ be the density of random vector (Y_{j-1}, Y_j, Y_{j+1}),

$r^j(y_1, y_2)$ be the density of the random vector (Y_j, Y_{j+1}).

We assume that

$$\mathfrak{d}_4(y_1, y_2, y_3, y_4) = \sum_{\substack{i,\,j:\\ i+1<j}} r_{i\,j}(y_1, y_2, y_3, y_4) < \infty \ (\lambda^4\text{-a.s.}),$$

$$\eth_3(s_1, s_2, s_3) = \sum_j r_j(s) < \infty \ \ (\lambda^3\text{-a.s.}),$$

and

$$\eth_2(y_1, y_2) = \sum_j r^j(y_1, y_2) < \infty \ \ (\lambda^2\text{-a.s.}).$$

For a nonnegative function $\psi(x), x = (x_1, x_2, x_3, x_4)$ and $x < z$ we have

$$\mathbf{E}\,\psi(L(x), R(x), L(z), R(z)) =$$

$$= \mathbf{E} \sum_{i,j} \psi(Y_i, Y_{i+1}, Y_j, Y_{j+1}) \mathbb{1}_{(Y_i, Y_{i+1}]}(x) \mathbb{1}_{(Y_j, Y_{j+1}]}(z) =$$

$$= \sum_j \mathbf{E}\,\psi(Y_j, Y_{j+1}, Y_j, Y_{j+1}) \mathbb{1}_{(Y_j, Y_{j+1}]}(x) \mathbb{1}_{(Y_j, Y_{j+1}]}(z) +$$

$$+ \sum_j \mathbf{E}\,\psi(Y_{j-1}, Y_j, Y_j, Y_{j+1}) \mathbb{1}_{(Y_{j-1}, Y_j]}(x) \mathbb{1}_{(Y_j, Y_{j+1}]}(z) +$$

$$+ \sum_{\substack{i,j: \\ i+1<j}} \mathbf{E}\,\psi(Y_i, Y_{i+1}, Y_j, Y_{j+1}) \mathbb{1}_{(Y_i, Y_{i+1}]}(x) \mathbb{1}_{(Y_j, Y_{j+1}]}(z).$$

Thus,

$$\mathbf{E}\,\psi(L(x), R(x), L(z), R(z)) =$$

$$+ \iint \psi(s_1, s_2, s_1, s_2) \eth_2(s_1, s_2) \mathbb{1}_{(s_1, s_2]}(x) \mathbb{1}_{(s_1, s_2]}(z)\, ds_1 ds_2 +$$

$$+ \iiint \psi(s_1, s_2, s_2, s_3) \eth_3(s_1, s_2, s_3) \mathbb{1}_{(s_1, s_2]}(x) \mathbb{1}_{(s_2, s_3]}(z)\, ds_1 ds_2 ds_3 +$$

$$+ \iiiint \psi(s_1, s_2, s_3, s_4) \times$$

$$\eth_4(s_1, s_2, s_3, s_4) \mathbb{1}_{(s_1, s_2]}(x) \mathbb{1}_{(s_3, s_4]}(z)\, ds_1 ds_2 ds_3 ds_4.$$

If we define $\nu-$measurable function $\eth(s|x, z), s = (s_1, s_2, s_3, s_4)$, by

$$\eth(s|x, z) = \mathbb{1}_{(s_1, s_2]}(x) \eth_*(s|z),$$

where

$$\eth_*(s \mid z) = \begin{cases} \eth_2(s_1, s_2) \mathbb{1}_{(s_1, s_2]}(z), & \text{if } s_1 = s_3 < s_2 = s_4 \\ \eth_3(s_1, s_2, s_3) \mathbb{1}_{(s_2, s_3]}(z), & \text{if } s_1 < s_2 = s_3 < s_4 \\ \eth_4(s_1, s_2, s_3, s_4) \mathbb{1}_{(s_3, s_4]}(z), & \text{if } s_1 < s_2 < s_3 < s_4 \\ 0, & \text{else} \end{cases} \quad (8)$$

then we obtain for $x < z$

$$\mathbf{E}\,\psi(L(x), R(x), L(z), R(z)) = \iiiint \psi(s)\mathfrak{d}(s \mid x, z)\, d\nu,$$

and therefore

$$q_{x,\,z}(s_1, s_2, s_3, s_4) = \mathbb{1}_{(s_1, s_2]}(x)\mathfrak{d}_*(s_1, s_2, s_3, s_4 \mid z). \tag{9}$$

5 The distribution of random vector $(L(X), R(X), L(Z), R(Z))$.

For the right truncated density function $f(x)$ we shall use the following notation

$$f_a(x) = \frac{f(x)}{\int\limits_{u \le a} f(u)\, du}\, \mathbb{1}_{(-\infty,\, a]}(x).$$

Now we suppose that for fixed z and fixed value of $\tau = t$, random variable X is taken from the truncated distribution with density $f_{\mathfrak{z}}(x)$. Here $\mathfrak{z} = \mathfrak{z}(t, z) = L(z)$. It follows from (9) that in that case the distribution P_z of random vector $(L(X), R(X), L(z), R(z))$ has density (with respect to the measure ν) $q(s_1, s_2, s_3, s_4 \mid z)$,

$$q(s_1, s_2, u, v \mid z) = \int q_{x,\,z}(s_1, s_2, u, v)\, f_u(x)\, dx,$$

and (see (8))

$$q(s_1, s_2, u, v \mid z) = \int\limits_{s_1}^{s_2} f_u(x)\, dx \,\times\, \mathfrak{d}_*(s_1, s_2, s_3, s_4 \mid z),$$

where for $s = (s_1, s_2, s_3, s_4)$

$$\mathfrak{d}_*(s|z) = \begin{cases} \mathfrak{d}_3(s_1, s_2, s_3)\mathbb{1}_{(s_2, s_3]}(z), & \text{if } s_1 < s_2 = s_3 < s_4 \\ \mathfrak{d}_4(s_1, s_2, s_3, s_4)\mathbb{1}_{(s_3, s_4]}(z), & \text{if } s_1 < s_2 < s_3 < s_4 \\ 0, & \text{else} \end{cases}$$

Therefore the distribution P_z is absolutely continuous with respect to the measure ν_*, which is defined for continuous nonnegative functions $\psi(s)$ by the relation

$$\iiiint \psi(s)\, d\nu_* =$$

$$= \iiint \psi(s_1, s_2, s_2, s_4)\, ds_1 ds_2 ds_4 + \iiiint \psi(s_1, s_2, s_3, s_4)\, ds_1 ds_2 ds_3 ds_4,$$

and

$$\frac{dP_z}{d\nu_*} = q(s \mid z).$$

Now suppose that Z is a random variable with density g, which is independent from the random covering $\vartheta(\cdot)$. For fixed values $Z = z$ and $\tau = t$, random variable X is taken from the truncated distribution with density $f_{\mathfrak{z}}(x)$, $\mathfrak{z} = \mathfrak{z}(t, z) = L(z)$. Denote by P_* the distribution of random vector $(L(X), R(X), L(Z), R(Z))$. It is clear that the distribution P_* has density $q(s)$ with respect to the measure ν_*,

$$q(s_1, s_2, u, s_4) =$$

$$\int_{s_1}^{s_2} f_u(x)\, dx \times \int \mathfrak{d}_*(s_1, s_2, u, s_4 \mid z) g(z)\, dz =$$

$$= \int_{s_1}^{s_2} f_u(x)\, dx \times \mathfrak{d}(s_1, s_2, u, s_4).$$

Now consider the random vector $W = (L(X), R(X), L(Z))$. Let μ be the measure on $\mathbb{R}^3$, defined for continuous nonnegative functions ψ by

$$\iiint \psi(s_1, s_2, s_3)\, d\mu =$$

$$= \iint \psi(s_1, s_2, s_2)\, ds_1 ds_2 + \iiint \psi(s_1, s_2, s_3)\, ds_1 ds_2 ds_3,$$

It is clear that the distribution P_W of random vector W is absolutely continuous with respect to the measure μ and

$$p(y) = p(y_1, y_2, y_3) = \frac{dP_z}{d\mu} = \int q(y_1, y_2, y_3, u) du, .$$

Therefore,

$$p(u, v, z) = \int_u^v f_z(x)\, dx \times r(u, v, z),$$

where

$$r(u, v, z) = \int \mathfrak{d}(u, v, z, x)\, dx.$$

6 Maximum likelihood estimators.

Let $W, W_1, \ldots, W_n$ be i.i.d. random vectors, $W = (L(X), R(X), L(Z))$, with unknown density

$$p(u,v,z) = r(u,v,z) \times \frac{\int_u^v f(x)\,dx}{\int_{x \le z} f(x)\,dx}$$

We assume that the baseline density r and density f belong to given sets $\mathcal{G}$ and $\mathcal{F}$ correspondingly, and specify these sets later. We set

$$\varphi(f;u,v,z) = \frac{\int_u^v f(x)\,dx}{\int_{x \le z} f(x)\,dx},$$

$$\mathcal{L} = \{p : \ p = r\,\varphi(f;\cdot), \ (r,f) \in \mathcal{G} \times \mathcal{F}\}$$

Denote by P_n the empirical measure,

$$P_n\{A\} = \frac{1}{n} \sum_{j=1}^{n} 1\!\!1_A(W_j).$$

Consider the maximum likelihood estimator $\widehat{p}_n$ for unknown $p \in \mathcal{L}$,

$$\int \ln \widehat{p}_n \, dP_n = \max_{q \in \mathcal{L}} \int \ln q \, dP_n.$$

It is clear, that $\widehat{p}_n = \widehat{r}_n \times \varphi(\widehat{f}_n;\cdot)$, where $\widehat{r}_n$ and $\widehat{f}_n$ are maximum likelihood estimators for r and f,

$$\int \ln \varphi(\widehat{f}_n;\cdot)\, dP_n = \max_{q \in \mathcal{F}} \int \ln \varphi(q;\cdot)\, dP_n,$$

$$\int \ln \widehat{r}_n \, dP_n = \max_{q \in \mathcal{G}} \int \ln q \, dP_n.$$

The estimator $\widehat{f}_n$ in general situation was suggested by Turnbull B.W. [Tur76], see also Finkelstein, D.M., Moore, D.F., Schoenfeld D.A. [FMD93].

6.1 The bracketing Hellinger ε−entropy

Let $(\mathscr{Y}, \mathscr{B}, \mu)$ be a measurable space and $Y_1, \ldots, Y_n$ be i.i.d. random elements of $\mathscr{Y}$ with common distribution $P \in \mathscr{P}$ and density f,

$$f(y) = \frac{dP}{d\mu}(y), \quad f \in \mathscr{F} = \left\{ f : \ f = \frac{dP}{d\mu}, \text{ for some } P \in \mathscr{P} \right\}.$$

For nonnegative f, g let $h(f,g)$ be the Hellinger distance,

$$h^2(f,g) = \frac{1}{2} \int_{\mathscr{Y}} \left(\sqrt{f} - \sqrt{g}\right)^2 d\mu.$$

For a pair of nonnegative functions $g^L \le g^R$ denote by $V(g^L, g^R)$ the set

$$V(g^L, g^R) = \left\{g : \; g^L \le g \le g^R\right\}.$$

Denote by $N(\varepsilon, \mathscr{F})$ the smallest value of m such that

$$\mathscr{F} \subset \bigcup_{j=1}^{m} V(g_j^L, g_j^R), \text{ where } h(g_j^L, g_j^R) \le \varepsilon, \; j = 1, \ldots, m.$$

The bracketing Hellinger ε–entropy $H(\varepsilon, \mathscr{F})$ is defined as:

$$H(\varepsilon, \mathscr{F}) = \ln N(\varepsilon, \mathscr{F}).$$

We assume that for a constant c

$$\int_{\varepsilon^2}^{\varepsilon} H^{1/2}(s, \mathscr{F})\, ds \le c\varepsilon^2 \sqrt{n}. \tag{10}$$

Theorem 1 (W.H.Wong and X.Shen). *Suppose that $f(x)$ is the true density. Then under condition (10), there exist positive constants c_1, c_2, C such that*

$$P\left\{ \sup_{\substack{h(g,f)\ge\varepsilon,\\ g\in\mathscr{F}}} \prod_{j=1}^{n} \frac{g(Y_j)}{f(Y_j)} \ge \exp\{-c_1 n\varepsilon^2\} \right\} \le C \exp\{-c_2 n \varepsilon^2\}. \tag{11}$$

Now suppose that $q_n = q_n(y; x)$ is a nonnegative function $q_n : \mathscr{Y}^n \times \mathscr{Y} \to \mathbb{R}^1$, such that

$$\int_{\mathscr{Y}} q_n(y; x)\, \mu(dx) = 1 \;\; (\mu^n\text{–a.s. on } y).$$

Here $y = (y_1 \ldots, y_n) \in \mathscr{Y}^n, x \in \mathscr{Y}, \; \mu^n = \underbrace{\mu \times \ldots \times \mu}_{n\text{–times}}$. We assume that

$$q_n(y; \cdot) \in \mathscr{F} \;\; (\mu^n\text{–a.s. on } y). \tag{12}$$

So, the random function $\widetilde{f}_n(x) = q_n(Y_1, \ldots, Y_n; x)$ may be considered as an estimator for f.

Suppose that $f(x)$ is the true density, and that function $q_n(y; \cdot)$ satisfies the condition

$$\prod_{j=1}^{n} \frac{q_n(y; y_j)}{f(y_j)} \ge \exp\left\{-c_1 n \varepsilon^2\right\} \;\; (\mu^n\text{–a.s. on } y). \tag{13}$$

Here $y = (y_1, \ldots, y_n)$.

Lemma 1. *Suppose that $f(x)$ is the true density, $\widetilde{f}_n(x)$ is an estimator for f with values in $\mathscr{F}$. Then under conditions (11),(13) for some positive constants c, C*

$$P\left\{h(\widetilde{f}_n, f) \geq \varepsilon\right\} \leq C \exp\{-cn\varepsilon^2\}. \tag{14}$$

Proof. Let $\widetilde{f}_n(\cdot) = q_n(Y_1, \ldots, Y_n; \cdot)$. We may assume that $q_n(y; \cdot) \in \mathscr{Y}$. Denote by $d(y)$ the Hellinger distance between $q_n(y; \cdot)$ and $f(\cdot)$,

$$d^2(y) = \frac{1}{2} \int\limits_{\mathscr{Y}} \left(\sqrt{q_n(y;x)} - \sqrt{f(x)}\right)^2 \mu(dx).$$

It is clear that

$$\{y : \, d(y) \geq \varepsilon, \} \subset \left\{ y : \sup_{\substack{h(g,f)\geq\varepsilon, \\ g\in\mathscr{F}}} \prod_{j=1}^{n} \frac{g(y_j)}{f(y_j)} \geq \exp\left\{-c_1 n\varepsilon^2\right\} \right\}.$$

Therefore,

$$P\left\{h(\widetilde{f}_n, f) \geq \varepsilon\right\} \leq P \left\{ \sup_{\substack{h(g,f)\geq\varepsilon, \\ g\in\mathscr{F}}} \prod_{j=1}^{n} \frac{g(Y_j)}{f(Y_j)} \geq \exp\left\{-c_1 n\varepsilon^2\right\} \right\},$$

and (14) follows from (11).

6.2 Hellinger and Kullback-Leibler distances.

Let P and Q be two measures both dominated by a σ−finite measure μ, $H^2(P, Q)$ be the Hellinger distance between P and Q,

$$H^2(P, Q) = h^2(f, q) = \frac{1}{2} \int \left(\sqrt{\frac{dP}{d\mu}} - \sqrt{\frac{dQ}{d\mu}}\right)^2 d\mu.$$

Here

$$f = \frac{dP}{d\mu}, \quad q = \frac{dQ}{d\mu}.$$

Consider the Kullback-Leibler distance

$$K(f, q) = \int\limits_{f>0} [\ln(f/q)] \, f \, d\mu = \int\limits_{f>0} \ln(f/q) \, dP.$$

Here P is the probability distribution with density f with respect to the measure μ.

Let $X_1, \ldots, X_n$ be i.i.d random variables with the common distribution $P \in \mathcal{P}$ and density $f \in \mathcal{F}$, P_n be the empirical distribution

$$P_n\{A\} = \sum_{j=1}^{n} \delta_{X_j}\{A\}, \quad \text{where } \delta_{X_j}\{A\} = \begin{cases} 1, \text{ if } X_j \in A \\ 0, \text{ if } X_j \notin A \end{cases}$$

Suppose we have to estimate the unknown density $f \in \mathcal{F}$ on the observations $X_1, \ldots, X_n, \ldots$ with common distribution $P \in \mathcal{P}$. R. Fisher suggested to minimize the functional $K(\cdot, f)$ on empirical data to choose estimator $\widehat{f}_n$. Namely, we put

$$K_n(f, g) = \int_{f>0} \ln(f/g)\, dP_n.$$

Here f is the true density. Let $\widehat{f}_n$ be a point of $\mathcal{F}$ which minimizes the functional $K_n(\cdot, f)$:

$$\int_{f>0} \ln\left(f/\widehat{f}_n\right) dP_n \le \int_{f>0} \ln(f/g)\, dP_n \ \text{ for all } g \in \mathcal{F}. \tag{15}$$

The estimator $\widehat{f}_n$ of f which is defined in (15), is called the maximum likelihood estimator since $\widehat{f}_n$ is a point of maximum, on $\mathcal{F}$, for the likelihood function $\mathcal{L}$ (function of g)

$$\mathcal{L}(g \,|\, X_1, \ldots, X_n) = \prod_{j=1}^{n} g(X_j).$$

Let $\mathscr{S}$ and $\mathscr{G}$ be two classes of nonnegative functions, such that for any $s \in \mathscr{S}$ and $g \in \mathscr{G}$

$$\int s(x)\, g(x)\, d\mu = 1.$$

We suppose that

$$\mathcal{F} = \{f : f = s\,g, \text{ for some } s \in \mathscr{S} \text{ and } g \in \mathscr{G}\}.$$

We denote by $P(f)$ the distribution with density f.

Now suppose we have to estimate unknown function $s \in \mathscr{S}$ on the observations $X_1, \ldots, X_n, \ldots$ with common distribution $P \in \mathcal{P}$ and density $f = s\,g \in \mathcal{F}$. The maximum likelihood estimator $\widehat{s}_n$ of s is defined by the relation

$$\int \ln(s/\widehat{s}_n)\, dP_n \le \int \ln(s/g)\, dP_n, \; g \in \mathscr{S}. \tag{16}$$

Lemma 2. *Suppose that P is the true distribution with density $f = s\,g$, then*

$$0 \leq \int\limits_{f>0} \ln \frac{s}{\widehat{s}_n}\, dP \leq \int\limits_{f>0} \ln \frac{\widehat{s}_n}{s}\, d\big(P_n - P\big).$$

Proof. It is clear that it is sufficient to prove that

$$\int\limits_{f>0} \ln \frac{\widehat{s}_n}{s}\, dP_n \geq 0.$$

It follows from (16).

Lemma 3. *Suppose that $\widetilde{g}_n$ is a nonnegative random function, such that*

$$\int \widehat{s}_n\, \widetilde{g}_n\, d\mu = \int s\, \widetilde{g}_n\, d\mu = 1,$$

$P(s\,\widetilde{g}_n)$ is the distribution with density $s\,\widetilde{g}_n$, then

$$0 \leq \int\limits_{f>0} \ln \frac{s}{\widehat{s}_n}\, dP(s\,\widetilde{g}_n) \leq \int\limits_{f>0} \ln \frac{\widehat{s}_n}{s}\, d\big(P_n - P(s\,\widetilde{g}_n)\big).$$

Proof. Lemma 4 can be proved in the same way as lemma 2.

Denote $\widehat{g}_n$ the maximum likelihood estimator of g,

$$\int \ln\,(g/\widehat{g}_n)\; dP_n \leq \int \ln\,(g/h)\; dP_n,\; h \in \mathscr{G}.$$

Corollary 1. *Let $P(s\,\widehat{g}_n)$ be the distribution with density $s\,\widehat{g}_n$, then*

$$0 \leq \int\limits_{f>0} \ln \frac{s}{\widehat{s}_n}\, dP(s\,\widehat{g}_n) \leq \int\limits_{f>0} \ln \frac{\widehat{s}_n}{s}\, d\big(P_n - P(s\,\widehat{g}_n)\big).$$

Corollary 2. *Let $P(s\,\widehat{g}_n)$ be the distribution with density $s\,\widehat{g}_n$, then*

$$\int |s - \widehat{s}_n|\, \widehat{g}_n\, d\mu \leq \sqrt{2 \int\limits_{f>0} \ln \frac{\widehat{s}_n}{s}\, d\big(P_n - P(s\,\widehat{g}_n)\big)}\,.$$

Proof. Corollary 1 follows from lemma 4 if we take $\widetilde{g}_n = \widehat{g}_n$.

Lemma 4 (Sara van de Geer). *Let P be the true distribution with density $f \in \mathfrak{F}$, and $\widehat{f}_n$ be the maximum likelihood estimator for f, then*

$$h^2(\widehat{f}_n, f) \leq \int\limits_{f>0} \left(\sqrt{\frac{\widehat{f}_n}{f}} - 1 \right) d(P_n - P).$$

Lemma 5. *Suppose that $\widetilde{g}_n$ is a nonnegative random function, such that*

$$\int s\, \widetilde{g}_n \, d\mu = 1,$$

$P(s\,\widetilde{g}_n)$ is the distribution with density $s\,\widetilde{g}_n$, then

$$\frac{1}{2} \int\limits_{f>0} \left(\sqrt{s} - \sqrt{\widehat{s}_n} \right)^2 \widetilde{g}\, d\mu \leq \int\limits_{f>0} \left(\sqrt{\frac{\widehat{s}_n}{s}} - 1 \right) d(P_n - P(s\,\widetilde{g}_n)). \tag{17}$$

Proof. We rewrite the proof of Sara van de Geer [VdG93].

$$0 \leq \frac{1}{2} \int\limits_{f>0} \ln \frac{\widehat{s}_n}{s} \, dP_n \leq \int\limits_{f>0} \left(\sqrt{\frac{\widehat{s}_n}{s}} - 1 \right) dP_n =$$

$$= \int\limits_{f>0} \left(\sqrt{\frac{\widehat{s}_n}{s}} - 1 \right) d(P_n - P(s\,\widetilde{g}_n)) + \int\limits_{f>0} \left(\sqrt{\frac{\widehat{s}_n}{s}} - 1 \right) dP(s\,\widetilde{g}_n).$$

Since

$$\int\limits_{f>0} \left(1 - \sqrt{\frac{\widehat{s}_n}{s}} \right) dP(s\,\widetilde{g}_n) = \frac{1}{2} \int\limits_{f>0} \left(\sqrt{s} - \sqrt{\widehat{s}_n} \right)^2 \widetilde{g}\, d\mu,$$

we obtain (17).

Corollary 3. *Let $P(s\,\widehat{g}_n)$ be the distribution with density $s\,\widehat{g}_n$, then*

$$\frac{1}{2} \int\limits_{f>0} \left(\sqrt{s} - \sqrt{\widehat{s}_n} \right)^2 \widehat{g}_n \, d\mu \leq \int\limits_{f>0} \left(\sqrt{\frac{\widehat{s}_n}{s}} - 1 \right) d(P_n - P(s\,\widehat{g}_n)).$$

6.3 Estimation in the presence of a nuisance parameter

Now we consider the following case

$$\mathcal{F} = \{ f : \, f(x) = f_{s,g}(x) = s(x)\, g(x), \text{ for some } s \in \mathscr{S} \text{ and } g \in \mathscr{G} \, \}.$$

Here s is the parameter of interest, g is the nuisance parameter. Let ρ be a metric on $\mathscr{S}$. Denote

$$\delta(\varepsilon) = \inf_{s,s_*,g} h(f_{s,g}, f_{s_*,g}), \tag{18}$$

where inf is taken on all $g \in \mathscr{D}$ and all $s, s_* \in \mathscr{S}$ such that $\rho(s, s_*) \geq \varepsilon$. It is clear that if

$$h(f_{s,g}, f_{s_*,g}) < \delta(\varepsilon),$$

then

$$\rho(s, s_*) < \varepsilon.$$

So, if for any $\varepsilon > 0$ the value

$$\delta(\varepsilon) > 0, \tag{19}$$

then from

$$h(f_{s,g}, f_{s_n,g}) \to 0$$

follows

$$\rho(s, s_n) \to 0.$$

But condition (19) is never carried out. Therefore we need to assume that inf in (18) is taken over all

$$g \in \mathscr{D} \text{ such that } g \in V(g_*)$$

and all

$$s, s_* \in \mathscr{S} \text{ such that } \rho(s, s_*) \geq \varepsilon.$$

Here g_* is a known point of $\mathscr{D}$, $V(g_*)$ is a neighborhood of g_*. It is clear that function $\delta(\varepsilon)$ depends on $V(g_*)$.

We denote P_f the distribution with density f.

Lemma 6. *Let s_n, g_n be estimators of s, g, and $V(g_n)$ a neighborhood of g_n such that*

$$\inf_{f=f_{s,g}\in\mathcal{F}} P_f \left\{g \in V(g_n)\right\} \to 1, \text{ as } n \to \infty,$$

and for any $\varepsilon > 0$

$$\delta(\varepsilon) = \inf_{\rho(s,s_n)\geq\varepsilon,\, g\in V(g_n)} h(f_{s,g}, f_{s_*,g}) > 0.$$

If for any $\varepsilon > 0$

$$\sup_{f=f_{s,g}\in\mathcal{F}} P_f \left\{h(f_{s,g}, f_{s_n,g_n}) > \varepsilon\right\} \to 0, \text{ as } n \to \infty,$$

Then

$$\sup_{f=f_{s,g}\in\mathcal{F}} P_f \left\{\rho(s, s_n) > \varepsilon\right\} \to 0, \text{ as } n \to \infty.$$

Let $W, W_1, \ldots, W_n$ be i.i.d. random vectors, $W = (L(X), R(X), L(Z))$, with unknown density

$$p(u,v,z) = p^{r,f}(u,v,z) = r(u,v,z) \times \frac{\int_u^v f(x)\,dx}{\int_{x\le z} f(x)\,dx}$$

We use notation $\widehat{f}_n$ for maximum likelihood estimator of f.

We suppose that the baseline density r and density f belong to given sets $\mathcal{G}$ and $\mathcal{F}$ correspondingly. And denote

$$\mathscr{P} = \left\{p : p = p^{r,f},\ r \in \mathcal{G},\ f \in \mathcal{F}\right\}.$$

We assume that the parametric set $\mathscr{P}$ is totally bounded in the Hellinger metric. Moreover, for a constant $C = C_{\mathscr{P}}$ and $\varepsilon > 0$ there exist finite coverings

$$V(\varepsilon) = \left\{V(f_i^L, f_i^R),\ i = 1, \ldots, m\right\} \text{ and } W(\varepsilon) = \left\{W(r_j^L, r_j^R),\ j = 1, \ldots, k\right\}$$

of sets $\mathcal{F}$ and $\mathcal{G}$:

$$\mathcal{F} \subset \bigcup_{i=1}^m V(f_i^L, f_i^R),\ \ \mathcal{G} \subset \bigcup_{j=1}^k W(r_j^L, r_j^R);$$

and finite covering

$$U(\varepsilon) = \left\{U(p_{i,j}^L, p_{i,j}^R),\ i = 1, \ldots, m;\ j = 1, \ldots, k\right\},$$

of the set $\mathscr{P}$: $\mathscr{P} \subset \bigcup_{i,j} U(p_{i,j}^L, p_{i,j}^R)$, such that

C$_1$ $\quad \left\{p : p = p^{r,f},\ \text{for some } r \in W(r_j^L, r_j^R),\ f \in V(f_i^L, f_i^R)\right\} \subset U_{i,j} = U(p_{i,j}^L, p_{i,j}^R);$

C$_2$ $\quad h(p_{i,j}^L, p_{i,j}^R) \le \varepsilon,\ h(f_i^L, f_i^R) \le \varepsilon;$

C$_3$ $\quad \int p_{i,j}^R\, d\mu < C_{\mathscr{P}}, \quad \int f_i^R\, dx < C_{\mathscr{P}};$

C$_4$ $\quad$ for any $\varepsilon > 0$, $z_0 > 0$

$$\inf_{p \in U_{i,j}} \int_{v-u \le \varepsilon,\ z \ge z_0} p(u,v,z)\, d\mu > 0;$$

Theorem 2 (consistency of the Non Parametric Maximum Likelihood estimate of f). *Under conditions* **C**$_1$ – **C**$_4$ *for any* $\varepsilon > 0$

$$\sup_{p = p^{r,f} \in \mathscr{P}} P\left\{h\left(\widehat{f}_n, f\right) > \varepsilon\right\} \to 0, \ \text{as } n \to \infty.$$

References

[BiM98] L. Birgé and P. Massart. Minimum contrast estimators on sieves:exponential bounds and rates of convergence. *Bernoulli*, 4:329–375, 1998.

[DeL01] L. Devroye and G. Lugosi *Combinatorial methods in density estimation.* Springer-Verlag, 2001.

[Fin04] J. P. Fine, M. R. Kosorok, and B. L.Lee Robust Inference for univariate proportional hazards frailty regression models. *AS*, 32, 4:1448–1491, 2004.

[FMD93] D. M. Finkelstein, D. F. Moore, and D. A.Schoenfeld A proportional hazard model for truncated aids data. *Biometrics*, 49:731–740, 1993.

[Tur76] B. W. Turnbull. The empirical distribution function with arbitrary grouped, censored and truncated data. *Journal of the Royal Statistical Siciety*, 38:290–295, 1976.

[HbV04] C. Huber-Carol and F. Vonta. Semiparametric Transformation Models for Arbitrarily Censored and Truncated Data . *in "Parametric and Semiparametric Models with Applications to Reliability, Survival Analysis and Quality of Life", Birkhauser ed.* ,167-176 , 2004.

[VdG93] S. Van de Geer. Hellinger–consistenscy of certain nonparametric maximum likelihood estimators. *The Annals of Statistics*, 21:14–44, 1993.

[WSh95] Wing Hung Wong and N. Xiatong Shen. Probability inequalities and convergence rates of sieve mles. *The Annals of Statistics*, **23**, 2:339–362, 1995.

[Sh97] N. Xiatong Shen. On methods sieves and penalization. *The Annals of Statistics*, 6:339–362, 1997.

[ACo96] A. Alioum and D. Commenges. A proportional hazard model for arbitrarily censored and truncated data. *Biometrics*, 52:512–524, 1996.

[Fry94] H. Frydman. A note on nonparametric estimation of the distribution function from interval-censored and truncated observations. *Journal of the Royal Statistical Society, Series B*, 56:71–74, 1994.

[NiS02] M. Nikulin and V. Solev. Testing problem for Increasing Function in a Model with Infinite Dimensional Parameter. In: C.Huber-Carol, N.Balakrishnan, M.Nikulin, M.Mesbah (eds) Goodness-of-fit Tests and Model Validity. Birkhauser: Boston, 477–494, 2002.

[NiS04] M. Nikulin and V. Solev. Probléme de l'estimation et ε-entropie de Kolmogorov. In : E.Charpentier, A.Lesne, N.Nikolski (eds) L'Héritage de Kolmogorov en mathématiques. Belin : Paris, 121–150, 2004.

Statistical Analysis of Some Parametric Degradation Models *

Waltraud Kahle and Heide Wendt

Otto-von-Guericke-University, Faculty of Mathematics,
D-39016 Magdeburg, Germany
waltraud.kahle@mathematik.uni-magdeburg.de

Summary. The applicability of purely lifetime based statistical analysis is limited due to several reasons. If the random event is the result of an underlying observable degradation process then it is possible to estimate the parameters of the resulting lifetime from observations of these process. In this paper we describe the degradation by a position-dependent marked doubly stochastic Poisson process. The intensity of such processes is a product of a deterministic function and a random variable Y which leads to an individual intensity for each realization. Our main interest consists in estimating the parameters of the distribution of Y under the assumption that the realization of Y is not observable.

1 Introduction

One of the simplest models for describing degradation is the Wiener process with linear drift. The design of the mathematical model is based on the assumption of an additive accumulation of degradation without any variation in the tendency of the degradation intensity. Some of such models and their parameter estimations are described in [KaL98]. A similar model and its application in medicine is described in [DoN96]. Several generalizations of this model were given. It is possible to include measurement errors [Whi95], or to transform the time scale [WhS97]. Some more general models have been developed in [BaN01] and [BBK02]. The advantages of using the Wiener process and its generalizations for describing the damage process are its simple form (at least for the univariate Wiener process) and, secondly, that a statistical analysis can be carried out for observations at any discrete time points. But these models have also disadvantages: It is possible that the damage is decreasing in any interval, which is difficult to interpret in practical applications. The second disadvantage is that these models become very complicated if a nonlinear drift is assumed. But for many products we can expect an increasing damage which becomes faster over time.

*This research was supported by DFG # Ka 1011/3-1

Actually, we consider a degradation process (Z_t) whose paths are monotone increasing step functions. For modeling it, we use marked point processes $\Phi = ((T_n, X_n))_{n\geq 1}$, presented in detail e.g. in [LaB95] or [ABG93]. The cumulative process (Z_t) is assumed to be generated by a position–dependent marking of a doubly stochastic Poisson process (T_n). The doubly stochastic Poisson process was introduced by Cox [Cox55]. Cramer [Cra66] applied it in risk theory, and Grandell [Gra91] gave a detailed discussion of these processes and their impact on risk theory. Further applications of the doubly stochastic Poisson process (T_n) may be found in reliability theory, medicine and queuing theory [Bre81], [ABG93], [Gra97]. Our aim is to describe suitable models for degradation accumulation. In section 2 the model is described. In section 1 and 4 maximum likelihood and moment estimates are found for the (in our view) most interesting parameters of the model. Section 5 contains some results of a simulation study.

2 A Degradation Model

We consider a shock model which is well known in reliability. The random variable T_n $(n \geq 1)$ is the time of the n–th shock. We suppose

$$T_n < T_{n+1} \quad \text{if } T_n < \infty \quad \text{and} \quad T_n = T_{n+1} = \infty \quad \text{otherwise}\,.$$

Every shock causes an random increment of degradation. The size of the n–th increment of the cumulative degradation process $(Z(t))_{t\geq 0}$ is given by a nonnegative random variable X_n $(n \geq 1)$. Thus,

$$Z(t) = \sum_{n=1}^{\infty} I(T_n \leq t) \cdot X_n$$

where $I(T_n \leq t)$ is an indicator function:

$$I(T_n \leq t) = \begin{cases} 1 & \text{if} \quad T_n \leq t \\ 0 & \text{if} \quad \text{otherwise} \end{cases}$$

describes the total amount of degradation at time t. The sequence $\Phi = ((T_n, X_n))$ is called a marked point process, and $\Phi(t)$ is defined as the random variable representing the number of events occurred up to time t. Frequently, it is of interest to discuss the first passage problem that the process (Z_t) exceeds a pre–specified constant threshold level $h > 0$ for the first time. This first passage time is the random lifetime of the item. It is also possible to regard a (random) state of first X_0 at time $T_0 := 0$. Then the corresponding first passage time Z^h is given as

$$Z^h = \inf\{t : \sum_{n=0}^{\infty} I(T_n \leq t) \cdot X_n \geq h\} \tag{1}$$

Let us mention Z^h coincide with some T_m for $m \in \mathbf{N}$.

Now we make some assumptions to specify the degradation model.

2.1 The distribution of (T_n)

The cumulated stochastic intensity $\overline{\nu}(t)$ of (T_n) is assumed to be given by $\overline{\nu}(t) = Y \cdot \eta(t)$, where $\eta(t)$ is a deterministic function with derivative $\xi(t) \geq 0$. Hence, given the outcome $Y = y$ the random number $\Phi(t)$ of shocks up to time t is Poisson distributed with mean $y \cdot \eta(t)$. Each realization of the degradation process has its own individual intensity. Consequently, it is possible to model different environmental conditions or different frailties for each individual. The unconditional distribution of $\Phi(t)$ is given by

$$p_k(t) = P(\Phi(t) = k) = E\left[\frac{[Y\eta(t)]^k}{k!}\exp(-Y\eta(t))\right]$$

$$= \int_0^\infty \frac{[y\eta(t)]^k}{k!} e^{-y\eta(t)} dF_Y(y)\ , \quad k = 0, 1, \ldots\ . \tag{2}$$

The sequence (T_n) is called a doubly stochastic Poisson process. Special cases of this process are the mixed Poisson process where $\eta(t) = t$ and the non homogeneous Poisson process where $P(Y = y_0) = 1$.

The following types belong to the most common models for η:

1. Weibull type: $\eta(t) = t^{\alpha+1} \quad (\alpha > -1)$
2. log–linear type: $\eta(t) = t \cdot e^{\alpha t^\gamma} \quad (\alpha \geq 0,\ \gamma \geq 0)$
3. logistic type: $\eta(t) = t \cdot [1 + \ln(1 + \alpha t^\gamma)] \quad (\alpha \geq 0,\ \gamma > -1)$.

The frailty variable Y is a nonnegative random variable which can be specified, too:

1. Y is rectangular distributed in $[a, b]$ with $0 \leq a < b$. Then we get from equation (2) for $\eta(t) > 0$ by partial integration and with the convention $0^0 := 1$

$$p_k(t) = \frac{1}{\eta(t)\,(b-a)} \sum_{u=0}^{k} \left[\frac{[a\,\eta(t)]^u}{u!} e^{-a\,\eta(t)} - \frac{[b\,\eta(t)]^u}{u!} e^{-b\,\eta(t)}\right]\ . \tag{3}$$

2. Let $Y - y_0$ be Gamma distributed with parameters $c > 0$ and $b > 0$ and pdf

$$f_Y(y) = I(y \geq y_0)\,\frac{c^b}{\Gamma(b)}(y - y_0)^{b-1} e^{-c(y-y_0)}\ .$$

Using $\int_{y_0}^\infty \frac{[c+\eta(t)]^{k-u+b}}{\Gamma(k-u+b)}(y - y_0)^{k-u+b-1} e^{-[c+\eta(t)]\cdot(y-y_0)} dy = 1$ we get

$$p_k(t) = \sum_{u=0}^{k} \frac{\Gamma(k-u+b)}{\Gamma(b)\,\Gamma(k+1)} \binom{k}{u} \left[\frac{c}{c+\eta(t)}\right]^b \left[\frac{\eta(t)}{c+\eta(t)}\right]^k \times$$

$$\times\,(y_0\,[c+\eta(t)])^u\, e^{-y_0\,\eta(t)}\ , \tag{4}$$

If we use the notations

$$q(t) := \frac{c}{c+\eta(t)} = \frac{c\eta^{-1}(t)}{c\eta^{-1}(t)+1}$$

and

$$y_0[c+\eta(t)] = y_0\eta(t)[c\eta^{-1}(t)+1]$$

then we get

$$p_k(t) = \sum_{u=0}^{k} \left(\frac{[y_0\eta(t)]^u}{u!} e^{-y_0\eta(t)} \right) \times$$
$$\times \left(\frac{\Gamma(k-u+b)}{\Gamma(b)\,\Gamma(k-u+1)}\, q(t)^b \,(1-q(t))^{k-u} \right) . \qquad (5)$$

Hence, $p_k(t)$ are the probabilities of the Delaport distribution with parameters $\frac{c}{\eta(t)}$, b and $y_0 \cdot \eta(t)$. From (5) the random number of shocks $\Phi(t)$ can be interpreted as a sum of two independent random variables $W_1(t)$ and $W_2(t)$ where $W_1(t)$ is Poisson distributed with expectation $y_0\eta(t)$ and $W_2(t)$ is negative binomial distributed with parameters $q(t) \in (0,1)$ and $b > 0$. In the special case of $y_0 = 0$ $\Phi(t) = W_2(t)$ is negative binomial distributed and if Y is exponential distributed ($b = 1$) we get the geometrical distribution for $W_2(t)$ in $\Phi(t) = W_1(t) + W_2(t)$.

3. Let Y be inverse Gaussian distributed with pdf

$$f_Y(y) = I(y \geq 0) \sqrt{\frac{\beta}{2\pi\, y^3}} \exp\left(-\frac{1}{2}\, \frac{\beta(y-\mu)^2}{\mu^2\, y} \right) .$$

From (2) we get

$$p_k(t) = \int_0^\infty \frac{(y\eta(t))^k}{k!} \sqrt{\frac{\beta}{2\pi\, y^3}}\, e^{-\frac{\beta\,[y\sqrt{1+2\eta(t)\mu^2/\beta}-\mu]^2}{2\,\mu^2\, y}} dy \times$$
$$\times e^{-\frac{\beta}{\mu}(\sqrt{1+2\eta(t)\mu^2/\beta}-1)}$$
$$= e^{-\frac{\beta}{\mu}(\sqrt{1+2\eta(t)\mu^2/\beta}-1)} \frac{\eta(t)^k}{k!\,\sqrt{(1+2\eta(t)\mu^2/\beta)^k}} \cdot E[W^k] \quad .$$

The moments of order k of the inverse Gaussian distribution are given by

$$E[W^k] = \mu^k \sum_{u=0}^{k-1} \frac{(k-1+u)!}{(k-1-u)!\,u!} \left[\frac{\mu}{2\beta\,\sqrt{1+2\eta(t)\mu^2/\beta}} \right]^u .$$

Finally we get

$$p_k(t) = \exp\left(-\frac{\beta}{\mu}(\sqrt{1+2\eta(t)\mu^2/\beta}-1) \right) \left[\frac{\mu\,\eta(t)}{\sqrt{1+2\eta(t)\mu^2/\beta}} \right]^k \frac{1}{k!} \times$$
$$\times \sum_{u=0}^{k-1} \frac{(k-1+u)!}{(k-1-u)!\,u!} \left[\frac{\mu}{2\beta\,\sqrt{1+2\eta(t)\mu^2/\beta}} \right]^u . \qquad (6)$$

2.2 Marking the sequence (T_n)

Next we consider a marking of the sequence (T_n). At every time point T_n a shock causes a random degradation. We describe the degradation increment at T_n by the mark X_n. $\Phi = ((T_n, X_n))$ is said to be a position–dependent G–marking of (T_n) if X_1, $X_2, \ldots$ are conditionally independent given (T_n):

$$P(X_n \in B \,|\, (T_n)) = G(T_n, B) \ . \tag{7}$$

Moreover, we assume that each mark X_n and Y are conditionally independent given (T_n), i.e. $P(X_n \in B \,|\, (T_n)\, , Y\,) = P(X_n \in B \,|\, (T_n))$. Note that the distribution of the $n - th$ degradation increment X_n depends on the random time of the n-th shock. With a position-dependent-marking it is possible to describe degradation processes where the degradation becomes faster (or slower) with increasing time. We want to give two simple examples.

1. Let $t_0 \geq 0$ be a fixed time and let (U_n), (V_n) be two sequences of iid random variables with cdf F_U and F_V, respectively. The sequence of degradation increments (X_n) is defined by

$$X_n := I(T_n \leq t_0)\, U_n + I(T_n > t_0)\, V_n$$

and $G(t, [0, x])$ is given by

$$G(t, [0, x]) = I(t \leq t_0)\, F_U(x) + I(t > t_0)\, F_V(x) \ . \tag{8}$$

That means that at time t_0 the distribution of degradation increments is changing. For $t_0 = 0$ we get the independent marking.

2. Let (U_n) be a sequence of non–negative iid random variables with the density f_U and let δ be a real number. We assume that the sequence (U_n) is independent on (T_n). The sequence (X_n) is defined by $X_n = U_n \cdot e^{\delta T_n}$. That means we get damage increments which tend to be increasing ($\delta > 0$) or decreasing ($\delta < 0$). The stochastic kernel G is given by

$$G(t, B) \;=\; \int_B f_U(x \cdot e^{-\delta t}) \cdot e^{-\delta t} \mathrm{d}x \quad .$$

Again, for $\delta = 0$ we have the special case of independent marking. In this case G defines a probability measure which is independent on the time t.

Last we have to specify the distribution of the marks. This can be any distribution law with nonnegative realizations, such as the exponential, gamma, Weibull, lognormal or Pareto.

For practical applications it is necessary to estimate the parameters of all considered distributions. That can be done or by likelihood theory or by the method of moments. The likelihood function for such degradation models and parameter estimates are given in detail in [WeK04]. Some characteristics of the process such as the cumulative degradation at any time t, the moments of the counting process, and others, are developed. Here we will restrict us to the estimation of the distribution parameters of Y by maximum likelihood and moment methods. We can have different levels of information observing the degradation process:

1. All random variables, Y, (T_n), and (X_n) are observable. Then the likelihood function is a product of three densities and the parameters can be estimated independently for each random variable. This case is not very realistic.
2. More interesting is the the assumption that we can observe each time point of a shock and each increment of degradation but cannot observe the random variable Y. This is a more realistic assumption because Y is a variable which describes the individual shock intensity for each item, a frailty variable.
3. In many situations it might be possible that a failure is the result of an degradation process but we cannot observe the underlying degradation. By the maximum likelihood method it is possible to estimate all parameters in the model because the distribution of the the first passage time contains all these parameters.

3 Maximum Likelihood Estimates

Let θ be given as $\theta = (\theta^Y, \theta^T, \theta^X) \in R^p$ with $p = u+v+w$. Here, $\theta^Y \in R^u$ is a parameter of the distribution function F_Y of Y, $\theta^T \in R^v$ denotes a parameter of the deterministic terms η and its derivative ξ, respectively. And $\theta^X \in R^w$ represents a parameter of the distribution of degradation increments. Under the assumptions of section 2 we get the following stochastic intensity of the marked point process $\Phi = ((T_n, X_n))$

$$\lambda(t, B; \theta) = Y \cdot \xi(t; \theta^T) \cdot G(t, B; \theta^X) , \qquad B \in \mathcal{B}^+ \quad .$$

If we have the full information about the degradation process then it can be shown that the likelihood function consists of three independent parts, each of them contains the full information about θ^Y, θ^T, and θ^X, respectively. If we want to estimate θ^Y, then we have the classical problem of estimating parameters from a sample of m iid observations [Wen99].
In [WeK04] it is shown that in the second case (Y is not observable) the intensity $\widetilde{\lambda}$ is given by

$$\widetilde{\lambda}(t, B; \theta) = \xi(t; \theta^T)\, G(t, B; \theta^X)\, \frac{\int_0^\infty y^{\Phi(t-)+1} e^{-y\,\eta(t;\theta^T)} F_Y(\mathrm{d}y; \theta^Y)}{\int_0^\infty y^{\Phi(t-)} e^{-y\,\eta(t;\theta^T)} F_Y(\mathrm{d}y; \theta^Y)} \,. \tag{9}$$

The essential part for estimating the parameter θ^Y is the last term

$$\lambda^* = \frac{\int_0^\infty y^{\Phi(t-)+1} e^{-y\,\eta(t;\theta^T)} F_Y(\mathrm{d}y;\theta^Y)}{\int_0^\infty y^{\Phi(t-)} e^{-y\,\eta(t;\theta^T)} F_Y(\mathrm{d}y;\theta^Y)}$$

which can be interpreted as the conditional expectation of Y given the history of observation. It is easy to see that this term depends only on θ^Y and θ^T. Our aim is to determine an estimator of θ^Y based on $m \geq 1$ independent copies of the process Φ. Let $\Phi_i(t)$ be the observed number of shocks in the i–th copy ($i = 1,\dots,m$). For the three special distributions of Y introduced in section 2 we get the following essential parts of the process intensity and resulting maximum likelihood estimates:

1. If Y is rectangular distributed in $[a,b]$:

$$\lambda^* = \frac{\Phi(t-)+1}{\eta(t;\theta^T)} \cdot \frac{\sum\limits_{u=0}^{\Phi(t-)+1} \left(\frac{[b\,\eta(t;\theta^T)]^u}{u!} e^{-b\,\eta(t;\theta^T)} - \frac{[a\,\eta(t;\theta^T)]^u}{u!} e^{-a\,\eta(t;\theta^T)} \right)}{\sum\limits_{u=0}^{\Phi(t-)} \left(\frac{[b\,\eta(t;\theta^T)]^u}{u!} e^{-b\,\eta(t;\theta^T)} - \frac{[a\,\eta(t;\theta^T)]^u}{u!} e^{-a\,\eta(t;\theta^T)} \right)} \,.$$

   For this distribution we get two likelihood equations which are linear dependent and which both leads to

$$\frac{1}{m} \sum_{i=1}^m \Phi_i(t) = \frac{\hat{a}+\hat{b}}{2} \cdot \eta(t;\hat{\theta}^T) \quad .$$

   Consequently, it is not possible to estimate both parameters a and b.

2. If $(Y - y_0)$ is gamma distributed:

$$\lambda^* = \frac{\Phi(t-)+1}{c+\eta(t;\theta^T)} \cdot \frac{\sum\limits_{u=0}^{\Phi(t-)+1} \frac{\Gamma(\Phi(t-)+1-u+b)}{u!\,(\Phi(t-)+1-u)!} [y_0\,(c+\eta(t;\theta^T))]^u}{\sum\limits_{u=0}^{\Phi(t-)} \frac{\Gamma(\Phi(t-)-u+b)}{u!\,(\Phi(t-)-u)!} [y_0\,(c+\eta(t;\theta^T))]^u} \qquad (0^0 := 1)\,.$$

   The likelihood equations can be found to be

$$0 = \sum_{i=1}^m \left\{ \frac{\hat{b}}{\hat{c}} - \frac{\Phi_i(t)+\hat{b}}{\hat{c}+\eta(t;\hat{\theta}^T)} + \hat{y}_0 \frac{\mathcal{U}_i(\,\Phi_i(t)-1\,)}{\mathcal{U}_i(\,\Phi_i(t)\,)} \right\} \tag{10}$$

$$0 = \sum_{i=1}^m \left\{ -\eta(t;\hat{\theta}^T) + (\hat{c}+\eta(t;\hat{\theta}^T)) \frac{\mathcal{U}_i(\,\Phi_i(t)-1\,)}{\mathcal{U}_i(\,\Phi_i(t)\,)} \right\} \tag{11}$$

$$0 = \sum_{i=1}^m \Bigg\{ \ln(\hat{c}) - \ln(\hat{c}+\eta(t;\hat{\theta}^T))$$

$$+ \frac{\mathcal{U}_i(\,\Phi_i(t)-1\,)}{\mathcal{U}_i(\Phi_i(t))} (\hat{y}_0\,[\hat{c}+\eta(t;\hat{\theta}^T)])^l \sum_{n=1}^{\Phi_i(t)-l} \frac{1}{n-1+\hat{b}} \Bigg\} \tag{12}$$

where

$$\mathcal{U}_i(n) = \sum_{l=0}^{n} \frac{\Gamma(n-l+\hat{b})}{\Gamma(n-l+1)\,l!} \cdot \left(\hat{y}_0\,[\hat{c}+\eta(t;\hat{\theta}^T)]\right)^l \quad .$$

These equations must be solved numerically. For the special case of $b = 1$ (two parametric exponential distribution) the two equations (10) and (11) have to be solved with $\hat{b} = 1$ and $\mathcal{U}_i(n) = \sum_{l=0}^{n} \frac{(\hat{y}_0\,[\hat{c}+\eta(t;\hat{\theta}^T)])^l}{l!}$. In the case of a two parametric Gamma distribution ($y_0 = 0$) we must consider the equations (10) and (12) $\hat{y}_0 = 0$ and $\mathcal{U}_i(n) = \frac{\Gamma(n+\hat{b})}{n!}$.

3. If Y is inverse Gaussian distributed:

$$\lambda^* = \frac{\mu}{\sqrt{1+2\mu^2\eta(t;\theta^T)/\beta}} \cdot \frac{\sum\limits_{u=0}^{\Phi(t-)} \frac{(\Phi(t-)+u)!}{(\Phi(t-)-u)!\,u!}\left[\frac{\mu}{2\,\beta\,\sqrt{1+2\mu^2\eta(t;\theta^T)/\beta}}\right]^u}{\sum\limits_{u=0}^{\Phi(t-)-1} \frac{(\Phi(t-)-1+u)!}{(\Phi(t-)-1-u)!\,u!}\left[\frac{\mu}{2\,\beta\,\sqrt{1+2\mu^2\eta(t;\theta^T)/\beta}}\right]^u} \; .$$

The parameter μ can be found from

$$\hat{\mu} = \frac{1}{m\cdot\eta(t;\hat{\theta}^T)} \sum_{i=1}^{m} \Phi_i(t)$$

and β is the solution of

$$0 = \sum_{i=1}^{m} \left\{ \frac{1}{\hat{\mu}}\left(1 - \frac{1+\eta(t;\hat{\theta}^T)\,\hat{\mu}^2/\hat{\beta}}{\sqrt{1+2\eta(t;\hat{\theta}^T)\,\hat{\mu}^2/\hat{\beta}}}\right) + \frac{\Phi_i(t)\cdot\eta(t;\hat{\theta}^T)\,\hat{\mu}^2/\hat{\beta}^2}{1+2\eta(t;\hat{\theta}^T)\,\hat{\mu}^2/\hat{\beta}} \right.$$
$$\left. - \mathcal{H}_i \cdot \frac{\hat{\mu}\cdot(1+\eta(t;\hat{\theta}^T)\,\hat{\mu}^2/\hat{\beta})}{2\hat{\beta}^2\,(1+2\eta(t;\hat{\theta}^T)\,\hat{\mu}^2/\hat{\beta})^{1.5}} \right\}$$

where

$$\mathcal{H}_i := \frac{\sum\limits_{k=0}^{\Phi_i(t)-2} \frac{(\Phi_i(t)+k)!}{(\Phi_i(t)-2-k)!\,k!}\left(\frac{\hat{\mu}}{2\hat{\beta}\,\sqrt{1+2\eta(t;\hat{\theta}^T)\,\hat{\mu}^2/\hat{\beta}}}\right)^k}{\sum\limits_{k=0}^{\Phi_i(t)-1} \frac{(\Phi_i(t)-1+k)!}{(\Phi_i(t)-1-k)!\,k!}\left(\frac{\hat{\mu}}{2\hat{\beta}\,\sqrt{1+2\eta(t;\hat{\theta}^T)\,\hat{\mu}^2/\hat{\beta}}}\right)^k}$$

The further restriction of information from observation to the knowledge of only failure times makes the problem more complicated. First it is necessary to find the distribution of the first passage time Z^h of the degradation process:

$$Z^h = \inf\left\{t : \sum_{n=0}^{\infty} I(T_n \le t)\cdot X_n \ge h\right\} = \inf\left\{t :\, Z(t) \ge h - X_0\right\}$$

where X_0 is a (possible random) state at time $T_0 := 0$. The explicit calculation is possible only for some special cases. Further, this distribution contains all parameters we considered and it is nearly impossible to find explicit estimates except for very simple assumptions. Nevertheless, the problem can be solved numerically.

4 Moment Estimates

Let us consider again the case of observable counting process and unobservable frailty variable Y. Let $\mathfrak{g}_k(W)$ and $\mathfrak{z}_k(W)$ be the empirical ordinary and central moments, respectively, of a random variable W:

$$\mathfrak{g}_k(W) = \frac{1}{m}\sum_{i=1}^{m} W_i^k \quad \text{and} \quad \mathfrak{z}_k(W) = \frac{1}{m}\sum_{i=1}^{m} [W_i - \mathfrak{g}_1(W)]^k \quad . \tag{13}$$

According to (2) we can express the k–th ordinary and central moments of $\Phi(t)$ as linear combinations of moments of Y multiplied by powers of the deterministic function $\eta(t)$. Actually, let $S(k,u)$ denote the Stirling numbers of second kind where $S(k,u)$ can be recursively determined

$$S(k,u) = S(k-1,u-1)+u\cdot S(k-1,u),\ 1 \le u \le k, \quad S(0,0) := 1,\ S(0,u) = 0.$$

We make use of

$$n^k = \sum_{u=1}^{k} S(k,u)\, n\,(n-1)\cdot \ldots \cdot (n-u+1)$$

and we consider the factorial moments of a Poisson distributed random variable with mean $y\,\eta(t)$. Some elementary calculations yield

$$\begin{aligned} E[\Phi(t)^k] &= \int_0^{\infty} \sum_{n=0}^{\infty} n^k \frac{[y\eta(t)]^n}{n!} e^{-y\eta(t)} \mathrm{d}F_Y(y) \\ &= \sum_{u=1}^{k} S(k,u)\cdot E[Y^u]\cdot \eta(t)^u\,. \end{aligned} \tag{14}$$

In particular, we find

$$\begin{aligned} E_\theta[\Phi(t)] &= E_\theta[Y]\cdot \eta(t;\theta^T) \\ \mu_2^\theta(\Phi(t)) &= E_\theta[Y]\cdot \eta(t;\theta^T) + \mu_2^\theta(Y)\cdot \eta(t;\theta^T)^2 \\ \mu_3^\theta(\Phi(t)) &= E_\theta[Y]\cdot \eta(t;\theta^T) + 3\mu_2^\theta(Y)\cdot \eta(t;\theta^T)^2 + \mu_3^\theta(Y)\cdot \eta(t;\theta^T)^3 \quad . \end{aligned}$$

where $\mu_k(\cdot)$ denotes the k–th central moment of a random variable. Let us further assume that the deterministic function $\eta(t)$ is known and that we are interested only in estimating the parameters of distribution of Y. The moments at the left hand site are replaced by its empirical moments. Further, the moments of Y can be expressed in dependence of the moments of $\Phi(t)$ and the function η:

$$E_\theta[Y] = \eta(t;\theta^T)^{-1} \cdot E_\theta[\Phi(t)] \tag{15}$$

$$\mu_2^\theta(Y) = \eta(t;\theta^T)^{-2} \left\{ \mu_2^\theta(\Phi(t)) - E_\theta[\Phi(t)] \right\} \tag{16}$$

$$\mu_3^\theta(Y) = \eta(t;\theta^T)^{-3} \left\{ \mu_3^\theta(\Phi(t)) - 3\,\mu_2^\theta(\Phi(t)) + 2\,E_\theta[\Phi(t)] \right\} \quad . \tag{17}$$

Now it is possible to find moment estimates for all parameters of the distribution of Y. Let us consider again the three previous examples:

1. If Y is rectangular distributed in $[a,b]$ we get from (15) and (16) and taking into account $0 \le a < b$

$$\hat{a} = \frac{\mathfrak{g}_1(\Phi(t))}{\eta(t;\hat{\theta}^T)} - \sqrt{3\mathcal{D}^2} \quad , \quad \hat{b} = \frac{\mathfrak{g}_1(\Phi(t))}{\eta(t;\hat{\theta}^T)} + \sqrt{3\mathcal{D}^2}$$

with

$$\mathcal{D}^2 := \frac{\mathfrak{z}_2(\Phi(t)) - \mathfrak{g}_1(\Phi(t))}{\eta(t;\hat{\theta}^T)^2} \ .$$

In difference to the maximum likelihood method an unique admissible estimator exists if $\hat{a} \ge 0$ and $\mathcal{D}^2 > 0$. The assumption $\mathcal{D}^2 > 0$ is fulfilled for sufficient large values of m because $\mathcal{D}^2$ is a consistent estimate of the variance $\mu_2^\theta(Y)$ of Y.

2. If $Y - y_0$ is gamma distributed than it has the first three moments

$$E_\theta[Y] = \frac{b}{c} + y_0 \quad , \quad \mu_2^\theta(Y) = \frac{b}{c^2} \quad \text{and} \quad \mu_3^\theta(Y) = 2\,\frac{b}{c^3} \ .$$

From (15), (16) and (17) we get the unique moment estimators

$$\hat{c} = 2\,\frac{\mathfrak{z}_2(\Phi(t)) - \mathfrak{g}_1(\Phi(t))}{\mathcal{J}}\,\eta(t;\hat{\theta}^T)$$

$$\hat{y}_0 = \frac{\mathfrak{g}_1(\Phi(t))}{\eta(t;\hat{\theta}^T)} - 2\,\frac{\left[\mathfrak{z}_2(\Phi(t)) - \mathfrak{g}_1(\Phi(t))\right]^2}{\mathcal{J} \cdot \eta(t;\hat{\theta}^T)}$$

$$\hat{b} = 4\,\frac{\left[\mathfrak{z}_2(\Phi(t)) - \mathfrak{g}_1(\Phi(t))\right]^3}{\mathcal{J}^2}$$

with

$$\mathcal{J} = \mathfrak{z}_3(\Phi(t)) - 3\,\mathfrak{z}_2(\Phi(t)) + 2\,\mathfrak{g}_1(\Phi(t)) \ .$$

3. For an inverse Gaussian distributed Y with $E_\theta[Y] = \mu$ and $\mu_2^\theta(Y) = \mu^3/b$ the equations (15) and (16) gives the unique estimators

$$\hat{\mu} = \frac{\mathfrak{g}_1(\Phi(t))}{\eta(t;\hat{\theta}^T)} \qquad \text{and} \qquad \hat{\beta} = \frac{\mathfrak{g}_1(\Phi(t))^3}{\eta(t;\hat{\theta}^T)\cdot[\mathfrak{z}_2(\Phi(t)) - \mathfrak{g}_1(\Phi(t))]} \quad .$$

For this distribution we get the same $\hat{\mu}$ from the moment method as from the maximum likelihood method in section 1.

The advantages of moment estimators in comparison to maximum likelihood estimators are its simple form and the fact that they can be found explicitely. But it is well known that in general maximum likelihood estimates have better properties. In the next section we compare the two methods concerning to its bias and variance.

5 Comparison of Maximum Likelihood and Moment Estimates

Let $Y - y_0$ Gamma distributed and let η be Weibull, it is $\eta(t;\alpha) = t^{\alpha+1}$ with $\alpha > -1$. We have considered sample sizes of $m = 50$, $m = 100$, $m = 250$ and $m = 500$. For each realization a path of the degradation process was simulated with true parameter $\theta^Y = (c, y_0, b) = (2.4, .5, 1.2)$. The observation is assumed to continue up to time $t = 10$. We have considered three different values of the parameter α in the deterministic part $\eta(t;\alpha)$. For $\alpha = -.3$ the derivative of $\eta(t;\alpha)$ is a decreasing function. The expected number of jumps up to time $t = 10$ is 5.01. If $\alpha = 0$ then we get a linear cumulative intensity or a constant hazard and expected number of jumps up to time $t = 10$ is 10. Last, for $\alpha = .3$ the derivative of $\eta(t;\alpha)$ is increasing and the expected number of jumps up to time $t = 10$ is 19.95. Such a simulation was repeated 750 times and from these 750 parameter estimates the mean and the variance of the estimator where calculated. The results are shown in table 1.

If the parameter α in the deterministic part is unknown, too, then it can be estimated by

$$\hat{\alpha} = \frac{\sum_{i=1}^{m} \Phi_i(t)}{\ln(t)\cdot\sum_{i=1}^{m}\Phi_i(t) - \sum_{i=1}^{m}\sum_{n=1}^{\Phi_i(t)}\ln(T_{n,i})} - 1 \, .$$

$\hat{\alpha}$ is a maximum likelihood estimator which does not contain other parameters [WeK04]. In table 2 the results of the same simulation are shown with the difference that now α is unknown and has to be estimated.

>From the simulation we get the following results:

1. Influence of α:
   In both cases we can see that for $\alpha = 0.3$ the variances of the estimator $\hat{y}_0$ (in both cases, MLE and ME) are smaller than for $\alpha = 0$ or $\alpha = -0.3$. The variances of the moment estimators for $\hat{b}$ and $\hat{c}$ are also smaller for $\alpha = 0.3$ than for $\alpha = 0$ or $\alpha = -0.3$, while α does not influence the variances of the maximum likelihood estimators.

Table 1. Empirical moments of maximum likelihood (MLE) and moment (ME) estimators ($\theta_0^Y = (2.4, 0.5, 1.2)$)

		$m = 50$		$m = 100$		$m = 250$		$m = 500$	
		Mean	Variance	Mean	Variance	Mean	Variance	Mean	Variance
$\alpha = -0.3$									
MLE	c	2.614	1.798	2.435	0.762	2.386	0.302	2.401	0.140
	y_0	0.599	0.018	0.561	0.016	0.520	0.008	0.508	0.004
	b	0.861	0.184	0.975	0.190	1.119	0.110	1.171	0.062
ME	c	2.910	3.902	2.824	2.208	2.826	1.475	2.758	0.893
	y_0	0.480	0.047	0.458	0.041	0.436	0.031	0.444	0.023
	b	1.680	2.552	1.708	2.026	1.754	1.507	1.650	1.052
$\alpha = 0.0$									
MLE	c	2.197	0.694	2.205	0.435	2.351	0.262	2.339	0.149
	y_0	0.574	0.013	0.560	0.010	0.532	0.005	0.536	0.003
	b	0.977	0.249	1.034	0.198	1.154	0.146	1.125	0.078
ME	c	3.189	2.533	2.846	1.325	2.771	0.876	2.613	0.427
	y_0	0.431	0.037	0.457	0.027	0.466	0.019	0.489	0.010
	b	1.982	2.030	1.716	1.257	1.621	0.855	1.431	0.375
$\alpha = 0.3$									
MLE	c	2.497	0.706	2.484	0.573	2.392	0.268	2.446	0.143
	y_0	0.521	0.010	0.519	0.007	0.523	0.004	0.497	0.002
	b	1.282	0.404	1.270	0.336	1.198	0.147	1.231	0.079
ME	c	3.361	2.066	3.133	1.399	2.758	0.672	2.602	0.347
	y_0	0.388	0.030	0.418	0.024	0.464	0.013	0.470	0.007
	b	2.276	2.110	2.004	1.457	1.596	0.625	1.417	0.279

2. The variances of the moment estimators are 2-4 times larger than the variances of the maximum likelihood estimators .
3. Both, bias and variance are particularly visible smaller if the parameter θ^T is known (with the exception of the independent of η moment estimate $\hat{b}$). The ratio of the variances of the maximum likelihood estimators and moment estimators, however, is the same for known and for unknown θ^T.

6 Conclusion

In the paper we have shown the advantages and disadvantages of maximum likelihood and moment estimators. The moment estimates are easy to calculate, where in many cases it is difficult to find maximum likelihood estimates. Moreover, there are problems, in which the maximum likelihood estimate of the parameters does not exist.

On the other hand the maximum likelihood estimators have a noticeable smaller variance as moment estimators. The ratio of the variances of the

Table 2. Empirical moments of maximum likelihood (MLE) and moment (ME) estimators ($\theta_0^Y = (2.4,\, 0.5,\, 1.2)$)

		$m = 50$		$m = 100$		$m = 250$		$m = 500$	
		Mean	Variance	Mean	Variance	Mean	Variance	Mean	Variance
$\alpha = -0.3$									
MLE	c	2.187	3.217	2.000	0.913	2.125	0.481	2.197	0.407
	y_0	0.683	0.040	0.666	0.040	0.623	0.027	0.607	0.024
	b	0.818	0.157	0.891	0.141	1.026	0.084	1.072	0.054
	α	-0.328	0.011	-0.337	0.010	-0.337	0.009	-0.333	0.009
ME	c	2.801	3.942	2.729	3.067	2.641	1.622	2.607	1.144
	y_0	0.548	0.066	0.533	0.066	0.523	0.049	0.527	0.041
	b	1.621	2.179	1.739	2.357	1.669	1.410	1.572	0.916
$\alpha = 0.0$									
MLE	c	2.194	1.156	2.177	0.695	2.209	0.625	2.299	0.403
	y_0	0.594	0.029	0.594	0.028	0.602	0.028	0.592	0.019
	b	1.072	0.304	1.157	0.207	1.194	0.147	1.137	0.084
	α	-0.031	0.014	-0.040	0.014	-0.048	0.012	-0.036	0.010
ME	c	3.128	3.691	2.800	1.645	2.660	1.400	2.503	0.678
	y_0	0.438	0.051	0.471	0.041	0.516	0.038	0.532	0.025
	b	2.082	2.271	1.885	1.264	1.721	0.826	1.485	0.394
$\alpha = 0.3$									
MLE	c	2.571	0.924	2.652	1.001	2.394	0.495	2.363	0.286
	y_0	0.510	0.014	0.520	0.014	0.546	0.011	0.552	0.010
	b	1.424	0.516	1.455	0.378	1.269	0.148	1.160	0.059
	α	0.286	0.010	0.280	0.009	0.277	0.007	0.278	0.006
ME	c	3.324	2.478	3.234	1.748	2.687	0.900	2.494	0.497
	y_0	0.394	0.038	0.421	0.032	0.489	0.020	0.517	0.015
	b	2.297	1.926	2.171	1.535	1.639	0.703	1.409	0.250

maximum likelihood estimators and moment estimators becomes smaller with increasing sample size.

References

[ABG93] Anderson, P., Borgan, Ø., Gill, R., Keiding, N.: Statistical Models Based on Counting Processes. Springer, New-York (1993)

[BaN01] Bagdonavičius, V., Nikulin, M.: Estimation in Degradation Models with Explanatory Variables. Lifetime Data Analysis, **7**, 85–103, (2001)

[BBK02] Bagdonavičius, V., Bikelis, A.,Kazakevičius, A., Nikulin, M.: Non-parametric Estimation from Simultaneous Degradation and Failure

Time Data. Comptes Rendus, Academie des Sciences de Paris, **335**, 183–188, (2002)

[BBK03] Bagdonavičius, V., Bikelis, A.,Kazakevičius, A., Nikulin, M.: Estimation from Simultaneous degradation and Failure Data, In: Lindquist, B.H., Doksum, K.A. (eds) Mathematical and Statistical Methods in Reliability. World Scientific Publishing Co., New Jersey London Singapore HongKong (2003)

[Bre81] Bremaud, P.: Point Processes and Queues. Springer, New York Berlin Heidelberg (1981)

[Cra66] Cramer, H.: The Elements of Probability Theory and Some of Its Applications. Krieger Publishing Co., Melbourne, Florida (1966)

[Cox55] Cox, D.R.: Some statistical methods connected with series of events. J. R. Statist. Soc B, **17**, 129–164 (1955)

[DoN96] Doksum, K.A., Normand, S.T.: Models for degradation processes and event times based on gaussian processes. In: Jewell, N.P. et al. (eds) Lifetime data: Models in reliability and survival analysis. Kluwer academic publishers, Dordrecht (1996)

[Gra91] Grandell, J.: Aspects of risk theory. Springer, New York (1991)

[Gra97] Grandell, J.: Mixed Poisson Processes. Chapman & Hall, London (1997)

[KaL98] Kahle, W., Lehmann, A.: Parameter Estimation in Damage Processes: Dependent Observations of Damage Increments and First Passage Time. In: Kahle, W. et al. (eds) Advances in Stochastic Models for Reliability, Quality and Safety. Birkhauser, Boston (1998)

[KaW04] Kahle, W., Wendt, H.: On a Cumulative Damage Process and Resulting First Passage Times. Applied Stochastic Models in Business and Industry, **20**, 17–26 (2004)

[LaB95] Last, G., Brandt, A.: Marked Point Processes on the Real Line. Springer, New York Berlin Heidelberg (1995)

[Wen99] Wendt, H.: Parameterschätzungen für eine Klasse doppelt–stochastischer Poisson Prozesse bei unterschiedlichen Beobachtungsinformationen. PhD-Thesis, Otto-von-Guericke-University, Magdeburg (1999)

[WeK04] Wendt, H., Kahle, W.: On Parameter Estimation for a Position-Dependent Marking of a Doubly Stochastic Poisson Process. In: Nikulin, N. et al. (eds) Parametric and Semiparametric Models with Applications to Reliability, Survival Analysis, and Quality of Life. Birkhauser, Boston Basel Berlin (2004)

[Whi95] Whitmore, G.A.: Estimating degradation by a Wiener diffusion process subject to measurement error. Lifetime data analysis, **1**, 307–319 (1995)

[WhS97] Whitmore, G.A., Schenkelberg, F.: Modeling accelerated degradation data using Wiener diffusion with a time scale transformation, Lifetime data analysis, **3**, 27–45, (1997)

Use of statistical modelling methods in clinical practice

Klyuzhev V.M., Ardashev V.N., Mamchich N.G., Barsov M.I.,Glukhova S.I.

Burdenko Main Military Clinical Hospital, Moscow, Russia name@email.address

1 Introduction

The necessity to generalize the great amount of information concerning the investigated physiological systems, the possibility to predict the body functional reserves have lead to the wide use of statistical modelling methods in the medical practice. The present paper based on the experience of collaborative work of medical specialists and statisticians is devoted to the review of some methods of multivariate statistics used in medicine. The statistical analysis of correlation matrices allowing to carry out the systemic approach to the phenomena under discussion underlies these methods. The data processing using factor and cluster analysis allows to gather the signs into groups identical to the concept of disease syndrome, to obtain the patient grouping, to reveal the connections between the signs, and, according to it, to form the new hypotheses about the revealed causes of dependence. At the stage of diagnostic decision the regression and discriminant analysis can be used.

2 Methods of statistical modelling

The main statistical methods used in diagnosis and prediction according to the problems and their clinical significance are shown in Table 1. Not dwelling upon the well-known Student t-test, Walsh t-test and Hotelling T2-test as they are discussed in the available literature, we shortly describe the multivariate statistical methods. More about these and other statistical methods one can see, for example, in [BS83], [GN96],[VN93],[VN96],[BR04],[KK02],[Zac71], etc...

Problem	Method	Purpose and Model
Comparison of statistically significant difference between the separate signs	Student t-test Walsh t-test	Testing of statistical significance of changes
Comparison of statistically significant difference by sign population	Hotelling T2-test	Testing of statistical significance of multi-variate measurements Syndrome approach model
Unification of signs into groups identical to the concept of disease ńsyndromeż	Factor and cluster analyses	Analysis of connections between the signs. Hypothesis forming
		Classification and diagnosis model
Hierarchical patient classification	Cluster and factor analyses	Diagnosis and prediction model
Evaluation of signs for diagnosis and prediction	Regression analysis	Differential diagnosis and prediction model
Differential evaluation of sign population for diagnosis and prediction	Discriminant analysis	Appropriate therapy choice model
Scheme of medicament choice	Logical programming	

Table 1. The main methods

3 Results

Factor analysis is based on the general idea according to which the values of all analyzed symptoms are under the influence of a rather small set of factors. These factors cannot be measured directly and that is why they are called *latent*. To a certain extent, the factors play the part of causes and the observed symptoms act as the consequences. As the number of latent factors is significantly lower than the number of analyzed signs the aim of factor analysis is to reduce the dimensionality of sign space.

The first factor accumulates maximal information about symptom interconnections and reveals the most distinct signs of the phenomenon under investigation. The second and the subsequent factors comprise the signs that supplement this information by giving some additional significant and individual disease features. The factors are not correlated and ordered according to the variance decrease (the highest is in the first one). It allows to gather the different clinical signs into groups similar to the concept of disease syndrome and to rank them according to significance degree.

The main information about the investigated phenomenon can be presented graphically as vectors in space, the axes of which are the values of first, second and subsequent factors. The use of factor analysis method allows to establish the connection between the diseases, to reveal the signs having no direct connection (or having slight connection) with the given disease.

We give the examples of statistical processing of the material in patient group with different variants of a coronary disease (CHD).

The clinical signs included in the factor which we named *cardiorrhexis* during myocardial infarction are presented in Table 2.

Signs	Factor load
The first myocardial infarction	0,8
Female sex	0,7
History of hypertensive disease	0,7
Severe, sometimes intolerable cardiac pain	0,5
Pain syndrome lasts more than 4 h, pain relapse	0,6
Arterial hypertension	0,7
Trinomial cardiac rhythm	0,6
Systolic murmur above the whole cardiac surface	0,5
Leukocytosis	0,5
Increase in sialic acid level	0,4
High activity of creatine phosphokinase	0,4

Table 2. The results of factor analysis in patients with acute myocardial infarction complicated by *cardiorrhexis*

One can see that these signs have factor loads ranging from 0,4 to 0,8. They show the acuity of disease manifestation and allow us to formulate the hypothesis of mechanical incompetence of myocardial connective stroma. *Cardiorrhexis* occurs during the first myocardial infarction with a background of pre-existent hypertensive disease. Pain severity, high fermentative activity, increased level of sialic acids, high leukocytosis reflect the severity of pathological process and probable myocardial stroma destruction.

So, with the help of factor analysis, it is possible to order the system volumes according to the levels and to create the hierarchical classification of the phenomenon under investigation.

The aim of **cluster analysis** is to partition a set of objects into preset or unknown number of classes based on a certain mathematical criterion of classification quality (a cluster is a group of elements characterized by any general feature). It can be used for disease class detection, patient attribution to appropriate groups and classification of disease symptoms. Based on the measurement of similarity and differences between the *patterns* (clinical profile) the clusters or groups of subjects investigated are selected. The selected clusters are compared with disease actual outcomes. Depending on their concurrence or difference the problem whether the clinical profile of the disease corresponds to the actual outcomes is solved. Such grouping based on the simple diagnostic principle, i.e. the similarity of one patient with another is the mathematical model of classification. The use of clinical signs allowed us to divide the investigated subjects into 4 groups. Each group corresponds to a disease functional class. The accuracy of classification obtained is 95%.

The most successful model of differential diagnosis is **discriminant analysis** [OW61]. Its aim is to include the subject (according to a certain rule) in one of the classes (k) depending on the parameters observed (p). This problem is solved with the help of step discriminant analysis. At every step the variable exercising the most significant influence on group division is entered into discriminant equation. As a result the following evaluation of discriminant function for i population is obtained

$$d_i = a_{i_1}x_1 + a_{i_2}x_2 + ... + a_{i_p}x_p + c_i,$$

where $i = 1, ..., k$.

When k equals 2 (two populations) the investigated subject belongs to group 1 if the following condition is carried out:

$$\sum_{i=1}^{p} a_j x_j > c_2 - c_1, \quad a_j = a_{1_j} + a_{2_j}, \quad j = 1, ..., p.$$

So the method based on the analysis of multiple correlation allows to reveal the most significant differential diagnostic signs and to obtain the decision rule of differential diagnosis. The discriminant analysis is successfully used in prediction of insult outcome when different methods of its treatment were used. It can also predict the patient survivability when operated for renal cancer and to determine the survival time for patients with renal cancer having metastases in different organs [For88]. The accuracy is within the range from 68-73% (with survivability prognosis) to 90% (with operation outcome prognosis).

There are cases in diagnosis when relying upon the number of indirect signs it is necessary to evaluate the most important sign to detect which is very difficult. It can be done with the use of regression analysis. Such approach helps to determine the degree of anatomical lesion based on the indirect diagnostic signs, to evaluate the complication probability, survival time, biological age.

The discriminant and regression analyses are based on the assumption that the statistical data correspond to the normal distribution law. Meanwhile there is a great number of data that either cannot be subjected to the analysis with the help of normal distribution curve or do not satisfy the main prerequisites necessary for its use. To analyze such data the multi-modal distribution laws [CZ85] and mathematical apparatus of catastrophe theory can be used allowing to reveal the most significant factors of multivariate population of statistical data and to detect geometrically the critical region where the qualitative changes in the investigated objects occur.

The process of penetration of mathematical methods into theory and practice of medicine is natural. The analysis of literature published during the last decades shows that the number of works devoted to this problem is still increasing. The wide and methodologically substantiated application of mathematical methods in different fields of health service makes it possible to put the medical information processing on principally new basis.

The most significant are information systems based on the principle of gathering the multiple case records into the large database. The database means the system of information storage, processing and analysis consisted of sign population among certain patients. For example, in CHD patients, the first stage for creating such a base provides the signs collection in the patient according to the formalized case record and input of this information into computer. The second stage is the information analysis with the help of mathematical techniques and sampling of decision rule of differential diagnosis, disease prognosis and patient treatment. The third stage is the decision making based on the created decision rules in CHD patient and diagnosis making with recommendations concerning the methods of adequate therapy. The fourth stage is the storage and updating of database in computer.

The result of database formation is the construction of mathematical methods capable to reflect the patient specific state. It is usually directed towards development of individual therapy and creation of algorithm of patient treatment methods and rehabilitation.

References

[CZ85] Cobb, L., Zacs, S. : Applications of catastrophe theory for statistical modeling in biosciences. IASA., **80, N 392**, 793–802, (1985)

[For88] Forges, F. et al.: Prognostic factors of metastatic renal carcinoma: a multivariate analysis . Seminars Oncol., **4, N 3**, 149–154 (1988)

[OW61] Overall, J.E., Williams, C.M.: Models for medical diagnosis . Behav. Science, **6, N 2**, 134–146 (1961)

[VMO93] Vander Poel, H.C., Mulders, P.F., Oosterhof, C.O. et al. : Prognostic value of karyometric and clinical characteristics in renal carcinoma. Cancer, **72. N 9**, 2667-2674 (1993)

[BR04] Balakrishnan, N., Rao, C.R.: Handbook in Statistics: Advances in Survival Analysis, **23**, Elsevier, New York (2004).

[VN93] Voinov, V.G., Nikulin, M. Unbiased Estimators and Their Applications, **1**, Univariate case. Kluwer Academic Publishres: Dordrecht, (1993).

[VN96] Voinov, V.G., Nikulin, M.: Unbiased Estimators and Their Applications, **2**, Multivariate case. Kluwer Academic Publishres: Dordrecht, (1996).

[GN96] Greenwood, P.E., Nikulin, M.: A Guide to Chi-Squared Testing. John Wiley and Sons, New York, (1996).

[BS83] Bolshev, L.N., Smirnov, N.V. Tables of Mathematical Statistics, Nauka, Moscow, (1983).

[Zac71] Zacks, Sh.: The Theory of Statistical Inference, Nauka, Moscow, (1971).

[KK02] Kleinbaum, D.G., Klein, M.: Lodistic Regression, Springer, New York, (2002).

Degradation-Threshold-Shock Models

Axel Lehmann

Otto–von–Guericke–University Magdeburg
Institute of Mathematical Stochastics
PF 4120, D-39016 Magdeburg, Germany
axel.lehmann@mathematik.uni-magdeburg.de

Summary. This paper deals with the joint modeling and simultaneous analysis of failure time data and degradation and covariate data. Many failure mechanisms can be traced to an underlying degradation process and stochastically changing covariates. We consider a general class of reliability models in which failure is due to the competing causes of degradation and trauma and express the failure time in terms of degradation and covariates. We compute the survival function of the resulting failure time and derive the likelihood function for the joint observation of failure time data and degradation data at discrete times.

Key words: Degradation process; degradation-threshold-shock model; dts-model; traumatic event; threshold; first passage time

1 Introduction

This paper deals with the joint modeling and simultaneous analysis of failure time data and covariate data like internal degradation and external environmental processes. Many failure mechanisms in engineering, medical, social, and economic settings can be traced to an underlying degradation process and stochastically changing covariates that may influence degradation and failure.

Most items under study degrade physically over time and a measurable physical deterioration almost always precedes failure. The level of deterioration of an item is represented by a degradation process. In engineering applications, degradation may involve chemical changes brought about by corrosion and electro-migration or physical changes due to wearing out and fracturing, whereas degradation may be characterized by markers of health status and quality of life data in medical settings. Frequently, an item is regarded as failed and switched off when degradation first reaches a critical threshold level.

Moreover, in most practical applications items or systems operate in heterogeneous environments and loads, environmental stresses, and other dynamically changing environmental factors may influence their failure rate.

When it is possible to measure degradation as well as covariates, an alternative approach to reliability and survival analysis is the modeling of degradation and environmental factors by stochastic processes and their failure-generating mechanisms. This stochastic-process-based approach shows great flexibility and can give rise to new or alternative time-to-failure distributions defined by the degradation model. It provides additional information to failure time observations and is particularly useful when the application of traditional reliability models based only on failure and survival data is limited due to rare failures of highly reliable items or due to items operating in dynamic environments.

Two relevant stochastic models relating failure to degradation and other covariates have evolved in the theoretical and applied literature, *threshold-models* and *shock-models*. A threshold-model supposes that the item or system fails whenever its degradation level reaches a certain critical deterministic or random threshold. In a shock-model the item or system is subjected to external shocks which may be survived or which lead to failure. The shocks usually occur according to a Poisson process whose intensity depends on degradation and environmental factors. It appears that Lemoine and Wenocur [LW85] may have been the first to combine both approaches by considering two competing causes of failure: degradation reaching a threshold, and occurrence of a traumatic event like a shock of large magnitude severe enough to destroy the item. So, the failure time of an item is the minimum of the moment when degradation first reaches a critical threshold and the moment when a censoring traumatic event occurs.

We call this class of reliability models which consider failure due to the competing causes of degradation and trauma *degradation-threshold-shock-models* (DTS-models). Singpurwalla [Sin95] and Cox [Cox99] give detailed reviews on stochastic-process-based reliability models including DTS-models.

In this paper, we derive an expression for the survival function of the failure time in a general DTS-model and consider certain classes of submodels. For the joint observation of failure time data and degradation data at discrete times the likelihood function is given.

The intension of this paper is to give a general framework for dealing with DTS-models. To apply the DTS-model to real data situations it is of course neccessary to specify the degradation and covariate processes. In applied literature degradation processes are frequently described by a general path model, i.e., by a stochastic process that depends only on a finite dimensional random variable (see [ME98]), or by a univariate process with stationary independent increments. In the context of DTS-models, Bagdonavičius, Haghighi and Nikulin [BHN05] consider a general path model with time dependent covariates and multiple traumatic event modes. Several processes with stationary independent increments have been used in degradation models. Frequently degradation is related to external covariates through a random time scale describing slowing or accelerating degradation in real time. The time scale and the intensity of traumatic events may depend on possibly time-varying covariates, for instance on different stress levels, to model the influence on failure of a dynamic operating environment of the item and to cover non-linear degradation behavior.

Wiener diffusion processes have found application in Doksum and Hoyland [DH92], Doksum and Normand [DN95], Lawless, Lu and Cao [LHC95], Whitmore [Whi95], Whitmore and Schenkelberg [WS97] and Whitmore, Crowder and Lawless [WCL98]. Degradation models based on the gamma process were considered by Wenocur [Wen89] and, in the context of a DTS-model, by Bagdonavičius and Nikulin [BN01] together with a random time scale depending on covariates. Lehmann [Leh04] considers a DTS-model with a Lévy degradation process and a random time scale and extends this DTS-model to the case of repairable items by a marked point process approach.

2 Degradation-Threshold-Shock-Models

Suppose that the degradation level of some item is described by a stochastic process $X = \{X(t) : t \in \mathbb{R}_+\}$, defined on a fixed probability space $(\Omega, \mathcal{F}, \mathrm{P})$. Let $X_t = \{X(s) : 0 \leq s \leq t\}$ denote the path of X on $[0, t]$ and $\mathcal{F}_t^X = \sigma(X_t)$ the history of all paths of X up to time t. For simplicity, we do not consider an external covariate process for the present.

An item is regarded as failed when the degradation process reaches a critical threshold level X^* which is possibly random but independent of X. Additionally to failures which are immediately related to degradation, an item can also fail when a traumatic event like a shock of large magnitude occurs although the degradation process has not yet reached the threshold. We model such a censoring traumatic event as the first point of a doubly stochastic Poisson process $\Psi = \{\Psi(t) : t \in \mathbb{R}_+\}$ with a stochastic intensity $\kappa(t, X(t))$ that may depend on time t and on the degradation level $X(t)$. That means, given a known path $x(\cdot)$ of X, Ψ is an nonhomogeneous Poisson process with intensity $\kappa(t, x(t))$. Hence, the failure time of an item is defined as the minimum

$$T = \min(D, C), \tag{1}$$

of the nontraumatic failure time $D = \inf\{t \geq 0 : X(t) \geq X^*\}$ caused by degradation and the traumatic failure time $C = \inf\{t \geq 0 : \Psi(t) = 1\}$ caused by a traumatic event. Given the degradation path X_t up to time t, the conditional survival function of C is

$$\mathrm{P}(C > t \,|\, X_t) = \exp\left(-\int_0^t \kappa(s, X(s))\, \mathrm{d}s\right). \tag{2}$$

Thus, the survival funtion of T is

$$\mathrm{P}(T > t) = \mathrm{E}\left[\mathbf{1}(D > t)\, \exp\left(-\int_0^t \kappa(s, X(s))\, \mathrm{d}s\right)\right], \tag{3}$$

where $\mathbf{1}(D > t)$ denotes the indicator function of the event $\{D > t\}$.

We call this model *degradation-threshold-shock-model* (DTS-model). Supposed that D has a failure rate $\lambda(t)$ and that C has a deterministic intensity $\kappa(t)$, (3) simplifies to

$$\mathrm{P}(T > t) = \exp\left(-\int_0^t \kappa(s)\, \mathrm{d}s\right) \mathrm{P}(D > t) = \exp\left(-\int_0^t (\kappa(s) + \lambda(s))\, \mathrm{d}s\right).$$

To find an expression of the survival function and the failure rate of T in the general case we use a theorem given by Yashin and Manton [YM97]:

Theorem 1 (Yashin, Manton). *Let ζ and ξ be stochastic processes influencing a failure rate $\alpha(t, \zeta, \xi)$ and satisfying measurability conditions such that, for $t \geq 0$*

$$\mathrm{E} \int_0^t \alpha(u, \zeta, \xi)\, \mathrm{d}u < \infty,$$

and let T be related to ζ_t and ξ_t by

$$\mathrm{P}(T > t \,|\, \zeta_t, \xi_t) = \exp\left(-\int_0^t \alpha(u, \zeta, \xi)\, \mathrm{d}u\right). \tag{4}$$

If the trajectories of ζ are observed up to t, then

$$\mathrm{P}(T > t \,|\, \zeta_t) = \exp\left(-\int_0^t \overline{\alpha}(u, \zeta_u)\, \mathrm{d}u\right),$$

where

$$\overline{\alpha}(t, \zeta_t) = \mathrm{E}[\alpha(t, \zeta, \xi) \,|\, \zeta_t, T > t].$$

The random failure rate $\alpha(t, \zeta, \xi)$ may depend on either the current values $\zeta(t)$ and $\xi(t)$ or on the trajectories ζ_t and ξ_t up to t. If the covariate process ζ does not appear in (4), i.e. $\alpha = \alpha(t, \xi)$, the statement of Theorem 1 obviously reads

$$\mathrm{P}(T > t) = \exp\left(-\int_0^t \mathrm{E}[\alpha(u, \xi) \,|\, T > u]\, \mathrm{d}u\right) \tag{5}$$

(see [Yas85]). Although we observe that

$$\mathrm{P}(T > t \,|\, X_t) = \mathrm{E}[\mathbf{1}(D > t)\mathbf{1}(C > t) \,|\, X_t] = \mathbf{1}(D > t)\mathrm{P}(C > t \,|\, X_t)$$

is not of the form (4), Theorem 1 can be used to compute $\mathrm{P}(T > t)$.

Theorem 2. *Let the traumatic failure time C has the stochastic failure rate $\kappa(t, X(t))$ with $\mathrm{E} \int_0^t \kappa(s, X(s))\, \mathrm{d}s < \infty$ for all $t \geq 0$ and assume that, given $X^* = x^*$, the nontraumatic failure time D have the conditional failure rate $\lambda(t, x^*)$ with $\mathrm{E} \int_0^t \lambda(s, X^*)\, \mathrm{d}s < \infty$ for all $t \geq 0$. Then,*

$$\mathrm{P}(C > t) = \exp\left(-\int_0^t \breve{\kappa}(s)\,\mathrm{d}s\right)$$

and

$$\mathrm{P}(D > t) = \exp\left(-\int_0^t \overline{\lambda}(s)\,\mathrm{d}s\right),$$

where the failure rates $\breve{\kappa}$ *and* $\overline{\lambda}$ *are given by* $\breve{\kappa}(t) = \mathrm{E}[\kappa(t, X(t)) \,|\, C > t]$ *and* $\overline{\lambda}(t) = \mathrm{E}[\lambda(t, X^*) \,|\, D > t]$. *The survival function of* T *can be expressed as*

$$\mathrm{P}(T > t) = \exp\left(-\int_0^t (\overline{\kappa}(s) + \overline{\lambda}(s))\,\mathrm{d}s\right),$$

where $\overline{\kappa}(t) = \mathrm{E}[\kappa(t, X(t)) \,|\, T > t]$ *is the failure rate of a traumatic event if a nontraumatic event has not occurred.*

Proof. Let $H(t) = \mathbf{1}(D \leq t)$ and $H_t = \{H(s) : 0 \leq s \leq t\}$. Since D is a $\mathcal{F}_t^X$-stopping time we have $\sigma(H_t) \subseteq \mathcal{F}_t^X$ for all $t \geq 0$ and, therefore,

$$\mathrm{P}(C > t \,|\, H_t, X_t) = \mathrm{P}(C > t \,|\, X_t) = \exp\left(-\int_0^t \kappa(s, X(s))\,\mathrm{d}s\right). \quad (6)$$

First, apply (5) with $\xi_t = X_t$ to the second equation of (6) to get the survival function $\mathrm{P}(C > t) = \exp\left(-\int_0^t \breve{\kappa}(s)\,\mathrm{d}s\right)$ and then, with $\xi_t = X^*$, to $\mathrm{P}(D > t \,|\, X^*) = \exp\left(-\int_0^t \lambda(s, X^*)\,\mathrm{d}s\right)$ to show $\mathrm{P}(D > t) = \exp\left(-\int_0^t \overline{\lambda}(s)\,\mathrm{d}s\right)$.

Moreover, applying Theorem 1 with $\zeta_t = H_t$ and $\xi_t = X_t$ to (6) we obtain

$$\mathrm{P}(C > t \,|\, H_t) = \exp\left(-\int_0^t \kappa^*(s, H_s)\,\mathrm{d}s\right)$$

where

$$\kappa^*(t, H_t) = \mathrm{E}[\kappa(t, X(t)) \,|\, H_t, C > t].$$

Obviously, on $\{D > t\}$, we have $\kappa^*(t, H_t) = \overline{\kappa}(t)$ and

$$\mathrm{P}(C > t \,|\, D > t) = \exp\left(-\int_0^t \overline{\kappa}(s)\,\mathrm{d}s\right).$$

Thus, we conclude

$$\mathrm{P}(T > t) = \mathrm{P}(C > t \,|\, D > t)\,\mathrm{P}(D > t) = \exp\left(-\int_0^t (\overline{\kappa}(s) + \overline{\lambda}(s))\,\mathrm{d}s\right).$$

□

In the following theorem an expression is derived for the density of the degradation process $X(t)$ conditioned on the event that no failure has occurred up to the moment t. We assume that $X(t)$ possesses a density $f_{X(t)}$ with respect to some dominating measure ν, usually the Lebesgue measure or the counting measure. However, in the following we will write $\mathrm{d}x$ instead of $\nu(\mathrm{d}x)$ regardless of the nature of ν. Further, we assume that for all $t \geq 0$ and $\mathbf{t}_k = (t_0, \ldots, t_k) \in \mathbb{R}^{k+1}$ with $0 \leq t_0 < \ldots < t_k \leq t$ and $\mathbf{x}_k = (x_0, \ldots, x_k) \in \mathbb{R}^{k+1}$ the conditional joint density $g(t, \mathbf{t}_k, \mathbf{x}_k; x^*)$ with

$$\mathrm{P}(D > t, \mathbf{X}_k \in \mathrm{d}\mathbf{x}_k \mid X^* = x^*) = g(t, \mathbf{t}_k, \mathbf{x}_k; x^*)\, \mathrm{d}\mathbf{x}_k \tag{7}$$

and $\mathbf{X}_k = (X(t_0), \ldots, X(t_k))$ is known.

Of course, g must satisfy $g(t, \mathbf{t}_k, \mathbf{x}_k; x^*) = 0$ if $\min(x_0, \ldots, x_k) \geq x^*$. If the paths of X are increasing then, obviously,

$$g(t, \mathbf{t}_k, \mathbf{x}_k; x^*) = f_{\mathbf{X}_k}(\mathbf{x}_k)\, \mathrm{P}(X(t) < x^* \mid \mathbf{X}_k = \mathbf{x}_k).$$

for $x_0 \leq \ldots \leq x_k < x^*$ and $g(t, \mathbf{t}_k, \mathbf{x}_k; x^*) = 0$ otherwise.

If X is a *Wiener process with drift* with drift coefficient μ, variance coefficient σ and initial value $X(0) = 0$, then

$$g(t, \mathbf{t}_k, \mathbf{x}_k; x^*) = \prod_{j=1}^{k} g_0(t_j - t_{j-1}, x_j - x_{j-1}; x^*)\, \overline{F_0}(t - t_k, x^* - x_k)$$

for $\min(x_1, \ldots, x_k) < x^*$ where $t_0 = x_0 = 0$ and

$$g_0(t, x; x^*) = \mathbf{1}(x^* > x) \frac{1}{\sigma\sqrt{t}} \varphi\left(\frac{x - \mu t}{\sigma\sqrt{t}}\right) \left[1 - \exp\left(-\frac{2x^*(x^* - x)}{\sigma^2 t}\right)\right]$$

for all $t > 0$ (see [KL98]). The function $\overline{F_0}$ is the survival function of an *Inverse Gaussian* distribution, i.e.,

$$\overline{F_0}(t, x) = N\left[\frac{x - \mu t}{\sigma\sqrt{t}}\right] - \mathrm{e}^{(2\mu\sigma^{-2}x)}\, N\left[\frac{-x - \mu t}{\sigma\sqrt{t}}\right]$$

where φ and N denote the pdf and the cdf of a standard normal random variable.

Theorem 3. *Let the traumatic failure time C has the stochastic failure rate $\kappa(t, X(t))$ with $\mathrm{E}\int_0^t \kappa(s, X(s))\, \mathrm{d}s < \infty$ for all $t \geq 0$ and assume (7) for all $t \geq 0$, $\mathbf{t}_k = (t_0, \ldots, t_k) \in \mathbb{R}^{k+1}$ with $0 \leq t_0 < \ldots < t_k \leq t$ and $\mathbf{x}_k = (x_0, \ldots, x_k) \in \mathbb{R}^{k+1}$. Then,*

$$\mathrm{P}(T > t, \mathbf{X}_k \in \mathrm{d}\mathbf{x}_k) = \exp\left(-\int_0^t \overline{\kappa}(s, \mathbf{x}_{k(s)})\, \mathrm{d}s\right) g(t, \mathbf{t}_k, \mathbf{x}_k)\, \mathrm{d}\mathbf{x}_k,$$

where $\overline{\kappa}(s, \mathbf{x}_{k(s)}) = \mathrm{E}[\kappa(s, X(s)) \mid T > s, \mathbf{X}_{k(s)} = \mathbf{x}_{k(s)}]$ with $k(s) = \max\{j \geq 0 : t_j \leq s\}$ for $0 \leq s \leq t$ and $g(t, \mathbf{t}_k, \mathbf{x}_k) = \mathrm{E}[g(t, \mathbf{t}_k, \mathbf{x}_k; X^*)]$.

Proof. For all $u \geq 0$, let $H(u) = \mathbf{1}(D \leq u)$ and $H_u = \{H(s) : 0 \leq s \leq u\}$. Set $t_{k+1} = \infty$ and let $X^{\Delta}(t) = \sum_{j=0}^{k} \mathbf{1}(t_j \leq t < t_{j+1}) X(t_j)$ denote the process of discrete observations of X Defining $\zeta_u = (H_u, X_u^{\Delta}) = \{(H(s), X^{\Delta}(s)) : 0 \leq s \leq u\}$ we have

$$\mathrm{P}(C > t \,|\, \zeta_t, X_t) = \mathrm{P}(C > t \,|\, X_t) = \exp\left(-\int_0^t \kappa(s, X(s))\, \mathrm{d}s\right) \tag{8}$$

since $\sigma(\zeta_t) \subseteq \mathcal{F}_t^X$. Applying Theorem 1 with $\xi_t = X_t$ to (8) we obtain

$$\mathrm{P}(C > t \,|\, \zeta_t) = \mathrm{P}(C > t \,|\, H_t, X_t^{\Delta}) = \exp\left(-\int_0^t \kappa^*(s, H_s, X_s^{\Delta})\, \mathrm{d}s\right)$$

where

$$\kappa^*(s, H_s, X_s^{\Delta}) = \mathrm{E}[\kappa(s, X(s)) \,|\, H_s, X_s^{\Delta}, C > s].$$

for all $0 \leq s \leq t$. Since $\kappa^*(s, H_s, X_s^{\Delta}) = \overline{\kappa}(s, \mathbf{X}_{k(s)})$ on $\{D > t\}$, we finally conclude

$$\begin{aligned}\mathrm{P}(T > t, \mathbf{X}_k \in \mathrm{d}\mathbf{x}_k) &= \mathrm{P}(C > t \,|\, D > t, \mathbf{X}_k = \mathbf{x}_k)\, \mathrm{P}(D > t, \mathbf{X}_k \in \mathrm{d}\mathbf{x}_k) \\ &= \exp\left(-\int_0^t \overline{\kappa}(s, \mathbf{x}_{k(s)})\, \mathrm{d}s\right) \mathrm{E}[g(t, \mathbf{t}_k, \mathbf{x}_k; X^*)]\, \mathrm{d}x.\end{aligned}$$

□

The class of DTS-models contains two important subclasses, *degradation-threshold-models* (DT-models) and *degradation-shock-models* (DS-models).

2.1 Degradation-Threshold-Models

In a degradation-threshold-model only nontraumatic failures can occur, i.e., the traumatic event intensity κ is equal to zero and the failure time $T = D = \inf\{t \geq 0 : X(t) \geq x^*\}$ is the first passage time of X to the threshold x^*, which is assumed to be nonrandom in this subsection.

If degradation is modeled by a one dimensional *Wiener process with drift* $X(t) = x + \mu t + \sigma W(t)$ where W denotes a standard Brownian motion, then it is well known, that $T \sim \mathsf{IG}\left(\frac{x^*-x}{\mu}, \frac{(x^*-x)^2}{\sigma^2}\right)$ is *Inverse Gaussian* distributed if $x < x^*$. In general, T has an upside bathtub failure rate, but it has an essentially increasing failure rate (IFR) if $(x^* - x)/\sigma \gg 1$ and a decreasing failure rate (DRF) if $(x^* - x)/\sigma \ll 1$ (see Fig. 1).

For an increasing degradation process X the survival function of T is given by $\mathrm{P}(T > t) = \mathrm{P}(X(t) < x^*)$. If, for instance, X is a *homogeneous Poisson process* with rate $\lambda > 0$ then $T \sim \mathsf{Ga}([x^*], \lambda)$ follows a *gamma* distribution, which has an increasing failure rate. If X is an increasing jump process, then T has always increasing failure rate average (see [SS88]):

Theorem 4 (Shaked, Shantikumar). *If X is an increasing jump process, then T has increasing failure rate average (IFRA). If X is an increasing Lévy process with a Lévy measure ν which has a decreasing density, then T has increasing failure rate (IFR).*

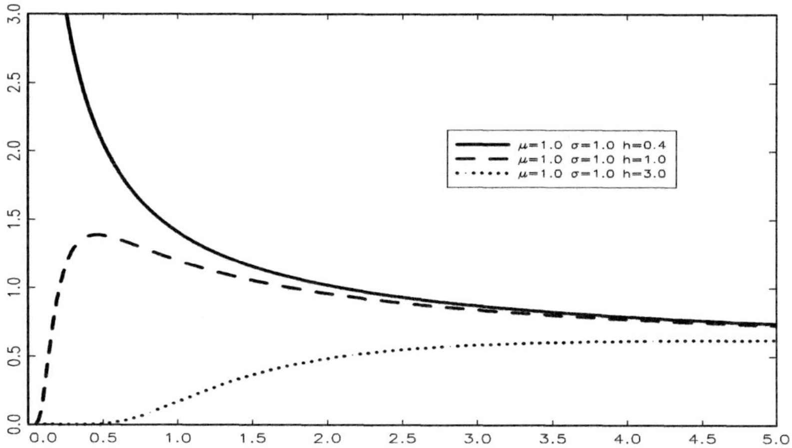

Fig. 1. Failure rates of Inverse Gaussian distribution ($x^* = h$)

2.2 Degradation-Shock-Models

The class of degradation-shock-models is characterized by the absence of a critical threshold X^* which can be described formally by $X^* = \infty$. Here the failure time is given by $T = C = \inf\{t \geq 0 : \Psi(t) = 1\}$, i.e., only traumatic failures can occur. Hence, by Theorem 2 we have

$$\mathrm{P}(T > t) = \mathrm{E}\left[\exp\left(-\int_0^t \kappa(s, X(s))\,\mathrm{d}s\right)\right] = \exp\left(-\int_0^t \overline{\kappa}(s)\,\mathrm{d}s\right).$$

with $\overline{\kappa}(t) = \breve{\kappa}(t) = \mathrm{E}\left[\kappa(s, X(s)) \,|\, C > t\right]$. For a positive increasing Lévy degradation process X with Lévy measure ν and drift rate μ and an intensity $\kappa(t, X(t)) = \gamma X(t)$ that depends proportionally on degradation, Kebir [Keb91] proved

$$\begin{aligned}\mathrm{P}(T > t) &= \mathrm{E}\left[\exp\left(-\gamma \int_0^t X(s)\,\mathrm{d}s\right)\right] \\ &= \exp\left(-\frac{\mu\gamma t^2}{2} - \gamma \int_0^t \int_0^\infty [1 - \mathrm{e}^{-sx}]\,\nu(dx)ds\right),\end{aligned}$$

i.e., T has the increasing failure rate $\overline{\kappa}(t) = \gamma(\mu t + \int_0^\infty [1 - \mathrm{e}^{-tx}]\,\nu(dx))$.

Applying Kebirs formula to a *homogeneous Poisson* process with rate $\lambda > 0$ we see that T follows a *Makeham* distribution with the survival function

$$\mathrm{P}(T > t) = \exp(-\lambda\gamma t + \lambda(1 - \mathrm{e}^{-\gamma t}))$$

and the failure rate $\overline{\kappa}(t) = \lambda\gamma(1 - \mathrm{e}^{-\gamma t})$.

For a DS-model with degradation modeled by a *Wiener process with drift* $X(t) = x + \mu t + \frac{\sigma}{\sqrt{2}} W(t)$ and a quadratic intensity $\kappa(t, X(t)) = (X(t))^2$, Wenocur [Wen86] computed the survival function of T as

$$\begin{aligned} \mathrm{P}(T > t) &= \mathrm{E}\left[\exp\left(-\int_0^t (X(s))^2\,\mathrm{d}s\right)\right] \\ &= \exp\left\{-\frac{\mu^2}{\sigma^2}t + \left(\frac{\mu^2}{\sigma^3} - \frac{x^2}{\sigma}\right)\tanh(\sigma t) + 2\frac{x\mu}{\sigma^2}(\mathrm{sech}(\sigma t) - 1)\right\}\sqrt{\mathrm{sech}(\sigma t)}. \end{aligned} \tag{9}$$

Some failure rates of T are shown in Fig. 2. If σ tends to zero the distribution

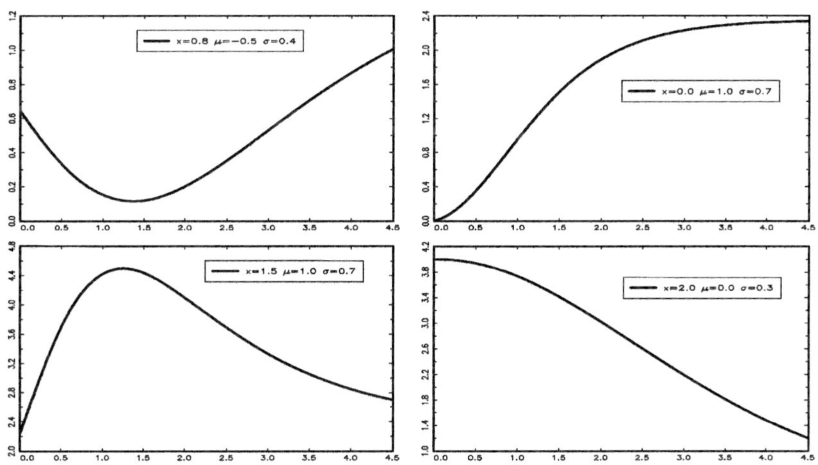

Fig. 2. Failure rates of the distribution (9)

of T converges to a *"generalized" Weibull* distribution with form parameter three:

$$\mathrm{P}(T > t) \quad \stackrel{\sigma \downarrow 0}{\longrightarrow} \quad \exp\left(-(\mu^2/3)t^3 - x\mu t^2 - x^2 t\right).$$

3 Maximum Likelihood Estimation

Suppose that n independent items are observed in $[0, t^*]$ with identically distributed degradation processes X_i, traumatic event processes Ψ_i, failure times T_i and thresholds X_i^*. We assume that X_i^* and (X_i, Ψ_i) are independent for all $i = 1, \ldots, n$. The ith item is observed at planned inspection times

$0 \le t_{i0} < t_{i1} < \ldots$ until t^*. Let $x_{ij} = X_i(t_{ij})$ denote the observed degradation levels. If a failure occurs in $[0, t^*]$, the observation of X_i will be stopped after this event. That means, in each interval $(t_{ij-1}, t_{ij}]$ we observe either a failure at $t_i \in (t_{ij-1}, t_{ij}]$ and the degradation level $X_i(t_i)$ or we observe the degradation level $x_{ij} = X_i(t_{ij})$ at t_{ij} (and $x_i = X_i(t^*)$ at t^*) under the condition that degradation has not yet exceeded the threshold. For the ith item let

$$l_i = l_i(t^*) = \max\{j \ge 0 : t_{ij} < \min(T_i, t^*)\},$$

i.e., $l_i + 1$ is the number of observed degradation levels without failure in $[0, t^*)$. Further, let $\widetilde{T_i} = \min(T_i, t^*)$ be the observable censored failure time and

$$V_i = V_i(t^*) = \begin{cases} 0, & \text{if } T_i > t^* \quad \text{(no failure in } [0,t^*]) \\ 1, & \text{if } D_i < C_i, D_i \le t^* \quad \text{(nontraumatic failure in } [0,t^*]) \\ -1, & \text{if } C_i \le \min(D_i, t^*) \quad \text{(traumatic failure in } [0,t^*]) \end{cases},$$

an observable failure mode indicator.

Hence, the data for the ith item in $[0, t^*]$ is

$$(\widetilde{t_i} = \widetilde{T_i}, v_i = V_i, \mathbf{x}_{il} = \mathbf{X}_{il}, x_i = X_i(\widetilde{t_i}))$$

with $\mathbf{X}_{il} = (X_i(t_{i0}), \ldots, X_i(t_{il_i}))$. For $k_i = l_i + 1$ set $t_{ik_i} = \widetilde{t_i}$, $x_{ik_i} = x_i$ and $\mathbf{X}_{ik} = (\mathbf{X}_{il}, X_i(\widetilde{t_i}))$. By $f_D(t \,|\, \mathbf{t}_l, \mathbf{x}_l; x^*)$ we denote the conditional density of the nontraumatic failure time given $\{X^* = x^*\}$ and given $l+1$ observations of the degradation process without reaching the threshold up to $t_l < t$:

$$f_D(t \,|\, \mathbf{t}_l, \mathbf{x}_l; x^*)\, \mathrm{d}t = \mathrm{P}(D \in \mathrm{d}t \,|\, D > t_l, \mathbf{X}_{\mathbf{t}_l} = \mathbf{x}_l; X^* = x^*)$$

and by f_{X^*} the density of the random threshold X^*. Dropping the subscript i the likelihood of the data is according to Theorem 3

$$\begin{aligned} \mathrm{P}(\widetilde{T} \in \mathrm{d}t, v = 0, \mathbf{X}_k \in \mathrm{d}\mathbf{x}_k) &= \mathbf{1}(t = t^*)\, \mathrm{P}(T > t^*, \mathbf{X}_k \in \mathrm{d}\mathbf{x}_k) \\ &= \mathbf{1}(t = t^*) \exp\left(- \int_0^{t^*} \overline{\kappa}(s, \mathbf{x}_{k(s)})\, \mathrm{d}s \right) g(t^*, \mathbf{t}_k, \mathbf{x}_k)\, \mathrm{d}\mathbf{x}_k, \end{aligned}$$

if no failure has occured in $[0, t^*]$,

$$\begin{aligned} \mathrm{P}(\widetilde{T} \in \mathrm{d}t, v = 1, \mathbf{X}_k \in \mathrm{d}\mathbf{x}_k) &= \mathrm{P}(C > t, D \in \mathrm{d}t, \mathbf{X}_k \in \mathrm{d}\mathbf{x}_k) \\ &= \mathrm{P}(C > t \,|\, D = t, \mathbf{X}_k = \mathbf{x}_k)\, \mathrm{P}(D \in \mathrm{d}t, \mathbf{X}_k \in \mathrm{d}\mathbf{x}_k) \end{aligned}$$

with

$$\begin{aligned} \mathrm{P}(C > t \,|\, D = t, \mathbf{X}_k = \mathbf{x}_k) &= \mathrm{P}(C > t \,|\, D = t, \mathbf{X}_k = \mathbf{x}_k, X^* = x_k) \\ &= \mathrm{P}(C > t \,|\, D \ge t, \mathbf{X}_k = \mathbf{x}_k) \\ &= \exp\left(- \int_0^t \overline{\kappa}(s, \mathbf{x}_{k(s)})\, \mathrm{d}s \right) \end{aligned}$$

and

$$\begin{aligned}\mathrm{P}(D\in \mathrm{d}t,\mathbf{X}_k\in \mathrm{d}\mathbf{x}_k) &= \mathrm{P}(D\in \mathrm{d}t\,|\,D>t_l,\mathbf{X}_l=\mathbf{x}_l,X^*=x_k)\\ &\qquad \mathrm{P}(D>t_l,\mathbf{X}_l\in \mathrm{d}\mathbf{x}_l\,|\,X^*=x_k)\,\mathrm{P}(X^*\in \mathrm{d}x_k)\\ &= f_D(t\,|\,\mathbf{t}_l,\mathbf{x}_l;x_k)\mathrm{d}t\; g(t_l,\mathbf{t}_l,\mathbf{x}_l;x_k)\mathrm{d}\mathbf{x}_l\; f_{X^*}(x_k)\mathrm{d}x_k,\end{aligned}$$

if a nontraumatic failure has occured first in $[0,t^*]$, and, finally,

$$\begin{aligned}&\mathrm{P}(\widetilde{T}\in \mathrm{d}t, v=-1,\mathbf{X}_k\in \mathrm{d}\mathbf{x}_k)\\ &\quad= \mathrm{P}(C\in \mathrm{d}t\,|\,D>t,\mathbf{X}_k=\mathbf{x}_k)\,\mathrm{P}(D>t,\mathbf{X}_k\in \mathrm{d}\mathbf{x}_k)\\ &\quad= \overline{\kappa}(t,\mathbf{x}_k)\,\exp\left(-\int_0^t \overline{\kappa}(s,\mathbf{x}_{k(s)})\,\mathrm{d}s\right)\mathrm{d}t\; g(t,\mathbf{t}_k,\mathbf{x}_k)\mathrm{d}\mathbf{x}_k,\end{aligned}$$

if a traumatic failure has occured first in $[0,t^*]$.

Thus, the complete likelihood function for the observation of n independent items in $[0,t^*]$ is given by

$$\begin{aligned}L_{t^*}(\widetilde{t}_i,v_i,\mathbf{X}_{ik}) = \prod_{i=1}^{n}\Bigg\{&\Big(g(\widetilde{t}_i,\mathbf{t}_{ik},\mathbf{x}_{ik})\Big)^{\mathbf{1}\{v_i<1\}}\\ &\times\Big(f_D(\widetilde{t}_i\,|\,\mathbf{t}_{il},\mathbf{x}_{il};x_i)\,g(t_{il_i},\mathbf{t}_{il},\mathbf{x}_{il};x_i)\,f_{X^*}(x_i)\Big)^{\mathbf{1}\{v_i=1\}}\\ &\times \overline{\kappa}(\widetilde{t}_i,\mathbf{x}_{ik})^{\mathbf{1}\{v_i=-1\}}\exp\left(-\int_0^{\widetilde{t}_i}\overline{\kappa}(s,\mathbf{x}_{ik(s)})\,\mathrm{d}s\right)\Bigg\}.\end{aligned}$$

Based on this complex likelihood structure the Maximum Likelihood estimators of model parameters have to be found numerically in general. Explicit estimators of the degradation parameters in a special DT-model based on the Wiener process were given in Lehmann [Leh01].

4 Concluding remarks

The DTS-model can be easily extended to the case that m different modes of traumatic events are considered such that traumatic failures of mode i occur due to a point process Ψ_i. If all these point processes are doubly stochastic Poisson processes conditionally independent given the degradation path $X(\cdot)$ and adapted to appropriate filtrations with intensities $\kappa_i(t,X(t))$, then the Theorems 2 and 3 remain valid if we replace $\overline{\kappa}(\cdot)$ by $\sum_{i=1}^m \overline{\kappa}_i(\cdot)$

Additionally to the degradation process, which is an internal covariate, one can consider an external covariate process $Z=\{Z(t):t\in\mathbb{R}_+\}$ which describes the dynamic environment and may influence degradation and the intensity of traumatic events. Since such covariate processes like loads, stresses or usage measures can often be completely observed, one is interested in the conditional distribution of degradation and failure time given the covariate history $Z_t = \{Z(s) : 0 \leq s \leq t\}$ up to some time t. If the failure rate $\lambda(t, Z_t, X^*)$ of D depends on the covariate Z and the threshold X^* and if the intensity of traumatic events $\kappa(t, Z(t), X(t))$ depends on the environment and on the degradation level, all concerned theorems and formulas remain valid if all probabilities and expectations are additionally conditioned on Z_t. For instance, the survival function of T given in Theorem 2, but conditioned on Z_t now, is

$$\mathrm{P}(T > t \,|\, Z_t) = \exp\left(-\int_0^t (\overline{\kappa}(s, Z_s) + \overline{\lambda}(s, Z_s))\, \mathrm{d}s\right)$$

with conditional failure rates $\overline{\kappa}(t, Z_t) = \mathrm{E}[\kappa(t, Z(t), X(t)) \,|\, Z_t, T > t]$ and $\overline{\lambda}(t, Z_t) = \mathrm{E}[\lambda(t, Z_t, X^*) \,|\, Z_t, D > t]$.

References

[Bag78] Bagdonavičius, V.: Testing the hypothesis of the additive accumulation of damage. Probab. Theory and its Appl., **23**, 403–408 (1978)

[BHN05] Bagdonavičius, V., Haghighi, F, Nikulin, M.S.: Statistical analysis of general degradation path model and failure time data with multiple failure modes. (accepted for publication in Communications in Statistics in 2005)

[BN01] Bagdonavičius, V., Nikulin, M.S.: Estimation in degradation models with explanatory variables. Lifetime Data Analysis, **7**, 85–103 (2001)

[Cox99] Cox, D.R.: Some remarks on failure-times, surrogate markers, degradation, wear and the quality of life. Lifetime Data Analysis, **5**, 307–314 (1999)

[DH92] Doksum, K.A., Hoyland, A.: Models for variable-stress accelerated life testing experiment based on a Wiener process and the inverse Gaussian distribution. Technometrics, **34**, 74–82 (1992)

[DN95] Doksum, K.A., Normand, S.L.T.: Gaussian models for degradation processes - part I: Methods for the analysis of biomarker data. Lifetime Data Analysis, **1**, 131–144 (1995)

[KL98] Kahle, W., Lehmann, A.: Parameter estimation in damage processes: Dependent observations of damage increments and first passage time. In: Kahle, W., von Collani, E., Franz, J., Jensen, U. (eds) Advances in Stochastic Models for Reliability, Quality and Safety, 139–152. Birkhauser, Boston (1998)

[Keb91] Kebir, Y.: On hazard rate processes. Naval Res. Logist. Quart., **38**, 865–876 (1991)

[LHC95] Lawless, J., Hu, J., Cao, J.: Methods for the estimation of failure distributions and rates from automobile warranty data. Lifetime Data Analysis, **1**, 227–240 (1995)

[Leh01] Lehmann, A.: A Wiener process based model for failure and degradation data in dynamic environments. Dresdner Schriften zur Mathemat. Stochastik, **4/2001**, 35–40 (2001)

[Leh04] Lehmann, A.: On a degradation-failure model for repairable items. In: Nikulin, M., Balakrishnan, N., Limnios, N., Mesbah, M. (eds) Semiparametric Models and its Applications for Reliability, Survival Analysis and Quality of Life, 65–79. Birkhauser, Boston (2004)

[LW85] Lemoine, A.J., Wenocur, M.L.: On failure modeling. Naval Res. Logist. Quart., **32**, 479–508 (1985)

[ME98] Meeker, W.Q., Escobar, L.A.: Statistical methods for reliability data. Wiley, New York (1998)

[SS88] Shaked, M., Shantikumar, J.G.: On the first-passage times of pure jump processes. J. Appl. Prob., **25**, 501–509 (1988)

[Sin95] Singpurwalla, N.D.: Survival in dynamic environments. Statistical Science, **10**, 86–103 (1995)

[Wen86] Wenocur, M.L.: Brownian motion with quadratic killing and some implications. J. Appl. Prob., **23**, 893–903 (1986)

[Wen89] Wenocur, M.L.: A reliability model based on the gamma process and its analytical theory. Adv. Appl. Prob., **21**, 899–918 (1989)

[Whi95] Whitmore, G.A.: Estimation degradation by a Wiener diffusion process subject to measurement error. Lifetime Data Analysis, **1**, 307–319 (1995)

[WS97] Whitmore, G.A., Schenkelberg, F.: Modelling accelerated degradation data using Wiener diffusion with a time scale transformation. Lifetime Data Analysis, **3**, 27–45 (1997)

[WCL98] Whitmore, G.A., Crowder, M.I., Lawless, J.: Failure inference from a marker process based on a bivariate Wiener model. Lifetime Data Analysis, **4**, 229–251 (1998)

[Yas85] Yashin, A.I.: Dynamics in survival analysis: conditional Gaussian property versus Cameron-Martin formula. In: Krylov, N.V., Liptser, R.S., Novikov, A.A. (eds) Statistics and Control of Stochastic Processes, 466–475. Springer, New York (1985)

[YM97] Yashin, A.I., Manton, G.M.: Effects of unobserved and partially observed covariate processes on system failure: A review a models and estimation strategies. Statistical Science, **12**, 20–34 (1997)

Comparisons of Test Statistics Arising from Marginal Analyses of Multivariate Survival Data

Qian H. Li[1] and Stephen W. Lagakos[2]

[1] Food and Drug Administration
Center for Drug and Evaluation Research, HFD-705
7500 Standish Place, Metro Park North (MPN) II,
Rockville, MD 20855
liq@cder.fda.gov

[2] Department of Biostatistics, Harvard School of Public Health
655 Huntington Avenue, Boston MA 02115
lagakos@hsph.harvard.edu

Summary. We investigate the properties of several statistical tests for comparing treatment groups with respect to multivariate survival data, based on the marginal analysis approach introduced by Wei, Lin and Weissfeld [WLW89]. We consider two types of directional tests, based on a constrained maximization and on linear combinations of the unconstrained maximizer of the working likelihood function, and the omnibus test arising from the same working likelihood. The directional tests are members of a larger class of tests, from which an asymptotically optimal test can be found. We compare the asymptotic powers of the tests under general contiguous alternatives for a variety of settings, and also consider the choice of the number of survival times to include in the multivariate outcome. We illustrate the results with two simulations and with the results from a clinical trial examining recurring opportunistic infections in persons with HIV.

Key words: Directional tests; Marginal model; Multivariate survival data; Omnibus test; Recurring events

1 Introduction

In some comparative clinical trials, each subject is followed for K failure-time events, each of which can be right censored. One example is recurring event data, where the K outcomes represent the times from the start of the trial until the occurrence of K clinical/biological events, such as recurring seizures or recurring opportunistic infections [HRL98]. Another is the repeated assessment, under different experimental conditions, of an infectious disease, as measured, for example, by the inhibitory concentration of drug needed to achieve a particular effect on the amount of virus [RGL90]. In the former example the K survival times for an individual are necessarily ordered in magnitude, but in the latter example they do not need to be.

Given the multivariate nature of these data, it is tempting to employ multivariate methods when comparing two treatment groups in the hope that this could provide a more meaningful assessment of their relative efficacy, or a more powerful statistical test than would be available from a univariate analysis, such as when examining the first survival time. Several semi-parametric approaches have been proposed for multivariate failure time data [PWP81], [AG81], [WLW89], [LW92], [LSC93], [CLN96], [CP95]. These methods each make certain assumptions, and their relative power characteristics are not well understood. In practice, however, these and other multivariate failure time methods do not appear to be used very often, and in most cases more familiar methods, such as the logrank test and Cox's proportional hazards model [COX72], are employed. One reason for this might be concerns about the additional assumptions that need to be made when employing most multivariate failure time methods, and lack of knowledge about the consequences of their violation. Another may be the lack of easily accessible software.

The goal of this paper is to assess the properties of statistical tests for comparing treatment groups based on the most popular of these approaches – the marginal analysis proposed by Wei, Lin, and Weissfeld (WLW). The WLW method derives its appeal from its avoidance of assumptions about the dependencies among an individual's K failure times and its simple computational aspects. However, use of the WLW method requires the choice from among several directional or omnibus tests whose relative performance are not fully understood. Additionally, in settings such as the first example of recurring events, one must also choose the number of outcomes, K, on which to base a test, and very little has been done to provide insight into the tradeoffs that arise. By investigating these issues, we aim to provide the analyst with guidelines on how best to utilize multivariate failure time data with this approach when comparing treatment groups.

The properties of the WLW method have also been examined by Hughes [HUG97], who approximated the power of the directional test and omnibus test proposed by WLW under a proportional hazards alternative to the null hypothesis of no treatment effect. Hughes uses the approximate power formulae to assess when a test based on $K = 1$ event is more or less powerful than an omnibus K df test based on K events, with special attention given to the comparison of using $K = 1$ versus $K = 2$ events. We build upon these initial results in several ways. In Section 3 we derive the asymptotic power of the two directional and one omnibus test that have been proposed by WLW and Lin [LIN94] under general alternative to the null hypothesis. In Section 4 we show that one of the directional tests proposed for the case of an equal treatment effects across the K failure times is, in general, inefficient relative to the other, and we derive the optimal directional test for an arbitrary alternative to the null hypothesis. We also provide a simple expression for the loss in power of the omnibus K df test relative to the optimal 1 df directional test. In Section 5 we consider the choice of K. When the treatment effect is homogeneous across the K failure times, we show that the power of the directional test proposed by WLW is increasing with K, and describe the relative efficiency of the omnibus test relative to this directional test. We also conduct simulations to examine the relative performance of the omnibus and directional tests for non-recurrent events with proportional hazards and recurrent events with non-proportional hazards alternative to the null hypothesis. We illustrate the methods in Section 6 using data from a HIV trial. Technical details are deferred to the Appendices. We note that a shorter version of this paper with fewer simulation results appears in [LL04].

2 The WLW Method and Definitions of Test Statistics

In this section, we describe the WLW approach, including the three statistics that have been proposed for comparing treatment groups.

Assume that each subject is followed for the occurrence of K survival times, denoted $T_1, T_2, \cdots, T_K$. The marginal hazard function associated with T_k is denoted by $\lambda_k(t|Z)$, where Z is a covariate which for simplicity we take to be binary, denoting treatment group. The null hypothesis that treatment group is not associated with any of the K failure times is given by

$$H_0 : \lambda_k(t|Z = 0) = \lambda_k(t|Z = 1)$$

where $t \geq 0$, $k = 1, 2, ..., K$. The WLW method is derived from the assumption that the marginal distributions for the two treatment groups have proportional hazard functions; that is, $\lambda_k(t|Z) = \lambda_k(t)exp(\beta_k Z)$, for $k = 1, \cdots, K$, where $\beta_1, \beta_2, \cdots, \beta_K$ are unknown parameters. The null hypothesis thus reduces to $H_0 : \beta_1 = \beta_2 = \cdots = \beta_K = 0$.

When the K survival times are not necessarily ordered, it is straightforward to show that there are proper $2K$ dimensional joint distributions which admit the K proportional hazards relationships represented above. Yang and Ying [YY01] show the existence of such joint distributions when T_1, $T_2, \cdots, T_K$ are ordered, as in the example of recurring events.

WLW allow noninformative right censoring of each T_k by introducing i.i.d. potential censoring times $C_1, C_2, \cdots, C_K$, which are assumed to be independent from the T_k. That is, the observation for a subject consists of (X_k, Δ_k), $k = 1, 2, \ldots, K$, where $X_k = min\{T_k, C_k\}$ is the observed portion of T_k and Δ_k is an indicator of whether T_k is uncensored ($\Delta_k = 1$) or right censored ($\Delta_k = 0$). Suppose that the data consist of n independent copies of $(Z, X_1, \Delta_1, \cdots, X_K, \Delta_K)$, the ith of which we denote by $(Z_i, X_{1i}, \Delta_{1i}, \cdots, X_{Ki}, \Delta_{Ki})$. Then if $L_k(\beta)$ denotes Cox's partial likelihood function based on the data $(Z_i, X_{ki}, \Delta_{ki})$ for $i = 1, \cdots, n$, WLW propose that the vector $\boldsymbol{\beta} = (\beta_1, \beta_2, \cdots, \beta_K)'$ be estimated by maximizing the working likelihood function

$$L(\boldsymbol{\beta}) = \prod_{k=1}^{K} L_k(\beta_k) \tag{1}$$

Denote the solution to this working likelihood by $\hat{\boldsymbol{\beta}} = (\hat{\beta}_1, \hat{\beta}_2, \cdots, \hat{\beta}_K)$. WLW show that when $\lambda_k(t|Z) = \lambda_k(t)exp(\beta_k Z)$, $\hat{\boldsymbol{\beta}}$ is consistent and asymptotically normal as $n \to \infty$; that is, $\hat{\boldsymbol{\beta}} \xrightarrow{p} \boldsymbol{\beta}$ and $\sqrt{n}(\hat{\boldsymbol{\beta}} - \boldsymbol{\beta}) \xrightarrow{\mathcal{L}} N(0, \Sigma)$, where $\Sigma = V_D^{-1} V V_D^{-1}$, V is a K-dimensional matrix obtained from the working likelihood (see Appendix 1), and V_D is the diagonal matrix with the same diagonal elements as V. WLW also provide a consistent sandwich estimate of Σ, which we denote by $\hat{\Sigma}$.

WLW propose a directional and omnibus test of H_0, which we denote by Q_2 and Q_3, respectively. Specifically,

$$Q_2 = \frac{n(c_2'\hat{\boldsymbol{\beta}})^2}{c_2' \Sigma c_2} = \frac{n(\mathbf{1}'\hat{\Sigma}^{-1}\hat{\boldsymbol{\beta}})^2}{\mathbf{1}'\hat{\Sigma}^{-1}\mathbf{1}},$$

where $c_2 = \frac{\hat{\Sigma}^{-1}\mathbf{1}}{\mathbf{1}'\hat{\Sigma}^{-1}\mathbf{1}}$ and $\mathbf{1}$ is a vector with elements all equal to 1, and

$$Q_3 = n\hat{\boldsymbol{\beta}}'\hat{\Sigma}^{-1}\hat{\boldsymbol{\beta}}.$$

Because of the asymptotic normality of $\hat{\beta}$, it follows that under H_0, $Q_2 \xrightarrow{\mathcal{L}} \chi_1^2$ and $Q_3 \xrightarrow{\mathcal{L}} \chi_K^2$ as $n \to \infty$. Q_2 is proposed by WLW for situations where it is felt that the components of $\boldsymbol{\beta}$ are approximately equal, whereas Q_3 is intended to be an omnibus test.

Another test of H_0, proposed by Lin [LIN94], arises from maximization of (1) under the constraint that $\beta_1 = \beta_2 = \cdots = \beta_K = \beta$; i.e., by maximizing

$$L(\beta) = \prod_{k=1}^{K} L_k(\beta) \quad . \tag{2}$$

If $\hat{\beta}$ denotes the maximizing value of β, Lin [LIN94] shows that the test statistic

$$Q_1 = n(\frac{\hat{\beta}}{\hat{\sigma}})^2$$

is asymptotically χ_1^2 under H_0, where $\hat{\sigma}^2$ is the estimate of $\sigma^2 = \frac{\mathbf{1}'V\mathbf{1}}{(\mathbf{1}'V_D\mathbf{1})^2}$ obtained by replacing V and V_D by the same estimators of these used by WLW to estimate Σ. Li [LI97] proves that Q_1 is asymptotically equivalent under H_0 (and under H_A defined below) to a test, Q_1^*, defined in the same way as Q_2, but with weight $c_1 = \frac{V_D\mathbf{1}}{\mathbf{1}'V_D\mathbf{1}}$. Thus, in studying the relative properties of directional tests of H_0, we restrict attention to those based on an arbitrary linear combination of $\hat{\boldsymbol{\beta}}$, say $Q_c = \frac{n(c'\hat{\boldsymbol{\beta}})^2}{c'\Sigma c}$. It is easy to see that under H_0, $Q_c \xrightarrow{\mathcal{L}} \chi_1^2$ as $n \to \infty$.

3 Asymptotic Properties of the Test Statistics under Contiguous Alternatives

In this section we derive the asymptotic distributions of the test statistics Q_1, Q_2, Q_3 and Q_c under a sequence of arbitrary contiguous alternatives to H_0. The results indicate the alternatives for which each test is asymptotically optimal, and provide the basis for their comparison in sections 5 and 6. Since Q_1 is asymptotically equivalent to the linear combination test Q_1^*, these tests are used interchangeably in this section.

The test statistics Q_2 and Q_3 introduced in the previous section were derived under the proportional hazard assumption $\lambda_k(t|Z) = \lambda_k(t)e^{\beta_k Z}$ for $k = 1, 2, \cdots, K$. We now consider their behavior under an arbitrary alternative to H_0. Consider the following sequence of alternatives to H_0:

$$H_A: \quad \lambda_k(t|Z) = \lambda_k(t)exp(\alpha_k g_k(t)Z), \qquad (3)$$

for $k = 1, 2, \cdots, K$. For simplicity we assume that the functions $g_k(t)$ are bounded and, without loss of generality, take $sup_{t\in[0,\infty]}|g_k(t)| = 1$. We further assume that the family of alternatives H_A is contiguous to H_0 by taking $\sqrt{n}\alpha_k \to \delta_k$, where $\delta_1, \delta_2, \cdots, \delta_K$ are fixed constants, where we suppress the dependency of α_k on n for simplicity of notation. The special case of proportional hazards alternatives is obtained by taking $g_k(t) \equiv 1$, in which case δ_k represents the limiting treatment group hazard ratio for the kth failure time.

Consider the estimator $\hat{\beta}$, which arises from model (2). The asymptotic distribution of $\hat{\beta}$ under H_A is shown in Appendix I to be $\sqrt{n}\hat{\beta} \xrightarrow{\mathcal{L}} N(\mu, (\sum_{k=1}^{K} \mathrm{v}_k^2)^{-2}\mathbf{1}'V\mathbf{1})$, where

$$\mu = \frac{\sum_{k=1}^{K} \delta_k \int_0^\infty g_k(t)v_k(0,t)s_k^{(0)}(0,t)\lambda_k(t)dt}{\sum_{k=1}^{K} \int_0^\infty v_k(0,t)s_k^{(0)}(0,t)\lambda_k(t)dt}, \qquad (4)$$

and where $v_k(0,t)$ and $s_k^{(0)}(0,t)$ are defined in Appendix 1. It follows that under H_A, $Q_1 \xrightarrow{\mathcal{L}} \chi_1^2(\xi_1)$ as $n \to \infty$, where the noncentrality parameter ξ_1 is given by $\xi_1 = \mu^2(\sum_{k=1}^{K} \mathrm{v}_k^2)^2/\mathbf{1}'V\mathbf{1}$. For proportional hazards alternatives, μ is seen to reduce to a linear combination of the δ_k, and ξ_1 reduces to $(1'V_D\boldsymbol{\delta})^2/1'V1$, where $\boldsymbol{\delta} = (\delta_1, \delta_2, \cdots, \delta_K)'$.

Next consider the vector estimator $\hat{\boldsymbol{\beta}}$ which arises from the model (1). The asymptotic distribution of $\hat{\boldsymbol{\beta}}$ under H_A is shown in Appendix I to satisfy $\sqrt{n}\hat{\boldsymbol{\beta}} \xrightarrow{\mathcal{L}} N(\boldsymbol{\mu}, \Sigma)$, where $\boldsymbol{\mu} = (\mu_1, \mu_2, \cdots, \mu_K)$, and

$$\mu_k = \frac{\delta_k \int_0^\infty g_k(t)v_k(0,t)s_k^{(0)}(0,t)\lambda_k(t)dt}{\int_0^\infty v_k(0,t)s_k^{(0)}(0,t)\lambda_k(t)dt}. \qquad (5)$$

It follows that $Q_2 \xrightarrow{\mathcal{L}} \chi_1^2(\xi_2)$ as $n \to \infty$, where $\xi_2 = \frac{(\mathbf{1}'\Sigma^{-1}\boldsymbol{\mu})^2}{\mathbf{1}'\Sigma^{-1}\mathbf{1}}$, and that $Q_3 \xrightarrow{\mathcal{L}} \chi_K^2(\xi_3)$ as $n \to \infty$, where $\xi_3 = \boldsymbol{\mu}'\Sigma^{-1}\boldsymbol{\mu}$. For proportional hazards alternatives, μ_k reduces to δ_k and the non-centrality parameters for Q_2 and Q_3 simplify to $\xi_2 = \frac{(\mathbf{1}'\Sigma^{-1}\boldsymbol{\delta})^2}{\mathbf{1}'\Sigma^{-1}\mathbf{1}}$ and $\xi_3 = \boldsymbol{\delta}'\Sigma^{-1}\boldsymbol{\delta}$.

Finally, consider the arbitrary linear combination test Q_c. It follows from the asymptotic normality of $\hat{\boldsymbol{\beta}}$ that Q_c converges to $\chi_1^2(\xi_c)$, where $\xi_c = \frac{(c'\boldsymbol{\mu})^2}{c'\Sigma c}$. The optimal test in this class, say Q_{opt}, is thus the one using the weight $c_{opt} = \frac{\Sigma^{-1}\boldsymbol{\mu}}{\mathbf{1}'\Sigma^{-1}\boldsymbol{\mu}}$, and has non-centrality parameter $\xi_{opt} = \boldsymbol{\mu}'\Sigma^{-1}\boldsymbol{\mu}$. Note the non-centrality parameter Q_{opt} is the same as that of the K df test Q_3. Furthermore, by comparing the optimal weight of c_{opt} with the weights forming Q_1 and Q_2, it is not hard to see that Q_2 is the optimal test when $\mu_1 = \mu_2 = \cdots = \mu_K$, and Q_1 is the optimal test when $\boldsymbol{\mu}$ is proportional to $\Sigma V_D \mathbf{1} = V_D^{-1} V \mathbf{1}$. We return to these results below.

4 Comparisons of Test Statistics

We now use the results of Section 3 to assess the relative power of Q_1, Q_2, Q_3, and Q_{opt} under variety of settings corresponding to homogeneous or heterogeneous treatment effects across failure times, and for special correlation structures among the failure times.

4.1 Equal $\mu_1, \mu_2, \cdots, \mu_K$

Suppose that the components of the mean of the asymptotic distribution of $\sqrt{n}\hat{\boldsymbol{\beta}}$ under H_A are equal, i.e., $\mu_1 = \mu_2 = \cdots = \mu_K = \mu_0$. This will result when the treatment groups have a common proportional hazards ratio for each k; that is, $g_1(t) \equiv \cdots \equiv g_k(t) = 1$ and $\delta_1 = \cdots = \delta_K$. However, this can also arise when non-proportional hazards relationships exist for various k, but in a way that the mean of the resulting asymptotic distribution of $\sqrt{n}\hat{\boldsymbol{\beta}}$ has equal components. As seen in Section 3, the non-centrality parameters of Q_1, Q_2, Q_3, and Q_{opt} are $\xi_1 = \mu_0^2 \frac{(\mathbf{1}'V_D\mathbf{1})^2}{\mathbf{1}'V\mathbf{1}}$, and $\xi_2 = \xi_3 = \xi_{opt} = \mu_0^2 \mathbf{1}'\Sigma^{-1}\mathbf{1}$. Thus, Q_2 is the optimal 1 df directional test for this setting and has the same non-centrality parameter as the K df omnibus test Q_3. It also follows that Q_2 has greater asymptotic power than Q_3 for all $K > 1$.

Figure 1 displays the asymptotic power of the directional test Q_{opt} and the omnibus test Q_3, based on a Type I error of 0.05, different values of ξ, and for $K = 2, 3, 4, 5, 6$. When $\mu_1 = \cdots = \mu_K$, $Q_2 = Q_{opt}$. Here we use this figure to discuss the choice of Q_2 and Q_3. The range of ξ was chosen to yield powers for Q_2 that range from 0.05 when $\xi = 0 (H_0)$ to 0.95. The successive lines beneath the top line represent the powers of Q_3 for $K = 2, 3, 4, 5, 6$. For example, when the power of Q_2 is 0.80, the power of Q_3 is $.71, .65, .60, .56, .52$ for $K = 2, 3, 4, 5, 6$. Thus with $K = 2$ failure times, the omnibus test Q_3 maintains reasonably good power against the Q_2, the optimal directional test when the treatment effects are homogeneous across failure times. This is consistent with the results found by Hughes [HUG97]. The relative power of the omnibus test remains relatively high when $K = 3$. For large K, however, the power advantage of using Q_2 becomes substantial. This potential advantage must be weighed against the possibility that the true alternative to H_0 may not correspond to homogeneity of treatment effects across failure times. For an arbitrary alternative, the asymptotic relative efficiency of Q_2 to Q_3 can be determined from ξ_2 and ξ_3, and K, and can be substantially lower than 1 for some alternatives, such as when one treatment is better for some K but worse for other K.

Because Q_1 and Q_2 each have 1 df, we can assess their relative powers by examining their asymptotic relative efficiency of Q_1 to Q_2, given by

$$ARE[Q_1, Q_2] = \frac{\xi_1}{\xi_2} = \frac{(\mathbf{1}'V_D\mathbf{1})^2}{(\mathbf{1}'V\mathbf{1})(\mathbf{1}'\Sigma^{-1}\mathbf{1})}.$$

Let I_D denote the diagonal matrix for which $I_D I_D = V_D$ and let C denote the corresponding correlation matrix, so that $V = I_D C I_D$ and $\Sigma^{-1} = I_D C^{-1} I_D$. Then

$$ARE[Q_1, Q_2] = \frac{(\mathbf{1}'I_D I_D\mathbf{1})^2}{(\mathbf{1}'I_D C I_D\mathbf{1})(\mathbf{1}'I_D C^{-1} I_D\mathbf{1})}.$$

From this representation, it can be seen that $ARE[Q_1, Q_2] \leq 1$, with equality when V is diagonal or of the form $aI + bJ$, where J is the KxK matrix of 1s. Thus, Q_1 can be as good as the optimal test Q_2 when $\mu_1 = \mu_2 = \cdots = \mu_K$, provided that correlation structure of $T_1, \cdots, T_K$ leads to an asymptotic covariance matrix for $\sqrt{n}\hat{\boldsymbol{\beta}}$ for which V has this form. This would arise, for example, when the T_k are uncorrelated.

For other correlation structures, however, the ARE of Q_1 to Q_2 can be very low. For example, suppose that $K = 4$ and that the survival times have the same variance and the correlation matrix

$$\begin{pmatrix} 1 & 0.7 & 0.7 & 0.4 \\ 0.7 & 1 & 0.1 & 0.1 \\ 0.7 & 0.1 & 1 & 0.1 \\ 0.4 & 0.1 & 0.1 & 1 \end{pmatrix},$$

then the asymptotic relative efficiency $ARE[Q_1, Q_2] = 0.20$, despite the homogeneity of treatment effects across strata. The fact that Q_1 can be substantially less efficient than Q_2 when the treatment hazard ratios are equal may seem counter-intuitive since Q_1 is derived from model (2), which assumes the β_k are equal. However, the working likelihood function (2) is created as if the failure times were uncorrelated. While this still leads to a consistent estimator of $\boldsymbol{\beta}$, the inefficient form of this working likelihood leads to an inefficient estimator. This can be seen analytically by noting that the asymptotically equivalent linear combination test corresponding to Q_1 does not use the optimal inverse-weighting that Q_2 uses when $\mu_1 = \mu_2 = \cdots = \mu_K$.

4.2 Unequal $\mu_1, \mu_2, \cdots, \mu_K$.

When the μ_k are not equal, the asymptotic expressions derived in Section 3 are not analytically difficult. However, because the relative efficiencies of the test statistics depend on the correlation structure among the T_k, the amount of censoring, and the magnitude of the treatment differences, the formula do not lend themselves to simple practical interpretations when the treatment differences are not homogeneous across failure times.

Since the optimal statistic Q_{opt} depends on the unknown parameter μ, use of Q_{opt} in practice is generally not feasible. However, when viewed as a "gold standard", this test provides insight into the choice and formation of test statistics. Because the noncentrality parameter of the 1 df optimal directional test is identical to that for Q_3, the top line in Figure 1 also represents the power of the optimal directional test for an arbitrary alternative to H_0. Thus, the power of the the omnibus test Q_3 for a particular choice of K can be compared to the maximum power achievable by a directional test. When $K = 2$ or 3, the power of Q_3 is surprisingly high compared to that achieved by the optimal directional test. Thus, one may not need to resort to a directional test when K is small. However, as K increases, the advantage of Q_{opt} to Q_3 increases substantially, so that the use of a directional test becomes more desirable when there is some confidence about the nature of the alternative to H_0. For example, if there is reason to believe that one treatment is uniformly superior to the other, but that the magnitude of benefit may decrease with k, then one may consider the weight $c = \frac{\Sigma^{-1} e_K}{\mathbf{1}'\Sigma^{-1} e_K}$, where $e'_K = (1, \frac{1}{2}, \cdots, \frac{1}{K})$. Such a directional test would be more powerful than Q_3 for alternatives that are reasonably approximated by $\mu_k \propto 1/k$ for $k = 1, 2, \cdots, K$.

4.3 Special Correlation Structures

Regardless of whether the treatment effect is homogeneous across failure times, the relative efficiency of Q_1 and Q_2 depends on the correlation structure of $T_1, T_2, \cdots, T_K$. Here we note two special cases for which Q_1 and Q_2 are equivalent. When the T_k are uncorrelated, we have $V = V_D$ and hence $\Sigma = V_D^{-1}$, from which it follows that Q_1 and Q_2 are equivalent with $\xi_1 = \xi_2 = \frac{(\mathbf{1}'V_D\boldsymbol{\mu})^2}{\mathbf{1}'V_D\mathbf{1}}$. In this case, $\xi_{opt} = \xi_3 = \boldsymbol{\mu}'V_D\boldsymbol{\mu}$, and thus the ARE of Q_1 or Q_2 to Q_{opt} is given by $\frac{(\mathbf{1}'V_D\boldsymbol{\mu})^2}{(\mathbf{1}'V_D\mathbf{1})(\boldsymbol{\mu}'V_D\boldsymbol{\mu})}$.

Another special covariance structure V is of the form $\mathrm{v}^2((1-\rho)I + \rho J)$, where I is identity matrix and $J = \mathbf{1}\mathbf{1}'$, i.e., all the marginal survival times have the same variance and the correlation between any two types of events are the same. It can be shown (see Appendix II) in this case that Q_1 and Q_2 are equivalent, with $\xi_1 = \xi_2 = \frac{\mathrm{v}^2 \sum_{k=1}^K \delta_k^2}{K+\rho(K-1)K}$, and that $\xi_{opt} = \xi_3 = v^2 \left(\frac{\boldsymbol{\mu}'\boldsymbol{\mu}}{1-\rho} - \frac{\rho(\mathbf{1}'\boldsymbol{\mu})^2}{(1+(K-1)\rho)(1-\rho)} \right)$.

5 Determining Sample Size and K

The focus of the previous section was a comparison of the test statistics for a particular choice of K. However, the design of a study also involves the choice of K, the number of failure times to analyze, and the sample size n. Suppose first that K and a test statistic have been selected for use and it is desired to determine the sample size or power for the study. Then the asymptotic results in Section 3 can be directly applied. To illustrate, suppose that we wish to use Q_2 and assume that the treatment groups have proportional hazards for each of the K failure times. That is, $g_k(t) = 1$ for each k. Then upon replacing δ_k by $\sqrt{n}\alpha_k$, equation (5) reduces to $\mu_k = \sqrt{n}\alpha_k$. Once a form for the censoring distribution is assumed, one can evaluate Σ and thus the noncentrality parameter ξ_2. Standard formula can then be used to determine the necessary sample size based on assuming that Q_2 has the 1 df chi-square distribution with noncentrality parameter ξ_2 under the alternative. Similar techniques can be applied to any of the other test statistics.

The choice of K can be complicated because the power of any test depends on the length of follow-up of subjects, the nature and magnitude of the treatment difference for the different failure times, and the correlation of the failure times. The asymptotic formula in Section 3 could be assessed to compare the relative powers of any particular test for various choices of K. However, one may not know enough about the amount of censoring of each failure time and of the relative treatment effects across failure times to evaluate these expressions before doing a trial.

In general, increasing K does not necessarily increase power of a test. One exception to this is the use of the directional test Q_2 when $\mu_1 = \mu_2 = \cdots = \mu_K$. Denote this test by $Q_2(K)$, and consider the use of the same statistic, say $Q_2(K')$ based on $K' < K$ failure times. Since $Q_2(K)$ is the optimal directional test in this setting, and $Q_2(K')$ can be expressed as a linear combination test based on the K-vector $\hat{\beta}$, it follows that the asymptotic power of $Q_2(K)$ is at least as great as that of $Q_2(K')$ for any $K' < K$. That is, when $\mu_1 = \mu_2 = \cdots = \mu_K$, the power of Q_2 increases with K regardless of the censoring distributions or the correlation structure of the failure time. The magnitude of the power gained with increasing K depends on censoring and the correlation structure, however, and in some cases could be small.

Similarly, the power of the optimal test Q_{opt} will increase with K for any alternative to H_0, any censoring pattern, and any correlation structure among the failure times. However, this result is of little practical value because one rarely is certain about the direction and magnitude of the treatment difference for each K. In contrast, the power of the omnibus test Q_3 need not increase with K. As we see below, while the noncentrality parameter of Q_3 increases with K, the increasing degrees of freedom can offset this and lead to less power as more failure times are included.

Hughes [HUG97] investigates in detail the choice between $K = 1$ (in which case the directional and omnibus tests are equivalent) and the use of the omnibus test when $K = 2$ in the setting of proportional hazards alternatives in which one treatment is superior to the other for each K. He shows that the more powerful of these tests depends on the amount of censoring as well as the correlation of the two failure times, and that the test based on a single failure time ($K = 1$) often is as or more powerful to the omnibus test with $K = 2$.

To get a practical sense of the trade-offs involved in choosing K for more complex settings, we conducted a simulation study. We generated $T_1, T_2 - T_1, T_3 - T_2, \cdots, T_K - T_{K-1}$ to be K independent exponential random variables, the kth with intensity h_k. Thus, the gap times between events are independent. T_j is the sum of j independent exponentials, so that the covariance between T_j and T_k for $j \leq k$ is just the variance of T_j. Two treatment groups are generated, each having a sample size of 200. For the first treatment group, we take $h_k = 1$ for all k, so that T_j has the Gamma distribution with parameters 1 and j. For the second treatment group, we use various choices for $h_1, h_2, \cdots, h_K$. We chose the h_k to be increasing, constant, decreasing, or non-monotone. The potential censoring time is taken to be an independent uniform $(0, \tau)$ random variable with τ chosen to give the desired censoring percentage. Each simulation is repeated 1000 times. Despite this simple correlation structure, the resulting failure time T_k for the treatment groups does not have a proportional hazard relationship for $k = 1, 2, \cdots, K$.

Table 1 displays the power of Q_2, Q_3 and Q_c, for $c = \frac{\Sigma^{-1} e_K}{\mathbf{1}' \Sigma^{-1} e_K}$ with $e_K = (1, 2, ..., K)$ or $(K, K-1, ..., 1)$, when choosing a univariate analysis ($K = 1$) or when analyzing $K = 2$, 3, or 4 events. The choice of e_K might be made when one expects the treatment difference to increase ($e_K = 1, 2, ..., K$) or decrease ($c = K, K-1, ..., 1$) with successive failure times. In these simulations we used $\tau = 4$, which leads to heavy censoring occurred in later events. The first column corresponds to gap time hazard ratios of $(h_1, h_2, h_3, h_4) = (1, .8, .6, .4)$, corresponding to strongly decreasing h_k with successive failure times. Since $h_1 = 1$, the univariate tests have power equal to the Type I error. In this case, the gain in power for $K > 1$ is evident for most of the test statistics, especially Q_3 and Q_c with increasing weight $e_K = (1, 2, ...K)$, despite of the increased rate of censoring of the second, third, and fourth failure times. Q_2 and the directional test based on the weight $e_K = (4, 3, 2, 1)$ do poorly here. The second column in Table 1 gives the power when the h_k equal 0.8 for each k. Here the power of the directional tests with weight $e_K = (1, 2, 3, 4)$ increases with K but the power of the omnibus test Q_3 does not, indicating that any gains in information from examining more survival times is more than offset by the increased censoring and degrees of freedom. The 3rd column of Table 1 gives the power of the tests for different K when the h_k increase slowly. None of the tests showed power gains by analyzing more events. This is not surprising as the treatment difference no longer exists for the third and fourth gap times. The last column gives the hazard rates of an inconsistent treatment effect among the gap time. The treatment effect of the first and the fourth gap time is just the opposite of the second and the third gap time. Here none of the directional tests are oriented towards such treatment differences and thus none perform well in comparison to Q_3.

We conducted another simulation study with $K = 2$, 3, and 4 unordered events and proportional hazards realtionships between the treatment groups. To do so, we extended Gumbel's bivariate exponential distribution [GUM60] to four exponential variables T_1, T_2, T_3, and T_4 with joint cumulative distribution:

$$
\begin{aligned}
F(t_1, t_2, t_3, t_4) = {} & (1 - e^{-h_1 t_1})(1 - e^{-h_2 t_2})(1 - e^{-h_3 t_3})(1 - e^{-h_4 t_4}) \\
& (1 + \alpha_{12} e^{-h_1 t_1 - h_2 t_2})(1 + \alpha_{13} e^{-h_1 t_1 - h_3 t_3})(1 + \alpha_{14} e^{-h_1 t_1 - h_4 t_4}) \\
& (1 + \alpha_{23} e^{-h_2 t_2 - h_3 t_3})(1 + \alpha_{24} e^{-h_2 t_2 - h_4 t_4})(1 + \alpha_{34} e^{-h_3 t_3 - h_4 t_4})
\end{aligned}
$$

where $h_i > 0$ $(i = 1, 2, 3, 4)$ and $-1 \leq \alpha_{ij} \leq 1$ $(1 \leq i < j \leq 4)$. The values of α_{ij} determine the correlations between T_i and T_j. For one treatment group, we set $h_1 = h_2 = h_3 = h_4 = 1$. We choose various values of h_i for the second treatment group. It follows that the treatment hazard ratio for survival time T_i is h_i. The potential censoring time is taken to have the uniform distribution on (0,4). For each simulation setting, we generate a sample size of 100 for each treatment group, and generate 1000 repetitions.

Four sets of α_{ij} values are used in these simulations, representing different correlation structures. For each set of α_{ij} value, three sets of hazard (h_1, h_2, h_3, h_4) are examined. Table 2 lists the power of Q_2 and Q_3 for $K = 2$ (the first 2 events), 3 (the first three events), and 4. When the marginal hazard ratios are the same for all the four survival times $(h_i = 1.25, i = 1, 2, 3, 4)$, it confirms that Q_2 is always superior to Q_3 for a given K, the power of Q_2 increases as K increases, and the power of Q_2 is about 10% higher than that of Q_3 when $K = 2$ and about 15% higher than that of Q_3 when $K = 3$. When the marginal hazard ratios are $h_1 = 1.67$, $h_2 = 1$, $h_3 = 1.67$, and $h_4 = 1$, there is no treatment group difference for T_2 and T_4. Here Q_2 has poorer power than Q_3 for almost all the correlation structures and choice of K in Table 2. Additionally, the power for Q_3 does not always increase with K. The power for Q_3 when $K = 3$ is similar to that when $K = 4$. It is interesting to see that there are not much power loss for Q_3 by adding the survival time T_4, which has no treatment difference. For one correlation structure, $\alpha_{11} = \alpha_{12} = \alpha_{13} = \alpha_{23} = \alpha_{24} = \alpha_{34} = 1$, the power Q_3 when $K = 4$ is slightly greater compared to that when $K = 3$. In these simulations, the correlation structure has a greater impact on the power of Q_2 than on that for Q_3. When the marginal hazard ratios are $h_1 = 1.25$, $h_2 = 0.8$, $h_3 = 0.8$, and $h_4 = 1.25$, the second treatment group has a higher hazard than the first one for T_1 and T_4 but a lower hazard for T_2 and T_3. In this case, Q_2 has almost no power to detect treatment differece irrespect of the correlation structures and the choices of K. This is merely because the treatment differences are in the opposit directions for the four survival times. Q_3 has better power than Q_2 and the power increases as K increases. The power does not change greatly for different correlation structures.

6 Example: Recurring Opportunistic Infections in HIV/AIDS

To illustrate the results presented in this paper, we consider an example from two companion AIDS clinical trials [DAF95], [KLR92]. Patients were followed for opportunistic infections, which could occur repeatedly, and for mortality. The primary endpoint in the trial was the time until a patient's first OI or death, whichever came first. A total of 1530 patients were randomized to receive either AZT (n=512) or ddI (1008). We exclude 10 patients due to missing or incorrect information about the times of subsequent events. Define T_j to be the time from randomization until the jth OI or death, whichever comes first, for $j = 1, \cdots, K$. Thus, if K=3 and a patient has 3 OIs, T_1, T_2, and T_3 will denote the elapsed times from randomization to these OIs. However, if a patient has 1 OI and then dies, T_1, T_2, T_3 are the time until the OI, the time until death, and the time until death, respectively. Extension of the primary endpoint in this way induces a correlation among the failure times.

In the ddI (AZT) group, the number of patients experiencing 1, 2, 3, and 4 distinct OIs were $342(204), 119(80), 26(15)$, and $4(1)$. A total of 199 and 95 patients died in the ddI and AZT groups, respectively. The percent of jth events that were censored for the ddI (AZT) group was $59(55), 74(72), 79(80)$ and $80(81)$. We define $Z = 1$ for the ddI group, so that β_j denotes the log relative hazard of ddI to AZT for T_j. The resulting estimates are

$$\hat{\beta}_1 = 0.163, \quad \hat{\beta}_2 = 0.106, \quad \hat{\beta}_3 = -0.022, \quad \hat{\beta}_4 = -0.069 \ ;$$

the corresponding estimated covariance matrix is

$$\begin{pmatrix} 0.0068 & 0.0061 & 0.0058 & 0.0057 \\ 0.0061 & 0.0107 & 0.0102 & 0.0101 \\ 0.0058 & 0.0102 & 0.0142 & 0.0140 \\ 0.0057 & 0.0101 & 0.0140 & 0.0151 \end{pmatrix} .$$

The estimators of the common β derived from (2) and used in Q_1 are $0.141, 0.104$, and 0.073, for $K = 2, 3$, and 4, respectively. The corresponding standard errors are $0.084, 0.087$, and 0.090.

Table 2 gives the p-values corresponding to Q_1, Q_1^*, Q_2, and Q_3 for $K = 1, 2, 3, 4$. As expected, the results for Q_1 and Q_1^* are very similar. Note that the estimated regression coefficients steadily decrease, suggesting that the initial advantage for the ddI group is only transient. Not surprisingly, the estimate of β, the common hazard ratio, also decreases but does not become negative, and the p-value corresponding to Q_2 steadily increases from 0.05

when analyzing just one event ($K = 1$) to 0.079 when analyzing all four. Despite the increased rate of censoring, the omnibus test Q_3 better detects a treatment difference than Q_2 when $K = 3$ or 4, but does less well when only analyzing the first ($K = 1$) or first two ($K = 2$) events. Q_1 and its asymptotic equivalent Q_1^* give the least significant p-values when $K = 3$ or 4.

7 Discussion

A goal of this paper was to build upon earlier work of Wei, Lin & Weissfeld [WLW89], Lin [LIN94], Hughes [HUG97], and shed light on whether and how to use the marginal analysis approach of WLW method for the analysis of multivariate survival data. We did so primarily by deriving and evaluating the asymptotic behavior of two directional tests, Q_1 and Q_2, and the omnibus test Q_3 that have been previously proposed, as well as a more general class of linear combination tests. Q_1 derives from solving a constrained working likelihood that assumes that the treatment group hazard ratios are the same across the K survival times. Q_2 is also motivated by the same assumption, but instead is based on a specific linear combination of the unconstrained estimators of the treatment hazard ratios. One noteworthy finding is that Q_2 is the optimal linear combination test under this assumption, and in general is asymptotically more powerful than Q_1, sometimes by a substantial amount. This stems from the fact that the working likelihood function used by WLW is inefficient, and as a result constrained estimators obtained from it are in general inefficient. We thus discourage the use of Q_1 and recommend Q_2 when it is believed that the marginal treatment effects are approximately equal for all events.

One can easily describe the power gains from using the optimal directional test as opposed to the omnibus test based on(Figure 1). These can be substantial when K is large, but are modest for $K = 2$ or 3. For $K = 2$, 3 the power of the omnibus test is often close to the maximal power achievable by a directional test, which suggests that the omnibus test should be considered when there is not a good sense of the alternatives to H_0 that are likely to occur.

When the treatment hazard ratios vary among the K survival times and the data are censored, the analytic formula for the asymptotic power of these tests do not lend themselves to simple guidelines on whether to use the omnibus test Q_3, the directional test Q_2, or some other directional test, say Q_c. The censoring settings we evaluated in simulations assumed a common censoring variable for all K survival times, as would commonly occur in a clinical trial when the survival times corresponded to recurring events. Here the amount of censoring necessarily increases with successive survival times, thus limiting the amount of information gained by analyzing larger K. In some instances, this along with the additional degrees of freedom on Q_3 with increasing K offset any gains in information.

In the HIV/AIDS example, the beneficial effect of the ddI treatment over AZT decreased with successive survival times. As a result, the omnibus test gave the smallest p-values when using $K = 3$ or 4 events. However, if $K = 1$ or $K = 2$ events had been analyzed, the directional test Q_2 would have given a more significant result. On the other hand, had the treatment effect increased with successive failure times, as illustrated in the simulation, use of larger K can increase power of a directional test. In many, if not most, settings, analysts would not know in advance how the treatment hazard ratio varies with successive failure times. If one expected a attenuated effect, as was seen in the HIV/AIDS example, then a univariate analysis may prove best, especially when later survival times are heavily censored. Alternatively, one might choose the linear combination test Q_c with weights c selected to reflect the suspected trend. On the other hand, if one expects treatment hazard ratios that are approximately equal, then use of Q_2 with $K > 1$ is justified. Exactly how to choose K is not simple, but if one had a sense of the degree of censoring, the noncentrality parameter $\xi \propto 1'\Sigma^{-1}1$ could be evaluated to approximate the power of Q_2 for various choices for K.

Acknowledgments

This research was supported by grants AI24643 from the National Institute of Allergy and Infectious Diseases. The authors are grateful to Drs. L. J. Wei, Michael Hughes, and the referees for their comments and suggestions.

Table 1. Asymptotic Power from a Simulation Study Using Independent Gap-time Exponential Distributions

		h_1, h_2, h_3, h_4			
K	$Test(e_k)$	1, .8, .6, .4	.8, .8, .8, .8	.8, .8, 1, 1	1.25, .8, .8, 1.25
1	$Q_2 = Q_3$	5.20	45.00	45.00	52.50
2	Q_2	6.10	57.30	57.30	30.40
	Q_3	25.00	55.30	55.30	61.10
	$Q_c(1,2)$	22.00	63.40	63.40	5.40
3	Q_2	8.00	62.20	56.70	24.60
	Q_3	68.50	58.00	49.10	65.30
	$Q_c(1,2,3)$	62.90	71.60	47.30	10.70
4	Q_2	9.30	62.40	56.80	26.00
	Q_3	91.80	56.20	44.10	63.30
	$Q_c(1,2,3,4)$	85.70	75.10	41.00	5.80
	$Q_c(4,3,2,1)$	8.30	33.00	44.00	47.70

Table 2. Asymptotic Power from a Simulation Study for Unordered Events Using Joint Distributions of Four Exponential Variables.

			h_1, h_2, h_3, h_4		
Correlation	K	*Test*	$1.67, 1, 1.67, 1$	$1.25, 1.25, 1.25, 1.25$	$1.25, .8, .8, 1.25$
$\alpha_{12} = 1$	2	Q_2	53.2	44.5	4.0
$\alpha_{13} = 1$		Q_3	84.8	34.4	43.9
$\alpha_{14} = 1$	3	Q_2	85.3	47.1	9.2
$\alpha_{23} = 1$		Q_3	95.1	33.5	56.1
$\alpha_{24} = 1$	4	Q_2	62.8	54.0	3.9
$\alpha_{34} = 1$		Q_3	97.0	33.8	74.2
$\alpha_{12} = 0.5$	2	Q_2	58.5	47.5	4.8
$\alpha_{13} = 0.5$		Q_3	83.8	38.0	40.8
$\alpha_{13} = 0.5$	3	Q_2	92.7	55.0	8.0
$\alpha_{23} = 0.5$		Q_3	96.0	40.5	49.7
$\alpha_{24} = 0.5$	4	Q_2	76.1	62.8	4.2
$\alpha_{34} = 0.5$		Q_3	95.4	44.7	60.7
$\alpha_{12} = -0.5$	2	Q_2	66.7	54.5	4.6
$\alpha_{13} = -0.5$		Q_3	83.3	45.3	38.5
$\alpha_{14} = -0.5$	3	Q_2	98.5	72.9	8.1
$\alpha_{23} = 0.5$		Q_3	99.0	53.4	43.7
$\alpha_{24} = 0.5$	4	Q_2	95.3	81.8	8.10
$\alpha_{34} = 0.5$		Q_3	98.8	62.3	62.8
$\alpha_{12} = -1$	2	Q_2	70.2	59.8	5.0
$\alpha_{13} = -1$		Q_3	83.5	51.3	38.9
$\alpha_{14} = -1$	3	Q_2	99.6	78.6	7.5
$\alpha_{23} = 1$		Q_3	99.4	62.1	41.4
$\alpha_{24} = 1$	4	Q_2	98.7	87.5	10.8
$\alpha_{34} = 1$		Q_3	99.3	68.2	72.5

Table 3. P-values for Different Test Statistics for HIV/AIDS Example

K	Q_1	Q_1^*	Q_2	Q_3
1	0.048	0.048	0.048	0.048
2	0.093	0.093	0.057	0.120
3	0.230	0.232	0.073	0.055
4	0.413	0.419	0.079	0.051

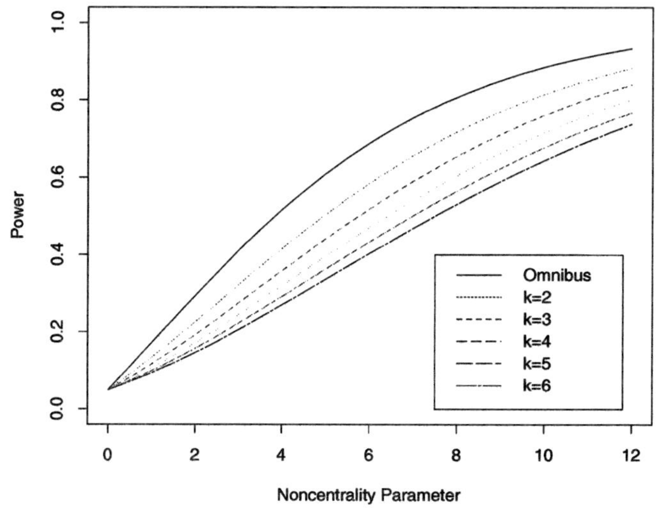

Fig. 1. Power of Q_{opt} versus Q_3 for Different K

References

[VKVN02] Andersen,P.K., Gill, R.D.: Cox's regression model for counting processes: a large sample study. The Annals of Statistics, **10**, 1100–20 (1981)

[VKVN02] Cai, J., Prentice, R.L.: Estimating Equations for Hazard Ratio Parameters Based on Correlated Failure Time Data. Biometrika, **82**, 151–64 (1995)

[VKVN02] Cook, R.J., Lawless, J.F., Nadeau, C.: Robust Tests for Treatment Comparisons Based on Recurrent Event. Biometrics, **52**, 557–71 (1996)

[VKVN02] Cox, D.R.: Regression Models and Life Tables. J. R. Statist. Soc. B, **34**, 187–220 (1972)

[VKVN02] Dolin, R., Amato, D.A., Fischl, M.A., et al.: Zidovudine Compared with Didanosine in Patients with Advanced HIV Type 1 Infection and Little or No Previous Experience with Zidovudine. Archives of Internal Medicine, **155** (9), 961–74 (1995)

[VKVN02] Gumbel,E.J.: Bivariate Exponential Distribtions. JASA **55**, 698–707 (1960)

[VKVN02] Hauser, W.A., Rich, S.S., Lee, J.R-J., et al.: Risk of Recurrent Seizures after Two Unprovoked Seizures. New England Journal of Medicine, **338**, 429–34 (1998)

[VKVN02] Hughes, M.: Power Considerations for Clinical Trials Using Multivariate Time to Event Data. Statistical in Medicine, **16**, 865–82 (1997)

[VKVN02] Kahn, J., Lagakos, S.W., Richman, D.D., et al.: A Controlled Trial Comparing Continued Zidovudine with Didanosine in Human Immunodeficiency Virus Infection. New England Journal of Medicine, **327**, 581–7 (1992)

[LI97] Li, Q.H.: The Relationship between Directional and Omnibus Tests for a Vector Parameter. Doctoral Thesis, Harvard University, Massachusetts (1997).

[VKVN02] Li, Q.H., Lagakos, S.W.: Comparisons of Test Statistics Arising from Marginal Analyses of Multivariate Survival Data. Lifetime Data Analysis, **10**, 389–405 (2004)

[VKVN02] Liang, K.Y., Self, S.G., Chang, Y.C.: Modeling Marginal Hazards in Multivariate Failure Time Data. J. Roy. Statist. Soc. B, **55**, 441–53 (1993)

[VKVN02] Lin, D.Y.: Cox Regression of Multivariate Failure Time Data: the Marginal Approach. Statistical in Medicine, **13**, 2233–47 (1994)

[VKVN02] Lin, J.S., Wei, L.J.: Linear Regression Analysis for Multivariate Failure Time Observations. JASA, **87**, 1091–7 (1992)

[VKVN02] Prentice, R.L., Williams, B.J., Peterson, A.V.: On the Regression Analysis of Multivariate Failure Time Data. Biometrika, **68**, 373–9 (1981)

[VKVN02] Richman, D., Grimes, J., Lagakos, S.: Effect of Stage of Disease and Drug Dose on Zidovudine Susceptibilities of Isolates of Human Immunodeficiency Virus. J. Acquired Immune Deficiency Syndromes, **3**, 743–6 (1990)

[VKVN02] Wei, L.J., Lin, D.Y., Weissfeld, L.: Regression Analysis of Multivariate Incomplete Failure Time Data by Modeling Marginal Distributions. JASA, **84**, 1065–73 (1989)

[VKVN02] Yang, Y., Ying, Z.: Marginal Proportional Hazards Models for MultipleEvent-time Data. Biometrika, **88**, 581–6 (2001)

Appendix I Asymptotic Normality of WLW Method under General Contiguous Alternatives

Let $Y_{ki}(t) = 1(X_{ki} \geq t)$ and $N_{ki}(t) = 1(X_{ki} \leq t, \Delta_{ki} = 1)$, and define the functions $S_k^{(r)}(\beta, t)$, $V_k(\beta, t)$, $s_k^{(r)}(\beta, t)$ and $v_k(\beta, t)$ for $r = 0, 1, 2$ by

$$S_k^{(r)}(\beta,t) = \frac{1}{n}\sum_{j=1}^{n} Y_{kj}(t)Z_j^r exp(\beta Z_j) \quad V_k(\beta,t) = \frac{S_k^{(2)}(\beta,t)}{S_k^{(0)}(\beta,t)} - \left(\frac{S_k^{(1)}(\beta,t)}{S_k^{(0)}(\beta,t)}\right)^2$$
$$s_k^{(r)}(\beta,t) = E[S_k^{(r)}(\beta,t)] \quad v_k(\beta,t) = \frac{s_k^{(2)}(\beta,t)}{s_k^{(0)}(\beta,t)} - \left(\frac{s_k^{(1)}(\beta,t)}{s_k^{(0)}(\beta,t)}\right)^2 .$$

Define the martingale $M_{ki}(t)$ as

$$M_{ki}(t) = N_{ki}(t) - \int_0^t Y_{ki}(\tau)\lambda_k(\tau|Z_i)d\tau .$$

First consider $\hat{\beta}$, the maximizing solution of equation (2). Under the regularity conditions, it can be shown that $\hat{\beta}$ converges to β, the solution of the following equation:

$$\sum_{k=1}^{K}\int_0^{\infty}\{s_k^{(1)}(\alpha_k g_k(t),t) - \frac{s_k^{(1)}(\beta,t)}{s_k^{(0)}(\beta,t)}s_k^{(0)}(\alpha_k g_k(t),t)\}\lambda_k(t)dt = 0.$$

Under contiguous alternatives, the above equation becomes

$$\sum_{k=1}^{K}\int_0^{\infty}\{s_k^{(1)}(0,t) - \frac{s_k^{(1)}(\beta,t)}{s_k^{(0)}(\beta,t)}s_k^{(0)}(0,t)\}\lambda_k(t)dt = 0,$$

which has its solution at $\beta = 0$. Let $\hat{\beta}^{(n)}$ denote the estimate of β when the alternatives are the sequence of general contiguous alternatives defined in Section 3. Using the above argument, it also follows that $\hat{\beta}^{(n)} \xrightarrow{p} 0$ as $n \to \infty$.

To determine the asymptotic distribution of $\sqrt{n}\hat{\beta}^{(n)}$ under the general contiguous alternative, we first derive the asymptotic distribution of the score statistic $U(\beta^{(n)})$, where

$$\beta^{(n)} = \frac{\sum_{k=1}^{K}\alpha_k\int_0^{\infty}g_k(t)v_k(0,t)s_k^{(0)}(0,t)\lambda_k(t)dt}{\sum_{k=1}^{K}\int_0^{\infty}v_k(0,t)s_k^{(0)}(0,t)\lambda_k(t)dt}.$$

Rewriting $U(\beta^{(n)})$, we have

$$\begin{aligned} n^{-\frac{1}{2}}U(\beta^{(n)}) &= n^{-\frac{1}{2}}\sum_{k=1}^{K}\sum_{i=1}^{n}\int_0^{\infty}\{z_i - \frac{S_k^{(1)}(\beta^{(n)},t)}{S_k^{(0)}(\beta^{(n)},t)}\}dN_{ki}(t) \\ &= n^{-\frac{1}{2}}\sum_{k=1}^{K}\sum_{i=1}^{n}\int_0^{\infty}\{z_i - \frac{S_k^{(1)}(\alpha_k g_k(t),t)}{S_k^{(0)}(\alpha_k g_k(t),t)}\}dN_{ki}(t) \\ &+ n^{-\frac{1}{2}}\sum_{k=1}^{K}\sum_{i=1}^{n}\int_0^{\infty}\{\frac{S_k^{(1)}(\alpha_k g_k(t),t)}{S_k^{(0)}(\alpha_k g_k(t),t)} - \frac{S_k^{(1)}(\beta^{(n)},t)}{S_k^{(0)}(\beta^{(n)},t)}\}dN_{ki}(t) \end{aligned}$$

The first term of right hand side of the second equal sign is martingale. It can be shown that this term converges to the $N(0, \mathbf{1}'V\mathbf{1})$ distribution and that the second term converge to 0 in probability as $n \to \infty$. Thus, as $n \to \infty$,

$$n^{-\frac{1}{2}}U(\beta^{(n)}) \longrightarrow N(0, \mathbf{1}'V\mathbf{1}) ,$$

where V is the KxK matrix with (k,k) element

$$\mathrm{v}_k^2(\beta_k) = \int_0^\infty v_k(\beta_k,t)s_k^{(0)}(\beta_k,t)\lambda_k(t)dt,$$

and with (k,w) element $(k = w)$

$$\mathrm{v}_{kw}(\beta_k,\beta_w) = E[\int_0^\infty \{z_i - \frac{s_k^{(1)}(\beta_k,t)}{s_k^{(0)}(\beta_k,t)}\}dM_{ki}][\int_0^\infty \{z_i - \frac{s_w^{(1)}(\beta_w,t)}{s_w^{(0)}(\beta_w,t)}\}dM_{wi}] \quad (k \neq w).$$

Thus, as $n \to \infty$

$$\sqrt{n}(\hat{\beta} - \beta^{(n)}) \xrightarrow{\mathcal{L}} N(0, (\sum_{k=1}^{K} \mathrm{v}_k^2)^{-2}\mathbf{1}'V\mathbf{1}).$$

Letting

$$\mu = \frac{\sum_{k=1}^{K} \delta_k \int_0^\infty g_k(t)v_k(0,t)s_k^{(0)}(0,t)\lambda_k(t)dt}{\sum_{k=1}^{K} \int_0^\infty v_k(0,t)s_k^{(0)}(0,t)\lambda_k(t)dt},$$

and noting that $\beta^{(n)} \to 0$, it follows that

$$\sqrt{n}\hat{\beta} \xrightarrow{\mathcal{L}} N(\mu, (\sum_{k=1}^{K} \mathrm{v}_k^2)^{-2}\mathbf{1}'V\mathbf{1})$$

as $n \to \infty$.

Next, consider the estimator $\hat{\boldsymbol{\beta}}^{(n)}$ arising from (1). Under the sequence of general contiguous alternatives, an approach similar to that for $\hat{\beta}^{(n)}$ can be applied, resulting in

$$\sqrt{n}\hat{\boldsymbol{\beta}} \xrightarrow{\mathcal{L}} N(\boldsymbol{\mu}, \Sigma)$$

as $n \to \infty$, where $\boldsymbol{\mu} = (\mu_1, \mu_2, \cdots, \mu_K)'$, and

$$\mu_k = \frac{\int_0^\infty \{\delta_k g_k(t)v_k(0,t)s_k^{(0)}(0,t)\}\lambda_k(t)dt}{\int_0^\infty v_k(0,t)s_k^{(0)}(\alpha_k g_k(t),t)\lambda_k(t)dt}.$$

Appendix II Equivalence of Q_1^* and Q_2 under the covariance structure $V = \sigma^2[(1-\rho)I + \rho J]$

Let

$$\hat{\gamma}_1 = \frac{\mathbf{1}'V_D\hat{\boldsymbol{\beta}}}{\mathbf{1}'V_D\mathbf{1}} \qquad \text{and} \qquad \hat{\gamma}_2 = \frac{\mathbf{1}'\Sigma^{-1}\hat{\boldsymbol{\beta}}}{\mathbf{1}'\Sigma^{-1}\mathbf{1}}$$

denote the linear combinations of $\hat{\boldsymbol{\beta}}$ on which Q_1^* and Q_2 are based.

When $V = \sigma^2(1-\rho)I + \sigma^2\rho J$, $\hat{\gamma}_1$ simplifies to

$$\frac{\mathbf{1}'V_D\hat{\boldsymbol{\beta}}}{\mathbf{1}'V_D\mathbf{1}} = \frac{\sum_{k=1}^{K}\hat{\beta}_k}{K}$$

Since $V^{-1} = \frac{1}{\sigma^2}(aI + bJ)$, where

$$a = \frac{1}{1-\rho}\ b = \frac{-\rho}{(1+(K-1)\rho)(1-\rho)},$$

$\hat{\gamma}_2$ simplifies to

$$\frac{\mathbf{1}'\Sigma^{-1}\hat{\boldsymbol{\beta}}}{\mathbf{1}'\Sigma^{-1}\mathbf{1}} = \frac{\mathbf{1}'V^{-1}\hat{\boldsymbol{\beta}}}{\mathbf{1}'V^{-1}\mathbf{1}} = \frac{\sum_{k=1}^{K}\hat{\beta}_k}{K} = \hat{\gamma}_1.$$

It follows that Q_1^* and Q_2 are equivalent, and hence that Q_1 and Q_2 are equivalent.

This article has appeared in the December 2004 issue of Lifetime Data Analysis

Nonparametric Estimation and Testing in Survival Models

Henning Läuter[1] and Hannelore Liero[2]

[1] Institute of Mathematics, University of Potsdam laeuter@rz.uni-potsdam.de
[2] Institute of Mathematics, University of Potsdam liero@rz.uni-potsdam.de

The aim of this paper is to demonstrate that nonparametric smoothing methods for estimating functions can be an useful tool in the analysis of life time data. After stating some basic notations we will present a data example. Applying standard parametric methods to these data we will see that this approach fails - basic features of the underlying functions are not reflected by their estimates. Our proposal is to use nonparametric estimation methods. These methods are explained in section 2. Nonparametric approaches are better in the sense that they are more flexible, and misspecifications of the model are avoided. But, parametric models have the advantage that the parameters can be interpreted. So, finally, we will formulate a test procedure to check whether a parametric or a nonparametric model is appropriate.

1 Stating the Problem

We consider life or failure times of individuals or objects belonging to a certain group, the so-called population of interest. Examples are: survival times of patients in a clinical trial, lifetimes of machine components in industrial reliability or times taken by subjects to complete specified tasks in psychological tests. We assume that these life times can be modelled by a random variable Y with a distribution F, that is, we assume that the probability that an individual of the underlying population dies (fails) before time point t can be expressed in the form

$$\mathsf{P}(Y \leq t) = F(t).$$

The probability that the individual survives the time point t is given by the survival function

$$S(t) = \mathsf{P}(Y > t) = 1 - F(t).$$

Other functions of interest are the density $f(t) = F'(t)$ and the hazard or failure rate

$$\lambda(t) = \lim_{s \downarrow 0} \frac{1}{s} \mathsf{P}(t < Y \leq t + s | Y \geq t)$$

describing the immediate risk attaching to an individual known to be alive at time point t.

Now, suppose that we have obtained data from the underlying population. How we can use these data to estimate the survival function or the hazard rate?

Assuming a parametric model for the distribution the survival times we have to estimate parameters. It is well-known, that the maximum likelihood method provides good estimates.

For example, if we assume that our data are realizations of exponential distributed random variables $Y_1, \ldots, Y_n$, that is, the survival function is given by

$$S(t) = \exp(-t\beta),$$

with parameter $\beta > 0$, then the problem of estimating the function S is simply the problem of estimating the parameter β. And the maximum likelihood estimator (m.l.e.) is given by

$$\hat{\beta} = \frac{1}{n} \sum_{i=1}^{n} Y_i.$$

Assuming a Weibull distribution with parameters β and ν, i.e

$$S(t) = \mathsf{P}(Y > t) = \exp(-(t/\beta)^{\nu}),$$

we obtain that the m.l.e. of the two-dimensional parameter is a solution of

$$\hat{\beta}^{\hat{\nu}} = \frac{1}{n} \sum_{i=1}^{n} Y_i^{\hat{\nu}}$$

$$\frac{\sum_{i=1}^{n} Y_i^{\hat{\nu}} \log Y_i}{\hat{\beta}^{\hat{\nu}}} = \frac{n}{\hat{\nu}} + \sum_{i=1}^{n} \log Y_i. \tag{1}$$

If the assumed parametric model is a good description of the of the underlying population, then parametric estimators and test procedures based on these estimators provide good results. But if the parametric model is not appropriate such an approach can lead to wrong conclusions. This is demonstrated in the following: Suppose that a mixture of two Weibull distributions is considered. The first group is characterized by parameters β_1, ν_1 and the second with β_2, ν_2, and let p be the portion of the first group. Then the survival function is given by

$$S^*(t) = (1-p)\exp(-(t/\beta_1)^{\nu_1}) + p\exp(-(t/\beta_2)^{\nu_2}) \tag{2}$$

For $\beta_1 = 1$, $\beta_2 = 4$, $\nu_1 = 2$, $\nu_2 = 4$ and $p = 0.05$ Figure 1 shows S^*, the density f^* and the hazard rate λ^* of the mixture (solid line). Further the

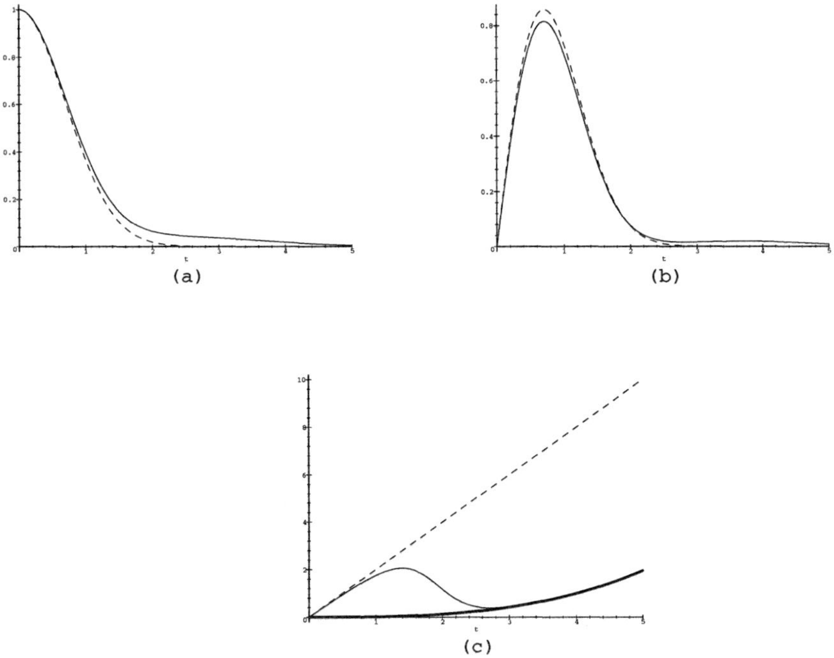

Fig. 1. (a) Survival functions, (b) Densities, (c) Hazard rates, for the main component (dashed line), for the mixture (thin solid line), in (c) the hazard rate for the minor component (bold solid line)

main part of the mixture, i.e. $\exp(-(t/\beta_1)^{\nu_1}$ is given in (a), in (b) and (c) you see not only this term of the mixture but also the minor one.
In such a case with a small p one can interprete the first Weibull distribution as a disturbation of the second one and one would hope that the fit with a single Weibull distribution is sufficiently well. Simulated data with 100 observations from the disturbed Weibull model were used to estimate the parameters β and ν in a single Weibull model with

$$S(t) = \exp(-(t/\beta)^{\nu}) \quad \text{and} \quad \lambda(t) = \frac{t^{\nu-1}}{\beta^{\nu}},$$

which was assumed neglecting the inhomogenity of the population.

The maximum likelihood estimates, computed according to (1), are: $\hat{\beta} = 1.057$ and $\hat{\nu} = 1.422$. Replacing these estimates into the functions S and λ we get Figure 2.

We see: The estimators using the single Weibull model are wrong estimators.

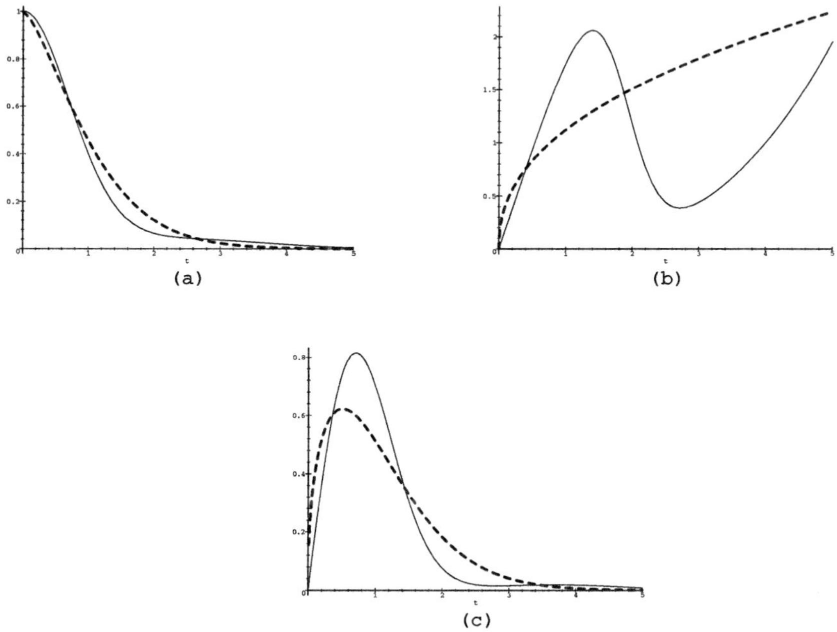

Fig. 2. a) Survival functions, (b) Hazard rates, (c) Densities, for the estimated single Weibull model (bold dashed line), for the mixture (thin solid line)

This model is unable to detect the features of the underlying functions!
Such a mixed distribution one meets if the underlying population is not homogenous. A latent factor, which is not observed divides the population into (for simplicity) two groups. Further, assume that both groups can be characterized by a Weibull distribution: the first with parameters β_1, ν_1 and the second with β_2, ν_2, and let p be the portion of the first group. Latent factors can be: a not observed underlying disease (depression), different litter in an animal experiment or different producer of a technical component.

2 Nonparametric Estimators

2.1 Model with censoring

Very often, in practical applications the life times Y_i's are subject to random right censoring, i.e. some individuals may not be observed for the full time to failure. Thus, our observations are values of r. v.'s T_i which are censored or uncensored. Here we assume a random censoring scheme characterized by i.i.d. r. v.'s C_i which are independent of the Y- sequence. Thus, we observe (T_i, δ_i), $i = 1, \ldots, n$ with

$$T_i = \min(Y_i, C_i) \quad \text{and} \quad \delta_i = 1(Y_i \le C_i).$$

The distribution of the observations is described by the distribution function and the subdistribution function of the uncensored observations

$$H(t) := \mathsf{P}(T_i \le t) \text{ and } H^U(t) := \mathsf{P}(T_i \le t, \delta_i = 1).$$

2.2 The Nelson-Aalen estimator for the cumulative hazard function

Starting point of the construction of an estimator for the hazard function λ and the survival function S is an estimator for Λ, the cumulative hazard function defined by

$$\Lambda(t) = \int_0^t \lambda(s)\,\mathrm{d}s.$$

Using standard transformations we can write this estimator in the following form

$$\Lambda(t) = \int_0^t \frac{\mathrm{d}F(y)}{S(y)} = \int_0^t \frac{\mathrm{d}H^U(y)}{1 - H(y)}. \tag{3}$$

The idea for the estimation of Λ goes back to [B81]. He proposed to replace the functions H and H^U in (3)by their empirical versions

$$\hat{H}_n^U(t) = \frac{1}{n}\sum_{i=1}^n 1(T_i \le t, \delta_i = 1), \qquad \hat{H}_n(t) = \frac{1}{n}\sum_{i=1}^n 1(T_i \le t). \tag{4}$$

The resulting estimator is the so-called Nelson-Aalen type estimator

$$\hat{\Lambda}_n(t) := \int_0^t \frac{\mathrm{d}\hat{H}_n^U(s)}{1 - \hat{H}_n(s_-)}.$$

The explicit formula of $\hat{\Lambda}_n$ is given by

$$\hat{\Lambda}_n(t) = \sum_{i=1}^n \frac{1(T_{(i)} \le t)\delta_{[i]}}{n - i + 1}.$$

Here $T_{(1)} \le \cdots \le T_{(n)}$ is the order statistic, and $\delta_{[i]} = \delta_j$ if $T_j = T_{(i)}$.

From this estimator we get the well-known Kaplan-Meier product limit estimator by the transformation

$$\hat{F}_n(t) = 1 - \exp(-\hat{\Lambda}_n(t)).$$

Asymptotic properties of these estimators were investigated by several authors, for example by [H81], [LS86] and [MR88].

2.3 A kernel estimator for the hazard function

The hazard function λ is the derivative of the cumulative hazard Λ. But the estimator $\hat{\Lambda}_n$ is not differentiable. So, we follow the same line as in the case of nonparametric density estimation. Let us estimate λ at point t. Consider a small interval $[t-b, t+b)$ of length $2b$ around t. We can approximate $\lambda(t)$ in the following way:

$$\lambda(t) \sim \frac{\int_{t-b}^{t+b} \lambda(s)\,\mathrm{d}s}{2b} = \frac{\Lambda(t+b) - \Lambda(t-b)}{2b} \sim \frac{\hat{\Lambda}_n(t+b) - \hat{\Lambda}_n(t-b)}{2b}. \tag{5}$$

The last term in (5) can be written in the form

$$\frac{1}{b} \sum_{i=1}^{n} K^* \left(\frac{t - T_{(i)}}{b} \right) \frac{\delta_{(i)}}{n - i + 1},$$

where

$$K^*(u) = \begin{cases} \frac{1}{2} & \text{for } -1 \le u \le 1 \\ 0 & \text{otherwise} \end{cases}.$$

The first approximation step in (5) yields a systematic error, which becomes small if the length of the interval is small. At the other hand, if b is small, then the second approximation error, the stochastic error, is large, because we have not enough observations for stability. To take these tendencies into account, we have to choose b depending on the sample size n, $b = b_n$, such that

$$b_n \to 0 \qquad \text{and} \qquad nb_n \to \infty. \tag{6}$$

Further, it is useful to take instead of the function K^* a more general function K, a function giving small weights to observations $T_{(i)}$ far away from the point t and large weights to observations very near to the point, at which we estimate. This is realized, for example, by taking a symmetric density function for K. So, finally we arrive at the following definition:

$$\hat{\lambda}_n(t) = \frac{1}{b_n} \sum_{i=1}^{n} K \left(\frac{t - T_{(i)}}{b_n} \right) \frac{\delta_{(i)}}{n - i + 1}. \tag{7}$$

Here $K : \mathbb{R} \to \mathbb{R}$ is the kernel function and $\{b_n\}$ the sequence of bandwidths satisfying (6). The estimator (7) can be written shortly as

$$\hat{\lambda}_n(t) = \frac{1}{b_n} \int K \left(\frac{t - s}{b_n} \right) \mathrm{d}\hat{\Lambda}_n(s).$$

Several properties of this estimator are known. Let us mention here papers [SW83], [TW83] and the results in [DS86]. In these papers conditions for consistency are derived and asymptotic expressions for the bias and the variance are given. Diehl and Stute considered an approximation for the difference between the estimator $\hat{\lambda}_n$ and a smoothed hazard rate by a sum of i.i.d. r.v.'s. On the basis of such a representation limit theorems can be derived.

The following picture shows a nonparametric kernel estimate for the data generated in the simulated model (2). Here the kernel function is the Gaussian kernel, the bandwidth is $b_n = 0.6$. We see, that this estimate reflects the features of the underlying hazard function much better than the parametric estimator.

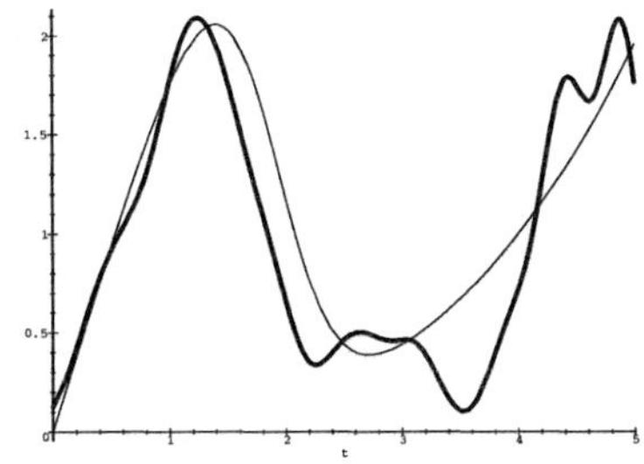

Fig. 3. True underlying hazard rate (thin) and nonparametric estimate (bold)

3 Testing the Hazard Rate

Nonparametric estimators of a curve are an appropriate tool in the analysis of data. But, sometimes in practical situations it seems to be useful to have a parametric model. The advantage of a parametric model is that the parameters have a some meaning, very often they can be interpreted. Of course, this holds only, if the chosen parametric model is appropriate. Thus, the question arises, whether the choice of a certain parametric model can be justified by the data. In this section we propose a test procedure for checking whether a hypothetical model fits the data, that is we consider the following hypothesis

$$\mathcal{H}: \ \lambda \in \mathcal{L} \quad \text{vs.} \quad \mathcal{K}: \ \lambda \notin \mathcal{L},$$

where $\mathcal{L}$ is the class of parametric hazard functions

$$\mathcal{L} = \{\lambda(\cdot;\vartheta) \,|\, \vartheta \in \Theta \subset \mathbb{R}^k\}$$

An example for such an parametric class $\mathcal{L}$ is the set of all Weibull hazards. Further parametric models are given in the book [BN02]. At the first view one would choose as test statistic the deviation of the nonparametric estimator $\hat{\lambda}_n$, which is a good estimator under the alternative, from a hypothetical hazard with estimated parameter $\hat{\vartheta}$, i.e. from $\lambda(t;\hat{\vartheta})$. Here $\hat{\vartheta}$ is an appropriate estimator of the unknown parameter. But the nonparametric $\hat{\lambda}_n$ is a result of smoothing procedure. Remember formulae (5) - it is an unbiased estimator of

$$\frac{1}{b_n}\int K\left(\frac{t-z}{b_n}\right)\lambda(z)\,\mathrm{d}z,$$

and not unbiased for the underlying hazard rate. So, it seems to be natural to compare $\hat{\lambda}_n$, which smoothes the data, with a smoothed version of the hypothesis. Thus, we will take the difference between $\hat{\lambda}_n$ and $\widetilde{\lambda}_n$ defined by

$$\widetilde{\lambda}_n(t;\hat{\vartheta}) = \frac{1}{b_n}\int K\left(\frac{t-z}{b_n}\right)\lambda(z;\hat{\vartheta})\,\mathrm{d}z.$$

Generally speaking, one can take as deviation measures L_p-distances for functions. Here we will consider a L_2-type distance, namely

$$\mathsf{Q}_n = \int\left(\hat{\lambda}_n(t) - \widetilde{\lambda}_n(t;\hat{\vartheta})\right)^2 a(t)\,\mathrm{d}t$$

The function a is a known weight function, it is introduced to control the region of integration.

3.1 An asymptotic α-test

To formulate a test based on this statistic we have to derive the distribution of Q_n, or at least the limiting distribution under the hypothesis. The theory about the asymptotical distributional behavior of quadratic forms yields the following limit statement. Under

- regularity conditions on the kernel K and the bandwidth b_n,
- smoothness of the functions $H$and H^U
- conditions on the function a such that the integrals given below exist and
- conditions ensuring that the estimator $\hat{\vartheta}_n$ is $\sqrt{n}$-consistent

the distribution of the standardized Q_n converges to the standard normal distribution, that is

$$\frac{nb_n^{1/2}}{\sigma}\left(\mathsf{Q}_n - \mu_n\right) \xrightarrow{\mathcal{D}} \mathsf{N}(0,1)$$

where

$$\mu_n = (nb_n)^{-1}\,\kappa_1 \int \frac{\lambda(t;\hat{\vartheta}_n)}{1-H(t)}\,a(t)\,\mathrm{d}t$$

$$\sigma^2 = 2\,\kappa_2 \int \left(\frac{\lambda(t;\hat{\vartheta}_n)}{1-H(t)}\right)^2 a^2(t)\,\mathrm{d}t \tag{8}$$

with $\kappa_1 = \int K^2(x)\,\mathrm{d}x$ and $\kappa_2 = \int (K*K)^2(x)\,\mathrm{d}x$ and "$*$" denotes the convolution.

The only unknown term in this limit statement is the distribution H of the observations. Replacing this by the empirical distribution $\hat{H}_n$ we obtain the following asymptotic α-test: Reject $\mathcal{H}$, iff

$$\mathrm{Q}_n \geq \frac{z_\alpha\,\hat{\sigma}_n}{nb_n^{1/2}} + \hat{\mu}_n. \tag{9}$$

Here z_α is the $(1-\alpha)$-quantile of the standard normal distribution and $\hat{\mu}_n$ and $\hat{\sigma}_n^2$ are defined as in (8), where H is replaced by $\hat{H}_n$. Note that one has to choose the function a such that regions where the kernel estimator of the hazard rate has a large variance are excluded.

3.2 Application to the example

Now, let us apply the proposed test to the example considered in Section 1. The nonparametric estimator of the hazard rate in the Weibull mixture model and the smoothed hypothetical hazard function, that is a hazard rate in a Weibull model with parameter $\hat{\vartheta} = (1.057, 1.422)$, are given in Figure 4. We compute the integrated quadratic distance over the interval $[0, 4]$. and get the following values for the test statistic and the standardizing terms

$$\mathrm{Q}_n = 2.8161$$
$$\hat{\mu}_n = 1.461$$
$$\hat{\sigma}_n^2 = 1853.717$$

With these values the test procedure yields for $\alpha = 0.05$: Reject $\mathcal{H}$. The p-value is 0.0025.

Conclusions

1. There are two possible points of view. The first is to consider the minor part of the mixture as a disturbation. That is, one is interested in the main part, for which the parametric model is justified. Then the nonparametric estimate of the hazard rate shows that the population is not homogenous, or in other words, our data are not appropriate for the estimation of both parameters. Further, we see that the hazard rate reflects this deviation

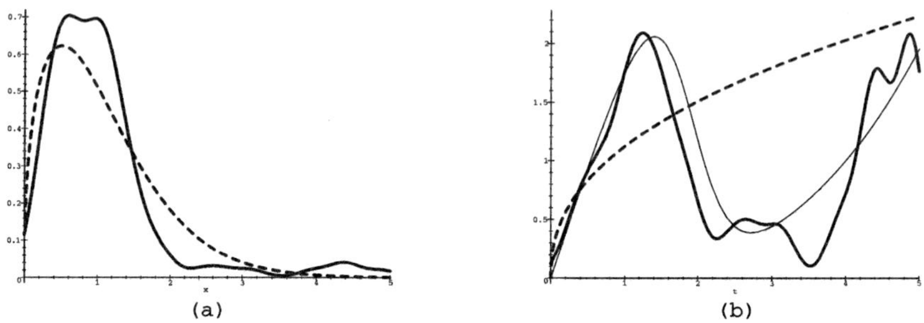

Fig. 4. (a) Densities, (b) Hazard rates. Hypothetical single Weibull model (dashed line), nonparametric estimate (bold solid line), in (b) true underlying mixture model (thin solid line)

much better then the survival function. Hence, in this case the application of a nonparametric estimator for the hazard rate is helpful for detecting outliers.

2. A second point of view is, that one is interested in the distribution of the population, that is the data are correct in the sense, that they are represent the population we are interested in. Then our nonparametric approach shows that the chosen parametric model is not appropriate. Thus, the nonparametric estimator can be helpful for stating a better parametric model. Of course a parametric mixture model with unknown parameter p is a complicated matter.
3. In both cases we see that the hazard rate is more sensitive. The deviation of a hazard rate from a hypothetical one, which can be seen very clearly, is smoothed away when we consider the corresponding survival functions.

4 Some further remarks

1. The proposed test is consistent, that is, if the distribution of the data does not belong to the hypothetical class, then the probability that the test rejects the hypothesis tends to one. This is not a very strong property. So, it seems to be useful to consider the power of the test under so-called local alternatives. For testing a density function nonparametrically such considerations were done in [LLK98]. The results for the hazard rate are similar. Roughly speaking one obtains, that the test is sensitive against alternatives tending to the hypothetical hazard function at the rate $\sqrt{nb_n^{1/2}}$.
2. The problem of the application of the nonparametric estimator and the test is the choice of the bandwidth b_n. If the bandwidth is chosen large,

the systematic error becomes large. At the first view this is not crucial, because we compare the smooth nonparametric estimator $\hat{\lambda}_n$ with the smoothed hypothetical function $\widetilde{\lambda}_n$. But the approximation of the distribution of the standardized test statistic Q_n by the normal distribution is worse for large b_n. Simulation results show that in this case the test has the tendency to accept the hypothesis. At the other hand, if b_n is chosen to small, then the resulting estimator is wiggly, and the power of the becomes worse.

5 About the Extension to the Model with Covariates

The approach described above can be generalized to the model with covariates. In applications often we observe in addition to the life times some covariates. These covariates can be e.g. the dosis of a drug, the temperature or other factors of influence. That is, we have observations (T_i, X_i, δ_i), where X_i is the covariate taking values in $\mathbb{R}$ or more general in $\mathbb{R}^k$. We can consider these covariates as fixed design points, or as random values. In both cases we are interested in statistical inference about the survival function $S(t|x)$, the density $f(t|x) = -\frac{\mathrm{d}S(t|x)}{\mathrm{d}t}$ and the hazard function $\lambda(t|x) = \frac{f(t|x)}{S(t|x)}$. Here $S(t|x)$ is the probability that an individuum or item survives the time point t given the covariate takes the value x. We do not want to go into further details, the basic idea is to estimate the distribution functions $H(\cdot|x)$ and $H^U(\cdot|x)$ not by the emprirical distribution functions given in (4), but by *weighted* empirical distribution functions

$$\hat{H}_n^U(t) = \sum_{i=1}^{n} w_{ni}(X, x; h_n) 1(T_i \le t, \delta_i = 1) \qquad \hat{H}_n(t) = \sum_{i=1}^{n} w_{ni}(X, x; h_n) 1(T_i \le t).$$

Here, the weights $w_{nj}(X, x)$ depend on the observed covariates $X = (X_1, \ldots, X_n)$, on x and on a smoothing parameter h_n. We assume $\sum_{i=1}^{n} w_{ni}(X, x; h_n) = 1$. They are chosen such that the T_j gets a large weight in counting all the T_i's, which are smaller or equal t, if the corresponding covariate X_j is near x. Appropriate weights are kernel weights of Gasser-Müller type for fixed covariates or Nadaraya-Watson kernel weights for random X_i's. The resulting estimator of the hazard rate has then the following form

$$\hat{\lambda}_n(t|x) = \frac{1}{b_n} \sum_{i=1}^{n} K\left(\frac{t - T_{(i)}}{b_n}\right) \frac{\delta_{[i]} w_{n[i]}(X, x; h_n)}{1 - \sum_{j=1}^{i-1} w_{n[j]}(X, x; h_n)}.$$

Properties of nonparametric estimators for the hazard rate, the cumulative hazard function and the survival functions for models with covariates are derived, for example, in papers [GMCS96] and [VKVN97], [VKVN01], [VKVN02].

For testing the hypothesis that $\lambda(\cdot|x)$ is equal to a given hazard function $\lambda^*(\cdot|x)$ we propose (for fixed covariates) the following test statistic

$$\mathrm{S}_n = \frac{1}{n}\sum_{k=1}^{n}\int\left(\hat{\lambda}_n(t|x_k) - \widetilde{\lambda}_n^*(t|x_k)\right)^2 a(t)\,\mathrm{d}t$$

Here $\widetilde{\lambda}_n^*(\cdot|x_k)$ is the smoothed hypothetical hazard function at fixed covariate x_k. In [L03a] it is shown that under certain conditions on K, b_n, the weights w_{ni} and h_n and on the smoothness of the underlying distribution functions that the (appropriate standardized) S_n is asymptotically normally distributed. Based on this limit statement a test procedure can be derived. Moreover, for testing the hypothesis, that $\lambda(\cdot|x)$ lies in a prespecified parametric class a test statistic with estimated parameters can be applied.

Appendix: Formulation of the Limit Theorem

This theorem is formulated not only for the behavior under the null hypothesis, but for general hazard rate λ. We define

$$\widetilde{\lambda}_n(t) := \int K_{b_n}(t-s)\,\lambda(s)\,\mathrm{d}s.$$

$$\mathrm{Q}_n = \int\left(\hat{\lambda}_n(t) - \widetilde{\lambda}_n(t)\right)^2 a(t)\,\mathrm{d}t$$

Further, let T_H be the right end point of the distribution H.

Theorem 1. *Suppose that*

(i) *K is a continuous density function vanishing outside the interval $[-L, L]$ for some $L > 0$.*
(ii) *λ and H are Lipschitz continuous.*
(iii) *The function a is continuous and $a(t) \equiv 0$ for all $t > T_H$ and the integrals defined below are finite.*
(iv) *$b_n \to 0$ and $nb_n^2 \to \infty$.*

Then for $n \to \infty$

$$\frac{nb_n^{1/2}}{\sigma}(\mathrm{Q}_n - \mu_n) \xrightarrow{\mathcal{D}} \mathsf{N}(0,1) \tag{10}$$

where

$$\mu_n = (nb_n)^{-1}\int\frac{\lambda(t)}{1-H(t)}\,a(t)\,\mathrm{d}t\,\kappa_1 \qquad \sigma^2 = 2\int\left(\frac{\lambda(t)}{1-H(t)}\right)^2 a^2(t)\,\mathrm{d}t\,\kappa_2$$

The proof of this theorem is given in [L03b].

References

[VKVN02] Beran, R.: Nonparametric regression with randomly censored survival times. Technical Report, Univ. California, Berkeley (1981)

[BN02] Bagdonavičius, V. and Nikulin, M.: Accelerated Life Models; Modeling and Statistical analysis. Boca Raton.; Chapman and Hall /CRC (2002)

[VKVN02] Diehl, S. and Stute, W.: Kernel density and hazard function estimation in the presence of censoring. J. Multivariate Anal., **25**, 299–310 (1988).

[VKVN02] González-Manteiga, W. and Cadarso-Suarez, C.: Asymptotic properties of a generalized Kaplan-Meier estimator with some applications. J. Nonparametic Statistics, **4**, 65–78 (1996).

[VKVN02] Liero, H. and Läuter, H. and Konakov, V. D.: Nonparametric versus parametric goodness of fit. Statistics, **31**, 115–149 (1998).

[VKVN02] Liero, H.: Goodness of fit tests of L_2-type. In: Statistical Inference for Semiparametric Models and Applications, Ed. Nikulin, M., Publisher Birkhäuser (2003a).

[VKVN02] Liero, H.: Testing the hazard rate. Preprint, Institut für Mathematik, Universität Potsdam. (2003b).

[VKVN02] Lo, S.-H. and Singh, K.: The product-limit estimator and the bootstrap: Some asymptotic representations. Probab. Theory Related Fields, **71**, 455–465 (1986).

[VKVN02] Major, P. and Rejtő: Strong embedding of the estimator of the distribution function under random censorship. Ann. Statist., **16**, 1113 – 1132 (1988).

[VKVN02] Singpurwalla, N. D. and Wong, M. Y.: Estimation of the failure rate - a survey of nonparametric methods, part I: Non- Baysian methods. Commun. Statist.- Theory and Meth., **12**, 559–588 (1983).

[VKVN02] Tanner, M. A. and Wong, W. H.: The estimation of the hazard function from randomly censored data by the kernel method. Ann. Statist., **11**, 989–993 (1983).

[VKVN02] Van Keilegom, I. and Veraverbeke, N.: Estimation and bootstrap with censored data in fixed design nonparametric regression. Ann. Inst. Statist. Math., **49**, 467–401 (1997).

[VKVN02] Van Keilegom, I. and Veraverbeke, N.: Hazard rate estimation in nonparametric regression with censored data. Ann. Inst. Statist. Math., **53**, 730–745 (2001).

[VKVN02] Van Keilegom, I. and Veraverbeke, N.: Density and hazard estimation in censored regression models. Bernoulli, **8**, 607–625 (2002).

Selecting a semi-parametric estimator by the expected log-likelihood

Benoit Liquet[1] and Daniel Commenges[2]

[1] Laboratoire de Statistique et Analyse des Données, BHSM, 1251 avenue centrale BP 47
38040 Grenoble Cedex 09, FRANCE
benoit.liquet@upmf-grenoble.fr
[2] INSERM E0338, Université Victor Segalen Bordeaux 2, 146 rue Léo Saignat, 33076 Bordeaux Cedex, FRANCE.
daniel.commenges@isped.u-bordeaux2.fr

A criterion for choosing an estimator in a family of semi-parametric estimators from incomplete data is proposed. This criterion is the expected observed log-likelihood (ELL). Adapted versions of this criterion in case of censored data and in presence of explanatory variables are exhibited. We show that likelihood cross-validation (LCV) is an estimator of ELL and we exhibit three bootstrap estimators. A simulation study considering both families of kernel and penalized likelihood estimators of the hazard function (indexed on a smoothing parameter) demonstrates good results of LCV and a bootstrap estimator called ELL_{bboot}. When using penalized likelihood an approximated version of LCV also performs very well. The use of these estimators of ELL is exemplified on the more complex problem of choosing between stratified and unstratified proportional hazards models. An example is given for modeling the effect of sex and educational level on the risk of developing dementia.

Key words: bootstrap, cross-validation, Kullback-Leibler information, proportional hazard model, semi-parametric, smoothing.

1 Introduction

The problem of model choice is obviously one of the most important in statistics. Probably one of the first solution to a model choice problem was given by Mallows [Mal73] who proposed a criterion (C_p) for selecting explanatory variables in linear regression problems. This problem of selection of variables was studied by many authors in more general regression models ([Cop83]; [Mil02]). The celebrated Akaike criterion [Aka74] brought a solution to the problem of parametric model selection. This criterion called AIC (An Information Criterion or Akaike Information Criterion) was based on an approximation of the Kullback-Leibler distance [Aka73]. Criterions improving AIC for small samples have been proposed: AIC_c [HT89] and EIC which is a bootstrap estimation [ISK97]. Finally in the case of missing data, Cavanaugh and Shumway [CS98] proposed a variant of AIC. A closely related, but more difficult problem, is that of choice of a smoothing parameter in smoothed semi-(or non-) parametric estimation of functions. These functions may be density function [Sil86], effect functions of an explanatory variable [HT90] or hazard functions ([O'S88], [JCL98]). Smoothing methods are in particular kernel smoothing methods and penalized likelihood. In simple regression problems, versions of AIC and AIC_c are available [HST98] and simple versions of the cross-validation criterion have been proposed: CV, GCV [CW79]. However in general problems only the likelihood cross-validation criterion (LCV) [O'S88] and bootstrap techniques, in particular, extension of EIC [LSC03] are available. In some problems approximations of the mean integrated square error (MISE) are available ([RH83], [MP87], [Fer99]).

Liquet et al. [LSC03] have introduced a general point of view which is to choose an estimator among parametric or semi-parametric families of estimators according to a criterion which is an approximation of the Kullback-Leibler distance; they have shown on some simulation studies that the best criterions were EIC and LCV. They treated a general multivariate regression problem. The aim of this paper is to extend this point of view to the case where incomplete data are observed. The data may be incomplete because of right or interval-censoring for instance. This is not a trivial extension: indeed, it becomes clear that for using relatively simply the bootstrap approach, the theoretical criterion to be estimated must be changed. The proposed criterion is the expectation of the (observed) log-likelihood (ELL) rather than the Kullback-Leibler distance. This paper uses much of the material of Liquet and Commenges [LC04] and applies the approach to the choice between stratified and unstratified proportional hazards models.

We define the ELL criterion in section 2 and give useful versions of it for use with right-censored data and with explanatory variables (where partial and conditional likelihood respectively are used). In section 3 we exhibit three bootstrap estimators of ELL and show that LCV also estimates ELL. Section 4 presents simulation studies for comparing the four estimators together with the Ramlau-Hansen approach for hazard functions using kernel smoothing methods or penalized likelihood.

In section 5, we show an application of these criterions to a more complex problem, which is to compare stratified and unstratified proportional hazards models. Our particular application is modeling onset of dementia as a function of sex and education level (coded as a binary variable). We could consider a proportional hazard for both variables or stratified models on one variable, or making four strata. No method has been proposed to our knowledge to compare such different semi-parametric models. We propose to compare them using the ELL criterion, in practice using LCV or a bootstrap estimator, and apply these methods to the data of the PAQUID study, a large cohort study on dementia [LCDBG94].

2 The expected log-likelihood as theoretical criterion

2.1 Definitions and notations

Let T be the time of the events of interest. Let f and F be the density function and the cumulative distribution function of T. The hazard function is defined by $\lambda(t) = \frac{f(t)}{S(t)}$ where $S = 1 - F$ is the survival function of T. However, we do not observe the realizations of T but only a sample $\mathcal{W} = \{W_1, \ldots, W_n\}$ of independent and identically distributed (i.i.d.) variables which bring information on the variable T. For instance, in the case of right-censored observations, the W_i's are copies of the random variable $W = (\widetilde{T}, \delta)$ where $\widetilde{T} = \min(T, C)$ and $\delta = I_{[T \leq C]}$ where C is a censoring variable independent of T. In the sequel, we denote by f_C the probability density functions and S_C the survival functions of C. Other cases of censoring are left and interval censoring. We denote by $\widehat{\lambda}_h^{\mathcal{W}}(\cdot)$ a family of estimators of $\lambda(\cdot)$, where h most often represents a smoothing parameter. To any particular estimator $\widehat{\lambda}_h^{\mathcal{W}}(\cdot)$ corresponds an estimator $\widehat{f}_h^{\mathcal{W}}(\cdot) = \widehat{\lambda}_h^{\mathcal{W}}(\cdot)\exp(-\int \widehat{\lambda}_h^{\mathcal{W}}(u)du)$. Our aim is to propose an information criterion to choose the smoothing parameter for a family of estimators and also to choose between different families of estimators.

2.2 The expected log-likelihood

For uncensored data, the useful part of the Kullback-Leibler information criterion, measuring the distance between $\widehat{f}_h^{\mathcal{T}}(\cdot)$ and f, is the conditional expectation of the log-likelihood of a future observation T' given $\mathcal{T}$

$$\mathrm{KL}(\mathcal{T}) = \mathrm{E}\left\{\log \widehat{f}_h^{\mathcal{T}}(T')|\mathcal{T}\right\} \tag{1}$$

where $\mathcal{T} = (T_1, \ldots, T_n)$ and T' an additional observation having the distribution F and being independent of the sample $\mathcal{T}$. Based on KL, Akaike [Aka74] (see also DeLeeuw, [DeL92]), in a parametric framework for complete data, defined the popular criterion AIC ($-2\log\mathcal{L} + 2p$; when $\mathcal{L}$ is the likelihood and p is the number of estimated parameter) as an estimator of the expectation of the Kullback-Leibler information EKL=E$[\mathrm{KL}(\mathcal{T})]$. In presence of incomplete data, even EKL is difficult to estimate. In particular, because $\mathcal{T}$ is not observed, it is not possible to directly estimate the different expectations by bootstrap.
Instead, we define a criterion as the expectation of the observed log-likelihood of a new sample which is a copy of the original sample, given the original sample:

$$\mathrm{ELL}(\widehat{\lambda}_h) = \mathrm{E}\left\{\log \mathcal{L}^{\widehat{\lambda}_h^{\mathcal{W}}}(\mathcal{W}')\right\}. \quad (2)$$

This criterion does not depend on $\mathcal{W}$ and judges a procedure of estimation $\widehat{\lambda}_h$ that can be applied to any $\mathcal{W}$ of same distribution. Thus the criterion that we propose is, in accordance with the pinciple of Akaike (see also DeLeeuw [DeL92]), the non-conditional expectation of the log-likelihood, ELL; ELL can be considered as equivalent to the expectation of the Kullback-Leibler information for observed data. Indeed it is relatively easy to show that for a parametric model, AIC defined as $-2\log\mathcal{L} + 2p$ (p being the number of parameters) is an estimator of ELL (more precisely of -2ELL) (see Cavanaugh and Shumway [CS98].

2.3 Case of right-censored data

In presence of right-censored data as defined in section 2.1, the likelihood $\mathcal{L}^{\widehat{\lambda}_h^{\mathcal{W}}}(\mathcal{W}')$ is:

$$\mathcal{L}^{\widehat{\lambda}_h^{\mathcal{W}}}(\mathcal{W}') = \prod_{i=1}^{n}\{\widehat{f}_h^{\mathcal{W}}(\widetilde{T}_i')\}^{\delta_i'}\{\widehat{S}_h^{\mathcal{W}}(\widetilde{T}_i')\}^{1-\delta_i'}\{f_C(\widetilde{T}_i')\}^{1-\delta_i'}\{S_C(\widetilde{T}_i')\}^{\delta_i'}$$

where $\widehat{f}_h^{\mathcal{W}}(\cdot)$ and $\widehat{S}_h^{\mathcal{W}}(\cdot)$, the estimators of f and S are deduced from $\widehat{\lambda}_h^{\mathcal{W}}(\cdot)$. The criterion defined in (2) can be decomposed in two parts:

$$\mathrm{ELL}(\widehat{\lambda}_h) = \mathrm{ELL}^p(\widehat{\lambda}_h) + \mathrm{E}\{\varphi(f_C, S_C, \mathcal{W}')\} \quad (3)$$

where

$$\mathrm{ELL}^p(\widehat{\lambda}_h) = \mathrm{E}\left\{\log \mathcal{L}_p^{\widehat{\lambda}_h^{\mathcal{W}}}(\mathcal{W}')\right\}$$

and

$$\mathcal{L}_p^{\widehat{\lambda}_h^{\mathcal{W}}}(\mathcal{W}') = \prod_{i=1}^{n}\{\widehat{f}_h^{\mathcal{W}}(\widetilde{T}_i')\}^{\delta_i'}\{\widehat{S}_h^{\mathcal{W}}(\widetilde{T}_i')\}^{1-\delta_i'}$$

which is the partial likelihood (in the sense of Andersen et al. [ABGK93]). The second term in (3) does not depend on $\widehat{\lambda}_h$; thus maximizing ELL is equivalent to maximize ELL^p. Finally our criterion is:

$$\mathrm{ELL}^p(\widehat{\lambda}_h) = \mathrm{E}\left\{\log \mathcal{L}_p^{\widehat{\lambda}_h^{\mathcal{W}}}(\mathcal{W}')\right\}$$

that is the ELL criterion applied to the partial likelihood; this is very fortunate because this avoids estimating the distribution of the censoring variable. Note however that the ELL criterion cannot be applied to the Cox partial likelihood, at least not directly: we need a smooth estimate of the hazard function to apply our criterion. Any non-smooth estimate has a value $-\infty$ and is rejected.

2.4 Case of explanatory variable

We consider the case of presence of explanatory variables. We note $W_i = (T_i, X_i)$ with T_i the survival time and X_i a vector of covariates for the ith individual. It is assumed that T_i has conditional density function $f(\cdot|x_i)$ given $X_i = x_i$. Our aim is to estimate $\lambda(\cdot|\cdot)$ the corresponding conditional hazard function. We note this estimator $\widehat{\lambda}_h^{\mathcal{W}}(\cdot|\cdot)$ and $\widehat{f}_h^{\mathcal{W}}(\cdot|\cdot)$ the corresponding density. The likelihood $\mathcal{L}^{\widehat{\lambda}_h^{\mathcal{W}}}(\mathcal{W}')$ is:

$$\mathcal{L}^{\widehat{\lambda}_h^{\mathcal{W}}}(\mathcal{W}') = \prod_{i=1}^{n} \{\widehat{f}_h^{\mathcal{W}}(T_i'|X_i')\}\{f_X(X_i')\}$$

where $\widehat{f}_X(\cdot)$ is the marginal density of X_i. With the same reasoning as in 2.3, the criterion in (2) can be decomposed in two parts:

$$\mathrm{ELL}(\widehat{\lambda}_h) = \mathrm{ELL}^c(\widehat{\lambda}_h) + \mathrm{E}\{\varphi(f_X, \mathbf{X}')\} \tag{4}$$

where $\mathbf{X}' = (X_1', \ldots, X_n')$,

$$\mathrm{ELL}^c(\widehat{\lambda}_h) = \mathrm{E}\left\{\log \mathcal{L}_c^{\widehat{\lambda}_h^{\mathcal{W}}}(\mathcal{W}')\right\}$$

and

$$\mathcal{L}_c^{\widehat{\lambda}_h^{\mathcal{W}}}(\mathcal{W}') = \prod_{i=1}^{n} \{\widehat{f}_h^{\mathcal{W}}(T_i'|X_i')\}$$

which is the conditionnal likelihood. The second term of (9) does not depend on $\widehat{\lambda}_h$; thus maximizing ELL is equivalent to maximizing ELL^c. Finally our criterion is:

$$\mathrm{ELL}^c = \mathrm{E}\left\{\log \mathcal{L}_c^{\widehat{\lambda}_h^{\mathcal{W}}}(\mathcal{W}')\right\}$$

that is the ELL criterion applied to the conditionnal likelihood; this is very fortunate because this avoids estimating the distribution of the explanatory variable. Both tricks can be applied when there are both explanatory variables and censoring.

3 Estimation of ELL

In order to obtain practical selection criteria, it is necessary to estimate ELL. Several estimators may be considered.

3.1 Likelihood cross-validation : LCV

Throughout this subsection, we index the sample $\mathcal{W}$ by its size n and thus use the notation $\mathcal{W}_n$. We recall that the likelihood cross-validation is defined as:

$$\mathrm{LCV}(\mathcal{W}_n) = \sum_{i=1}^{n} \log \mathcal{L}^{\widehat{\lambda}_h^{\mathcal{W}-i}}(W_i)$$

where $\mathcal{L}^{\widehat{\lambda}_h^{\mathcal{W}-i}}(W_i)$ is the likelihood contribution of W_i for the estimator defined on the sample $\mathcal{W}^{-i}$ in which W_i is removed. The LCV choice for $\widehat{\lambda}_h^{\mathcal{W}}$ is the estimator which maximizes LCV. An important property of LCV is that the expectation of LCV is approximatively equal to ELL and it is shown in Liquet and Commenges (2004) that when $n \longrightarrow \infty$,

$$\frac{\mathbf{E}\left[\mathrm{LCV}(\mathcal{W}_n)\right]}{\mathrm{ELL}(\widehat{\lambda}_h(n))} \longrightarrow 1,$$

where $\widehat{\lambda}_h(n)$ is an estimator applied to a sample of size n. If n is large, the computation of LCV is intensive. An approximation based on a first-order expansion of $\log \mathcal{L}^{\widehat{\lambda}_h^{\mathcal{W}-i}}(W_i)$ around $\log \mathcal{L}^{\widehat{\lambda}_h^{\mathcal{W}}}(W_i)$ can be used. This leads to an expression of the form

$$\mathrm{LCVa}(\mathcal{W}_n) = \sum_{i=1}^{n} \log \mathcal{L}^{\widehat{\lambda}_h^{\mathcal{W}}}(W_i) - mdf,$$

where the term mdf can be interpreted as the model degrees of freedom, and this expression is analogous to an AIC criterion. For instance, in the spline approximation of the penalized likelihood, we have $mdf = \mathrm{trace}([\widehat{H} - 2h\Omega]^{-1}\widehat{H})$ where $\widehat{H}$ is the converged Hessian matrix of the log-likelihood, and Ω is the penalized part of the converged Hessian matrix, see Joly et al. [JCL98] for more details.

3.2 Direct bootstrap method for estimating ELL (ELL$_{boot}$ and ELL$_{iboot}$)

We can directly estimate by bootstrap the expectation of the log-likelihood (ELL). We define this bootstrap estimator as

$$\mathrm{ELL}_{boot} = \mathrm{E}_*\left\{\log \mathcal{L}^{\widehat{\lambda}_h^{\mathcal{W}^*}}(\mathcal{W}'^*)\right\}$$

where $\mathcal{W}^* = (W_1^*, \ldots, W_n^*)$, $W_j^* \sim \widehat{F}_W$, $\mathcal{W}'^* = (W_1'^*, \ldots, W_n'^*)$ and $W_j'^* \sim \widehat{F}_W$, $\widehat{F}_W$ being the empirical distribution of W_i based on $\mathcal{W}$. We use the notation E_* to remind that the expectation is taken relatively to the estimated distribution $\widehat{F}_W$. In practice, the expectation is approximated by a mean of B repetitions of bootstrap samples ($\mathcal{W}^j \stackrel{d}{=} \mathcal{W}'^j \stackrel{d}{=} \mathcal{W}^*$)

$$\mathrm{ELL}_{boot} \simeq \frac{1}{B}\sum_{j=1}^{B} \log \mathcal{L}^{\widehat{\lambda}_h^{\mathcal{W}^j}}(\mathcal{W}'^j)$$

To improve this criterion, we can iterate the bootstrap method [Hal92]. We define this new estimator as:

$$\mathrm{ELL}_{iboot} = \mathrm{E}_{**}\left\{\log \mathcal{L}^{\widehat{\lambda}_h^{\mathcal{W}^{**}}}(\mathcal{W}'^{**})\right\}$$

where $\mathcal{W}^{**} = (W_1^{**}, \ldots, W_n^{**})$, $W_j^{**} \sim \widehat{F}_{\mathcal{W}^*}$, $\mathcal{W}'^{**} = (W_1'^{**}, \ldots, W_n'^{**})$ and $W_j'^{**} \sim \widehat{F}_{\mathcal{W}^*}$, $\widehat{F}_{\mathcal{W}^*}$ being the empirical distribution of W_i^* based on $\mathcal{W}^*$. E_{**} is calculated with respect to the distribution $\widehat{F}_{\mathcal{W}^*}$. The expectation is also approximated by a mean of B repetitions of bootstrap samples ($\mathcal{W}^j \stackrel{d}{=} \mathcal{W}'^j \stackrel{d}{=} \mathcal{W}^{**}$)

$$\mathrm{ELL}_{iboot} \simeq \frac{1}{B}\sum_{j=1}^{B} \log \mathcal{L}^{\widehat{\lambda}_h^{\mathcal{W}^j}}(\mathcal{W}'^j)$$

More explicitly, for each j, we take a bootstrap sample from $\mathcal{W}^*$ from $\mathcal{W}$, then we take $\mathcal{W}^j$ a bootstrap sample from $\mathcal{W}^*$; we obtain $\mathcal{W}'^j$ by the same way; $\widehat{\lambda}_h^{\mathcal{W}^j}$ is the estimator of λ for fixed h based on $\mathcal{W}^j$.

3.3 Bias corrected bootstrap estimators

To construct this estimator, we first propose the log-likelihood as naive estimator of the ELL criterion and then correct it by estimating its bias by bootstrap [Hal92]. This approach is similar to that used for deriving the EIC [LSC03] criterion available for complete data.
Our corrected estimator of ELL is:

$$\mathrm{ELL}_{bboot} = \log \mathcal{L}^{\widehat{\lambda}_h^{\mathcal{W}}}(\mathcal{W}) - \widehat{b}(\mathcal{W}). \tag{5}$$

where $\widehat{b}(\mathcal{W}) \simeq \frac{1}{B}\sum_{j=1}^{B}\left\{\log \mathcal{L}^{\widehat{\lambda}_h^{\mathcal{W}^j}}(\mathcal{W}^j) - \log \mathcal{L}^{\widehat{\lambda}_h^{\mathcal{W}^j}}(\mathcal{W})\right\}$ is the bootstrap estimate of the bias ($bias = \mathrm{E}\{\log \mathcal{L}^{\widehat{\lambda}_h^{\mathcal{W}}}(\mathcal{W})\} - \mathrm{ELL}$), B is the number of bootstrap sample $\mathcal{W}^j$ taken at random from the distribution of $\mathcal{W}^*$. For more details see Liquet and Commenges (2004).
Remark : for all the bootstrap methods when treating right-censored observations the bootstrap expectations have to be conditionned on having at least one uncensored observation because the estimator are not defined otherwise.

4 Simulation

We have compared ELL_{boot}, ELL_{iboot}, ELL_{bboot} and LCV using both families of kernel and penalized likelihood estimators of hazard functions. We have included the Ramlau-Hansen method when using kernels, a popular method for estimating hazard functions [ABGK93]. We compare the criteria when using kernel smoothing in 4.1 and penalized likelihood in 4.2.

4.1 Kernel estimator

The smoothed Nelson-Aalen estimator is

$$\widehat{\lambda}(t) = \frac{1}{h}\int K\left(\frac{t-u}{h}\right) d\widehat{A}(u)$$

where $K(\cdot)$ is a kernel function, $\widehat{A}(\cdot)$ is the Nelson-Aalen estimator of $A(\cdot)$, the cumulative hazard function, and h is the bandwith parameter. Ramlau-Hansen [RH83] has proposed an estimator of the MISE (mean integrated square error) based on an approximated cross-validation method for estimating h; we call it the RH method. We apply gaussian kernels to allow the use of the different criteria. Indeed, if we used a kernel with compact support, we risk for small h to have LCV criteria equal to $-\infty$. For the criteria based on bootstrap, kernels with compact support are prohibited since the bootstrap expectations are theoretically equal to $-\infty$ for bandwidth lower than the range of the observed event times. We consider problems where the density near zero is very low so there is no edge effect near zero.
The data were generated from a mixture of gamma distributions. We generated random samples $T_1, ..., T_n$ of i.i.d. failure times and $C_1, ..., C_n$ of i.i.d. censoring times; the C_i were independent of the T_i. So the observed samples were $(\widetilde{T}_1, \delta_1), ..., (\widetilde{T}_n, \delta_n)$ where $\widetilde{T}_1 = \min(T_i, C_i)$ and $\delta_i = \mathrm{I}_{[T_i \leq C_i]}$. The density of T was a a mixture of Gamma $\{0.4\Gamma(t; 40, 1) + 0.6\Gamma(t; 80, 1)\}$, with the probability density functions $\Gamma(t; \alpha, \gamma) = \frac{\alpha^\gamma t^{\gamma-1} e^{-\alpha t}}{\Gamma(\gamma)}$. The probability density function of C_i was a simple Gamma: $\Gamma(t; 90, 1)$, $\Gamma(t; 90, 1.1)$ and $\Gamma(t; 90, 1.3)$ corresponding to a percentage of censoring around 15%, 25% and 50% respectively. Samples of sizes 30, 50 and 100 were generated. Figure 1 displays the smoothed Nelson-Aalen estimate chosen by ELL_{bboot} and the true hazard function for one simulated example from a mixture of gamma.

Each bootstrap estimator was computed using B=400 samples. Each simulation involved 100 replications. For each replication we computed the useful part of the Kullback-Leibler information (KL) between the true density function f and the estimators chosen by each criterion

$$\mathrm{KL}(f; \widehat{f}_h^{\mathcal{W}}) = \int_J \log \widehat{f}_h^{\mathcal{W}}(t) f(t) dt$$

where $J =]0; T_{max}]$. We do not take T_{max} equal to $+\infty$, because for large times t when there is censoring, we do not have enough information to determine $\widehat{f}_h^{\mathcal{W}}(t)$. T_{max} was chosen for each simulation such as

$$\Pr\{\mathrm{E}(n_{T_{max}}) \geq 1\} = 0.95$$

where $n_{T_{max}}$ represents the risk set at time T_{max}. We computed, for each simulation presented, the average of KL and its standard error. Since KL generally takes negative values we give in tables 1, 2, 4 the values of -KL: low

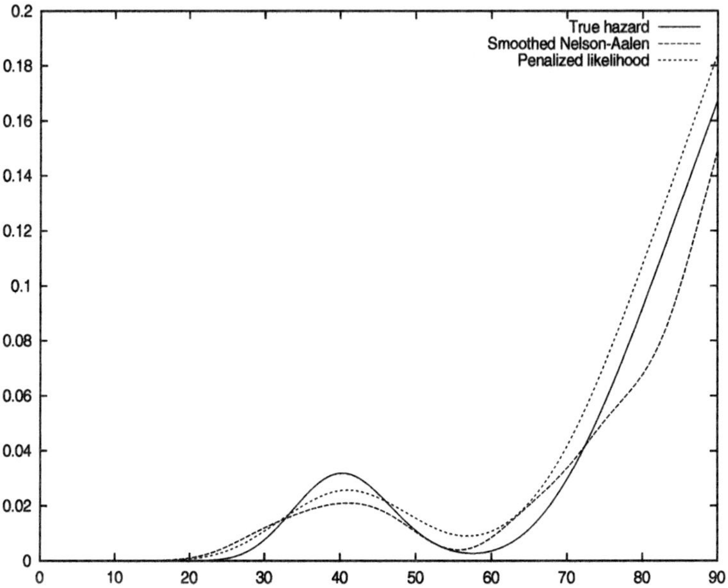

Fig. 1. True hazard function (solid line), smoothed Nelson-Aalen estimator (dashed line) and penalized likelihood estimate (dotted line) chosen by ELL_{bboot} for a simulated example. The sample size is 50, with 15% right-censored observations.

values then correspond to estimators close to the true distribution. First we present in table 1 the results of the simulation comparing the optimal criterion KL and the new criterion ELL. The two theoretical criteria give practically the same results. We note some differences only when there is little information (small sample size and high censoring level). The average of -KL obtained for the pratical criteria are given in table 2. These averages can be compared to an optimal value, the value of KL when estimators are chosen using the true ELL.

We may note that RH yielded in all cases much higher (worse) values of -KL than the other criteria. The ELL_{boot} criterion, although better than RH, had in pratically all the cases higher values than the other criteria. The differences were very small between LCV, ELL_{iboot} and ELL_{bboot} although for high censoring level and small sample sizes, LCV tended to perform not as well as the bootstrap methods. For all the simulations, the three competitive criteria had values of KL quite close to the values given by ELL. Although the simulations were based on only 100 replications some differences were large comparatively to the standard errors. To make the comparisons more rigorous we performed paired t-tests for comparing the criteria in the case of 25% of censoring. All the tests (except one) of ELL_{boot} and RH versus LCV, ELL_{iboot} and ELL_{bboot} were significant at the 0.001 level; the three tests comparing ELL_{boot} with RH were also significant. This confirms that

Table 1. Average Kullback-Leibler information $-\mathrm{KL}(\widehat{\lambda}_h^{\mathcal{W}})$ for the kernel estimator for estimating the hazard function of the mixture of gamma $(0.4\Gamma(t,a,b) + 0.6\Gamma(t,c,d))$ for bandwith chosen by ELL and KL, based on 100 replications. Standard errors are given in parentheses.

	$-\mathrm{KL}(\widehat{\lambda}_h^{\mathcal{W}})$ for kernel estimators	
n	KL	ELL
15% censoring		
30	3.96(0.005)	3.96(0.005)
50	3.98(0.003)	3.99(0.003)
100	4.01(0.002)	4.01(0.002)
25% censoring		
30	3.89(0.004)	3.91(0.005)
50	3.93(0.004)	3.93(0.004)
100	3.95(0.002)	3.95(0.002)
50% censoring		
30	3.81(0.02)	3.92(0.04)
50	3.80(0.009)	3.84(0.02)
100	3.80(0.005)	3.80(0.005)

Table 2. Average Kullback-Leibler information $-\mathrm{KL}(\widehat{\lambda}_h^{\mathcal{W}})$ for the kernel estimator for estimating the hazard function of the mixture of gamma $\{0.4\Gamma(t,a,b) + 0.6\Gamma(t,c,d)\}$ for each criterion based on 100 replications. Standard errors are given in parentheses.

	$-\mathrm{KL}(\widehat{\lambda}_h^{\mathcal{W}})$ for kernel estimators					
n	ELL	ELL_{bboot}	LCV	ELL_{iboot}	ELL_{boot}	RH
15% censoring						
30	3.96(0.005)	4.00(0.009)	4.01(0.02)	3.98(0.005)	4.04(0.01)	4.19(0.06)
50	3.99(0.003)	4.00(0.006)	4.00(0.008)	4.00(0.005)	4.04(0.01)	4.22(0.06)
100	4.01(0.002)	4.02(0.002)	4.02(0.002)	4.02(0.002)	4.05(0.005)	4.12(0.02)
25% censoring						
30	3.91(0.005)	3.94(0.009)	3.96(0.01)	3.92(0.006)	3.98(0.01)	4.26(0.08)
50	3.93(0.004)	3.95(0.007)	3.96(0.01)	3.94(0.006)	3.99(0.01)	4.2(0.06)
100	3.95(0.002)	3.96(0.002)	3.96(0.002)	3.96(0.002)	3.99(0.007)	4.10(0.03)
50% censoring						
30	3.92(0.04)	3.99(0.07)	4.04(0.07)	4.01(0.07)	4.02(0.08)	4.36(0.1)
50	3.84(0.02)	3.85(0.02)	3.91(0.03)	3.85(0.02)	3.88(0.03)	4.18(0.09)
100	3.80(0.005)	3.81(0.005)	3.83(0.02)	3.80(0.005)	3.84(0.008)	3.95(0.03)

the criteria can be classified in three groups ordered from best to worst : 1) LCV, ELL_{iboot} and ELL_{bboot}; 2) ELL_{boot}; 3) RH.

We also compared the different criteria in term of MISE (mean integrated squared error). The result of this simulation are summarized in table 3. Although the RH criterion was based on minimizing the MISE, it gave the worst result. Since Marron and Padgett [MP87] proved an optimality property of cross-validation for bandwidth choice, this may be due to the approximation done for obtaining the RH criterion.

Table 3. Comparison of the criteria by the MISE distance for the kernel estimator for estimating the hazard function of the mixture of gamma $\{0.4\Gamma(t,a,b) + 0.6\Gamma(t,c,d)\}$ based on 100 replications. Standard errors are given in parentheses.

			MISE for kernel estimators		
n	ELL_{bboot}	LCV	ELL_{iboot}	ELL_{boot}	RH
25% censoring					
30	0.039(0.005)	0.044(0.008)	0.034(0.004)	0.043(0.006)	0.095(0.02)
50	0.038(0.004)	0.039(0.005)	0.035(0.004)	0.041(0.006)	0.110(0.02)
100	0.034(0.004)	0.033(0.004)	0.034(0.004)	0.046(0.004)	0.073(0.012)

4.2 Penalized likelihood estimator

Another approach to estimate the hazard function is to use penalized likelihood:

$$p\mathcal{L}_h(\mathcal{W}) = \log \mathcal{L}_p^{\lambda}(\mathcal{W}) - h \int \lambda''^2(u)du \qquad (6)$$

where $\mathcal{L}_p^{\lambda^{\mathcal{W}}}$ is the partial log-likelihood (in the sense of section 2.3) and h is a positive smoothing parameter which controls the tradeoff between the fit of the data and the smoothness of the function. Maximization of (6) over the desired class of functions defines the maximum penalized likelihood estimator (MPLE) $\widehat{\lambda}_h^{\mathcal{W}}$. The solution is then approximated on a basis of splines. The main advantage of the penalized likelihood approach over the kernel smoothing method is that there is no edge problem; the drawback is that it is more computationally demanding. The method of likelihood cross-validation (LCV) may be used to select h. To circumvent the computational burden of the LCV a one-step Newton-Raphson expansion has been proposed by O'Sullivan [O'S88] and adapted by Joly et al. [JCL98]; we denote this approximation by LCV_a.
ELL_{bboot} and ELL_{iboot} are also applicable to select the smoothing parameter for penalized likelihood estimators. Figure 1 displays the penalized likelihood estimate chosen by ELL_{bboot} and the true hazard function for one simulated example. We have compared LCV_a, LCV, ELL_{bboot} and ELL_{iboot} to ELL in

a short simulation study (penalized likelihood estimators require more computation than kernel estimators). We used the sample with size $n = 50$, generated in section 4.1. The results of the simulation are summarized in table 4. For penalized likelihood estimators, the differences were small between LCV, LCV_a and ELL_{bboot}; ELL_{iboot} seemed to be less satisfactory.

Table 4. Average Kullback-Leibler information $-KL(\widehat{\lambda}_h^{\mathcal{W}})$ for penalized likelihood estimator for each criterion. Standard errors are given in parentheses.

	$-KL(\widehat{\lambda}_h^{\mathcal{W}})$
n=50 and 15% censoring	
ELL	3.99(0.003)
LCV	4.00(0.005)
LCV_a	4.00(0.005)
ELL_{bboot}	4.00(0.006)
ELL_{iboot}	4.06(0.01)
n=50 and 25% censoring	
ELL	3.93(0.006)
LCV	3.98(0.006)
LCV_a	3.99(0.009)
ELL_{bboot}	4.01(0.02)
ELL_{iboot}	4.08(0.02)
n=50 and 50% censoring	
ELL	3.96(0.008)
LCV	4.00(0.02)
LCV_a	4.07(0.03)
ELL_{bboot}	4.06(0.03)
ELL_{iboot}	4.32(0.06)

5 Choosing between stratified and unstratified survival models

5.1 Method

The estimators of ELL can be used to choose between stratified and unstratified survival models. Consider right-censored data as defined in section 2.1 and let $\mathbf{X} = (X_1, \ldots, X_n)$ a vector of binary variable (coded 0/1). Finally, we note $\mathcal{W} = (W_1, \ldots, W_n)$ with $W_i = (\widetilde{T}_i, \delta_i, X_i)$ the observed data. We propose to use the ELL_{bboot} or the LCV_a criteria, to choose between a proportional hazards model and a stratified model. We define by

$$\lambda(t|X_i) = \lambda^0(t) \exp \beta X_i \quad i = 1, \dots, n$$

the proportional hazards model ([COX72]) and by

$$\lambda(t|X_i) = \begin{cases} \lambda^0(t) & \text{if } X_i = 0 \\ \lambda^1(t) & \text{if } X_i = 1 \end{cases}$$

the stratified model. To estimate these two models, we may use the penalized likelihood approach. In the proportional hazards regression model, $\widehat{\lambda}_h^0(\cdot)$ and $\widehat{\beta}$ maximize the penalized log-likelihood:

$$p\mathcal{L}_h(\mathcal{W}) = \log \mathcal{L}_p^{\lambda^0,\beta}(\mathcal{W}) - h \int \lambda^{0''^2}(u)du$$

In the stratified model, $\widehat{\lambda}_h^0(\cdot)$ and $\widehat{\lambda}_h^1(\cdot)$ maximize:

$$p\mathcal{L}_h(\mathcal{W}) = \log \mathcal{L}_p^{\lambda^0,\lambda^1}(\mathcal{W}) - h \int \left\{ \lambda^{0''^2}(u) + \lambda^{1''^2}(u) \right\} du$$
$$= \log \mathcal{L}_p^{\lambda^0}(\mathcal{W}^0) - h \int \lambda^{0''^2}(u)du + \log \mathcal{L}_p^{\lambda^1}(\mathcal{W}^1) - h \int \lambda^{1''^2}(u)du$$

where $\mathcal{W}^0 = (W_1^0, \dots, W_{n_0}^0)$ with $W_i^0 = (\widetilde{T}_i, \delta_i, X_i = 0)$ and $\mathcal{W}^1 = (W_1^1, \dots, W_{n_1}^1)$ with $W_i^1 = (\widetilde{T}_i, \delta_i, X_i = 1)$. We can remark that, we do not estimate separately $\lambda^0(\cdot)$ and $\lambda^1(\cdot)$ on the sample $\mathcal{W}^0$ and $\mathcal{W}^1$. $\lambda^0(\cdot)$ and $\lambda^1(\cdot)$ are estimated using the same smoothing parameter; thus the family of estimators $\widehat{\lambda}_h(\cdot|\cdot)$ of the proportional hazards model and the family of estimator $\widehat{\lambda}_h(\cdot|\cdot)$ of the stratified model have both just one hyper-parameter h. Therefore, we can discriminate between these two models (we return on this theoretical issue in the discussion). The LCV_a criterion could be applied to select h in the two models and thus to choose between them. It is appealing to apply in addition to the condition of the remark of section 3.3, the stronger condition $\sum X_i^{'} = n_1$. This has the advantage on conditioning on an ancillary statistic (the sample sizes in the strata, which does not carry information) and to yield the addition formula (7) below. The conditional criterion is thus:

$$\mathrm{ELL}_c(\widehat{\lambda}_h) = \mathrm{E}\left[\log \mathcal{L}_p^{\widehat{\lambda}_h^{\mathcal{W}}}(\mathcal{W}^{'}) | \sum X_i^{'} = n_1\right].$$

where $\mathcal{W}^{'} \overset{d}{=} \mathcal{W}$,$\mathcal{W}^{'} = (W_1^{'}, \dots, W_n^{'})$ with $W_i^{'} = (\widetilde{T}_i^{'}, \delta_i^{'}, X_i^{'})$.
To calculate $\mathrm{ELL}_c(\widehat{\lambda}_h)$ we use ELL_{bboot} defined in (5) with each bootstrap sample j that satisfies the condition $\sum_{i=1}^n X_i^j = n_1$. For the stratified estimator, we note that:

$$\mathrm{ELL}_c(\widehat{\lambda}_h) = \mathrm{ELL}(\widehat{\lambda}_h^0) + \mathrm{ELL}(\widehat{\lambda}_h^1) \tag{7}$$

So, in practice for each h we estimated $\mathrm{ELL}(\widehat{\lambda}_h^0)$ and $\mathrm{ELL}(\widehat{\lambda}_h^1)$ by (5) applied separately to $\mathcal{W}^0$ and $\mathcal{W}^1$ then computed $\mathrm{ELL}_c(\widehat{\lambda}_h)$ by (7). To minimize the different selection criteria we use a golden section search.

5.2 Example

We analysed data from the Paquid study [LCDBG94], a prospective cohort study of mental and physical aging that evaluates social environment and healh status. The Paquid study is based on a large cohort randomly selected in a population of subjects aged 65 years or more, living at home in two departments of southwest France (Gironde and Dordogne). There were 3675 non demented subjects at entry in the cohort and each subject has been visited six times or less, between 1988 and 2000; 431 incident cases of dementia were observed during the folow up. The risk of developing dementia was modeled as a function of age. As prevalent cases of dementia were excluded, data were left-truncated and the truncation variable was the age at entry in the cohort (for more details see Commenges et al., [CLJ+98]). Two explanatory variables were considered: sex (noted S) and educational level (noted E). In the sample, there were 2133 women and 1542 men. Educational level was classified into two categories: no primary school diploma and primary school diploma [LGC+99]. The pattern of observations involved interval censoring and left truncation. It is straightfoward to extent the theory described above to that case. For the sake of simplicity, we kept here the survival data framework, treating death as censoring rather than the more adapted multistate framework (Commenges ,2002). We were first interested in the effect of sex. The penalized likelihood estimate was used to compare the risk of dementia for men and women with a stratified model (model A) (figure 2) using ELL_{bboot} for choosing the smoothing parameter.

The penalized likelihood estimate using the LCV_a criterion was very close to the one obtained with ELL_{bboot}. It appears that women tend to have a lower risk of dementia than men before 78 years and a higher risk above that age and shows a non proportional hazard model. Indeed the proportional hazards model (model B) had lower value for both LCV_a and ELL_{bboot} than the stratified model (table 5).

Another important risk factor for dementia is educational level. As the proportional hazards assumption does not hold, we performed several analyses on the educational level stratified on sex. We considered three models. The stratified proportional hazards model (model C):

$$\lambda(t|S_i,E_i) = \begin{cases} \lambda_h^0(t)\exp\beta E_i \text{ if } S_i = 0 \text{ (women)} \\ \lambda_h^1(t)\exp\beta E_i \text{ if } S_i = 1\text{(men)} \end{cases}$$

the proportional hazard model performed separately (model D):

$$\lambda(t|S_i,E_i) = \begin{cases} \lambda_h^0(t)\exp\beta_0 E_i \text{ if } S_i = 0 \text{ (women)} \\ \lambda_h^1(t)\exp\beta_1 E_i \text{ if } S_i = 1\text{(men)} \end{cases}$$

the model stratified on both sex and educational level (model E):

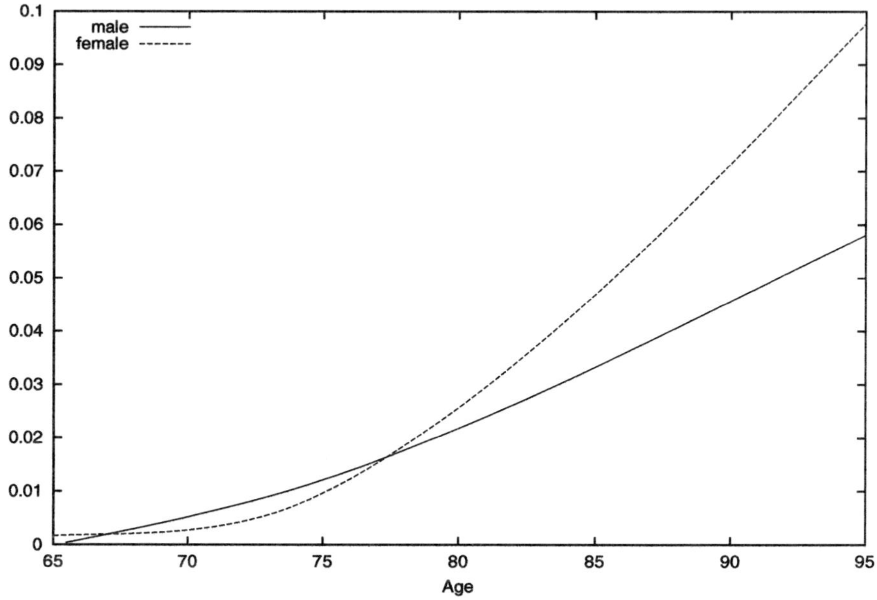

Fig. 2. Estimates of the hazard function of dementia for male (solid line) and female (dotted line) chosen by ELL_{bboot} criterion.

$$\lambda(t|S_i, E_i) = \begin{cases} \lambda_h^{0,0}(t) & \text{if } S_i = 0 \text{ and } E_i = 0 \\ \lambda_h^{1,0}(t) & \text{if } S_i = 1 \text{ and } E_i = 0 \\ \lambda_h^{0,1}(t) & \text{if } S_i = 0 \text{ and } E_i = 1 \\ \lambda_h^{1,1}(t) & \text{if } S_i = 1 \text{ and } E_i = 1 \end{cases}$$

Table 5 presents the results of the different models. The two criteria give the same conclusion: the best model is the stratified proportional hazard model (highest values; model C). Subjects with no primary school diploma have an increased risk of dementia. For this model (model C), the estimated relative risk for educational level is equal to 1.97; the corresponding 95% confidence interval is [1.63; 2.37].

6 Conclusion

We have presented a general criterion for selection of semi-parametric models from incomplete observations. This theoretical criterion, the expectation of the observed log-likelihood (ELL) performs nearly as well as the optimal KL distance (which is very difficult to estimate in this setting) as soon as there is enough information. We have shown that LCV estimates ELL. LCV and two proposed bootstrap estimators yield nearly equivalent results; ELL_{bboot} seems the best bootstrap estimator. The approximate version of LCV (for

Table 5. Comparison of the stratified and proportional hazards models according ELL_{bboot} and LCV_a criterion; A and B: unstratified and stratified models on sex; C, D, E: 3 models stratified on sex with educational level as new covariable (see text).

	ELL_{bboot}	LCV_a
model A	-1515.61	-1517.45
model B	-1517.71	-1519.92
model C	-1492.61	-1496.28
model D	-1493.51	-1497.18
model E	-1495.48	-1498.42

penalized likelihood) also performs very well and thus appears as the method of choice for this problem, due to the short computation time it requires. When no approximation of LCV is available, bootstrap estimators such as ELL_{bboot} are competitive because the amount of computation can be more flexibly tuned than for LCV.

ELL can be used for choosing a model in semi-parametric families. An important example is the choice beetween stratified and unstratified survival models. We have shown that this could be done using LCV or a bootstrap estimator of ELL in the case where all the models are indexed by a single hyper-parameter. This raises a completely new problem which is how to compare families of models of different complexities, i.e indexed by a different number of hyper-parameters. For instance this problem would arise if we compared a proportionnal hazards model (1 hyper-parameter) to a stratified model with one hyper-parameter for each stratum. We conjecture that there is a principle of parsimony at the hyper-parameter level, similar to that known for the ordinary parameters.

References

[ABGK93] P. K. Andersen, R.D. Borgan, R.D. Gill, and D. Keiding. *Statistical models based on counting processes.* Springer-Verlag, New-York, 1993.

[Aka73] H. Akaike. Information theory and an extension of the maximum likelihood principle. In B.N. Petrov and F. Csaki, editors, *Second International Symposium on Information Theory*, pages 267–281, Budapest, 1973. Akademiai kiado.

[Aka74] H. Akaike. A new look at the statistical model identification. *IEEE Transactions on Automatic Control*, 19:716–723, 1974.

[CLJ+98] D. Commenges, L. Letenneur, P. Joly, A. Alioum, and J.F. Dartigues. Modelling age-specific risk: application to dementia. *Statistics in Medicine*, 17:1973–1988, 1998.

[Com02] D. Commenges. Inference for multistate models from interval-censored data. *Statistical Methods in Medical Research*, 11:1–16, 2002.

[Cop83] J. B. Copas. Regression, prediction and shrinkage (with discussion). *Journal of the Royal Statistical Society B*, 45:311–354, 1983.

[COX72] D.R. Cox. Regression models and life tables (with discussion). *Journal Royal Statistical Society B*, 34:187–220, 1972.

[CS98] J. E. Cavanaugh and R. H. Shumway. An Akaike information criterion for model selection in the presence of incomplete data. *Journal of Statistical Planning and Inference*, 67:45–65, 1998.

[CW79] P. Craven and G. Wahba. Smoothing noisy data with spline functions: estimating the correct degree of smoothing by the method of generalized cross-validation. *Num. Math.*, 31:377–403, 1979.

[DeL92] J. DeLeeuw. *Breakthroughs in statistics*, volume 1, chapter Introduction to Akaike (1973) information theory and an extension of the maximum likelihood principle, pages 599–609. Springer-Verlag, London, 1992. Kotz, S. and Johnson, N. L.

[Fer99] J. D. Fermanian. A new bandwitdth selector in hazard estimation. *Nonparametric Statistics*, 10:137–182, 1999.

[Hal92] P. Hall. *The bootstrap and Edgeworth expansion.* Springer-Verlag, New york, 1992.

[HST98] C. M. Hurvich, J.S. Simonoff, and C.L Tsai. Smoothing parameter selection in nonparametric regression using an improved Akaike information criterion. *Journal of the Royal Statistical Society B*, 60:271–293, 1998.

[HT89] C. M. Hurvich and C.L Tsai. Regression and time series model selection in small samples. *Biometrika*, 76:297–307, 1989.

[HT90] T. J. Hastie and R. J. Tibshirani. *Generalized Additive Models.* Chapman and Hall, London, 1990.

[ISK97] M. Ishiguro, Y. Sakamoto, and G. Kitagawa. Bootstrapping log likelihood and EIC, an extension of AIC. *Ann. Inst. Statist. Math*, 49:411–434, 1997.

[JCL98] P. Joly, D. Commenges, and L. Letenneur. A penalized likelihood approach for arbitrarily censored and truncated data: application to age-specific incidence of dementia. *Biometrics*, 54:185–194, 1998.

[LC04] B. Liquet and D. Commenges. Estimating the expectation of the log-likelihood with censored data for estimator selection. *LIDA*, 10:351–367, 2004.

[LCDBG94] L. Letenneur, D. Commenges, J.F. Dartigues, and P. Barberger-Gateau. Incidence of dementia and alzheimer's disease in elderly community residents of south-western france. *Int. J. Epidemiol.*, 23:1256–1261, 1994.

[LGC+99] L. Letenneur, V. Gilleron, D. Commenges, C. Helmer, J.M. Orgogozo, and J.F. Dartigues. Are sex and educational level independent predictors of dementia and alzheimer's disease? incidence data from the paquid project. *J. Neurol. Neurosurg. Psychiatry.*, 66:177–183, 1999.

[LSC03] B. Liquet, C. Sakarovitch, and D. Commenges. Bootstrap choice of estimators in non-parametric families: an extension of EIC. *Biometrics*, 59:172–178, 2003.

[Mal73] C.L. Mallows. Some comments on C_p. *Technometrics*, 15:661–675, 1973.

[Mil02] A.J. Miller. *Subset Selection in Regression (Second Edition)*. Chapman and Hall, London, 2002.

[MP87] J. S. Marron and W. J. Padgett. Asymptotically optimal bandwidth selection for kernel density estimators from randomly right-censored samples. *The Annals of Statistics*, 15:1520–1535, 1987.

[O'S88] F. O'Sullivan. Fast computation of fully automated log-density and log-hazard estimators. *SIAM J. Sci. Stat. Comput.*, 9:363–379, 1988.

[RH83] H. Ramlau-Hansen. Smoothing counting process intensities by means of kernel functions. *The Annals of Statistics*, 11:453–466, 1983.

[Sil86] B.W. Silverman. *Density estimation for statistics and data analysis*. Chapman and Hall, London, 1986.

Imputing responses that are not missing

Ursula U. Müller[1], Anton Schick[2], and Wolfgang Wefelmeyer[3]

[1] Fachbereich 3, Universität Bremen, Postfach 330 440, 28334 Bremen, Germany uschi@math.uni-bremen.de
[2] Department of Mathematical Sciences, Binghamton University, Binghamton, NY 13902-6000, USA anton@math.binghamton.edu
[3] Mathematisches Institut, Universität zu Köln, Weyertal 86–90, 50931 Köln, Germany wefelmeyer@math.uni-koeln.de

We consider estimation of linear functionals of the joint law of regression models in which responses are missing at random. The usual approach is to work with the fully observed data, and to replace unobserved quantities by estimators of appropriate conditional expectations. Another approach is to replace all quantities by such estimators. We show that the second method is usually better than the first.

1 Introduction

Let (X, Y) be a random vector. We want to estimate $E[h(X, Y)]$, the expectation of some known square-integrable function h. If we are able to sample from (X, Y), we can use the empirical estimator $\frac{1}{n}\sum_{i=1}^n h(X_i, Y_i)$. If nothing is known about the distribution of (X, Y), this estimator is efficient. We are interested in the situation where we always observe X, but Y only if some indicator Z equals one. We assume that Z and Y are conditionally independent given X. Then one says that Y is *missing at random*. In this case the empirical estimator is not available unless all Z_i are one. Let $\pi(X) = E(Z \mid X) = P(Z = 1 \mid X)$. If π is known and positive, we could use the estimator $\frac{1}{n}\sum_{i=1}^n Z_i h(X_i, Y_i)/\pi(X_i)$. If π is unknown, one could replace π by an estimator $\hat{\pi}$, resulting in

$$\frac{1}{n}\sum_{i=1}^{n} \frac{Z_i}{\hat{\pi}(X_i)} h(X_i, Y_i). \tag{1}$$

Surprisingly, even if π is known, replacing π by an estimator can decrease the asymptotic variance. Such an improvement is given by Schisterman and Rotnitzky [SR01]. A similar result, on average treatment effects, is in Hirano, Imbens and Ridder [HIR03]. Another estimator for $E[h(X, Y)]$ is the *partially imputed estimator*

$$\frac{1}{n}\sum_{i=1}^{n}\Big(Z_i h(X_i, Y_i) + (1 - Z_i)\hat{\chi}(X_i)\Big), \tag{2}$$

where $\hat{\chi}(X_i)$ is an estimator of the conditional expectation

$$\chi(X_i) = E(h(X_i, Y_i) \mid X_i).$$

An alternative to the partially imputed estimator is the *fully imputed* estimator

$$\frac{1}{n}\sum_{i=1}^{n}\hat{\chi}(X_i). \tag{3}$$

An extreme case would be that the conditional distribution of Y given X is *known*. It is easy to see that then the fully imputed estimator $\frac{1}{n}\sum_{i=1}^{n}\chi(X_i)$ is at least as good as the partially imputed estimator, and strictly better unless $Z(h(X,Y) - \chi(X))$ is zero almost surely.

We show that the fully imputed estimator (3) is usually better than the partially imputed estimator (2). We restrict attention to the situation where π is bounded away from zero but otherwise completely unknown. We also impose no structural assumptions on the covariate distribution. We consider four different models for the conditional distribution of Y given X.

Suppose first that the conditional distribution $Q(X, dy)$ of Y given X is completely unknown. For the case $h(X,Y) = Y$, Cheng [Che94] shows that the partially and fully imputed estimators are asymptotically equivalent, and obtains their asymptotic distribution. He estimates $E(Y \mid X)$ by a truncated kernel estimator. Wang and Rao [WR02] obtain a similar result with a differently truncated kernel estimator. Cheng and Chu [CC96] study estimation of the response distribution function and quantiles. We generalize Cheng's result to arbitrary functions h and prove efficiency.

Suppose now that we have a parametric model $Q_\vartheta(X, dy)$ for the conditional distribution of Y given X. In this case the conditional expectation is of the form $\chi_\vartheta(x) = \int h(x,y)\, Q_\vartheta(x, dy)$. This suggests estimating χ_ϑ by $\chi_{\hat{\vartheta}}$. The natural estimator for ϑ is the conditional maximum likelihood estimator. We show that the fully imputed estimator $\frac{1}{n}\sum_{i=1}^{n}\chi_{\hat{\vartheta}}(X_i)$ is efficient, and better than the corresponding partially imputed estimator except in degenerate cases. This is related to Tamhane [Tam78] who assumes a parametric model for the *joint* distribution of X and Y. Then $E[h(X,Y)]$ is a smooth function of ϑ; hence it can be estimated efficiently by plugging in an efficient estimator, such as the maximum likelihood estimator.

Next we consider a model between the fully nonparametric and parametric ones for Q, a linear regression model with covariates and errors independent. For simplicity we take $Y = \vartheta X + \varepsilon$. We do not assume that ε has mean zero but require X to have positive variance for identifiability. Here $Q(x, dy) = f(y - \vartheta x)\, dy$, where f is the (unknown) density of the errors. Then $\chi(x) = \int h(x, \vartheta x + u) f(u)\, du$. Exploiting this representation, we estimate $\chi(x)$ by $\sum_{j=1}^{n} Z_j h(x, \hat{\vartheta}x + Y_j - \hat{\vartheta}X_j)/\sum_{j=1}^{n} Z_j$. We show that the corresponding fully imputed estimator is efficient if an efficient estimator for ϑ is used. Again the partially imputed estimator will not be efficient in general, even if an efficient estimator for ϑ is used.

Finally we consider a linear regression model *without* assuming independence between covariates and errors. For simplicity we take $Y = \vartheta X + \varepsilon$ with $E(\varepsilon \mid X) = 0$. This can be written as a constraint on the conditional distribution of Y given X, namely $\int y\, Q(X, dy) = \vartheta X$. For $h(X,Y) = Y$ this suggests the estimator $\hat\vartheta \frac{1}{n}\sum_{i=1}^n X_i$, which happens to be the fully imputed estimator. Matloff [Mat81] has shown that such an estimator improves upon the partially imputed estimator for his choice of $\hat\vartheta$. We show that the fully imputed estimator of $E[h(X,Y)]$ for general h is efficient if an appropriate estimator for χ is used. This requires an efficient estimator $\hat\vartheta$ for ϑ and a correction term to the nonparametric estimator of χ. An efficient estimator of ϑ can be obtained as a weighted least squares estimator with estimated optimal weights, based on the fully observed pairs. Efficient estimation of ϑ for more general regression models and various models for π has been studied in Robins, Rotnitzky and Zhao [RRZ94], Robins and Rotnitzky [RbRt95], and Rotnitzky and Robins [RtRb95], among others. Efficient score functions for ϑ are calculated by Nan, Emond and Wellner [NEW04] and Yu and Nan [YN03]. The partially imputed estimator will not be efficient, in general. In view of this, partially imputed estimators such as the one by Wang, Härdle and Linton [WHL04] for $E[Y]$ in a partly linear model are not efficient.

The paper is organized as follows. In Section 2 we characterize efficient estimators for linear functionals of arbitrary regression models with responses missing at random; in particular for the four cases above. Our results show that the model is adaptive in the sense that we can estimate $E[h(X,Y)]$ as well not knowing π as knowing π. In Section 3 we construct efficient fully imputed estimators of $E[h(X,Y)]$ in these four models.

2 Efficient influence functions

In this section we calculate the efficient influence function for estimating the expected value $E[h(X,Y)]$ with observations (X, ZY, Z) as described in the Introduction. The joint distribution $P(dx, dy, dz)$ of the observations depends on the marginal distribution $G(dx)$ of X, the conditional probability $\pi(x)$ of $Z = 1$ given $X = x$, and the conditional distribution $Q(x, dy)$ of Y given $X = x$. More precisely, we have

$$P(dx, dy, dz) = G(dx) B_{\pi(x)}(dz)\big(zQ(x, dy) + (1-z)\delta_0(dy)\big),$$

where $B_p = p\delta_1 + (1-p)\delta_0$ denotes the Bernoulli distribution with parameter p and δ_t the Dirac measure at t. Consider perturbations G_{nu}, Q_{nv} and π_{nw} of G, Q and π that are *Hellinger differentiable* in the following sense:

$$\int \Big(n^{1/2}(dG_{nu}^{1/2} - dG^{1/2}) - \frac{1}{2}u\,dG^{1/2}\Big)^2 \to 0,$$

$$\iint \Big(n^{1/2}(dQ_{nv}^{1/2}(x,\cdot) - dQ^{1/2}(x,\cdot)) - \frac{1}{2}v(x,\cdot)dQ^{1/2}(x,\cdot)\Big)^2 G(dx) \to 0,$$

$$\iint \Big(n^{1/2}(dB_{\pi_{nw}(x)}^{1/2} - dB_{\pi(x)}^{1/2}) - \frac{1}{2}(\cdot - \pi(x))w(x)dB_{\pi(x)}^{1/2}\Big)^2 G(dx) \to 0.$$

This requires that u belongs to

$$L_{2,0}(G) = \{u \in L_2(G) : \int u\,dG = 0\};$$

that v belongs to

$$V_0 = \{v \in L_2(M) : \int v(x,y)\,Q(x,dy) = 0\}$$

with $M(dx,dy) = Q(x,dy)G(dx)$; and that w belongs to $L_2(G_\pi)$, where $G_\pi(dx) = \pi(x)(1-\pi(x))\,G(dx)$.

We have *local asymptotic normality*: With P_{nuvw} denoting the joint distribution of the observations (X, ZY, Z) under the perturbed parameters G_{nu}, Q_{nv} and π_{nw},

$$\sum_{i=1}^n \log \frac{dP_{nuvw}}{dP}(X_i, Z_iY_i, Z_i) = n^{-1/2}\sum_{i=1}^n t_{uvw}(X_i, Z_iY_i, Z_i) - \frac{1}{2}E[t_{uvw}^2(X, ZY, Z)] + o_p(1),$$

where $t_{uvw}(X,ZY,Z) = u(X) + Zv(X,Y) + (Z-\pi(X))w(X)$ and

$$E[t_{uvw}^2(X,ZY,Z)] = E[u^2(X)] + E[Zv^2(X,Y)] + E[(Z-\pi(X))^2 w^2(X)] = \int u^2\,dG + \iint \pi(x)v^2(x,y)\,Q(x,dy)G(dx) + \int w^2\,dG_\pi.$$

If we have models for the parameters G, Q and π, then, in order for the perturbations G_{nu}, Q_{nv} and π_{nw} to be within these models, the functions u, v and w must be restricted to subsets U of $L_{2,0}(G)$, V of V_0, and W of $L_2(G_\pi)$. The choices $U = L_{2,0}(G)$ and $V = V_0$ correspond to fully nonparametric models for G and Q. Parametric models for G and Q result in finite-dimensional U and V. In what follows the spaces U, V and W will be assumed to be closed and linear.

Let now κ be a functional of G, Q and π. The functional is *differentiable* with *gradient* $g \in L_2(P)$ if, for all $u \in U$, $v \in V$ and $w \in W$,

$$n^{1/2}\big(\kappa(G_{nu}, Q_{nv}, \pi_{nw}) - \kappa(G,Q,\pi)\big) \to E[g(X,ZY,Z)t_{uvw}(X,ZY,Z)].$$

The gradient g is not unique. The *canonical gradient* is g_*, where $g_*(X,ZY,Z)$ is the projection of $g(X,ZY,Z)$ onto the *tangent space*

$$T = \{t_{uvw}(X, ZY, Z) : u \in U, v \in V, w \in W\}.$$

Since T is a sum of orthogonal spaces

$$T_1 = \{u(X) : u \in U\},$$
$$T_2 = \{Zv(X,Y) : v \in V\},$$
$$T_3 = \{(Z - \pi(X))w(X) : w \in W\},$$

the random variable $g_*(X, ZY, Z)$ is the sum

$$g_*(X, ZY, Z) = u_*(X) + Zv_*(X,Y) + (Z - \pi(X))w_*(X),$$

where $u_*(X)$, $Zv_*(X,Y)$ and $(Z - \pi(X))w_*(X)$ are the projections of the random variable $g(X, ZY, Z)$ onto T_1, T_2 and T_3, respectively. We assume that $E[g_*^2(X, ZY, Z)]$ is positive.

An estimator $\hat{\kappa}$ for κ is *regular* with *limit* L if L is a random variable such that, for all $u \in U$, $v \in V$ and $w \in W$,

$$n^{1/2}\big(\hat{\kappa} - \kappa(G_{nu}, Q_{nv}, \pi_{nw})\big) \Rightarrow L \quad \text{under } P_{nuvw}.$$

The Hájek–Le Cam convolution theorem says that L is distributed as the sum of a normal random variable with mean zero and variance $E[g_*^2(X, ZY, Z)]$ and some independent random variable. This justifies calling an estimator $\hat{\kappa}$ *efficient* if it is regular with limit such a normal random variable.

An estimator $\hat{\kappa}$ for κ is *asymptotically linear* with *influence function* $\psi \in L_{2,0}(P)$ if

$$n^{1/2}\big(\hat{\kappa} - \kappa(G, Q, \pi)\big) = n^{-1/2}\sum_{i=1}^{n} \psi(X_i, Z_iY_i, Z_i) + o_p(1).$$

As a consequence of the convolution theorem, a regular estimator is efficient if and only if it is asymptotically linear with influence function g_*. A reference for the convolution theorem and the characterization is Bickel, Klaassen, Ritov and Wellner [BKRW98].

We are interested in estimating

$$\kappa(G, Q, \pi) = E[h(X,Y)] = \iint h(x,y)\, Q(x, dy)G(dx) = \int h\, dM.$$

Let $M_{nuv}(dx, dy) = Q_{nv}(x, dy)G_{nu}(dx)$. Then M_{nuv} is Hellinger differentiable in the following sense:

$$\int \Big(n^{1/2}\big(dM_{nuv}^{1/2} - dM^{1/2}\big) - \frac{1}{2} t\, dM^{1/2}\Big)^2 \to 0$$

with $t(x,y) = u(x) + v(x,y)$. If M_{nuv} satisfies $\limsup_n \int h^2\, dM_{nuv} < \infty$, then

$$n^{1/2}\Big(\int h\, dM_{nuv} - \int h\, dM\Big) \to E\big[h(X,Y)\big(u(X) + v(X,Y)\big)\big];$$

see e.g. Ibragimov and Has'minskiˇi [IH81], p. 67, Lemma 7.2.

Thus the canonical gradient of $E[h(X,Y)]$ is determined by

$$E[u_*(X)u(X)] + E[Zv_*(X,Y)v(X,Y)] + E[(Z-\pi(X))^2 w_*(X)w(X)] \\ = E\big[h(X,Y)\big(u(X)+v(X,Y)\big)\big]$$

for all $u \in U$, $v \in V$ and $w \in W$. Setting first $u = 0$ and $v = 0$, we see that $w_* = 0$. Setting $v = 0$, we see that $u_*(X)$ is the projection of $h(X,Y)$ onto T_1. Taking $u = 0$, we see that the projection of $Zv_*(X,Y)$ onto $\widetilde{V} = \{v(X,Y) : v \in V\}$ must equal the projection of $h(X,Y)$ onto $\widetilde{V}$.

We are mainly interested in a fully nonparametric model for G, for which $U = L_{2,0}(G)$. Then $u_*(X) = \chi(X) - E[\chi(X)]$. We now give explicit formulas for v_*, and hence for the canonical gradient of $E[h(X,Y)]$, in four cases: fully nonparametric conditional distribution, with $V = V_0$; parametric conditional distribution, with V finite-dimensional; and two semiparametric models, namely linear regression with and without independence of covariate and error.

1. Nonparametric conditional distribution. If $V = V_0$, then the projections of $h(X,Y)$ and $Zv_*(X,Y)$ onto $\widetilde{V}$ are $h(X,Y)-\chi(X)$ and $\pi(X)v_*(X,Y)$. Thus

$$v_*(X,Y) = \frac{h(X,Y)-\chi(X)}{\pi(X)}.$$

Hence, if $U = L_{2,0}(G)$, the canonical gradient of $E[h(X,Y)]$ is

$$\psi_{np}(X,ZY,Z) = \chi(X) - E[\chi(X)] + \frac{Z}{\pi(X)}\big(h(X,Y)-\chi(X)\big).$$

For the important special case $h(X,Y) = Y$ we obtain

$$\psi_{np}(X,ZY,Z) = E(Y \mid X) - E[Y] + \frac{Z}{\pi(X)}(Y - E(Y \mid X)).$$

2. Parametric conditional distribution. Let $Q(x,dy) = q_\vartheta(x,y)\,dy$, where ϑ is an m-dimensional parameter. In this case, V will be the span of the components of the score function ℓ_ϑ, the Hellinger derivative of the parametric model q_ϑ at ϑ:

$$\iint \Big(q_{\vartheta+t}^{1/2}(x,y) - q_\vartheta^{1/2}(x,y) - \frac{1}{2}t^\top \ell_\vartheta(x,y)q_\vartheta^{1/2}(x,y)\Big)^2\,dy\,G(dx) = o(t^2).$$

We also assume that $E[Z\ell_\vartheta(X,Y)\ell_\vartheta(X,Y)^\top]$ is positive definite. If q_ϑ is differentiable in ϑ, then $\ell_\vartheta = \dot{q}_\vartheta/q_\vartheta$, where $\dot{q}_\vartheta$ is the derivative of q_ϑ with respect to ϑ. If we set $L = \ell_\vartheta(X,Y)$, then $\widetilde{V} = \{c^\top L : c \in \mathbb{R}^m\}$. Thus v_* is of the form $c_*^\top L$. Since the projections of $h(X,Y)$ and $Zv_*(X,Y)$ onto $\widetilde{V}$ are $a^\top L$ and

$b^\top L$ with $a = (E[LL^\top])^{-1}E[Lh(X,Y)]$ and $b = (E[LL^\top])^{-1}E[ZLL^\top]\,c_*$, we obtain $c_* = (E[ZLL^\top])^{-1}E[Lh(X,Y)]$. Thus, if $U = L_{2,0}(G)$, the canonical gradient of $E[h(X,Y)]$ is

$$\psi_p(X, ZY, Z) = \chi(X) - E[\chi(X)] + Zc_*^\top \ell_\vartheta(X,Y).$$

3. Linear regression with independence. We consider the linear regression model $Y = \vartheta X + \varepsilon$ with ε and X independent. We assume that ε has an unknown density f with finite Fisher information J for location and X has finite and positive variance. We do *not* assume that ε has mean zero. In this model, $Q(x, dy) = f(y - \vartheta x)\,dy$. Write F for the distribution function of f. As shown in Bickel [Bic82],

$$\widetilde{V} = \{\alpha X\ell(\varepsilon) + \beta(\varepsilon) : \alpha \in \mathbb{R}, \beta \in L_{2,0}(F)\}.$$

Here ℓ denotes the score function $\ell(y) = -f'(y)/f(y)$ for location. The space $\widetilde{V}$ can be written as the orthogonal sum of the spaces $\widetilde{V}_1 = \{\alpha\xi : \alpha \in \mathbb{R}\}$ with

$$\xi = (X - E[X])\ell(\varepsilon),$$

and $\widetilde{V}_2 = \{\beta(\varepsilon) : \beta \in L_{2,0}(F)\}$. The projection of $h(X,Y)$ onto $\widetilde{V}_1$ is $c_h\xi/E[\xi^2]$ with $c_h = E[h(X,Y)\xi]$, and the projection of $h(X,Y)$ onto $\widetilde{V}_2$ is $\bar{h}(\varepsilon) - E[\bar{h}(\varepsilon)]$ with $\bar{h}(\varepsilon) = E(h(X,Y) \mid \varepsilon)$. For $b \in L_2(F)$, the projection of $Zb(\varepsilon)$ onto $\widetilde{V}_1$ is $c\xi/E[\xi^2]$ with

$$c = E[Zb(\varepsilon)\xi] = E[Z](E(X|Z=1) - E[X])E[b(\varepsilon)\ell(\varepsilon)],$$

and the projection of $Zb(\varepsilon)$ onto $\widetilde{V}_2$ is $E[Z](b(\varepsilon) - E[b(\varepsilon)])$. Let

$$\xi_* = \big(X - E(X \mid Z = 1)\big)\ell(\varepsilon).$$

Then $Z\xi_*$ is orthogonal to $\widetilde{V}_2$, and its projection onto $\widetilde{V}_1$ is $a_*\xi/E[\xi^2]$ with $a_* = E[Z\xi_*\xi] = E[Z\xi_*^2]$. Since

$$c_h = E[h(X,Y)\xi] = E[h(X,Y)\xi_*] + (E(X|Z=1) - E[X])E[h(X,Y)\ell(\varepsilon)],$$

it follows that

$$v_*(X,Y) = \frac{E[h(X,Y)\xi_*]}{E[Z\xi_*^2]}\,\xi_* + \frac{1}{E[Z]}\big(\bar{h}(\varepsilon) - E[\bar{h}(\varepsilon)]\big).$$

Thus, if $U = L_{2,0}(G)$, the canonical gradient of $E[h(X,Y)]$ is

$$\psi_I(X, ZY, Z) = \chi(X) - E[\chi(X)] + Z\Big(\frac{E[h(X,Y)\xi_*]}{E[Z\xi_*^2]}\,\xi_* + \frac{1}{E[Z]}\big(\bar{h}(\varepsilon) - E[\bar{h}(\varepsilon)]\big)\Big).$$

For $h(X,Y) = Y$ we can use the identity $E[\varepsilon\ell(\varepsilon)] = 1$ to simplify the canonical gradient to

$$\vartheta(X - E[X]) + \frac{Z(E[X] - E(X|Z=1))}{E[Z\xi_*^2]}\xi_* + \frac{Z(\varepsilon - E[\varepsilon])}{E[Z]}.$$

4. Linear regression without independence. Now we consider the linear regression model $Y = \vartheta X + \varepsilon$ with $E(\varepsilon \mid X) = 0$. We write $\sigma^2(X) = E(\varepsilon^2 \mid X)$ and $\rho_h(X) = E(h(X,Y)\varepsilon \mid X)$. In this model, we have only the constraint $\int y\, Q(x, dy) = \vartheta x$ on the transition distribution Q. In this case, the space $\widetilde{V}$ is the sum of the two orthogonal spaces

$$\widetilde{V}_1 = \{a\sigma^{-2}(X)X\varepsilon : a \in \mathbb{R}\},$$
$$\widetilde{V}_2 = \{v(X,Y) : v \in V_0, E(v(X,Y)\varepsilon \mid X) = 0\}.$$

For details see Müller, Schick and Wefelmeyer [MSW04]. The projection of $h(X,Y)$ onto $\widetilde{V}_1$ is $a_h\sigma^{-2}(X)X\varepsilon$ with

$$a_h = E[h(X,Y)\sigma^{-2}(X)X\varepsilon]/E[\sigma^{-2}(X)X^2],$$

while the projection onto $\widetilde{V}_2$ is $\widetilde{h}_2 = h(X,Y) - \chi(X) - E[\rho_h(X)]\sigma^{-2}(X)\varepsilon$. It is now easy to check that $v_*(X,Y) = a_*\sigma^{-2}(X)X\varepsilon + \widetilde{h}_2/\pi(X)$. Thus, if $U = L_{2,0}(G)$, the canonical gradient of $E[h(X,Y)]$ is

$$\psi_{II}(X, ZY, Z) = \chi(X) - E[\chi(X)] + \frac{Z}{\pi(X)}(h(X,Y) - \chi(X))$$
$$- \frac{Z\varepsilon}{\sigma^2(X)}\Big(\frac{\rho_h(X)}{\pi(X)} - a_* X\Big).$$

Note that $\psi_{II} = \psi_{np} - \psi_{II}^*$ with

$$\psi_{II}^*(X, ZY, Z) = \frac{Z\varepsilon}{\sigma^2(X)}\Big(\frac{\rho_h(X)}{\pi(X)} - a_* X\Big).$$

3 Efficient estimators

In this section we indicate that the fully imputed estimators are efficient in the four models discussed at the end of Section 2. Throughout we assume that we have no structural information on the covariate distribution G.

1. Nonparametric conditional distribution. In this model, Q is completely unspecified. The usual partially imputed estimators for $E[h(X,Y)]$ are of the form

$$\hat{H}_1 = \frac{1}{n}\sum_{i=1}^{n}\Big(Z_i h(X_i, Y_i) + (1 - Z_i)\hat{\chi}(X_i)\Big),$$

where $\hat{\chi}$ is a nonparametric estimator for χ of the form

$$\hat{\chi}(X_i) = \sum_{j=1}^{n} W_{ij} Z_j h(X_j, Y_j)$$

with weights W_{ij} depending on $X_1, \dots, X_n, Z_1, \dots, Z_n$ only. This includes kernel-type estimators and linear smoothers. Under appropriate smoothness conditions on χ and π, and for properly chosen weights W_{ij}, the estimator $\hat{H}_1$ has the stochastic expansion

$$\hat{H}_1 = \frac{1}{n}\sum_{i=1}^{n} \chi(X_i) + \frac{1}{n}\sum_{i=1}^{n} \frac{Z_i}{\pi(X_i)}(h(X_i, Y_i) - \chi(X_i)) + o_p(n^{-1/2}). \tag{4}$$

In the case $h(X, Y) = Y$, such conditions are given by Cheng [Che94] and Wang and Rao [WR02]. These authors use weights W_{ij} corresponding to truncated kernel estimators. Cheng [Che94] also shows that $\hat{H}_1$ is asymptotically equivalent to the fully imputed $\hat{H}_2 = \frac{1}{n}\sum_{i=1}^{n} \hat{\chi}(X_i)$. It follows from (4) that $\hat{H}_1$ and $\hat{H}_2$ have influence function $\psi = \psi_{np}$ and are therefore efficient by Section 2.

2. Parametric conditional distribution. In this model, $Q = Q_\vartheta$, with ϑ an m-dimensional parameter. Then

$$\chi(x) = \chi_\vartheta(x) = \int h(x, y)\, Q_\vartheta(x, dy).$$

Here we use an estimator $\hat{\vartheta}$ of ϑ and obtain for $E[h(X, Y)]$ the partially and fully imputed estimators

$$\hat{H}_3 = \frac{1}{n}\sum_{i=1}^{n} \Big(Z_i h(X_i, Y_i) + (1 - Z_i)\chi_{\hat{\vartheta}}(X_i)\Big) \quad \text{and} \quad \hat{H}_4 = \frac{1}{n}\sum_{i=1}^{n} \chi_{\hat{\vartheta}}(X_i).$$

For the following discussion, we assume again Hellinger differentiability of Q_ϑ as in Section 2 and write ℓ_ϑ for the score function. A natural estimator for ϑ is the conditional maximum likelihood estimator, which solves $\frac{1}{n}\sum_{i=1}^{n} Z_i \ell_\vartheta(X_i, Y_i) = 0$. Under some additional regularity conditions, this estimator has the expansion

$$\hat{\vartheta} = \vartheta + I_\vartheta^{-1} \frac{1}{n}\sum_{i=1}^{n} Z_i \ell_\vartheta(X_i, Y_i) + o_p(n^{-1/2})$$

with $I_\vartheta = E\big[\pi(X)\ell_\vartheta(X, Y)\ell_\vartheta(X, Y)^\top\big]$. One can show that $\hat{\vartheta}$ is efficient for $\vartheta = \kappa(G, Q_\vartheta, \pi)$. Moreover, under regularity conditions, for any $n^{1/2}$-consistent $\hat{\vartheta}$,

$$\frac{1}{n}\sum_{i=1}^{n} Z_i\chi_{\hat\vartheta}(X_i) = \frac{1}{n}\sum_{i=1}^{n} Z_i\chi_{\vartheta}(X_i) + D_1^\top(\hat\vartheta - \vartheta) + o_p(n^{-1/2}),$$

$$\frac{1}{n}\sum_{i=1}^{n}(1 - Z_i)\chi_{\hat\vartheta}(X_i) = \frac{1}{n}\sum_{i=1}^{n}(1 - Z_i)\chi_{\vartheta}(X_i) + D_0^\top(\hat\vartheta - \vartheta) + o_p(n^{-1/2}),$$

where

$$D_1 = E[Zh(X,Y)\ell_\vartheta(X,Y)] \quad \text{and} \quad D_0 = E[(1-Z)h(X,Y)\ell_\vartheta(X,Y)].$$

Thus, if we use the conditional maximum likelihood estimator for ϑ, we have the expansions

$$\begin{aligned}\hat H_3 &= \frac{1}{n}\sum_{i=1}^{n}\Big(Z_i h(X_i,Y_i) + (1 - Z_i)\chi_\vartheta(X_i) + D_0^\top I_\vartheta^{-1} Z_i \ell_\vartheta(X_i,Y_i)\Big)\\ &\quad + o_p(n^{-1/2}),\\ \hat H_4 &= \frac{1}{n}\sum_{i=1}^{n}\Big(\chi_\vartheta(X_i) + (D_0 + D_1)^\top I_\vartheta^{-1} Z_i\ell_\vartheta(X_i,Y_i)\Big) + o_p(n^{-1/2}).\end{aligned}$$

Since $D_0 + D_1 = E[h(X,Y)\ell_\vartheta(X,Y)]$, we see that $\hat H_4$ has influence function $\psi = \psi_p$ and is therefore efficient. The difference between the estimators is

$$\hat H_3 - \hat H_4 = \frac{1}{n}\sum_{i=1}^{n} Z_i\Big(h(X_i,Y_i) - \chi_\vartheta(X_i) - D_1^\top I_\vartheta^{-1}\ell_\vartheta(X_i,Y_i)\Big) + o_p(n^{-1/2}).$$

Hence $\hat H_3$ is asymptotically equivalent to $\hat H_4$, and therefore also efficient, if and only if $Z\big(h(X,Y) - \chi_\vartheta(X) - D_1^\top I_\vartheta^{-1}\ell_\vartheta(X,Y)\big)$ is zero almost surely. Since this is usually not the case, the partially imputed estimator $\hat H_3$ is typically inefficient.

3. Linear regression with independence. In this model, $Q(x,dy) = Q_{\vartheta,f}(x,dy) = f(y - \vartheta x)\,dy$. We assume that f has finite Fisher information J for location and X has finite and positive variance. Now

$$\chi(x) = \chi(x,\vartheta,f) = \int h(x, \vartheta x + u) f(u)\,du.$$

This suggests the estimator

$$\hat\chi(x,\hat\vartheta) = \frac{\frac{1}{n}\sum_{j=1}^{n} Z_j h(x, \hat\vartheta x + Y_j - \hat\vartheta X_j)}{\overline{Z}},$$

where $\overline{Z} = \frac{1}{n}\sum_{j=1}^{n} Z_j$. Then the partially and fully imputed estimators for $E[h(X,Y]$ are

$$\hat{H}_5 = \frac{1}{n}\sum_{i=1}^n \Big(Z_i h(X_i, Y_i) + (1 - Z_i)\hat{\chi}(X_i, \hat{\vartheta})\Big) \quad \text{and} \quad \hat{H}_6 = \frac{1}{n}\sum_{i=1}^n \hat{\chi}(X_i, \hat{\vartheta}).$$

Let

$$S = \frac{1}{n^2}\sum_{i=1}^n \sum_{j=1}^n \frac{Z_j}{E[Z]} h(X_i, \vartheta X_i + \varepsilon_j).$$

Then $E[S] = E[h(X,Y)] = \kappa$. By the Hoeffding decomposition,

$$S = \kappa + \frac{1}{n}\sum_{i=1}^n (\chi(X_i) - \kappa) + \frac{1}{n}\sum_{j=1}^n \Big(\frac{Z_j \bar{h}(\varepsilon_j)}{E[Z]} - \kappa\Big)$$

with $\bar{h}(\varepsilon) = E(h(X,Y) \mid \varepsilon)$. Using this we obtain

$$\begin{aligned}\frac{1}{n}\sum_{i=1}^n \hat{\chi}(X_i, \vartheta) &= \frac{E[Z]}{\bar{Z}} S \;=\; S - \frac{\bar{Z} - E[Z]}{E[Z]}\kappa + o_p(n^{-1/2}) \\ &= \frac{1}{n}\sum_{i=1}^n \chi(X_i) + \frac{1}{n}\sum_{j=1}^n \frac{Z_j}{E[Z]}(\bar{h}(\varepsilon_j) - \kappa) + o_p(n^{-1/2}).\end{aligned}$$

Under additional assumptions,

$$\begin{aligned}\hat{H}_6 &= \frac{1}{n^2}\sum_{i=1}^n \sum_{j=1}^n \frac{Z_j}{\bar{Z}} h\big(X_i, \vartheta X_i + \varepsilon_j + (\hat{\vartheta} - \vartheta)(X_i - X_j)\big) \\ &= \frac{1}{n^2}\sum_{i=1}^n \sum_{j=1}^n \frac{Z_j}{\bar{Z}} h(X_i, \vartheta X_i + \varepsilon_j) + D(\hat{\vartheta} - \vartheta) + o_p(n^{-1/2})\end{aligned}$$

with

$$\begin{aligned}D &= \frac{1}{E[Z]} E\big[h(X_1, X_1 + \varepsilon_2) Z_2 (X_1 - X_2)\ell(\varepsilon_2)\big] \\ &= E\big[h(X,Y)(X - E(X|Z=1))\ell(\varepsilon)\big].\end{aligned}$$

In the linear regression model *without* missing responses, efficient estimators for ϑ have been constructed by Bickel [Bic82], Koul and Susarla [KS83], and Schick [Sch87, Sch93]. Their influence function is $\xi/E[\xi^2]$ with $\xi = (X - E[X])\ell(\varepsilon)$. An analogous construction based on the observations (X_i, Y_i) with $Z_i = 1$ yields an estimator for ϑ with influence function $Z\xi_*/E[Z\xi_*^2]$ with $\xi_* = \big(X - E(X \mid Z = 1)\big)\,\ell(\varepsilon)$. One can show that $\hat{\vartheta}$ is efficient for $\vartheta = \kappa(G, Q_{\vartheta,f}, \pi)$. If we use an estimator $\hat{\vartheta}$ with this influence function, then $\hat{H}_6$ has the stochastic expansion

$$\begin{aligned}\hat{H}_6 = \frac{1}{n}\sum_{i=1}^n \Big(\chi(X_i) &+ \frac{Z_i}{E[Z]}(\bar{h}(\varepsilon_i) - \kappa) \\ &+ \frac{D}{E[Z\xi_*^2]}\, Z_i\big(X_i - E(X \mid Z = 1)\big)\,\ell(\varepsilon_i)\Big) + o_p(n^{-1/2}).\end{aligned}$$

Thus this estimator has influence function $\psi = \psi_I$ and is therefore efficient by Section 2. Note that in general the partially imputed estimator $\hat{H}_5$ is different from $\hat{H}_6$ and therefore inefficient. If $h(X,Y) = Y$, our estimator becomes $\hat{\vartheta}\overline{X} + \frac{1}{n}\sum_{i=1}^n Z_i(Y_i - \hat{\vartheta}X_i)/\overline{Z}$.

4. Linear regression without independence. In this model, Q satisfies the constraint $\int y\, Q(x,dy) = \vartheta x$. We estimate ϑ by a weighted least squares estimator based on (X_i, Y_i) with $Z_i = 1$,

$$\hat{\vartheta} = \frac{\sum_{i=1}^n Z_i \hat{\sigma}^{-2}(X_i) X_i Y_i}{\sum_{i=1}^n Z_i \hat{\sigma}^{-2}(X_i) X_i^2},$$

with $\hat{\sigma}^2(x)$ an estimator of $\sigma^2(x) = E(\varepsilon^2 \mid X = x)$. Such estimators have been studied without missing responses by Carroll [Car82], Müller and Stadtmüller [MS87], Robinson [Rob87], and Schick [Sch87]. In view of their results, we get under appropriate conditions that

$$\hat{\vartheta} = \vartheta + \frac{\frac{1}{n}\sum_{i=1}^n Z_i \sigma^{-2}(X_i) X_i \varepsilon_i}{E[Z\sigma^{-2}(X)X^2]} + o_p(n^{-1/2}).$$

This estimator can be shown to be efficient for ϑ.

A possible estimator for χ is the nonparametric estimator $\hat{\chi}$ introduced above for the nonparametric model. Here, however, we have the constraint $\int y\, Q(x,dy) = \vartheta x$ and use the estimator

$$\hat{\chi}_{II}(X_i) = \sum_{i=1}^n W_{ij} Z_j h(X_j, Y_j) - \hat{c}$$

with

$$\hat{c} = \frac{1}{n}\sum_{i=1}^n \frac{Z_i \hat{\rho}_h(X_i)}{\hat{\pi}(X_i)\hat{\sigma}^2(X_i)}(Y_i - \hat{\vartheta}X_i),$$

where $\hat{\pi}(x)$ and $\hat{\rho}_h(x)$ are nonparametric estimators of $\pi(x)$ and $\rho_h(x) = E(h(X,Y)\varepsilon \mid X = x)$. Note that $\hat{c}$ is of order $n^{-1/2}$. Hence $\hat{\chi}_{II}(x)$ is asymptotically equivalent to the nonparametric estimator $\hat{\chi}$. Nevertheless, it leads to a better estimator for $E[h(X,Y)]$. Under appropriate assumptions, $\hat{c}$ has the expansion

$$\hat{c} = \frac{1}{n}\sum_{i=1}^n \frac{Z_i \rho_h(X_i)}{\pi(X_i)\sigma^2(X_i)}\varepsilon_i - d(\hat{\vartheta} - \vartheta) + o_p(n^{-1/2})$$

with $d = E[Z\rho_h(X)X/\pi(X)\sigma^2(X)] = E[h(X,Y)\sigma^{-2}(X)X\varepsilon]$. Using the expansion for the weighted least squares estimator $\hat{\vartheta}$, we see that

$$\begin{aligned}\hat{c} &= \frac{1}{n}\sum_{i=1}^n \frac{Z_i\varepsilon_i}{\sigma^2(X_i)}\Big(\frac{\rho_h(X_i)}{\pi(X_i)} - \frac{dX_i}{E[Z\sigma^{-2}(X)X^2]}\Big) + o_p(n^{-1/2}) \\ &= \frac{1}{n}\sum_{i=1}^n \psi_{II}^*(X_i, Z_iY_i, Z_i) + o_p(n^{-1/2}).\end{aligned}$$

Using this and the stochastic expansion of the nonparametric estimator $\hat{\chi}$, we obtain that the estimators $\hat{H}_1 - \hat{c}$ and $\hat{H}_2 - \hat{c}$ have influence functions $\psi = \psi_{II}$ and are therefore efficient by Section 2. Of course, $\hat{H}_2 - \hat{c}$ is the fully imputed estimator based on $\hat{\chi}_{II}$. Both $\hat{H}_1 - \hat{c}$ and $\hat{H}_2 - \hat{c}$ are better than the partially imputed estimators $\hat{H}_1$ based on the estimator $\hat{\chi}$, and $\hat{H}_1 - (1 - \overline{Z})\hat{c}$ based on the estimator $\hat{\chi}_{II}$.

Simpler estimators are possible for certain functions h, such as $h(x, y) = y$, which is the function usually treated in the literature. Since $E(Y \mid X) = \vartheta X$, we can use the fully imputed estimator $\hat{\vartheta}\overline{X}$, with $\overline{X} = \frac{1}{n}\sum_{i=1}^{n} X_i$. As smooth function of the two efficient estimators $\hat{\vartheta}$ and $\overline{X}$, the estimator $\hat{\vartheta}\overline{X}$ is efficient for $E(Y \mid X)$. Matloff [Mat81] has recommended an estimator of this form, but with a simpler, in general inefficient, estimator for ϑ.

Achnowledgment

Anton Schick was supported in part by NSF Grant DMS 0072174.

References

[Bic82] Bickel, P.J.: On adaptive estimation. Ann. Statist. **10**, 647–671 (1982)

[BKRW98] Bickel, P.J., Klaassen, C.A.J., Ritov, Y., Wellner, J.A.: Efficient and Adaptive Estimation for Semiparametric Models. Springer, New York (1998)

[Car82] Carroll, R.J.: Adapting for heteroscedasticity in linear models. Ann. Statist. **10**, 1224–1233 (1982)

[Che94] Cheng, P.E.: Nonparametric estimation of mean functionals with data missing at random. J. Amer. Statist. Assoc. **89**, 81–87 (1994)

[CC96] Cheng, P.E., Chu, C.K.: Kernel estimation of distribution functions and quantiles with missing data. Statist. Sinica **6**, 63–78 (1996)

[HIR03] Hirano, K., Imbens, G.W., Ridder, G.: Efficient estimation of average treatment effects using the estimated propensity score. Econometrica **71**, 1161-1189 (2003).

[IH81] Ibragimov, I.A., Has'minski˘i , R.Z.: Statistical Estimation. Asymptotic Theory. Applications of Mathematics 16, Springer, New York (1981)

[KS83] Koul, H.L., Susarla, V.: Adaptive estimation in linear regression. Statist. Decisions **1**, 379–400 (1983)

[Mat81] Matloff, N.S.: Use of regression functions for improved estimation of means. Biometrika **68**, 685–689 (1981)

[MS87] Müller, H.-G., Stadtmüller, U.: Estimation of heteroscedasticity in regression analysis. Ann. Statist. **15**, 610–625 (1987)

[MSW04] Müller, U.U., Schick, A., Wefelmeyer, W.: Estimating functionals of the error distribution in parametric and nonparametric regression. J. Nonparametr. Statist. **16**, 525–548 (2004)

[NEW04] Nan, B., Emond, M., Wellner, J.A.: Information bounds for Cox regression models with missing data. Ann. Statist. **32**, 723–753 (2004)

[RbRt95] Robins, J.M., Rotnitzky, A.: Semiparametric efficiency in multivariate regression models with missing data. J. Amer. Statist. Assoc. **90**, 122–129(1995)

[RRZ94] Robins, J.M., Rotnitzky, A., Zhao, L.P.: Estimation of regression coefficients when some regressors are not always observed. J. Amer. Statist. Assoc. **89**, 846–866 (1994)

[Rob87] Robinson, P.M.: Asymptotically efficient estimation in the presence of heteroskedasticity of unknown form. Econometrica **55**, 875–891 (1987)

[RtRb95] Rotnitzky, A., Robins, J.M.: Semi-parametric estimation of models for means and covariances in the presence of missing data. Scand. J. Statist. **22**, 323–333 (1995)

[Sch87] Schick, A.: A note on the construction of asymptotically linear estimators. J. Statist. Plann. Inference **16**, 89–105 (1987)

[Sch93] Schick, A.: On efficient estimation in regression models. Ann. Statist. **21**, 1486–1521 (1993). Correction and addendum: **23**, 1862–1863 (1995)

[SR01] Schisterman, E., Rotnitzky, A.: Estimation of the mean of a K-sample U-statistic with missing outcomes and auxiliaries. Biometrika **88**, 713–725 (2001)

[Tam78] Tamhane, A.C.: Inference based on regression estimator in double sampling. Biometrika **65**, 419–427 (1978)

[WHL04] Wang, Q., Härdle, W., Linton, O.: Semiparametric regression analysis under imputation for missing response data. J. Amer. Statist. Assoc. **99**, 334–345 (2004)

[WR01] Wang, Q., Rao, J.N.K.: Empirical likelihood for linear regression models under imputation for missing responses. Canad. J. Statist. **29**, 597–608 (2001)

[WR02] Wang, Q., Rao, J.N.K.: Empirical likelihood-based inference under imputation for missing response data. Ann. Statist. **30**, 896–924 (2002)

[YN03] Yu, M., Nan, B.: Semiparametric regression models with missing data: the mathematics in the work of Robins et al. Technical Report, Department of Biostatistics, University of Michigan (2003). `http://www.sph.umich.edu/~bnan/`

Bivariate Decision Processes

Martin Newby

Centre for Risk Management, Reliability and Maintenance
City University
LONDON EC1V 0HB

Summary. Models are developed for decision making where a system's evolution is described by a general stochastic process. The general structure of the problem includes many statistical tests such as treatment comparisons, regression models and likelihood ratio tests. The process is monitored and decisions are made in response to the observed system state. The decision process is simplified by using an associated process as well as the underlying state as decision variables; in many situations a functional of the underlying process defines a statistic. The approach is motivated by the idea of a performance metric based on the system state. The bivariate approach allows a wide class of models to be considered and can incorporate long term memory within a simple probability structure. The decisions in this study are based on an average cost and a life-cycle cost. The approach can deal with decisions that entail restarting the process as new or continuing the process after an intervention which changes the system state. The resulting optimization problem solved either by a renewal-reward argument or by a dynamic programming formulation.

Key words: Wiener process; Lévy process; regenerative process; renewal-reward; dynamic programming; statistical testing; health monitoring

1 Introduction

We consider the problem of inspecting and controlling a system. The system can be physiological, medical, technological, or an experiment such a clinical trial. The system state is described by a stochastic process X_t. The arguments are developed with few restrictions on the process. The arguments allow systems with monotone or non-monotone trajectories to be analyzed. The trajectories can be status measurements from clinical trials where the decision may be to stop the trial, choose between competing treatments [Bet98], or to change the treatment regimes. For an individual whose health status is being monitored the decision may be to make an intervention or to adjust the treatment regime. Typically if a status indicator crosses a critical level an intervention will be required. The extent of the intervention determines the future development of the system. We address the problem with two tools, firstly, the standard method of seeking the regeneration points of the stochastic process, secondly, by considering an associated stochastic real valued process. We thus begin with the underlying process X_t and work with a bivariate process (X_t, Y_t).

The process Y_t is generally constructed by applying a functional to the basic process, $Y_t = A(X_t)$, examples are the construction of a statistic or making a measurement. Within the models Y_t can also describe a covariate process, an imperfectly observed version of X_t, or a associated process used as a surrogate. The advantages of the approach are that decisions can be based the entry of the pair (X_t, Y_t) into a critical region, or on the associated process Y_t alone. The model structure allows X_t to be unobservable. Depending on the context of the problem, either of X_t or Y_t can be integrated out of the final results to give any desired marginal distribution.

The aim of the article is to show how optimal policies for a system described by the stochastic process X_t. Initially we make no strong assumptions about the process, only that it has the properties required to allow the necessary computations. The properties of the process are made explicit in particular examples. Examples arise in many ways, from the construction of statistics, for example to assist treatment comparisons [Bet98]; thc consideration of marker processes in HIV infected patients [JK96]; in risk analysis for engineering projects [Noo96]; fatigue crack growth [Sob87, New91, New98]. The approach in this paper differs through the use of the bivariate process (X_t, Y_t) motivated by an extension of the methodology in an earlier paper [New04]. The process can be monotone where the state is the level of a drug in the blood, whereas a measurement such as heart rate or blood pressure will be non-monotone. A simple example of the use of the maximum process is provided by temperature or blood pressure monitoring. It is a compromise between continuous monitoring and monitoring only at certain times. The associated process is frequently defined by a functional of the underlying process, namely $Y_t = A(X_t)$. Decision making is simplified if the process Y_t has monotone sample paths; the map from X_t to Y_t may be in two stages, the first an aggregation followed by taking the maximum of the aggregated process to guarantee monotonicity. Natural examples of an associated process are:

(a) The maximum process, $Y_t = \sup_{0 \le s \le t} X_s$;
(b) in a multivariate process $Y_t = \|X_t\|$;
(c) an accumulation process $Y_t = \int_{0 \le s \le t} X_s \mathrm{d}s$;
(d) a usage measure $Y_t = \int_{0 \le s \le t} \|X_s\| \mathrm{d}s$;
(e) errors in measurement, a distribution $F_{Y_t|X_t}$ describes the dependence of the observed process on the true process;
(f) covariate processes, a distribution $F_{X_t\,|\,Y_t}$ describes the dependence of X_t on covariate Y_t.

When the underlying process X_t is a Wiener process (a) above is well known, (b) is a Bessel process, and (c) is the Kolmogorov diffusion [McK63] which arises in a regression model [GJW99].

2 The Structure of the Model

The bivariate process (X_t, Y_t) is defined on a product space $\Omega \times \mathbb{R}$, the state transitions are described by a transition density $f_t^{x,y}(u, w)$ where

$$f_t^{x,y}(u,w)\mathrm{d}u\mathrm{d}w = \mathrm{vec}P\left[X_t \in (u, u + \mathrm{d}u), Y_t \in (w, w + \mathrm{d}w)\,|X_0 = x, Y_0 = y\right] .$$

The development of this transition density is the key to adapting the general approach to particular cases.

The system stops when the basic process X_t enters a critical set, that is the system continues if $X_t \in \mathcal{G}$ and stops at time t, the time of first entry into $\mathcal{G}^c$; $t = \inf\{t \geq 0\,|\,X_t \in \mathcal{G}^c\}$. By using the bivariate model more possibilities are available. The times at which the system stops are defined by the excursions of the bivariate process$(X_t, Y_t) \in \Omega \times \mathbb{R}$. Decisions can be made by partitioning the state space. For example define the partitions $\Omega = A_0 \cup \{\cup_{i=1..m} A_j\} \cup A^\star$ and $\mathbb{R} = B_0 \cup \{\cup_{j=1..n} B_i\} \cup B^\star$ where A_0 and B_0 indicate perfect condition and $A^\star$ and $B^\star$ are critical sets. Inspection reveals the system state as $(X_t, Y_t) \in A_i \times B_j$ and this determines the action. With each action there is an associated cost. The cost can represent a true cost or another measure of the benefit or harm incurred as a result of the chosen action.

3 Inspection Policies

The decision maker inspects the system according to a policy Π. The policy is a list of inspection epochs $\Pi = \{\tau_1, \tau_2, \ldots, \tau_n\}$ and we assume for the moment that inspection is perfect. The decision chooses an action determined by the system state (X_t, Y_t) which may change the system state. The actions are assumed to be instantaneous. We consider two cases, a renewal approach where the system is returned to its original state at each intervention, and one where the intervention returns to system to an arbitrary state [SZ91, SZ92]. In the renewal approach if the revealed state on inspection, (X_t, Y_t), falls in an interval $A_i \times B_j$, the system is completely restored to the original state with cost $C_{i,j}$. In the context of this chapter the renewal approach corresponds to changing, starting, or stopping a treatment and restoring the system to its initial state; the restoration to an arbitrary state corresponds to bringing the system to some state between the current state and the initial state. The system found with state $(X_{t-}, Y_{t-}) = (x, y)$ is restored to the state $(X_{t+}, Y_{t+}) = (x', y')$ with $(x', y') = D(x, y)$. The function D describes the decision maker's action. Usually the new state lies somewhere between the original state and the present state so that $0 \leq |x'| \leq |x|$. In most cases $y' = 0$ because the decision variable will be reset; in the case when Y_t is the maximum $y' = x'$. Clearly $x' = 0$ corresponds to restoration to the initial state and $x' = x$ implies no change. The planned inspection period ends normally with the planned inspection or is terminated by the entry into a critical set.

The hitting time of the critical set, starting from $(X_t, Y_t) = (x, y)$, is $T^{x,y}$. Candidates for the hitting times are:

$$T^{x,y}_{\Omega\times\mathbb{R}} = \inf\{t \,|\, (X_t, Y_t) \in A^\star \times B^\star\};$$

$$T^{x,y}_{\mathbb{R}} = \inf\{t \,|\, Y_t \in B^\star\};$$

$$T^{x,y}_{\Omega} = \inf\{t \,|\, X_t \in A^\star\}.$$

where $A^\star$ and $B^\star$ are critical sets. We shall write the hitting distribution starting from (x, y) as $G^{x,y}(t)$ and assume it possesses a density $g^{x,y}(t)$.

The state probabilities are

$$p^{x,y}_{i,j} = \text{vec}P\left[(X_t, Y_t) \in A_i \times B_j\right] = \text{vec}E\left[\text{vec}1_{\{(X_t,Y_t)\in A_i\times B_j\}}\right] = \int_{A_i}\int_{B_j} f^{x,y}_t(u, w)\text{d}u\text{d}v$$

and the hitting time distribution

$$p^{x,y}_F = G^{x,y}(t)\ .$$

4 The Inspection Cycle

The inspection and actions are assumed to occur at the beginning of each interval. This choice allows linking of the chain of decisions required in the dynamic programming solutions later.

4.1 System Renewal

The policies are determined by the intervals between inspections. We consider first the the simplest case in which after inspection or entry into a critical set the system is restored to the original state. Consider a single cycle where the policy is the time to the next inspection, τ. The decision makers actions are:

1. do nothing if the system is in a "good" state, $(X_t, Y_t) \in A_0 \times B_0$;
2. the system state is $A_i \times B_j$, the probability of this state is $p_{i,j}$, restoration to the initial state costs $C_{i,j}$;

3. the system enters the critical state with probability $p_F^{x,y}$ and is restored with cost C_F.

The expected cost of the planned and unplanned actions is

$$c_\tau(x,y) = \sum_{i,j} C_{i,j} p_{i,j}^{x,y} + C_F p_F^{x,y} .$$

Considering the whole cycle, if the total cost starting in state (x,y) is $V_\tau^{x,y}$ then

$$\text{vec}E\left[V_\tau^{x,y}\right] = \text{vec}E\left[V_\tau^{x,y}\text{vec}1_{\{X_t,Y_t)\in A_0\times B_0\}}\right] + c_\tau(x,y)$$

where the term $\text{vec}E\left[V_\tau^{x,y}\text{vec}1_{\{X_t,Y_t)\in A_0\times B_0\}}\right]$ arises because the system state is left unchanged when the system is found in the "good" state $A_0 \times B_0$. Writing $v_\tau(x,y) = \text{vec}E\left[V_\tau^{x,y}\right]$ it is clear that

$$v_\tau(x,y) = c_\tau(x,y) + \int_{A_0}\int_{B_0} v_\tau(u,w) f_\tau^{x,y}(u,w)\text{d}u\text{d}w .$$

4.2 Arbitrary Restoration

The state space is subdivided more simply, there are now only two states, "non-critical" and "critical" and the decision maker acts to change the state to $D(x,y)$ on finding the system in state (x,y), $D(x,y) \mapsto (x',y')$. The cost of this repair is $c(D(x,y))$.

Using a similar argument to above, it is clear that

$$\begin{aligned} v_\tau(x,y) =& c(D(x,y)) + \{v_\tau(0,0) + C_F\} p_F^{D(x,y)} \dots \\ & \dots + \int_{A_0}\int_{B_0} v_\tau(u,w) f_\tau^{D(x,y)}(u,w)\text{d}u\text{d}w \end{aligned}$$

where $v_\tau(0,0)$ arises from the restoration to the initial state.

5 Optimal Policies

The interval costs and the expected length of an interval can be used to construct an optimal average cost solution for a fixed maintenance interval with policy $\Pi = \{k\tau | k = 1, 2, \dots, n\}$.

5.1 Average Cost Criterion

If we take a fixed policy with $\Pi = \{k\tau | k = 1, 2, \ldots, n\}$ the sequence of entries into the critical set defines an embedded renewal process and the average cost per cycle can be obtained using the renewal-reward theorem. For this we need the expected length of an interval. For perfect restoration the expected interval length satisfies

$$\begin{aligned}
\ell_\tau(x,y) &= \int_0^\tau [1 - G^{x,y}(s)]\, ds + \int_{A_0}\int_{B_0} \ell_\tau(u,w) f_\tau^{x,y}(u,w) \mathrm{d}u\mathrm{d}w \\
&= \int_0^\tau s g^{x,y}(s) ds + \int_{A_0}\int_{B_0} \ell_\tau(u,w) f_\tau^{x,y}(u,w) \mathrm{d}u\mathrm{d}w
\end{aligned}$$

and for partial restoration

$$\ell_\tau(x,y) = \int_0^\tau \left[1 - G^{D(x,y)}(s)\right] ds + \int_{A_0}\int_{B_0} \ell_\tau(u,w) f_\tau^{D(x,y)}(u,w) \mathrm{d}u\mathrm{d}w$$

$$\ell_\tau(x,y) = \int_0^\tau s g^{D(x,y)}(s) ds + \int_{A_0}\int_{B_0} \ell_\tau(u,w) f_\tau^{D(x,y)}(u,w) \mathrm{d}u\mathrm{d}w$$

where $1 - G^{D(x,y)}(s)$ is the interval survival function.

On applying the renewal-reward theorem [Bat00], the average cost per unit time is

$$\mathcal{C}(x,\tau) = \frac{v_\tau(x,y)}{\ell_\tau(x,y)} \ .$$

The optimum policy for a system starting in state $X_0 = x$ can then be determined as

$$\tau^\star = \underset{\tau}{\mathrm{argmin}} \left\{\mathcal{C}(x,\tau)\right\} \ .$$

5.2 Total Cost Criterion

If the inspection intervals are allowed to change with the evolution of the system state, a non-periodic policy $\Pi = \{\tau_1, \tau_2, \ldots, \tau_n\}$ can be defined. The construction of the interval cost functions with the inspection and action at the beginning, makes the step from one interval to the next straightforward. With this construction the function $v_\tau(x,y)$ is the value function for a dynamic programming problem [Bat00]. The optimality equation for the perfect repair version is

$$v_\tau(x,y) = \inf_{\tau>0}\left\{c_\tau(x,y) + \int_{A_0}\int_{B_0} v_\tau(u,w) f_\tau^{x,y}(u,w)\mathrm{d}u\mathrm{d}w\right\} .$$

For partial restoration the programming problem becomes

$$v_\tau(x,y) = \inf_{\tau>0}\left\{\begin{array}{l} c(D(x,y)) + \{v_\tau(0,0) + C_F\}\, p_F^{D(x,y)} \ldots \\ \ldots + \int_{A_0}\int_{B_0} v_\tau(u,w) f_\tau^{D(x,y)}(u,w)\mathrm{d}u \end{array}\right\} .$$

If costs are discounted with rate r the value function is modified and the dynamic programming problem becomes

$$v_\tau(x,y) = \inf_{\tau>0}\left\{\begin{array}{l} e^{-r\tau}c(D(x,y)) + \{v_\tau(0,0) + C_F\}\, p_{r,F}^{D(x,y)} \ldots \\ \ldots + \int_{A_0}\int_{B_0} e^{-ru}\, v_\tau(u,w) f_\tau^{D(x,y)}(u,w)\,\mathrm{d}u\mathrm{d}w \end{array}\right\}$$

where

$$p_{r,F}^{u,w} = \int_{B_0} e^{-rs} g^{u,w}(s)\mathrm{d}s .$$

5.3 Obtaining Solutions

The optimization problems above contain integral equations of the Volterra type so that discretization of the state space and application of quadrature rules produce equivalent matrix equations with the general form

$$\mathrm{vec}v = \mathrm{vec}c + \mathrm{vec}Mv$$

which are readily solved numerically as long as care is taken in dealing with singularities [PTVF92]. The dynamic programming problems translate in the same way and allow a policy improvement algorithm [Bat00] to be applied to develop the optimal policy. Convergence proofs for the algorithms are given by Dagg in his thesis [Dag00].

6 Lévy Processes as Degradation Models

Many degradation models are based on the concept of accumulated damage. Noortwijk [Noo96] points out that in systems subject to shocks, the order in which the damage (i.e. the shocks) occurs is often immaterial so that the random deterioration incurred in equal time intervals forms a set of exchangeable random variables [BS92]. This also implies that the distribution of the degradation incurred is independent of the time scale, i.e. the process has stationary increments. Exchangeable and stationary increments are similar to the stronger properties of stationary and independent increments of Lévy processes [Bre68].

The restriction to stationary increments is outweighed by the analytical advantages of using Lévy processes. Amongst Lévy processes are compound Poisson process, the Wiener process and the gamma process, shot noise process, for which many results are readily available. The Lévy-Khinchine decomposition [Bre68, Ch 9,14] expresses any Lévy process as the sum of a Wiener process and a jump process with the consequence that any degradation model based on Lévy process is either a Wiener process, a jump process or the sum of these two processes. The Wiener process is the only Lévy process with continuous sample paths. Thus by insisting that a system whose degradation is continuous is modelled by a Lévy process with continuous sample paths restricts the choice to the Wiener process. Similarly, insisting on monotonicity allows only the jump processes within the class of Lévy processes [RW94]. Lehmann [Leh04, Leh01] develops a bivariate approach based on the Lévy-Khinchine decomposition. Diffusions also arise naturally [Sob87, New91, New98] from the stochastic analogue of the simplest growth laws

$$x'(t) = \alpha x(t)^{\lambda+1} ,$$

namely,

$$\mathrm{d}X_t = \alpha X_t^{\lambda+1}\mathrm{d}t + \beta \mathrm{d}B_t .$$

7 Examples

The models depend on obtaining the joint transition density $f_t^{u,w}(x,y)$ of the process starting from $X_0 = u$ and $Y_0 = v$. The examples give some instances of the way in which they can be derived.

7.1 Maximum Process

An illustration of the bivariate process is obtained by taking a basic process X_t and constructing the bivariate process (X_t, Y_t) with the maximum process $Y_t = \sup\{X_s \,|\, 0 \le s \le t\}$. A non-monotonic process with continuous sample paths is defined by the Wiener process $X_t = \sigma B_t + \mu t$ with drift μ and variance parameter σ and where B_t is a standard Brownian motion. The state space is $\Omega = \mathbb{R}$ and is divided into intervals at points $-\infty < s_1 < \ldots, s_j, \ldots, s_n$ so that $Y_t \ge s_n$ or $X_t \ge s_n$ indicate entry into the critical set. The required densities and distributions can be deduced from results in [RW94]; the joint density of the process and its maximum with $X_0 = u$ is

$$f_\tau^u(x,y) = \frac{2(2y-x-u)}{\sqrt{2\pi\sigma^6\tau^3}} \exp\left\{-\frac{(x-u-\mu\tau)^2}{2\sigma^2\tau}\right\} \exp\left\{-\frac{2(y-u)(y-x)}{\sigma^2\tau}\right\} .$$

The marginal distribution of Y_τ is

$$F_\tau^u(y) = \Phi\left(\frac{y-u-\mu\tau}{\sigma\sqrt{\tau}}\right) - \exp\left\{\frac{2\mu(y-u)}{\sigma^2}\right\} \Phi\left(\frac{-y+u-\mu\tau}{\sigma\sqrt{\tau}}\right)$$

which gives an inverse Gaussian distribution [CF89] as the hitting time distribution.

7.2 The Integrated Process

When the underlying process X_t is a Wiener the two dimensional Kolmogorov diffusion (X_t, Y_t) arises on setting $Y_t = \int_{0\le s\le t} X_s \mathrm{d}s$. For a Brownian motion B_t the transition density of $\left(B_t, \int B_s ds\right)$ starting from (u,w) is [McK63]

$$f_t^{u,w}(x,y,t) = \frac{\sqrt{3}}{\pi t^2} \exp\left(-6\frac{(v-y-tu)^2}{t^3} + 6\frac{(x-v)(y-v-tu)}{t^2} - 2\frac{(u-y)^2}{t}\right) .$$

The linearity of the integral shows that Y_t is also Gaussian and its moments are easily obtained. When the basic process X_t has drift μ and volatility σ the derived moments are

$$\text{vec}E\left[Y_t\right] = \tfrac{1}{2}\mu t^2 \qquad \text{vec}V\left[Y_t\right] = \tfrac{1}{3}\sigma^2 t^3$$

allowing the joint density to be written

$$f_t^{0,0}(x,y) = \frac{\sqrt{3}}{\pi\sigma^2 t^2} \exp\left(-6\frac{(y-\frac{1}{2}\mu t^2)^2}{\sigma^2 t^3} + 6\frac{(x-\mu t)(y-\frac{1}{2}\mu t^2)}{\sigma^2 t^2} - 2\frac{(x-\mu t)^2}{\sigma^2 t}\right) .$$

7.3 The Absolute Value

For the one dimensional Wiener process, X_t, the distribution of the absolute value of the process, $Y_t = |X_t|$, is relatively easy and mimics the arguments for the maximum value. The distribution is clearly

$$\begin{aligned} F_{Y_t}(y) &= P\left[|X| \le y\right] = P\left[-y \le X \le y\right] \\ &= \Phi\left(\frac{y-\mu t}{\sigma\sqrt{t}}\right) - \Phi\left(\frac{-y-\mu t}{\sigma\sqrt{t}}\right) . \end{aligned}$$

The moments are

$$\begin{aligned} \text{vec}E\left[Y_t,\right] &= 2\sigma\sqrt{t}\varphi\left(\frac{\mu\sqrt{t}}{\sigma}\right) + \mu t\left\{2\Phi\left(\frac{\mu\sqrt{t}}{\sigma}\right) - 1\right\} , \\ \text{vec}E\left[Y_t^2\right] &= \mu^2 t^2 + \sigma^2 t . \end{aligned}$$

7.4 Bessel Processes

Betensky [Bet98] uses Bessel processes in the comparison of treatment regimes. The Bessel process can be used directly, but in many cases the squared Bessel process allows simpler decision making and produces equivalent decisions. The squared Bessel processes are handled effectively through the well known properties of chi-squared and Wishart distributions.

The simplest approach is illustrated by the norm functional $Y_t = \|X_t\|$ which generates a Bessel process from a multivariate Wiener process. The basic process is based on an N-dimensional Brownian motion,

$$\mathrm{vec}B_t = [B_1(t), B_1(t), \ldots, B_\delta(t)] \, .$$

The process $Y_t = ||\mathrm{vec}B_t||$ or more explicitly

$$Y_t = \left[\sum_{i=1}^{n} B_i^2(t)\right]^{\frac{1}{2}}$$

is a Bessel process. For simplicity, the squared Bessel process

$$Z_t = Y_t^2 = \|\mathrm{vec}B_t\|^2 = \sum_{i=1}^{n} B_i^2(t)$$

is also used. The excursions of the two processes produce equivalent decision rules, and the second is simpler to handle. The Bessel process is a diffusion [Oks00] and Z_t, a squared Bessel process denoted $BESQ^{\delta}_{x_0}$, [RY99], is a solution of

$$Z_t = z_0 + \delta t + 2\int_0^t \sqrt{|Z_s|}\mathrm{d}B_s$$

with $\delta \geq 0$ and $z_0 \geq 0$. If $\delta \geq 2$, the process never reaches 0 for $t > 0$.

The squared Bessel process, is the square of the Euclidean norm of a δ-dimensional Brownian motion. Because the distributions of the $B_i(t)$ are normal, the probability density function is a chi-squared distribution

if $z_0 = 0$ and $\delta > 0$ the density is the chi-square

$$f_t^{\delta}(0, z) = \frac{z^{\frac{\delta}{2}-1}}{(2t)^{\frac{\delta}{2}}\Gamma(\frac{\delta}{2})} e^{-\frac{z}{2t}} \mathrm{vec}1_{\{z>0\}}$$

with δ degrees of freedom.

if $z_0 \neq 0$ and $\delta > 0$, the density is a non-central chisquare written in terms of modified Bessel functions of the first kind [AS72, result 9.6.7]

$$f_t^{\delta}(z_0, z) = \frac{1}{2t}\left(\frac{z}{z_0}\right)^{\frac{\nu}{2}} e^{-\frac{z0+z}{2t}} I_\nu\left(\frac{\sqrt{z_0 z}}{t}\right) \mathrm{vec}1_{\{z>0\}}$$

The model is more realistic if the Brownian motion is replaced by a Wiener process with drift

$$\mathrm{vec}W_t = [W_1(t), W_2(t), \ldots, W_\delta(t)]\,, \quad W_i(t) = \mu_i t + \sigma_i B_t$$

The distribution of $\mathrm{vec}W_t$ is $\mathrm{N}(\mathrm{vec}M_t, \Sigma_t)$ where $\mathrm{vec}M_t$ is the vector of means and Σ_t is the covariance matrix; $S_t = (\mathrm{vec}W_t - \mathrm{vec}M_t)^{\mathrm{T}}(\mathrm{vec}W_t - \mathrm{vec}M_t)$ thus has a Wishart distribution with parameters $\delta/2$ and $(1/2)\Sigma_t$. The distribution of the "non-standard" squared Bessel process

$$Y_t^\delta = \sum_{i=1}^{\delta} W_i^2(t)\,.$$

can thus be obtained.

7.5 Models for Imperfect Inspection

Imperfect inspection provides another example of associated processes. In this case Y_t is simply the observed level of degradation subject to error. The simplest model of imperfect inspection is when the system degradation X_t is observable, but with error. here a simple independent additive error is assumed

$$Y_t = X_t + \varepsilon_t$$

where ε_t represents the error.

The increments and the error are distributed

$$X_{t_j} - X_{t_i} \sim G(x_j - x_i)\,,$$

and

$$\varepsilon_t \sim H(\varepsilon) \qquad \forall\, t\,.$$

The distributions G and H have densities g and h, and assume that h is symmetrical about zero.

Since $Y_t - X_t = \varepsilon_t$ it is clear that

$$f_{Y_t|X_t}(y|x) = h(y-x)$$

and by symmetry

$$f_{X_t|Y_t}(x|y) = h(y-x)\,.$$

From the definition of the process

$$f_{X_t|X_0}(x|u) = G(x-u)\,.$$

These results combine to give the joint distribution of the future observed and true values of the degradation

$$f_{Y_t,X_t|X_0}(y,x|u) = f_{Y_t|X_0,X_t}(y|u,x) f_{X_t|X_0}(x|u) = f_{Y_t|X_t}(y|x) f_{X_t|X_0}(x|u)$$
$$= h(y-x)g(x-u) .$$

To be specific assume here that the $\varepsilon_t \sim N(0,\sigma^2)$ and that underlying process is the Gamma process with increments [Abd75]

$$X_{t_j} - X_{t_i} \sim Ga(\,\alpha(t_j - t_i), \beta\,) .$$

Plugging in the densities yields the joint density of the observed and true level of degradation, conditional on the true initial level of degradation,

$$f^u_\tau(y,x) = \frac{1}{\sigma\sqrt{2\pi}} \exp\left\{ -\frac{1}{2\sigma^2}(y-x)^2 \right\} \frac{\beta^{\alpha\,\tau}(x-u)^{\alpha\,\tau - 1} \exp\{\, -\beta(x-u)\}}{\Gamma(\alpha\,\tau)} .$$

It follows from the monotonicity of the gamma process that the distribution of the hitting time of the critical set from an initial degradation u is

$$p^x_F = P(T^u_c < h) = \frac{\Gamma(\alpha\,h, \beta(c-u))}{\Gamma(\alpha\,h)} .$$

8 Summary

The models proposed have developed a unified structure for decision making where the degradation is described by a process X_t has an associated process Y_t that is given or is constructed. The models require a bivariate transition density and some simple examples have been given. The formulation of the models depends on identifying the instants of perfect repair or replacement where the probability laws are reset to time zero. The construction of the intervals also allows the sequential version of the models to be formulated as a dynamic programming problem.

References

[Abd75] Abdel-Hameed, M. A.: A Gamma Wear Process, IEEE Trans. Reliab., **R-24**, 152–154 (1975)

[AS72] Abramowitz, M. and Stegun, I. Handbook of Mathematical Functions, Dover Publications (1972)

[Bat00] Bather, J.: Decision Theory: an Intoduction to Dynamic Programming and Sequential Decisions, Wiley (2000)

[BS92] Bernardo, J.M. and Smith, A.F.M.: Bayesian Theory, John Wiley and Sons, New York (1992)

[Bet98] Betensky, R. A.: A Boundary Crossing Probability for the Bessel Process, Advances in Applied Probability, **30**, 807–830 (1998).

[Bre68] Breiman, L. (1968) Probability, Addison-Wesley, Reading, Mass., 1968

[CF89] Chhikara, R.S. and Folks, J. L.: The Inverse Gaussian Distribution. Theory, Methodology and Applications, Marcel Dekker, New York (1989)

[Dag00] Dagg, R. Optimal Inspection and Maintenance for Stochastically Deteriorating Systems,PhD thesis, City University, London, (2000).

[GJW99] Groeneboom, P. Jongbloed, G, Wellner, J. A.: Integrated Brownian Motion, Conditioned To Be Positive, The Annals of Probability, **27**, No. 3, 1283–1303 (1999)

[JK96] Jewell, N. P. and Kalbfleisch, J. D.: Marker Processes in Survival Analysis, Lifetime Data Analysis, **2**, 15–29 (1996)

[Leh04] Lehmann, A.: Degradation Based Reliability Models with Imperfect Repair, International Workshop: "Longevity, Aging and Degradation Models in Reliability, Medicine and Biology", St. Petersburg, Russia, June 7–9, (2004)

[Leh01] Lehmann, A.: A Wiener process based model for failure and degradation data in dynamic environments. Dresdner Schriften zur Mathemat. Stochastik **4/2001** , 35–40 (2001)

[McK63] McKean, H.P. Jr.: A winding problem for a resonator driven by a white noise. J. Math. Kyoto Univ. **2**, 227–235. (1963)

[New91] Newby, M.J.: Estimating of Paris-Erdogan law parameters and the influence of environmental factors on crack growth, Int. J. Fatigue, **13**, 187–198 (1991)

[New98] Newby, M.J.: Analysing Crack Growth, Proc. ImechE. Part G – Aero. Eng., **212**, 157–166 (1998)

[New04] Newby, M J and Dagg, R, Optimal Inspection and Perfect Repair, IMA Journal of Management Mathematics, **15**, 17 192, 2004

[PTVF92] Press, W. H., Teukolsky, S. A., Vetterling, W. T. and Flannery, B. P.: Numerical Recipes in C, 2nd Edition, Cambridge University Press, Cambridge, UK (1992)

[Oks00] Øksendal, B.: Stochastic Differential Equations, Springer-Verlag, 2000

[RY99] Revuz, D., and Yor, M.: Continuous Martingales and Brownian Motion, Springer-Verlag; 3rd edition 1999

[RW94] Rogers, L. C. G. and Williams, D.: Diffusions, Markov Processes and Martingales, Volume 1: Foundations (2nd Edition), John Wiley and Sons,Chichester (1994)

[Sob87] Sobczyk, K.: Stochastic Models for Fatigue Damage of Materials, Adv. Appl. Prob., **19**, 652-673 (1987)

[SZ91] Stadje, W. and Zuckerman, D.: Optimal Maintenance Strategies for Repairable Systems with General Degree of Repair, J. Appl. Prob., **28**, 384–396 (1991)

[SZ92] Stadje, W. and Zuckerman, D.: Optimal repair policies with general degree of repair in two maintenance models, Operations Research Letters, **11**, 77–80 (1992)

[Noo96] Van Noortwijk, J M, Optimal Maintenance Decisions for Hydraulic Structures Under isotropic Deteriroation, PhD Thesis, Technical University Delft, Netherlands(1996)

[WCL98] Whitmore, G. A., Crowder, M. J. and Lawless, J. F.: Failure Inference from a Marker Process Based on a Bivariate Wiener Model, Lifetime Data Analysis, **4**, 229–251 (1998)

Weighted Logrank Tests With Multiple Events

C. Pinçon[1] and O. Pons[2]

[1] EA 3614 - Laboratoire de Biomathématiques
3, rue du Professeur Laguesse - 59006 Lille cédex - France.
cpincon@pharma.univ-lille2.fr
[2] Département MIA - INRA
Domaine de Vilvert - 78352 Jouy-en-Josas cédex - France.
Odile.Pons@jouy.inra.fr

Summary. We focus on the two group comparison when subjects of the sample may experience multiple distinct events, possibly censored. Because of the correlation arising between failure times, the sum of the marginal test statistics is not accurate. We propose a multivariate version of weighted logrank tests derived from the marginal logrank statistics, and we study their asymptotic distribution under null hypothesis; we construct a consistent estimator of their covariance using martingales properties. We present a simulation study and an application of this method to a study aimed to prove the association between retinopathy and diabetes.

Key words: Censoring; Correlated failure times; Martingales; Weighted logrank test statistics

1 Introduction and notations

It is often of high interest to study simultaneously the times to several events; for example, in breast cancer, one can be interested in studying the time until a tumor appears for both breasts, and in this case, the times to failure for a female patient are correlated. Suppose that the sample of size n is composed of two groups, A and B, with respective sizes n_A and n_B, and that each subject may experience $K > 1$ events, possibly censored. For an easier understanding, let $K = 2$, but generalization to $K > 2$ is immediate. For $k = 1, 2$ and $j = A, B$, let Λ_k^j be the cumulative hazard function for event k in group j. The null hypothesis $\{\Lambda_k^A = \Lambda_k^B\}$ can be tested with rank tests when there is no censoring, or with weighted logrank statistics in the censored case ([GEH65]; [MAN66]; [COX72]; [PETO72]; [PRE78]; [HF82]). But these test statistics can not be used if we are interested in testing H_0 : $\{\Lambda_1^A = \Lambda_1^B,\, \Lambda_2^A = \Lambda_2^B\}$ because of the dependence between the failure times. In 1984 Wei, L. J. and Lachin, J. M. [WL84] proposed a test statistic of H_0 based on the marginal weighted logrank statistics, say LR_k, for $k = 1, 2$; they

proved that the vector $(LR_1, LR_2)'$ converges in distribution under H_0 to a normal law, with null expectation and with a variance covariance matrix for which they constructed a consistent estimator. But their simulations study showed a rather high first type error rate. In 1987, Pocock [PGT87] developed a family of tests for treatment comparison with multiple endpoints, including failure times, in this latter case using Wei and Lachin covariance estimator found in [WL84]. Then Wei, L. J., Lin, D. Y. and Weissfeld, L. [WLW89] extended in 1989 this marginal approach to construct proportional hazards model for multiple events, their method allowing multiple hypotheses testing on regression parameter estimates with Wald test statistics.
We aimed to propose an improvement of weighted logrank test statistics for multiple events, by constructing a consistent estimator of the variance covariance matrix of $(LR_1, LR_2)'$ using martingales properties. We study the asymptotic distribution of the vector $(LR_1, LR_2)'$ under H_0; then we produce the results of a simulations study performed to compare the observed first type error rate of our test statistic to those of [WL84] and [WLW89]. The last section reports the application of our method to data from a clinical trial studying efficacy of laser treatment on retinopathy depending on type of diabetes (juvenile or adult).

In group j $(j = A, B)$, let (T_{1ji}, T_{2ji}) be the times to failure for subject i, independent and identically distributed for $i = 1, ..., n_j$ with joint survival function $\overline{F}^j$ where $\overline{F}^j(x_1, x_2) = \Pr\{T_{1ji} \geq x_1, T_{2ji} \geq x_2\}$, and with density f^j. For k and j fixed, we also suppose the random variables T_{kji} independent and identically distributed, with marginal survival function $\overline{F}_k^j$ and with cumulative hazard function Λ_k^j.
Let (C_{1ji}, C_{2ji}) be the censoring times, independent of the failure times (T_{1ji}, T_{2ji}), independent and identically distributed with joint survival function $\overline{G}$ and with marginal survival function $\overline{G}_k$ equal in both groups A and B.
Then, for subject i, we only observe $(X_{1ji}, X_{2ji}, \delta_{1ji}, \delta_{2ji})$ where for $k = 1, 2$ $X_{kji} = (T_{kji} \wedge C_{kji})$ and $\delta_{kji} = I\{T_{kji} < C_{kji}\}$.

Let

$$Y_{kji}(x) = I\{X_{kji} \geq x\},$$
$$N_{kji}(x) = I\{T_{kji} < C_{kji}, T_{kji} \leq x\},$$

and

$$\overline{Y}_{kj}(x) = \sum_{i=1}^{n_j} Y_{kji}(x),$$
$$\overline{Y}_k(x) = \overline{Y}_{kA}(x) + \overline{Y}_{kB}(x),$$
$$\overline{N}_{kj}(x) = \sum_{i=1}^{n_j} N_{kji}(x),$$
$$\overline{N}_k(x) = \overline{N}_{kA}(x) + \overline{N}_{kB}(x).$$

The maximum follow-up time is $\tau < \infty$, such that $\overline{Y}_{kA}(\tau) > 0$, $\overline{Y}_{kB}(\tau) > 0$.

The means $n_A^{-1}\overline{Y}_{kA}$, $n_B^{-1}\overline{Y}_{kB}$ and $n^{-1}\overline{Y}_k$ converge uniformly to $\overline{y}_{kA} = \overline{F}_k^A\overline{G}_k$, $\overline{y}_{kB} = \overline{F}_k^B\overline{G}_k$ and $\overline{y}_k = \rho_A\overline{y}_{kA} + \rho_B\overline{y}_{kB}$ respectively, where ρ_A and ρ_B are the limits of n_A/n and n_B/n.

For $k = 1, 2$, consider the weighted logrank test statistic for event k

$$LR_k = n^{-1/2} \int_0^\tau W_k(u) \frac{\overline{Y}_{kA}\overline{Y}_{kB}}{\overline{Y}_k}(u) \left\{ \frac{d\overline{N}_{kA}}{\overline{Y}_{kA}} - \frac{d\overline{N}_{kB}}{\overline{Y}_{kB}} \right\}(u), \qquad (1)$$

where W_k is a weight function that converges uniformly to a function w_k on $[0, \tau]$ (for example W_k constant and equals to 1 represents the logrank test, and $W_k(x) = \overline{Y}_k(x)$ the Gehan test statistic).

2 Asymptotic distribution of $(LR_1, LR_2)'$ under H_0 in a copula model

In this section, we assume that the joint survival function of the variables $\{(T_{1ji}, T_{2ji}), i = 1, ..., n_j\}$ is formulated by a copula model, that is $\overline{F}^j(x_1, x_2) = C_\alpha(\overline{F}_1^j(x_1), \overline{F}_2^j(x_2))$ for any $(x_1, x_2) \in [0, \tau]^2$.
The null hypothesis H_0 we wish to test is equality in both groups of the marginal survival distributions, that is $H_0 : \{\Lambda_1^A = \Lambda_1^B = \Lambda_1,\ \Lambda_2^A = \Lambda_2^B = \Lambda_2$ on $[0, \tau]\}$.
Then, under H_0, the $\{T_{kji}, i = 1, ..., n_j\}$ have marginal survival function $\overline{F}_k$ and cumulative hazard function Λ_k for $k = 1, 2$, and the couples $\{(T_{1ji}, T_{2ji}),\ i = 1, ..., n_j\}$ have joint survival function $\overline{F}$ and density f.

We introduce the following martingales, for $k = 1, 2$, $j = A, B$ and $i = 1, ..., n_j$:

$$M_{kji}(x) = N_{kji}(x) - \int_0^x Y_{kji}(u)d\Lambda_k(u),$$

$$\overline{M}_{kj}(x) = \sum_{i=1}^{n_j} M_{kji}(x),$$

$$\overline{M}_k(x) = \overline{M}_{kA}(x) + \overline{M}_{kB}(x).$$

Under H_0, (1) is equal to

$$LR_k = n^{-1/2} \int_0^\tau W_k(u) \left\{ \frac{\overline{Y}_{kB}}{\overline{Y}_k} d\overline{M}_{kA} - \frac{\overline{Y}_{kA}}{\overline{Y}_k} d\overline{M}_{kB} \right\}(u), \tag{2}$$

and the asymptotic distribution of the vector $(LR_1, LR_2)'$ will be derived from the asymptotic properties of the martingales vector $(n_A^{-1/2}\overline{M}_{1A}, n_B^{-1/2}\overline{M}_{1B}, n_A^{-1/2}\overline{M}_{2A}, n_B^{-1/2}\overline{M}_{2B})'$.

2.1 Preliminary results for the martingales under H_0

Rebolledo's theorem ensures the convergence of the vector $(n_A^{-1/2}\overline{M}_{1A}, n_B^{-1/2}\overline{M}_{1B}, n_A^{-1/2}\overline{M}_{2A}, n_B^{-1/2}\overline{M}_{2B})'$ to a Gaussian process $(\overline{m}_{1A}, \overline{m}_{1B}, \overline{m}_{2A}, \overline{m}_{2B})'$, with null expectation and with variances $v_{kj}(x)$ where

$$\begin{aligned} v_{kj}(x) &= \lim_{n_j \to \infty} E\left\{ n_j^{-1} \overline{M}_{kj}^2(x) \right\} \\ &= \lim_{n_j \to \infty} E\left\{ n_j^{-1} \int_0^x \overline{Y}_{kj}(u) d\Lambda_k(u) \right\} \\ &= \int_0^x \overline{y}_{kj}(u) d\Lambda_k(u). \end{aligned}$$

The covariance between $\overline{m}_{kA}$ and $\overline{m}_{kB}$ is null because subjects in group A are independent of those in group B. But we have to express the covariance between $\overline{m}_{1j}$ and $\overline{m}_{2j}$ for $j = A, B$. Let

$$v_{12j}(x_1, x_2) = E\left\{ \overline{m}_{1j}(x_1)\overline{m}_{2j}(x_2) \right\}.$$

By definition,

$$\begin{aligned} v_{12j}(x_1, x_2) &= \lim_{n_j \to \infty} E\left\{ n_j^{-1} \overline{M}_{1j}(x_1)\overline{M}_{2j}(x_2) \right\} \\ &= \int_0^{x_1} \int_0^{x_2} \lim_{n_j \to \infty} E\left\{ n_j^{-1} \sum_{i=1}^{n_j} dM_{1ji}(u) dM_{2ji}(v) \right\} \\ &= \int_0^{x_1} \int_0^{x_2} E\left\{ dM_{1ji}(u) dM_{2ji}(v) \right\}. \end{aligned}$$

A development of $dM_{1ji}(u)dM_{2ji}(v)$ shows that $E\{dM_{1ji}(u)dM_{2ji}(v)\}$ is equal to

$$E\{dN_{1ji}(u)dN_{2ji}(v)\} - d\Lambda_1(u)E\{Y_{1ji}(u)dN_{2ji}(v)\}$$
$$-d\Lambda_2(v)E\{Y_{2ji}(v)dN_{1ji}(u)\} + d\Lambda_1(u)d\Lambda_2(v)E\{Y_{1ji}(u)Y_{2ji}(v)\}.$$

The first term is

$$\Pr\{T_{1ji} \in [u, u+du], T_{2ji} \in [v, v+dv], C_{1ji} > T_{1ji}, C_{2ji} > T_{2ji}\},$$

that is equal to $f(u,v)\overline{G}(u,v)dudv$.

The second term is

$$-d\Lambda_1(u)\Pr\{T_{1ji} \geq u, C_{1ji} \geq u, T_{2ji} \in [v, v+dv], C_{2ji} > T_{2ji}\}$$

that equals $\overline{G}(u,v)\overline{F}(u,dv)d\Lambda_1(u)$, and symmetrically the third term equals $\overline{G}(u,v)\overline{F}(du,v)d\Lambda_2(v)$.

The last term is simply $\overline{G}(u,v)d\Lambda_1(u)d\Lambda_2(v)\overline{F}(u,v)$.

Finally,

$$E\{dM_{1ji}(u)dM_{2ji}(v)\} = \overline{G}(u,v) \times \left\{ \begin{array}{l} f(u,v)dudv + d\Lambda_1(u)\overline{F}(u,dv) \\ +d\Lambda_2(v)\overline{F}(du,v) + d\Lambda_1(u)d\Lambda_2(v)\overline{F}(u,v) \end{array} \right\}. \quad (3)$$

Let $\overline{\pi}(u,v) = \overline{F}(u,v)\overline{G}(u,v) = E\{Y_{1ji}(u)Y_{2ji}(v)\}$. Then (3) can be written as

$$E\{Y_{1ji}(u)Y_{2ji}(v)\} \times \left\{ \begin{array}{l} \frac{f(u,v)\overline{G}(u,v)dudv}{\overline{\pi}(u,v)} + d\Lambda_1(u)\frac{\overline{F}(u,dv)\overline{G}(u,v)}{\overline{\pi}(u,v)} \\ +d\Lambda_2(v)\frac{\overline{F}(du,v)\overline{G}(u,v)}{\overline{\pi}(u,v)} + d\Lambda_1(u)d\Lambda_2(v) \end{array} \right\},$$

that allows to establish a consistent estimator of each fraction in the above expression. The first numerator can be consistently estimated by

$$n^{-1}\sum_{j=A}^{B}\sum_{i=1}^{n_j} dN_{1ji}(u)dN_{2ji}(v);$$

an estimator of the second one is

$$-n^{-1}\sum_{j=A}^{B}\sum_{i=1}^{n_j} Y_{1ji}(u)dN_{2ji}(v),$$

and symmetrically $\overline{F}(du,v)\overline{G}(u,v)$ has for consistent estimator

$$-n^{-1}\sum_{j=A}^{B}\sum_{i=1}^{n_j} Y_{2ji}(v)dN_{1ji}(u);$$

last, the denominator can be estimated by

$$n^{-1}\sum_{j=A}^{B}\sum_{i=1}^{n_j} Y_{1ji}(u)Y_{2ji}(v) = n^{-1}r(u,v).$$

Finally, noticing that $dM_{1ji}(u)dM_{2ji}(v)$ is null if subject i is not at risk in u for event 1 and in v for event 2, a consistent estimator of (3) is

$$\begin{aligned}&\widehat{E}\left\{dM_{1ji}(u)dM_{2ji}(v)\right\}\\ &= \frac{Y_{1ji}(u)Y_{2ji}(v)}{r(u,v)}\sum_{j=A}^{B}\sum_{l=1}^{n_j}\left\{\begin{array}{l} dN_{1jl}(u)dN_{2jl}(v) - d\widehat{\Lambda}_1(u)Y_{1jl}(u)dN_{2jl}(v) \\ -d\widehat{\Lambda}_2(v)Y_{2jl}(v)dN_{1jl}(u) + d\widehat{\Lambda}_1(u)d\widehat{\Lambda}_2(v)r(u,v)\end{array}\right\},\end{aligned}$$

with $\widehat{\Lambda}_k$ the Nelson-Aalen estimate ([NEL69], [AAL78]) of the common cumulative hazard function for event k, $k = 1, 2$:

$$\widehat{\Lambda}_k(x) = \int_0^x \frac{d\overline{N}_k(u)}{\overline{Y}_k(u)}.$$

Introducing the martingale residuals

$$\widehat{M}_{kjl}(x) = N_{kjl}(x) - \int_0^x Y_{kjl}(u)d\widehat{\Lambda}_k(u), \tag{4}$$

the above estimator is simply

$$\widehat{E}\left\{dM_{1ji}(u)dM_{2ji}(v)\right\} = \frac{Y_{1ji}(u)Y_{2ji}(v)}{r(u,v)}\sum_{j=A}^{B}\sum_{l=1}^{n_j} d\widehat{M}_{1jl}(u)d\widehat{M}_{2jl}(v).$$

Consequently, an appealing estimator of $v_{12j}(du, dv)$ is

$$\begin{aligned}\widehat{v}_{12j}(du,dv) &= n_j^{-1}\sum_{i=1}^{n_j}\widehat{E}\left\{dM_{1ji}(u)dM_{2ji}(v)\right\} \\ &= n_j^{-1}\frac{r_j(u,v)}{r(u,v)}\sum_{j=A}^{B}\sum_{l=1}^{n_j} d\widehat{M}_{1jl}(u)d\widehat{M}_{2jl}(v),\end{aligned} \tag{5}$$

where

$$r_j(u,v) = \sum_{i=1}^{n_j} Y_{1ji}(u)Y_{2ji}(v).$$

2.2 Asymptotic distribution of $(LR_1, LR_2)'$ under H_0

The above results will help us to establish the asymptotic distribution of the vector $(LR_1, LR_2)'$ under H_0, and more particularly to express the covariance between its two components. But first consider

$$LR_k^* = n^{-1/2} \int_0^{\tau} w_k \left\{ \rho_B \frac{\overline{y}_{kB}}{\overline{y}_k} d\overline{M}_{kA} - \rho_A \frac{\overline{y}_{kA}}{\overline{y}_k} d\overline{M}_{kB} \right\};$$

under H_0, LR_k^* is an asymptotic equivalent to (2) because of the convergence in distribution of the martingales to Gaussian processes, and because of uniform convergence of $W_k \overline{Y}_k^{-1} \overline{Y}_{kj} - w_k \rho_j \overline{y}_k^{-1} \overline{y}_{kj}$ to 0 for $k = 1, 2$ and $j = A, B$. Then the asymptotic distribution of $(LR_1, LR_2)'$ will be the asymptotic distribution of $(LR_1^*, LR_2^*)'$, which is easier to derive because the only random terms in LR_1^* and LR_2^* are the martingales.

LR_k^* can be written

$$\int_0^{\tau} w_k \left\{ \rho_B \left(\frac{n_A}{n}\right)^{1/2} \frac{\overline{y}_{kB}}{\overline{y}_k} n_A^{-1/2} d\overline{M}_{kA} - \rho_A \left(\frac{n_B}{n}\right)^{1/2} \frac{\overline{y}_{kA}}{\overline{y}_k} n_B^{-1/2} d\overline{M}_{kB} \right\}.$$

Since $(n_A^{-1/2}\overline{M}_{1A}, n_B^{-1/2}\overline{M}_{1B}, n_A^{-1/2}\overline{M}_{2A}, n_B^{-1/2}\overline{M}_{2B})'$ converges in distribution to the Gaussian process $(\overline{m}_{1A}, \overline{m}_{1B}, \overline{m}_{2A}, \overline{m}_{2B})'$ previously defined, and since n_A/n and n_B/n have limits ρ_A and ρ_B, we can conclude that $(LR_1^*, LR_2^*)'$ converges in distribution to the Gaussian vector

$$\begin{pmatrix} \int_0^{\tau} w_1 \left\{ \rho_B \rho_A^{1/2} \frac{\overline{y}_{1B}}{\overline{y}_1} d\overline{m}_{1A} - \rho_A \rho_B^{1/2} \frac{\overline{y}_{1A}}{\overline{y}_1} d\overline{m}_{1B} \right\} \\ \int_0^{\tau} w_2 \left\{ \rho_B \rho_A^{1/2} \frac{\overline{y}_{2B}}{\overline{y}_2} d\overline{m}_{2A} - \rho_A \rho_B^{1/2} \frac{\overline{y}_{2A}}{\overline{y}_2} d\overline{m}_{2B} \right\} \end{pmatrix}.$$

This vector has null expectation because $(\overline{m}_{kA}, \overline{m}_{kB})'$ is centered for $k = 1, 2$.

Let $\Sigma^* = (\sigma_{kk'}^*)$, $k = 1, 2$, $k' = 1, 2$, be the variance covariance matrix of $(LR_1^*, LR_2^*)'$. The variance of LR_k^* is:

$$\begin{aligned} \sigma_{kk}^* &= \int_0^{\tau} w_k^2 \left\{ \rho_B^2 \rho_A \frac{\overline{y}_{kB}^2}{\overline{y}_k^2} dv_{kA} + \rho_A^2 \rho_B \frac{\overline{y}_{kA}^2}{\overline{y}_k^2} dv_{kB} \right\} \\ &= \int_0^{\tau} w_k^2 \left\{ \rho_B^2 \rho_A \frac{\overline{y}_{kB}^2}{\overline{y}_k^2} \overline{y}_{kA} d\Lambda_k + \rho_A^2 \rho_B \frac{\overline{y}_{kA}^2}{\overline{y}_k^2} \overline{y}_{kB} d\Lambda_k \right\} \\ &= \int_0^{\tau} w_k^2 \rho_A \rho_B \frac{\overline{y}_{kA} \overline{y}_{kB}}{\overline{y}_k^2} \left\{ \rho_B \overline{y}_{kB} + \rho_A \overline{y}_{kA} \right\} d\Lambda_k \\ &= \int_0^{\tau} w_k^2 \rho_A \rho_B \frac{\overline{y}_{kA} \overline{y}_{kB}}{\overline{y}_k} d\Lambda_k. \end{aligned}$$

The covariance between LR_1^* and LR_2^* is:

$$\sigma_{12}^* = \int_0^\tau \int_0^\tau w_1(u)w_2(v)\rho_B^2\rho_A \frac{\overline{y}_{1B}}{\overline{y}_1}(u)\frac{\overline{y}_{2B}}{\overline{y}_2}(v)v_{12A}(du,dv) \qquad (6)$$
$$+ \int_0^\tau \int_0^\tau w_1(u)w_2(v)\rho_A^2\rho_B \frac{\overline{y}_{1A}}{\overline{y}_1}(u)\frac{\overline{y}_{2A}}{\overline{y}_2}(v)v_{12B}(du,dv).$$

So the asymptotic distribution of the vector $(LR_1^*, LR_2^*)'$ is a Gaussian law, with null expectation and with variance covariance matrix Σ^*. A consistent estimator of the variance σ_{kk}^* is simply

$$\widehat{\sigma}_{kk} = n^{-1}\int_0^\tau W_k^2(u)\frac{\overline{Y}_{kA}\overline{Y}_{kB}}{\overline{Y}_k}d\widehat{\Lambda}_k(u),$$

which is the classical marginal variance estimator of the logrank statistic. And inserting (5) into (6), we obtain a consistent estimator of σ_{12}^*:

$$\widehat{\sigma}_{12} = \frac{n_A}{n}\int_0^\tau \int_0^\tau W_1(u)W_2(v)\frac{\overline{Y}_{1B}}{\overline{Y}_1}(u)\frac{\overline{Y}_{2B}}{\overline{Y}_2}(v)\widehat{v}_{12A}(du,dv)$$
$$+\frac{n_B}{n}\int_0^\tau \int_0^\tau W_1(u)W_2(v)\frac{\overline{Y}_{1A}}{\overline{Y}_1}(u)\frac{\overline{Y}_{2A}}{\overline{Y}_2}(v)\widehat{v}_{12B}(du,dv)$$
$$= n^{-1}\int_0^\tau \int_0^\tau \left\{ \begin{array}{l} \frac{W_1(u)W_2(v)}{\overline{Y}_1(u)\overline{Y}_2(v)} \frac{\sum_{j=A}^B \sum_{l=1}^{n_j} d\widehat{M}_{1jl}(u)d\widehat{M}_{2jl}(v)}{r(u,v)} \times \\ \left[\overline{Y}_{1B}(u)\overline{Y}_{2B}(v)r_A(u,v) + \overline{Y}_{1A}(u)\overline{Y}_{2A}(v)r_B(u,v)\right] \end{array} \right\}.$$

Notice that the above formula allows to find the variance estimator $\widehat{\sigma}_{kk}$. Actually, write $\widehat{\sigma}_{kk'}$ as

$$n^{-1}\int_0^\tau \int_0^\tau \left\{ \begin{array}{l} \frac{W_k(u)W_{k'}(v)}{\overline{Y}_k(u)\overline{Y}_{k'}(v)} \frac{\sum_{j=A}^B \sum_{l=1}^{n_j} d\widehat{M}_{kjl}(u)d\widehat{M}_{k'jl}(v)}{r(u,v)} \times \\ \left[\overline{Y}_{kB}(u)\overline{Y}_{k'B}(v)r_A(u,v) + \overline{Y}_{kA}(u)\overline{Y}_{k'A}(v)r_B(u,v)\right] \end{array} \right\} \qquad (7)$$

and let $k = k'$; in this case, $\sum_{j=A}^B \sum_{l=1}^{n_j} d\widehat{M}_{kjl}(u)d\widehat{M}_{kjl}(v)$ is null unless $u = v$, and then $r_j(u,u) = \overline{Y}_{kj}(u)$ and $r(u,u) = \overline{Y}_k(u)$. So

$$\widehat{\sigma}_{kk} = n^{-1}\int_0^\tau \frac{W_k^2}{\overline{Y}_k^2}\left\{\overline{Y}_{kB}^2\overline{Y}_{kA} + \overline{Y}_{kA}^2\overline{Y}_{kB}\right\}\frac{\sum_{j=A}^B \sum_{l=1}^{n_j} d\widehat{M}_{kjl}^2}{\overline{Y}_k}.$$

Since

$$\sum_{j=A}^{B}\sum_{l=1}^{n_j} d\widehat{M}_{kjl}^2 = \overline{Y}_k d\widehat{\Lambda}_k\left\{1 - d\widehat{\Lambda}_k\right\},$$

we obtain

$$\widehat{\sigma}_{kk} = n^{-1} \int_0^{\tau} \frac{W_k^2}{\overline{Y}_k^2} \left\{ \overline{Y}_{kB}^2 \overline{Y}_{kA} + \overline{Y}_{kA}^2 \overline{Y}_{kB} \right\} d\widehat{\Lambda}_k \left\{ 1 - d\widehat{\Lambda}_k \right\} \tag{8}$$

$$\simeq n^{-1} \int_0^{\tau} W_k^2 \frac{\overline{Y}_{kA} \overline{Y}_{kB}}{\overline{Y}_k} d\widehat{\Lambda}_k,$$

as we supposed Λ_k continuous on $[0, \tau]$.

Therefore, the asymptotic distribution of the vector $(LR_1, LR_2)'$ is a Gaussian law, with null expectation and with variance covariance matrix consistently estimated by $\widehat{\Sigma} = (\widehat{\sigma}_{kk'})$, $k = 1, 2$, $k' = 1, 2$, where $\widehat{\sigma}_{kk'}$ is defined by (7). A test statistic for $H_0 : \left\{ \Lambda_1^A = \Lambda_1^B = \Lambda_1, \Lambda_2^A = \Lambda_2^B = \Lambda_2 \text{ on } [0, \tau] \right\}$ is then

$$K = (LR_1, LR_2)\, \widehat{\Sigma}^{-1} (LR_1, LR_2)', \tag{9}$$

which asymptotically follows under H_0 a chi-square distribution with two degrees of freedom.

2.3 What if the joint censoring distributions or the joint survival functions differ in groups A and B under H_0 ?

If one suspects that the censoring distribution in group A differs from the one in group B, or that a copula model for the joint survival function is not appropriate, the above results have to be modified. For $j = A, B$, let $\overline{F}^j$ and $\overline{G}^j$ be the joint survival functions of $\{(T_{1ji}, T_{2ji}), i = 1, ..., n_j\}$ and $\{(C_{1ji}, C_{2ji}), i = 1, ..., n_j\}$ respectively, and let $\overline{F}_k^j$ and $\overline{G}_k^j$ be the respective marginal survival function of $\{T_{kji}, i = 1, ..., n_j\}$ and $\{C_{kji}, i = 1, ..., n_j\}$ for event k. Let also f^j denote the density function of $\{(T_{1ji}, T_{2ji}), i = 1, ..., n_j\}$. Then $(n_A^{-1/2}\overline{M}_{1A}, n_B^{-1/2}\overline{M}_{1B}, n_A^{-1/2}\overline{M}_{2A}, n_B^{-1/2}\overline{M}_{2B})'$ converges in distribution to a centered Gaussian process with variances

$$v_{kj}(x) = \int_0^x \overline{y}_{kj}(u) d\Lambda_k(u),$$

and with covariance between $n_j^{-1/2}\overline{M}_{kj}$ and $n_j^{-1/2}\overline{M}_{k'j}$ for $k = 1, 2$ and $j = A, B$ now equal to

$$v_{12j}(x_1, x_2) = \int_0^{x_1} \int_0^{x_2} E\{dM_{1ji}(u) dM_{2ji}(v)\}$$

$$= \int_0^{x_1} \int_0^{x_2} \overline{G}^j(s, u) \left\{ \begin{array}{l} f^j(s, u) ds du + d\Lambda_1(s) \overline{F}^j(s, du) \\ + d\Lambda_2(u) \overline{F}^j(ds, u) + d\Lambda_1(s) d\Lambda_2(u) \overline{F}^j(s, u) \end{array} \right\}.$$

Estimating consistently $E\{dM_{1ji}(u)dM_{2ji}(v)\}$ by

$$\widehat{E}\{dM_{1ji}(u)dM_{2ji}(v)\} = \frac{Y_{1ji}(u)Y_{2ji}(v)}{r_j(u,v)} \sum_{l=1}^{n_j} d\widehat{M}_{1jl}(u)d\widehat{M}_{2jl}(v),$$

a consistent estimator of $v_{12j}(du, dv)$ is now

$$\begin{aligned}\widehat{v}_{12j}(du,dv) &= n_j^{-1}\sum_{i=1}^{n_j}\frac{Y_{1ji}(u)Y_{2ji}(v)}{r_j(u,v)}\sum_{l=1}^{n_j} d\widehat{M}_{1jl}(u)d\widehat{M}_{2jl}(v)\\ &= n_j^{-1}\sum_{l=1}^{n_j} d\widehat{M}_{1jl}(u)d\widehat{M}_{2jl}(v),\end{aligned}$$

with the martingale residuals $\widehat{M}_{kjl}$ defined by (4).

Then the vector $(LR_1, LR_2)'$ converges in distribution to a Gaussian law, with null expectation and with variance covariance matrix $\Sigma = (\sigma_{kk'})$, $k = 1, 2$, $k' = 1, 2$, being consistently estimated by $\widehat{\Sigma} = (\widehat{\sigma}_{kk'})$, $k = 1, 2$, $k' = 1, 2$, with $\widehat{\sigma}_{kk'}$ now expressed as

$$n^{-1}\int_0^\tau\int_0^\tau \frac{W_k(u)W_{k'}(v)}{\overline{Y}_k(u)\overline{Y}_{k'}(v)}\left\{\begin{array}{l}\overline{Y}_{kB}(u)\overline{Y}_{k'B}(v)\sum_{i=1}^{n_A} d\widehat{M}_{kAi}(u)d\widehat{M}_{k'Ai}(v)\\ +\overline{Y}_{kA}(u)\overline{Y}_{k'A}(v)\sum_{i=1}^{n_B} d\widehat{M}_{kBi}(u)d\widehat{M}_{k'Bi}(v)\end{array}\right\}.$$

Notice that for $k = k'$, $\sum_{i=1}^{n_j} d\widehat{M}_{kji}(u)d\widehat{M}_{k'ji}(v)$ is no more equal to zero if $u \neq v$; more precisely, for $u < v$, it equals

$$\sum_{i=1}^{n_j} d\widehat{M}_{kji}(u)d\widehat{M}_{k'ji}(v) = d\widehat{\Lambda}_k(u)\overline{Y}_{kj}(v)\left\{d\widehat{\Lambda}_k(v) - d\widehat{\Lambda}_{kj}(v)\right\}$$

where

$$\widehat{\Lambda}_{kj}(x) = \int_0^x \frac{d\overline{N}_{kj}(u)}{\overline{Y}_{kj}(u)}$$

is the Nelson-Aalen estimate of the cumulative hazard function Λ_k^j in group j for event k. Then, with finite sample size, $\widehat{\sigma}_{kk}$ is not the classical variance estimator (8); nevertheless, since, under H_0, for $j = A, B$, $\widehat{\Lambda}_{kj}$ converges in probability to Λ_k as does $\widehat{\Lambda}_k$, we can conclude that $\widehat{\sigma}_{kk}$ converges in probability to σ_{kk}^*.

A test statistic for $H_0 : \{\Lambda_1^A = \Lambda_1^B = \Lambda_1,\ \Lambda_2^A = \Lambda_2^B = \Lambda_2 \text{ on } [0,\tau]\}$ is

$$(LR_1, LR_2)\,\widehat{\Sigma}^{-1}\,(LR_1, LR_2)',$$

whose asymptotic distribution under H_0 is a chi-square distribution with two degrees of freedom.

3 Simulations study

We performed simulations to study the first type error rate of the test statistic K and to compare it to those of the statistics developed by Wei and Lachin ([WL84]) and of the Wald test statistic proposed by Wei, Lin and Weissfeld ([WLW89]) in their generalization of the proportional hazards model for multivariate failure time data. 10000 samples were generated as following: we first created a dummy variable z with value 1 for group B and with value 0 otherwise; the sample sizes were either equal ($n_A = n_B = 50$) or unequal ($n_A = 25$, $n_B = 75$). The marginal distributions of failure times were supposed to have the same exponential survival function in both groups. Dependence between the two failure times was governed by a Clayton-Oakes model for a strong correlation and by a Gumbel model for a slight correlation; in the first case, the joint survival distribution was

$$\Pr\{T_1 > t_1, T_2 > t_2\} = \{\overline{F}_1(t_1)^{1-\alpha} + \overline{F}_2(t_2)^{1-\alpha} - 1\}^{\frac{1}{1-\alpha}},$$

with $\alpha > 1$, and where $\alpha \to 1$ leads to independence between the two events. We chose $\alpha = 5$.
In the Gumbel model, the joint survival distribution was

$$\Pr\{T_1 > t_1, T_2 > t_2\} = \overline{F}_1(t_1)\overline{F}_2(t_2)\{1 + \alpha F_1(t_1)F_2(t_2)\},$$

for $\alpha \in [-1; 1]$, and we let $\alpha = -1$.
We then generated one single censoring time, that means $C_1 = C_2$, from an uniform distribution on $[0; \tau]$, with τ being chosen to obtain the desired censoring proportion.

For a weak dependence between events (Table 1), we remark that the sum of the marginal logrank statistics has a first type error rates close to 5%. The Wei and Lachin's test shows the highest first type error rate, specially in case of unequal size groups; the Wald test proposed by Wei, Lin and Weissfeld and the proposed test (9) have similar results for equal groups sizes, but the test statistic (9) resists better to imbalance between groups sizes.
For a strong dependence between events (Table 2), the sum of the marginal logrank statistics is no more suitable, as expected. In case of equal size groups, Wei and Lachin's test first type error rate is higher than Wei, Lin and Weissfeld's and the proposed test's, that have similar results, with first type error rates close to 5%. In case of unequal size groups, Wei and Lachin's test is the one with the first type error rate the more distant from 5%; the first type error rates of Wei, Lin and Weissfeld's test and of test (9) are similar with heavy censoring, but the proposed test (9) is more appropriate with no or few censoring.

Table 1. First type error rate (in %) for the two group comparison with nominal level 5% in a Gumbel model with slightly negative dependence; sample is composed of two groups A and B either with equal sizes or unequal sizes. LR: sum of the marginal logrank test statistics; WL: Wei and Lachin's test statistic; WLW: Wei, Lin and Weissfeld' Wald test statistic; K: the proposed test statistic (9).

	Groups with equal sizes				Groups with unequal sizes			
	LR	*WL*	*WLW*	*K*	*LR*	*WL*	*WLW*	*K*
No censoring	5.61	6.27	5.82	5.94	6.09	7.51	7.63	6.40
25% censoring	5.53	6.16	5.74	5.69	5.60	7.20	6.94	5.78
50% censoring	5.41	5.88	5.61	5.70	5.47	7.37	6.79	5.80
75% censoring	5.33	5.40	5.03	5.11	5.75	7.19	6.06	5.65

Table 2. First type error rate (in %) for the two group comparison with nominal level 5% in a Clayton-Oakes model with strong positive dependence; sample is composed of two groups A and B either with equal sizes or unequal sizes. LR: sum of the marginal logrank test statistics; WL: Wei and Lachin's test statistic; WLW: Wei, Lin and Weissfeld' Wald test statistic; K: the proposed test statistic (9).

	Groups with equal sizes				Groups with unequal sizes			
	LR	*WL*	*WLW*	*K*	*LR*	*WL*	*WLW*	*K*
No censoring	8.02	5.69	4.94	5.04	8.53	6.19	6.28	5.83
25% censoring	7.22	4.77	4.52	4.52	8.16	5.89	5.78	5.35
50% censoring	7.01	5.35	5.11	5.10	7.60	6.30	5.49	5.57
75% censoring	6.50	4.81	4.38	4.78	6.62	7.08	4.97	5.40

4 Application

Data issue from a study concerned by the association between diabetes and retinopathy ([DRSRG76]). For each of the 197 diabetic patients, one eye was subject to a laser treatment, the other one remaining untreated, and time to retinopathy was recorded for both eyes, so that two events were studied: retinopathy for the treated eye, and retinopathy for the untreated eye. The patient type of diabetes was furthermore known as juvenile if detected before age of 20 and adult otherwise. Censoring rates according type of diabetes for each event are described in Table (3).

We wished to compare types of diabetes. Let T_1 and T_2 be the time to blindness for the treated eye and the untreated eye, with respective marginal survival function $\overline{F}_{1j}$ and $\overline{F}_{2j}$ in group j (j =juvenile diabetes or adult diabetes). The null tested hypothesis is:

$$H_0 : \begin{cases} \overline{F}_{1,\text{juvenile}} = \overline{F}_{1,\text{adult}} \\ \overline{F}_{2,\text{juvenile}} = \overline{F}_{2,\text{adult}} \end{cases} .$$

Results of H_0 testing against an unspecified alternative are produced in Table (4).

Table 3. Description of the sample of 197 diabetic patients - sizes and censoring proportions according type of diabetes and eye treatment.

		n	Censoring in %
Treated eye	Adult diabetes	83	78.3
	Juvenile diabetes	114	68.4
	All	197	72.6
Untreated eye	Adult diabetes	83	39.8
	Juvenile diabetes	114	55.3
	All	197	48.7

Table 4. Test of comparison of marginal survival functions for treated and untreated eyes according type of diabetes with nominal level 5%. p-values are expressed in %; LR: sum of the marginal logrank test statistics; WL: Wei and Lachin's test statistic; WLW: Wei, Lin and Weissfeld' Wald test statistic; K: the proposed test statistic (9).

	LR	WL	WLW	K
Test statistic value	5.953	8.146	8.089	7.141
p-value	0.051	0.017	0.018	0.028

We observe that the sum of the marginal logrank statistics p-value is close to 5%, that could question about not rejecting H_0; Wei and Lachin's and Wei, Lin and Weissfeld's test statistics values are equal, and greater than the value of the test statistic (9), as noticed with the simulations study.

5 Discussion

The proposed method is based on martingales properties. Our results for estimating the martingales covariance under null hypothesis are similar to those of [PRE92] when there is no censoring, but differ in the censored case, because their estimator is expressed with an estimator of the joint survival function of the two times to failure. In particular, their estimator of $E\left\{dM_{kji}(u)dM_{k'ji}(v)\right\}$ for $k = k'$ leads to

$$\widehat{E}\left\{dM_{ki}^2(u)\right\} = \widehat{\overline{F}}_k(u)d\widehat{\Lambda}_k(u),$$

that tends to under-estimate the classical variance of a martingale as the survival function estimate for event k is necessarily lower than 1.
We also would like to point out that the Wei and Lachin variance covariance matrix estimator of vector $(LR_1, LR_2)'$ under null hypothesis does not use the fact that under H_0 the marginal survival functions are equal in groups A and B, since their estimator is derived from averaging martingale residuals over subjects in each group and not over the whole sample.

References

[GEH65] Gehan, E. A. : A generalized Wilcoxon test for comparing arbitrarily singly-censored samples. Biometrika, **52**, 203–23 (1965)

[MAN66] Mantel, N. : Evaluation of survival data and two new rank order statistics arising in its consideration. Cancer Chemother. Rep., **50**, 163–170. (1966)

[COX72] Cox, D. R. :(1972). Regression models and life tables (with discussion). J. Roy. Statist. Soc., **B, 34**, 187–202 (1972)

[PETO72] Peto, R., and Peto, J. : Asymptotically efficient rank invariant test procedures (with discussion). J. Roy. Statist. Soc., **A, 135**, 185–206 (1972)

[PRE78] Prentice, R. L. : Linear rank tests with right censored data. Biometrika, **65**, 167–79 (1978)

[HF82] Harrington, D. P. and Fleming, T. R. : A class of rank test procedures for censored survival data. Biometrika, **69**, 553–66 (1982)

[WL84] Wei, L. J. and Lachin, J. M. : Two-sample asymptotically distribution-free tests for incomplete multivariate observations. J. Am. Statist. Ass., **79**, 653–61 (1984)

[PGT87] Pocock, S. J., Geller, N. L. and Tsiatis, A. A. : The analysis of multiple endpoints in clinical trials. Biometrics, **43**, 487–98 (1987)

[WLW89] Wei, L. J., Lin, D. Y. and Weissfeld, L. : Regression Analysis of Multivariate Incomplete Failure Time Data by Modeling Marginal Distributions.J. Am. Statist. Ass., **84**, 1065–73 (1989)

[NEL69] Nelson, W. : Hazard plotting for incomplete failure data. J. Qual. Technol., **1**, 27–52 (1969)

[AAL78] Aalen, O. O. : Nonparametric inference for a family of counting processes. Ann. Statist., **6**, 701–26 (1978)

[DRSRG76] The Diabetic Retinopathy Study Research Group : Preliminary report on effects of photocoagulation therapy. Am. J. Ophthalmol., *81*, 383–96 (1976)

[PRE92] Prentice, R. L. and Cai, J. : Covariance and survivor function estimation using censored multivariate failure time data. Biometrika, *79*, 495-512 (1992)

Explained Variation and Predictive Accuracy in General Parametric Statistical Models: The Role of Model Misspecification

Susanne Rosthøj[1] and Niels Keiding[1]

Department of Biostatistics, University of Copenhagen, Blegdamsvej 3, DK-2200 Copenhagen N, Denmark S.Rosthoej@biostat.ku.dk

1 Introduction

One of the purposes of a multivariate regression analysis is to determine covariates of importance for the outcome of interest. When the estimated effect of a covariate is highly statistically significant it is easy to be lead to the conclusion that such a covariate has a substantial effect on the outcome under study. However, this might not necessarily be the case. Quantities assessing the extent to which the covariates actually determine the outcome are needed to avoid overinterpretation of the effect. Another purpose of a multivariate regression analysis is to enable prediction of the outcome of interest and in this case a quantity assessing the accuracy of the predictions based on the regression model is needed.

Measures of explained variation and predictive accuracy can be used to address these questions. We carefully distinguish between the two concepts. Korn and Simon [KS91] provide a general framework and their approach is adopted and elaborated here. To asses the importance of the covariates, the *explained variation* is defined on a population level. This quantity is also related to the chosen class of regression models and if the model is misspecified, it cannot necessarily be considered as a measure of the ability of the covariates to determine the outcome. To quantify the ability of the covariates and the regression model to determine or rather predict the outcome the *predictive accuracy* is defined on a population level. A high predictive accuracy requires a useful prediction rule as well as informative covariates. Whether the model is misspecified or not, the predictive accuracy is a meaningful quantity.

In the linear normal model, the estimator of the explained variation is asymptotically equal to the estimator of the predictive accuracy and is better known as the R^2-statistic or the coefficient of determination. This statistic has become standard output from the statistical software packages. Outside of the linear model the estimators of the explained variation and the predictive accuracy usually differ and due to the possibility of the model being misspecified, the predictive accuracy has often been preferred instead of the explained variation. However, there seems to be some confusion in the literature on the distinction between the two measures and how they are actually affected by misspecification of the regression model. We here provide a more detailed discussion. In our exposition we put more weight on explicitly formulating the various underlying statistical models than in most of the literature in the area. We only consider parametric models.

Our interest is motivated by the use of explained variation and predictive accuracy in failure time models since here the estimation becomes complicated due to censoring of the outcome of interest. There is no unique generalization of the R^2-statistic to survival data and several authors have proposed other measures and estimators in the simple failure time model, see e.g. Schemper and Stare [SS96], O'Quigley and Xu [QX01], Graf et al. [GSSS99] and Schemper and Henderson [SH00]. Some authors, e.g. Graf et al. and Schemper and Henderson, are inspired by the approach of Korn and Simon whereas others have different approaches, of which some are only defined in the Cox regression model. So far none of the measures have been widely accepted.

Section 2 contains an introduction to the approach of Korn and Simon including a detailed discussion of consistency of the estimators. In Section 3 the concept of model misspecification is introduced and the effect of misspecification on the estimators is discussed. Furthermore, the simple failure time model is discussed shortly in Section 4, namely the estimation procedures proposed by Graf et al. and Schemper and Henderson. We do not give a detailed introduction to their work but only present their ideas. Finally we provide in Section 5 some concluding remarks.

2 Measures of explained variation

Suppose (Z, Y) is a random variable, Z being a q-dimensional vector of covariates and Y being a p-dimensional response variable. A parametric regression model indexed by a finite dimensional parameter $\theta \in \Theta \subseteq \mathbb{R}^d$ is proposed for the conditional distribution of Y given Z. The distribution of the vector of covariates is assumed not to depend on the parameter θ and left completely unspecified. Expectations with respect to the conditional distribution of Y given Z determined by θ, the marginal distribution of Z and the marginal distribution of Y determined by the parameter θ will be denoted by $\mathrm{E}_\theta\big[\cdot \mid Z\big]$, $\mathrm{E}\big[\cdot\big]$ and $\mathrm{E}_\theta\big[\cdot\big] = \mathrm{E}\big[\mathrm{E}_\theta\big[\cdot \mid Z\big]\big]$, respectively. As a starting point it is assumed that the true distribution of Y conditional on Z belongs to the model, that is the existence of a true parameter $\theta_0 \in \Theta$ is required.
Let V denote the variable of interest. It is assumed that V is a one-dimensional transformation of the response variable Y, i.e. $V = f(Y)$ for some function $f : \mathbb{R}^p \to \mathbb{R}$. Unless coarsened data is considered (see Section 4 on survival analysis below), f is usually the identity function. Using the regression model,

our two purposes are to determine how much of the variation in V is explained by the covariates and to make accurate predictions of V.

2.1 Definition of the explained variation

Following the loss function approach as proposed by Korn and Simon [KS91] a loss function L has to be defined. Then, $L(v, \hat{v})$ denotes the loss incurred when making the prediction $\hat{v}$ of an observation v of the variable of interest V. The loss function L is assumed to be bounded below by 0 and to attain the value 0 when the correct value $\hat{v} = v$ of v is predicted. Quadratic loss $L(v, \hat{v}) = (v - \hat{v})^2$, absolute loss $L(v, \hat{v}) = |v - \hat{v}|$ and entropy loss $L(v, \hat{v}) = -(v \log \hat{v} + (1 - v) \log(1 - \hat{v}))$ are the most commonly used loss functions (the latter only when predicting binary variables), see e.g. Korn and Simon [KS90] and Korn and Simon [KS91].

A prediction of the variable of interest V based on the vector of covariates Z can be defined by any function $\hat{v} : \mathbb{R}^q \to \mathbb{R}$ $(z \mapsto \hat{v}(z))$, since such a function determines a prediction rule. For every $\theta \in \Theta$, a measure of the ability of the covariates and the prediction rule $\hat{v}$ to predict the variable of interest V is the prediction error defined as the expected loss $\mathrm{E}[\mathrm{E}_\theta[L(V, \hat{v}(Z)) \mid Z]]$. Since interest is in making accurate predictions, the focus will be on the prediction rules giving rise to the smallest possible prediction error: For every $\theta \in \Theta$ the θ-optimal prediction rule is defined as the prediction rule $\hat{v}_\theta$ minimising the prediction error, i.e.

$$\mathrm{E}[\mathrm{E}_\theta[L(V, \hat{v}_\theta(Z)) \mid Z]] \leq \mathrm{E}[\mathrm{E}_\theta[L(V, \hat{v}(Z)) \mid Z]] \quad \text{for all } \hat{v} : \mathbb{R}^q \to \mathbb{R}.$$

Note that the θ-optimal prediction rule indeed depends on the choice of loss function: Using quadratic, absolute and entropy loss the θ-optimal prediction rules are given by the means, the medians and the means, respectively, of the conditional distributions of $V = f(Y)$ given $Z = z \in \mathbb{R}^q$ determined by the parameter θ.
The prediction error corresponding to the θ-optimal prediction rule will be denoted π_θ in the following, i.e. $\pi_\theta = \mathrm{E}[\mathrm{E}_\theta[L(V, \hat{v}_\theta(Z)) \mid Z]]$.

Since the prediction error is a positive number, it is difficult to determine whether it is small or large corresponding to whether the covariates and the prediction rule are good or bad in predicting the variable of interest. It may here be helpful to compare it to the prediction error based on a prediction rule not depending on the covariate values. Thus, consider a prediction rule of the form $z \mapsto \hat{v}^0$ for a fixed $\hat{v}^0 \in \mathbb{R}$. Such a prediction rule will be termed a marginal prediction rule. In this case the marginal prediction error is $\mathrm{E}_\theta[L(V, \hat{v}^0)]$. The θ-optimal marginal prediction rule is similarly defined as the prediction rule $(z \mapsto \hat{v}^0_\theta)$ minimising the marginal prediction error, i.e.

$$\mathrm{E}_\theta[L(V, \hat{v}^0_\theta)] \leq \mathrm{E}_\theta[L(V, \hat{v}^0)] \text{ for all } \hat{v}^0 \in \mathbb{R}.$$

The prediction error corresponding to the θ-optimal marginal prediction rule is denoted π_θ^0, i.e. $\pi_\theta^0 = \mathrm{E}_\theta[L(V, \hat{v}_\theta^0)]$.

When considering the θ-optimal prediction rules the prediction error based on the covariates and the marginal prediction error might be compared by the *explained variation*

$$\mathcal{V}_\theta = 1 - \frac{\mathrm{E}[\mathrm{E}_\theta\left[L(V, \hat{v}_\theta(Z)) \mid Z\right]]}{\mathrm{E}_\theta[L(V, \hat{v}_\theta^0)]} = 1 - \frac{\pi_\theta}{\pi_\theta^0} \tag{1}$$

for every $\theta \in \Theta$. This quantity attains values between zero and one. Values close to zero correspond to the prediction errors being almost equal, i.e. that the covariates and the prediction rule do not determine the variable of interest particularly accurately since the marginal prediction rule is almost as accurate. Values close to one on the other hand correspond to the covariates and the prediction rule determining the variable of interest to a large extent. Since the explained variation compares the best possible rules of prediction it becomes a measure of the degree to which the covariates determine the variable of interest.

When squared error loss is considered, $\mathcal{V}_\theta$ reduces to the variance of the conditional mean divided by the marginal variance of the variable of interest: $\mathcal{V}_\theta = \mathrm{Var}\mathrm{E}_\theta(V|Z)/\mathrm{Var}_\theta(Y)$. In this case it thus measures the reduction in the variance of the variable of interest when the information on the covariates is included in the model.

In this context the explained variation $\mathcal{V}_\theta$ is the quantity of interest. However, another quantity measuring the accuracy of a non-optimal prediction rule based on the covariates turns out to be of interest too. We postpone the introduction of this quantity, the population concept of predictive accuracy, until we have discussed estimation of the explained variation and misspecification of the model.

2.2 Estimation of the explained variation

Suppose $(Z_1, Y_1), \ldots, (Z_n, Y_n)$ is a sample of independent random variables distributed as (Z, Y). Based on this sample, the distribution of Y conditional on Z is estimated by a parameter $\hat{\theta}_n$ whereas the marginal distribution of the vector of covariates Z is estimated by the empirical distribution of $Z_1, \ldots, Z_n$.

Korn and Simon [KS91] suggest two estimators of the explained variation. Obviously, the explained variation of the estimated model might be used as an estimator, that is

$$\mathcal{V}_{\hat{\theta}_n} = 1 - \frac{\frac{1}{n}\sum_{i=1}^n \mathrm{E}_{\hat{\theta}_n}\left(L(V, \hat{v}_{\hat{\theta}_n}(Z)) \mid Z = Z_i\right)}{\mathrm{E}_{\hat{\theta}_n} L(V, \hat{v}_{\hat{\theta}_n}^0)} = 1 - \frac{\pi_{\hat{\theta}_n}}{\pi_{\hat{\theta}_n}^0} \tag{2}$$

This estimator is termed the *estimated explained variation.* Note that the estimated explained variation indeed is based on the estimated model since it is a function of the expected losses in the distribution determined by $\hat{\theta}_n$ whereas it only depends on the values of the sample through the estimated parameter $\hat{\theta}_n$ and the covariate values $Z_1, \ldots, Z_n$.

Korn and Simon [KS91] also consider the *explained residual variation,*

$$\widehat{\mathcal{V}}_{\hat{\theta}_n} = 1 - \frac{\sum_{i=1}^n L(V_i, \hat{v}_{\hat{\theta}_n}(Z_i))}{\sum_{i=1}^n L(V_i, \hat{v}^0_{\hat{\theta}_n})}. \tag{3}$$

This estimator only depends on the model through the $\hat{\theta}_n$-optimal prediction rules $z \mapsto \hat{v}_{\hat{\theta}_n}(z)$ and $z \mapsto \hat{v}^0_{\hat{\theta}_n}$. In the numerator, the values of the variable of interest are compared to the predicted values based on the covariates by the loss function L. Similarly, the values of the variable of interest are compared to the marginal predicted value $\hat{v}^0_{\hat{\theta}_n}$ in the denominator. The explained residual variation is therefore, besides being a measure of explained variation, also a measure of how accurate the predictions based on the $\hat{\theta}_n$-optimal prediction rule and the covariates actually are compared to the $\hat{\theta}_n$-optimal marginal prediction rule.

Korn and Simon [KS91] do not formulate conditions under which the two estimators are to be considered as consistent estimators of the explained variation $\mathcal{V}_{\theta_0}$ of the true model. In the Appendix we provide a theorem stating sufficient conditions. This theorem ensures that it is possible to obtain consistent estimators by averaging terms which are dependent through their common dependence on the estimated parameter $\hat{\theta}_n$ as is the case for the numerators and the denominators of the above estimators. How the theorem is used to guarantee the consistency of the two estimators above is also demonstrated in the Appendix.

When considering quadratic loss in the normal linear regression model, the explained variation is equal to the squared multiple correlation coefficient. The two estimators of the explained variation, the estimated explained variation and the explained residual variation, are almost identical. Traditionally the explained residual variation is used as the estimator of the explained variation (for reasons to be described below in Section 3 on misspecification of the model) and is probably better known as the R^2-statistic. However, it is well known that this estimator for small samples has a positive bias as an estimator of the explained variation and therefore the adjusted R^2-statistic R^2_{adj} is used instead (Helland [Hel87]):

$$R^2_{\mathrm{adj}} = 1 - \frac{\frac{1}{n-q-1}\sum_{i=1}^n (V_i - \hat{v}_{\hat{\theta}_n}(Z_i))^2}{\frac{1}{n-1}\sum_{i=1}^n (V_i - \hat{v}^0_{\hat{\theta}_n})^2}.$$

In the normal linear model, the adjusted R^2-statistic is exactly the estimated explained variation. Also in other regression models, the explained residual variation might be an inflated estimator of the explained variation when small samples are considered. Mittlböck and Waldhör [MW00] propose a similar adjustment of the explained residual variation for the Poisson regression model whereas Mittlböck and Schemper [MS02] propose similar and other adjustments for the logistic regression model.

3 Misspecification and definition of the predictive accuracy

The model is said to be misspecified if the true distribution of the response Y conditional on the covariate vector Z does not belong to the proposed regression model indexed by $\theta \in \Theta$. That is, a true $\theta_0 \in \Theta$ does not exist. In the case of the model being misspecified it is not possible to consider the estimated explained variation and the explained residual variation as estimators of the explained variation, namely the degree to which the covariates determine the value of the variable of interest. It is however, by appropriate use of the theorem provided in the Appendix, possible to state which quantities the two estimators estimate consistently.

White [Whi89] proves that the maximum likelihood estimator in a misspecified model indexed by a finite dimensional parameter $\theta \in \Theta$ under appropriate regularity conditions is a consistent estimator of the parameter $\theta^* \in \Theta$ minimising the Kullback-Leibler divergence. For every $\theta \in \Theta$, the Kullback-Leibler divergence is a measure of the distance from the true unknown density to the density determined by the parameter θ. In this sense, the maximum likelihood estimator suggests the distribution among the proposed distributions that agrees best with the true distribution and the parameter θ^* is therefore termed the least false or most fitting parameter.

By appropriate use of the theorem in the Appendix it then follows, when the prescribed conditions are fulfilled, that the estimated explained variation is a consistent estimator of $\mathcal{V}_{\theta^*}$. This quantity is a measure of the degree to which the actual covariates would determine the value of the variable of interest, if the distribution of this variable were described by the distribution determined by θ^*.

The explained residual variation is similarly, under appropriate conditions, a consistent estimator of

$$\mathcal{W}_{\theta^*} = 1 - \frac{\mathrm{E}[L(V, \hat{v}_{\theta^*}(Z))]}{\mathrm{E}[L(V, \hat{v}^0_{\theta^*})]}$$

where the mean in the numerator is with respect to the true unknown distribution of $(V, Z) = (f(Y), Z)$ whereas the mean in the denominator is with respect to the true distribution of V. The numerator is the prediction error of the prediction rule $z \mapsto \hat{v}_{\theta^*}(z)$ in the true distribution of (Z, Y) whereas the denominator is the marginal prediction error corresponding to the marginal prediction rule $(z \mapsto \hat{v}^0_{\theta^*})$. Thus, $\mathcal{W}_{\theta^*}$ is the *predictive accuracy* of the predictions based on the least false model and the covariates compared to the marginal predictions based on the least false model.
Some authors define the above quantity as the explained variation. Since it is based on a non-optimal prediction rule we prefer to think of it as the predictive accuracy instead.

From the above it follows that both estimators of the explained variation might be biased estimators of the true explained variation in case of the model being misspecified. The explained variation of the least false model estimated by the estimated explained variation is a measure related to the chosen, misspecified model and the covariates whereas the predictive accuracy estimated by the explained residual variation is a measure of the ability of the model and the covariates to describe, namely predict, the values of the variable of interest. According to the interpretation of these two quantities, the explained residual variation appears to be the most rational estimator since it still has a relevant interpretation when the model is misspecified. This is probably the reason why most papers on explained variation for uncoarsened data do not even consider the estimated explained variation as an estimator of the explained variation, e.g. Mittlböck and Waldhör [MW00] and Mittlböck and Schemper [MS02]. Others argue that an estimator of the explained variation should compare the observed and the predicted values directly as is the case for the explained residual variation but not always for the estimated explained variation.
It is our experience however that there will only be small, if any, differences between the quantities estimated by the two estimators and that these quantities will be rather close to the true explained variation. Korn and Simon [KS91] claim the opposite, namely that there might be considerable differences between the population measures estimated by the two estimators if the model is 'grossly' misspecified. We found that this is usually not the case provided the proposed regression model is defined in a sensible way, namely that the parameter space Θ is not unnecessarily restricted. Korn and Simon [KS91] base their statement on an example of a misspecified regression model for which the parameter space Θ consists of one point θ, i.e. $\Theta = \{\theta\}$. In this case the least false parameter θ^* equals θ. Using this distribution, they determine the explained variation $\mathcal{V}_{\theta^*}$ and the predictive accuracy $\mathcal{W}_{\theta^*}$. The difference between these two quantities turns out to be considerable as well as they both differ considerably from the true explained variation of the model considered. However, it must be obvious that it is not reasonable to pick an arbitrary distribution, determine the explained variation and the

predictive accuracy in the true distribution of (Z, Y) of this distribution and then expect these two quantities to be equal as well as equal to the true explained variation. If instead the parameter space Θ is allowed to be large as possible, the explained variation $\mathcal{V}_{\theta^*}$ and the predictive accuracy $\mathcal{W}_{\theta^*}$ of the least false model are equal and close to the true explained variation. The example of Korn and Simon [KS91] is given below:

Consider the logistic regression model where the true distribution of the binary response Y conditional on the covariate Z is Bernoulli with parameter $p(Z)$ where $\operatorname{logit} p(Z) = Z$ and Z is uniform on $\{-1.5, -1, -0.5, 0.5, 1, 1.5\}$. Using a quadratic loss function, the explained variation of this distribution, the true model, is 0.2256.

The model is misspecified by assuming Y conditional on Z to be Bernoulli with parameter $\widetilde{p}(Z) = 0.1I(Z < 0) + 0.9I(Z > 0)$. The distribution of the covariate Z remains unchanged. The proposed model is indexed by the single parameter $\theta \in \Theta = \{(0.1, 0.9)\}$ and has an explained variation of $\mathcal{V}_\theta = 0.64$. On the other hand, the predictive accuracy of this model in the true distribution of (Z, Y) is $\mathcal{W}_\theta = 0.0758$. On the basis of this example they conclude that there might be large differences between the quantities estimated by the estimated explained variation and the explained residual variation. However, if instead the parameter space is allowed to be as large as possible, i.e. $\Theta = (0, 1)^2$, then $\theta^* = (0.2763, 0.7237)$ resulting in an explained variation of $\mathcal{V}_{\theta^*} = 0.2002$ and a predictive accuracy of $\mathcal{W}_{\theta^*} = 0.2002$, that is the two quantities are equal.

We have furthermore considered examples of misspecification of the linear predictor in the normal and the logistic regression model but did not succeed in finding examples for which the explained variation and the predictive accuracy differed appreciably. These examples were examined analytically and by simulation studies.

4 The failure time model

In failure time analysis the response variable Y is a failure time. Measures of explained variation and predictive accuracy may be defined as above but when censoring occurs, as is usual in survival analysis, the estimation procedure gets complicated: The explained residual variation cannot be determined because the loss corresponding to a censored failure time is unavailable. Since the estimated explained variation has not been accepted as an estimator of the explained variation other estimation methods are needed. Graf et al. [GSSS99] and Schemper and Henderson [SH00] have proposed estimators of the explained variation in the failure time model. Graf et al. base their estimator on inverse probability weighting of the available losses whereas Schemper and Henderson use the proposed regression model

to determine a loss for the unavailable failure times too. Both estimators can be considered as generalizations of the explained residual variation and coincide with the explained residual variation in the case of no censoring.
How the estimators of Graf et al. and Schemper and Henderson are defined is not described in detail here. Graf et al. consider a model consisting of one parameter $\theta \in \Theta = \{\theta\}$ and estimate the predictive accuracy $\mathcal{W}_\theta$ of this model in the true distribution of (Z, Y) and Schemper and Henderson focus on the Cox regression model. However, it is not complicated to generalize their estimators to the above setting. We here shortly discuss the choice of the variable of interest and the properties of the estimation procedures proposed by these authors.

As noted in several papers on explained variation and predictive accuracy in failure time models, e.g. Korn and Simon [KS90] and Henderson [Hen95], the variable of interest is not necessarily the failure time Y. This is due to the nature of failure time data. From a medical point of view other variables of interest arise but the censoring mechanism may also influence the choice of variable of interest: If the individuals are followed until time point t it is not relevant to focus on how long they will survive further.
In many cases, when considering the failure time of a patient, the actual failure time is important for patients who are expected to die soon whereas the actual failure time for long-term survivors is of less interest than the fact that they will live for a long time. If long-term survivors are defined as the individuals surviving a specified time point t, this leads to the at time point t censored failure time as the variable of interest, i.e. $V = \min\{Y, t\}$. A prediction of $\hat{v} = t$ corresponds to the long-term prediction 'survival greater than or equal to t' and is to be considered successful if the individual survives time point t. No loss should be incurred in this case. When the predictions used attain the same values as the variable of interest, i.e. belong to the interval $[0, t]$, standard loss function like quadratic and absolute loss incorporate this feature. See Henderson [Hen95] for a more elaborate discussion of the choice of loss functions.
In some cases focus is on whether an individual is alive at a specified time point t. This may be the case if a patient can be considered cured if the patient survives this time point. Thus focus is on whether the patient will be cured or not and the variable of interest therefore becomes the survival status at time point t, i.e. $V = I(Y \leq t)$ where $I(\cdot)$ denotes the indicator function. In this case the variable of interest is binary and the standard loss functions are applicable.
A possible generalization of the survival status at time point t as the variable of interest can be obtained by considering the survival status as a process, i.e. $\big(I(y \leq s) : s \in [0, t]\big)$. In this case the prediction is also a process and a loss can be determined by averaging the loss for each time point in $[0, t]$. The average may be constructed by integration on $[0, t]$ with respect to the Lebesgue measure or another suitable measure (see e.g. Graf and

Schumacher [GS95], Graf et al. [GSSS99] or Schemper and Henderson [SH00]). The concept of explained variation and predictive accuracy can be defined in the same manner as above. However, it is not straightforward to prove consistency of the estimators when integrating the losses of the survival status process.
Note that when considering these variables of interest, the loss becomes available for some of the censored failure times: If a failure time is censored after time point t, the at time point t censored survival time is $\min\{Y, t\} = t$ whereas the survival status is $I(Y \leq t) = 0$.

The estimator proposed by Graf et al. is constructed using inverse probability weighting where the weights are based on the Kaplan-Meier estimator of the censoring distribution. Their estimator therefore resembles the explained residual variation in the sense that it estimates the predictive accuracy in case of model misspecification. The estimator of Schemper and Henderson is instead based on the proposed model and in case of model misspecification, it can neither be interpreted as the predictive accuracy nor the explained variation of the least false model but is rather an estimator of a quantity in between these two measures.
We have compared the three available estimation procedures available for survival data in simulation studies: The estimated explained variation, the estimator based on Graf et al. and the estimator based on Schemper and Henderson. We studied the exponential failure time model using survival status as the variable of interest and a quadratic loss function. We first considered the case where the model is not misspecified in order to study the efficiency. Here the estimator of Graf et al. turned out to be the least efficient whereas the estimated explained variation wass the most efficient estimator of the explained variation. Misspecifying the model (by leaving out covariates and still using an exponential model), we did not succed in finding examples where the quantities estimated by the three estimators differed appreciably.

5 Which estimation method to choose - model based or not?

When uncoarsened data are considered, there are two available estimators of the explained variation. Traditionally, the explained residual variation is preferred due to the nice interpretation as an estimator of predictive accuracy in the case of model misspecification. However, we have not been able to demonstrate considerable differences in the quantities estimated by the two estimators and therefore we do not consider the question of misspecification as a big problem as is the case in part of the literature on this area.

When considering survival data, the explained residual variation is undefined due to censoring and therefore other estimation procedures have been proposed by Graf et al. [GSSS99] and Schemper and Henderson [SH00]. When choosing one of the estimators in preference to the other two, the efficiency may be compared to the potential bias of the estimators in case of model misspecification. Since the model based estimators have a higher efficiency and the bias of these estimators is not necessarily large in case of model misspecification, we are tempted to prefer the model based approaches. Another criterion might also influence the choice of estimator: The estimated explained variation is, at least when quadratic loss is used, very simple to determine. The estimator based on the method of Graf et al. is also rather simple to calculate whereas the calculation of the estimator based on the method of Schemper and Henderson requires a bit more programming.

We have studied the issue of misspecification by several examples. It is possible however, that there exist examples for which the bias in misspecified models is more pronounced than for our examples.

6 Acknowledgement

This work was supported by Public Health Services Grant R01-CA54706-07 from the National Cancer Institute.

7 Appendix

The following result on consistency is easily proved using Lemma 2.8 and Lemma 2.13 of Pakes and Pollard [PP89].

Let $(Z_1, Y_1), \cdots, (Z_n, Y_n)$ be a sample of n independent random variables with the same distribution as the random variable (Z, Y) taking values in $\mathbb{R}^q \times \mathbb{R}^p$.

Theorem 1. *Assume $\hat{\theta}_n \to \overline{\theta} \in \Theta$ when n tends to infinity, the convergence being almost sure or in probability. Let $\{h_\theta : \mathbb{R}^q \times \mathbb{R}^p \to \mathbb{R} \mid \theta \in \Theta\}$ be a family of functions with $\mathrm{E}|h_\theta(Z, Y)| < \infty$ for all θ belonging to a bounded neighborhood of $\overline{\theta}$ (E denoting expectation with respect to the true distribution of (Z, Y)).*
Assume further that there exists an $\alpha > 0$ and a nonnegative function $\varphi : \mathbb{R}^q \times \mathbb{R}^p \to \mathbb{R}$ with $\mathrm{E}\varphi(Z, Y) < \infty$ for which

$$|h_\theta(z, y) - h_{\theta'}(z, y)| \leq \varphi(z, y)\|\theta - \theta'\|^\alpha$$

for some norm $\|\cdot\|$ on Θ, all (z, y) and all θ, θ' belonging to the bounded neighborhood of $\overline{\theta}$. Then

$$\frac{1}{n}\sum_{i=1}^{n} h_{\hat{\theta}_n}(Z_i, Y_i) \to \mathrm{E}h_{\overline{\theta}}(Z, Y) \tag{4}$$

for $n \to \infty$, $\to$ being the same convergence as above.

Note that the conditions are fulfilled if the functions $\theta \to h_\theta(z, y)$ are continously differentiable for every (z, y), h_θ and the derivatives of h_θ with respect to θ are integrable with respect to the true distribution of (z, y) in a bounded neighbourhood of $\overline{\theta}$. This will usually be the case for quadratic and entropy loss.

According to the theorem, the numerator of the estimated explained variation (2) is a consistent estimator of π_{θ_0} if $\hat{\theta}_n$ converges to θ_0 in probability or almost surely and the functions $\theta \mapsto \mathrm{E}_\theta(L(V, \hat{v}_\theta(Z)) \mid Z = z)$ fulfill the prescribed conditions for all z and θ in a bounded neighborhood of θ_0.
Similarly the numerator of the explained residual variation (3) is a consistent estimator of π_{θ_0} if $\hat{\theta}_n$ converges to θ_0 in probability or almost surely and the functions $\theta \mapsto L(v, \hat{v}_\theta(z)) = L(f(y), \hat{v}_\theta(z))$ fulfill the conditions for all (z, y) and θ in a bounded neighborhood of θ_0.

The theorem cannot be applied directly to the denominators of the two estimators to ensure that these are consistent estimators of the marginal prediction error $\pi^0_{\theta_0}$. The marginal prediction rule $(z \to \hat{v}^0_{\theta_0})$ is usually a simple function of the observed values $(Z_i, Y_i), i = 1, \ldots, n$, rather than a function of the estimated parameter $\hat{\theta}_n$. That is, $\hat{v}^0_{\hat{\theta}_n} = g((Z_1, Y_1), \cdots, (Z_n, Y_n))$ for some function $g : \mathbb{R}^{2n} \to \mathbb{R}$. If for example quadratic or entropy loss is used, g determines the average of $V_i = f(Y_i), i = 1, \ldots, n$.
The denominator of the estimated explained variation (2) cannot typically be written as an average but is often a simple function of the marginal prediction $\hat{v}^0_{\hat{\theta}_n}$ (see e.g. Korn and Simon [KS91] for some examples). Since it is often possible to use Theorem 1 or even simpler methods (for example the law of large numbers) to guarantee that $\hat{v}^0_{\hat{\theta}_n}$ is a consistent estimator of $\hat{v}^0_{\theta_0}$, the consistency of the denominator can be obtained.
The denominator of the explained residual variation (3) has the form $\sum_{i=1}^{n} L(V_i, \hat{v}^0_{\hat{\theta}_n}) = \sum_{i=1}^{n} L\big(V_i, g((Z_1, Y_1), \cdots, (Z_n, Y_n))\big)$ and hence does not have a form as the average in (4). It is however often possible to rewrite the denominator into a form for which it is possible to use Theorem 1 or simpler methods to guarantee the consistency.

References

[GS95] Graf, E., Schumacher, M.: An investigation on measures of explained variation in survival analysis, The Statistician, **44**, 497–507 (1995)

[GSSS99] Graf, E., Schmoor, C., Sauerbrei, W., Schumacher, M.: Assessment and comparison of prognostic classification schemes for survival data, Statistics in Medicine, **18**, 2529–2545 (1999)

[Hel87] Helland, I.S.: On the interpretation and use of R^2 in regression analysis, Biometrics, **43**, 61–69 (1987)

[Hen95] Henderson, R.: Problems and prediction in survival–data analysis, Statistics in Medicine, **14**, 161–184 (1995)

[KS90] Korn, E.L., Simon, R.: Measures of explained variation for survival data, Statistics in Medicine, **9**, 487–503 (1990)

[KS91] Korn, E.L., Simon, R.: Explained residual variation, explained risk and goodness of fit, The American Statistician, **45**, 201–206 (1991)

[MS02] Mittlböck, M., Schemper, M.: Explained variation for logistic regression – small sample adjustments, confidence intervals and predictive precision, Biometrical Journal, **44** (3), 263–272 (2002)

[MW00] Mittlböck, M., Waldhör, T.: Adjustments for R^2–measures for Poisson regression models, Computational Statistics and Data Analysis, **34**, 461–472 (2000)

[QX01] O'Quigley, J., Xu, R.: Explained variation in proportional hazards regression. In: Crowley, J. (ed) Handbook of Statistics in Clinical Oncology. Marcel Dekker, Inc., New York (2001)

[PP89] Pakes, A., Pollard, D.: Simulation and the asymptotics of optimization estimators, Econometrica , **57**, 1027–1057 (1989)

[SH00] Schemper, M., Henderson, R.: Predictive accuracy and explained variation in cox regression, Biometrics, **56**, 249–255 (2000)

[SS96] Schemper, M., Stare J.: Explained variation in survival analysis, Statistics in Medicine, **15**, 1999–2012 (1996)

[Whi89] White, H.: Maximum likelihood estimation of misspecified models, Econometrica, **50**,1–25 (1989)

This article has appeared in the December 2004 issue of Lifetime Data Analysis

Optimization of Breast Cancer Screening Modalities

Yu Shen[1] and Giovanni Parmigiani[2]

[1] Department of Biostatistics and Applied Mathematics
M. D. Anderson Cancer Center, University of Texas
Houston, TX 77030
yshen@mdanderson.org

[2] Departments of Oncology, Biostatistics and Pathology
Johns Hopkins University, Baltimore, MD 21205
gp@jhu.edu

ABSTRACT

Mathematical models and decision analyses based on microsimulations have been shown to be useful in evaluating relative merits of various screening strategies in terms of cost and mortality reduction. Most investigations regarding the balance between mortality reduction and costs have focused on a single modality, mammography. A systematic evaluation of the relative expenses and projected benefit of combining clinical breast examination and mammograpphy is not at present available. The purpose of this report is to provide methodologic details including assumptions and data used in the process of modeling for complex decision analyses, when searching for optimal breast cancer screening strategies with the multiple screening modalities. To systematic evaluate the relative expenses and projected benefit of screening programmes that combine the two modalities, we build a simulation model incorporating age-specific incidence of the disease, age-specific pre-clinical duration of the disease, age-specific sensitivities of the two screening modalities, and competing causes of mortality. Using decision models, we can integrate information from different sources into the modeling processes, and assess the cost-effectiveness of a variety of screening strategies while incorporating uncertainties.

1 Introduction

Breast cancer is the most frequently diagnosed cancer among women. Its rate of incidence in the United States has continued to increase since 1986

[WTHR03], while breast cancer mortality has decreased overall in the United States, Canada and the United Kingdom [SEER98, IARC99, WTHR03]. Plausible explanations for this decrease in mortality include progress in treatment, as well as widespread participation in early detection programs that contribute to increased cure rates and reduced disease-specific mortality. Many studies have indicated that early detection through screening can lead to more advantageous treatment options, and often leads to an increase in survival rates and improvement in the quality of life for women who develop breast cancer [FE03, Wan03]. The development of new technologies and further improvement of the existing modalities for disease detection may increasingly make screening for cancer a routine part of secondary prevention.

The goals of early detection are to reduce breast cancer morbidity and mortality. Optimal screening strategies are expected to carefully balance these goals against the associated burden to women and cost to health care systems. Several issues regarding the optimal choice of breast cancer screening strategies remain open. For example, debate surrounds the question of whether regular mammographies are beneficial to women in their forties. Evidence of benefit varies across the relevant randomized clinical trials [Ber9], and there is controversy on the relevance of the suggested benefits for individual women. Consensus panels [GBC97, CTF01] who reviewed the evidence did not find it sufficiently strong to make general recommendations, emphasizing that "women should be informed of the potential benefits and risks of screening mammography and assisted in deciding at what age they wish to initiate the manoeuvre" [CTF01]. In addition to the issue of the appropriate age at which screening should begin, complex open issues include the appropriate frequency of screening examinations; whether women who are at increased risk of breast cancer would benefit from more frequent screening; and what would be the impact of combining multiple screening modalities.

Evaluating alternative screening strategies is difficult because the benefits of screening depend on complex interaction among several factors, including the ability of various screening tests to detect cancer sufficiently early; the time window during which such detection can take place, and its relation to the interval between screening exams; the relative advantage of an early detection compared to waiting for symptoms to arise; the age distribution of onset of pre–symptomatic cancer; competing causes of mortality; and others.

Simulation-based decision models have proved to be an effective way to evaluate health care interventions whose consequences are complex and depend on the interaction of many factors. They can provide a formal structure for supporting optimal choice of screening strategies, cost–effectiveness analysis of specific interventions, and formal optimization of utility functions of interest. These models often generate simulated individual histories by drawing evidence from several sources, including epidemiology and genetic risk factors, relevant clinical trials of secondary prevention and treatment, and studies of tumor growth. A decision model can also support realistic assessments of uncertainty about the relative merits of alternative choices, an aspect that is often underappreciated in policy making [Par02]. The literature on model-based evaluation of screening strategies is now extensive [VBH95, PARM02].

In this article we consider the model by Parmgiani [Par93, Par02], and generalize it by incorporating the possibility of using two breast cancer screening modalities in concert: mammography (MM) and clinical breast examination (CBE). We also update the model inputs to reflect recent contributions to the literature. Existing investigations regarding the balance between mortality reduction and costs have focused on mammography only, and have paid less attention to the combined use of periodic mammography with clinical breast examination. [Dek00, LR95, Bro92, MF92, VVD93, Bro92, BF93, Eli91, Cla92, PK93, EHM89, CGV93, SKP97, BBE01, Fet01, YRP03, KSR03]. Recent studies have shown that periodic clinical breast examinations combined with mammograms improve the overall sensitivity of the screening exam compared with mammography alone, [BHF99, BMB99, BLT00, SZ01], and can be particularly valuable among younger women for whom the sensitivity of mammography alone is relatively low. Logistically, a regular clinical breast examination is easy to administer as part of a routine physical examination, and is less expensive compared to mammography.

To promote more efficient and cost-effective breast cancer early detection programs, we will explore optimal screening strategies in terms of the costs and the quality-adjusted years of life saved. The analyses focus on strategies that combine the use of both mammography and clinical breast examination. The other factors we investigate include age group and screening interval. The primary objectives of this article are to discuss modeling issues arising in optimization of screening strategies with multiple modalities, and to provide methodologic justifications for models and sources of data used in the analyses reported Shen and Parmigiani [SP05].

The results from our investigation will help in the design of more efficient and near optimal early detection programs, thereby maximizing the survival benefit for breast cancer patients while also considering the associated societal costs. This study focuses on breast cancer, but the methods are also applicable to early detection programs for other types of cancer. The proposed research will provide a basis to guide health policy makers in designing optimal and cost-effective screening programs, and in extending such benefits to a large population.

2 Model

2.1 Natural History of Breast Cancer

The basis of our investigation of optimal combinations of screening modalities is a simulation model that can generate individual health histories. It is useful to distinguish the natural history model, which refers to the health histories of women without early detection screening, from the intervention model, which refers to the effects of screening. For a patient with preclinical disease, the natural history model provides a way of simulating age of onset and preclinical sojourn time, or, equivalently, growth rate. Conditional on these, it then simulates the age of the woman and the tumor size at the time of diagnosis. These variables can then be used in turn as covariates in predicting a woman's survival and quality adjusted survival. This multi-stage prediction can be repeated for various screening strategies, by superimposing a history of examinations to the natural history, appropriately simulating results of screening tests based on assumed sensitivity, and appropriately adjusting age and size at detection when early detection takes place. Thus, given women's risk factors, a decision model using Monte Carlo simulations can be employed to jointly model the disease histories and screening interventions, and predict the outcomes of interest.

In the natural history model, breast cancer events are simulated according to the age-specific incidence of preclinical disease and mortality from other causes. For a woman with breast cancer, the natural history model also provides a way of generating her history of disease over time. The natural history of the disease over time requires a description of the transition between different states of the disease. Using the same notation as in Parmigiani [PARM02], we assume that there are four relevant states: H, women who are either disease-free or asymptomatic; P, women who have detectable pre-clinical disease; C, women with clinical manifestation of the disease; and D, women who have died. For women in the cohort who have breast cancer, we generate their ages at the onset of pre-clinical breast cancer, P, ages at the onset of clinical breast cancer, C, via tumor growth, and ages at death D according to corresponding models.

Because the age-specific incidence of pre-clinical disease cannot be directly observed, we have to estimate such a quantity from the sojourn time distribution and age-specific incidence of the clinical disease. Specifically, we can derive the incidence of pre-clinical breast cancer backward from the following deconvolution formula:

$$I_c(y) = \int_0^y w_{hp}(t) w_{pc}(y-t|t) dt, \tag{1}$$

where $I_c(y)$ is the age-specific incidence of clinical breast cancer, w_{hp} is the instantaneous probability of making a transition from H to P, and w_{pc} is age-specific sojourn time density.

Note that the age-specific incidence of clinical breast cancer can be observed and is often well documented in cancer registries or from the control arms of early detection trials. We use the age-cohort-specific breast cancer incidence estimates developed by Moolgavkar et al. [MSL79]. With a given distribution for the sojourn time of the pre-clinical disease state, the age-specific incidence of pre-clinical breast cancer (w_{hp}) can be estimated using the method of Parmigiani and Skates [PS01].

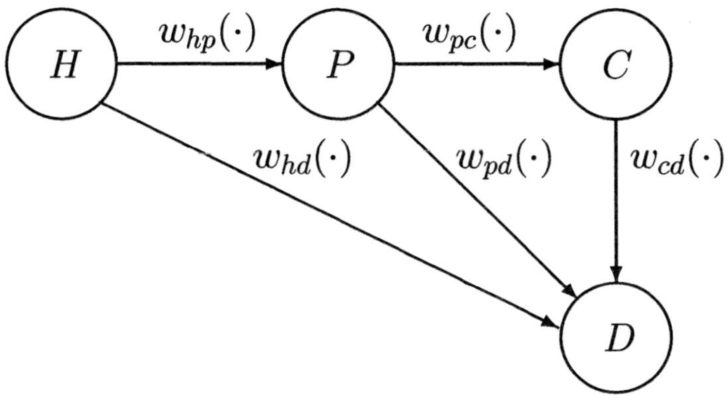

Fig. 1. Summary of states, possible transitions, and transition densities for the natural history model. This scheme describes the progress of breast cancer in the absence of screening. All instantaneous probabilities of transition are indicated next to the corresponding transition. The two subscripts correspond to the origin and destination states, respectively.

However, the estimation of the sojourn time distribution is not straightforward in general [AGL78, DW84, BDM86, ES97, SPV97, SZ99]. In this study, we focus on three commonly used parametric distributions for the sojourn time of the preclinical disease state, which are further modified to incorporate the effect of age at the onset of the preclinical disease.

We first consider a smoothed age-specific exponential sojourn time distribution:

$$w_{pc}(x|\lambda(t)) = \lambda^{-1}(t)\exp(-\lambda^{-1}(t)x),$$

where the mean sojourn time $\lambda(t)$ depends on the woman's age. To incorporate the uncertainty of the parameter into the model, we introduce an inverse gamma prior to the parameter λ, where the two parameters of the inverse gamma distribution are age-specific and are chosen to match the mean and standard deviation of sojourn times estimated from the Canadian National Breast Screening Studies (CNBSS) trials in Shen and Zelen [SZ01].

An alternative assumption for the sojourn time distribution is the log-normal assumption. We consider a modified version taking into account the womanŠs age at the onset of the preclinical disease, while generalizing the model by Spratt et al [SGH86]:

$$w_{pc}(x|t) = \frac{1}{\sqrt{2\pi}\sigma(t)x}\exp\left\{-\frac{1}{2\sigma^2}(\log(x) - \mu(t))^2\right\},$$

where the logarithm of the mean, $\mu(t)$ is specified to be a linear function of the womanŠs age, t. An inverse gamma prior is used to incorporate the uncertainty for the variance, which does not show an age effect [PARM02].

The parameters of the inverse gamma are chosen to match the moments of the reported age-specific variances in [SGH86].

A modified tumor growth distribution of Peer et al [PVH93, PVS96] is also used in our simulation for sensitivity analyses. Specifically, the sojourn time of the preclinical duration is modeled by the tumor growth rate or, equivalently, by the tumor doubling time. In particular, the relationship between sojourn time and tumor doubling time can be expressed as

$$X = \ln(V_p/V_c)DT/\ln(2),$$

where X is the sojourn time, DT is the tumor doubling time, and V_p and V_c are the volumes of a tumor at onset of detectable preclinical disease and at onset of clinical disease, respectively. We assume that the smallest tumor detectable by screening exam is 5mm, and that the average diameter at which breast cancer manifests is 20 mm [PARM02]. Tumor doubling times are assumed to follow an age-dependent log-normal distribution. Note that there is a direct relationship between the tumor growth rate (or doubling time) and the sojourn time in the preclinical duration. The parameters in the sojourn time distribution are estimated to match with the median and 95% quantile of the tumor doubling time based on findings from the Nijmegen trial [PVH93].

2.2 Survival Distributions and Mortality

One primary interest of the study is to evaluate the length of survival after diagnosis of breast cancer with various screening strategies. The survival distribution for women with breast cancer is determined by their age and tumor characteristics at diagnosis, and by the treatment they receive following diagnosis. Women in the cohort who receive periodic screenings are more likely to have breast tumors detected early and thus are more likely to have better prognoses than women who do not receive such screening. However, due to imperfect screening sensitivities and heterogeneity in pre-clinical durations, some breast cancer may still be clinically diagnosed between exams (interval cases). The survival distribution depends on screening only through the tumor characteristics and age at diagnosis.

Based on the natural history model, the tumor size and age at diagnosis are generated for a woman diagnosed to have breast cancer in the cohort. It is well known that lymph node involvement (nodal status) and the estogen receptor (ER) status of the tumor (positive or negative) are also important risk factors, and are related to treatment options and survival. To estimate the number of positive nodes at diagnosis, a predictive model was developed using the data of a womanŠs age and tumor size at diagnosis from the SEER registries [PARM02, SEER98]. A constraint via the truncated Poisson distribution is given to ensure that the number of positive nodes for a screening- detected breast tumor is less than or equal to that for the same woman if her tumor is clinically detected. Without enough evidence to connect ER status with other risk factors, the ER status of a womanŠs breast tumor is simulated independently of the other risk factors, but according to the distribution for the general population. It is estimated that roughly 70% of breast tumors are ER positive [NIH02].

As expected, the tumor characteristics at diagnosis will determine the treatment received thereafter. We assume that women in the cohort are treated according to the guidelines established by the NIH Consensus Conference on Early Breast Cancer (1991), given their risk factors including age, tumor ER status, tumor size, and nodal status at diagnosis. Whether a woman receives tamoxifen depends on her age and tumor ER status. The survival distribution for length with quality of life adjustment after diagnosis of breast cancer is estimated using a Cox regression model with covariates of treatment, age, tumor ER status, primary tumor size, and number of nodes involved. The predictive survival model was established based on a combined analysis of four CALGB trials [PARM02, WWT85, PNK96, WBK94], as described in [PBW99].

For a woman in the cohort, her age-specific mortality due to causes other than breast cancer is obtained from actuarial tables, using a 1960 birth cohort from the census database. If the breast-cancer-specific survival time for a woman is shorter than her simulated natural lifetime, then we assume that she died from breast cancer and contributed to the breast cancer mortality. Otherwise, we assume that she died from a competing cause.

2.3 Sensitivities of Mammography and Clinical Breast Examinations

The sensitivity of a screening program for the early detection of breast cancer plays a critical role in its potential for the reduction of disease-specific mortality. When a screening program involves more than one modality, it is important to obtain the sensitivity of each individual screening modality and the dependence structure among the multiple diagnostic tests [SZ99, SWZ01]. This knowledge provides a basis to guide health policy makers in designing optimal and cost-effective screening programs.

Some recent studies reveal that the sensitivity of a screening exam is likely to depend on tumor size and age at the time of diagnosis [PVS96, SZ01]. Based on literature in the area of breast cancer screening and the estimates of screening sensitivities for both MM and CBE, we consider a model to relate the sensitivity of each modality with age and tumor size at diagnosis, respectively [PARM02, SZ01]. In particular, a logit function is employed to model the effects of age and tumor size at diagnosis on the sensitivities of mammography and clinical breast exam, respectively. We assume the sensitivity of each modality satisfying the following equation:

$$\beta_k(t,d) = \frac{\exp\{\alpha_{k0} + \alpha_{k1}(t-45) + \alpha_{k2}(d-2)\}}{1+\exp\{\alpha_{k0} + \alpha_{k1}(t-45) + \alpha_{k2}(d-2)\}},$$

where t is the age at diagnosis, d is the diameter in centimeters of the primary tumor at diagnosis, $k = 1$ corresponds to mammography, and 2 is for CBE.

The coefficients in the logit models are determined based on the corresponding sensitivity estimates from the CNBSS trials [SZ01] as follows. A sensitivity of mammography of 0.61 corresponds to a woman at age 45 with a tumor diameter of 2cm; a sensitivity of 0.1 corresponds to a woman at the same age but with a tumor size of 0.1 cm; and a sensitivity of 0.66 corresponds to a woman of age 55 with a tumor size of 2cm: $\beta_1(45, 2) = 0.61$, $\beta_1(45, 0.05) = 0.1$ and $\beta_1(55, 2) = 0.66$. Thus, the coefficients in the logit model are solved to be, $\alpha_{10} = 0.447$, $\alpha_{11} = 0.216$ and $\alpha_{12} = 1.36$ for mammography. In the same vein, we can solve the coefficients for the sensitivity of CBE: $\alpha_{20} = 0.364$, $\alpha_{21} = -0.077$ and $\alpha_{22} = 1.31$. Moreover, because the sensitivity can vary from subject to subject even when given the same age and tumor size [KGB98], we use a beta distribution to reflect such a random variation for each sensitivity, while matching the corresponding mean and variance for the estimated sensitivity from the CNBSS trials, as reported in Shen and Zelen [SZ01].

The Health Insurance Plan of Greater New York (HIP) trial and the CNBSS both offered independent annual clinical breast exams and mammograms to women in their study arms, which gave us an opportunity to assess the dependence between the two screening modalities. The analyses based on data from these trials indicate that mammography and clinical breast examinations contribute independently to the detection of breast cancer [SWZ01]. Therefore, given the sensitivity of each individual screening modality, the overall sensitivity of a screening program using both MM and CBE is as follows:

$$\beta(t, d) = \beta_1(t, d) + \beta_2(t, d) - \beta_1(t, d)\beta_2(t, d),$$

when the two modalities are independent to each other.

2.4 Costs of Screening Programs

As expected in screening practices, the primary costs of a screening program is proportional to the total number of mammograms and clinical breast examinations given. Although there are additional costs related to follow-up confirmative tests such as a biopsy, and costs for the treatment of breast cancer at various stages after diagnosis, we will focus only on the cost of screening examinations in the current study. On its website, the National Cancer Institute lists the estimated cost of mammography in 2002 at $100-200, and acknowledges that the cost can vary widely among different centers and hospitals. Since it is frequently part of a routine physical examination, the cost of a CBE is often less than that of mammography. In a public website promoting cancer prevention, the estimated cost for an annual CBE is $45-55, whereas the cost of MM is $75-150 [PRE02]. In the decision analysis, it is clear that the cost ratio of MM and CBE determines the results in the comparison of different screening strategies. Therefore, we investigate the effects of two cost ratios (1.5 and 2) between MM and CBE, and allow the cost for a CBE to be $100. For simplification, we will not adjust for the type of currency, or for inflation over the years.

3 Optimization of Screening Strategies and Sensitivity Analyses

The focus of this investigation is to compare the effects of different breast cancer screening policies and the costs directly related to these policies, based on the models introduced in the last sections. The health outcome of interest is the expected gain in quality-adjusted survival. We interpret this quality adjustment to be relative to a typical health history rather than that of a state of perfect health [PBW99]. Quality adjustments are important because they allow, with certain limitations, to account for the effects of medical intervention on morbidity as well as mortality. In screening this is especially important becasue of the so-called overdiagnosis problem. While beneficial to many women, screening leads to discovering cancer that would have not otherwise affected certain womens health. While lenght of life may be unaffected, this is a considerable loss of quality of life. Also, early detection can prolong the portion of one's life spent as a cancer survivor. The specific quality adjustments used in our model are the same as Parmigiani [PARM02].

The marginal effectiveness for each screening strategy is calculated based on the difference between the expected quality-adjusted life in years for women in a cohort undergoing screening versus the same cohort of women without screening. The summaries of interest are the expected gain in quality life years (QALYS) and the expected total monetary cost for each screening strategy. Marginal cost is the difference in total cost between the screened and unscreened cohorts. The marginal effectiveness for each screening strategy is the difference between the expected QALYS in the screened and unscreened cohorts. The ratio is marginal cost per year of quality-adjusted life saved (MCYQLS).

Three important issues to consider for screening policies are the age at which a woman should start a screening program, the screening frequency, and what screening modalities are to be used. In this study, we will evaluate a total of 48 screening strategies with the following combinations:

- The age to begin and end periodic screening: 40-79, 45-79, and 50-79 years;
- The interval between consecutive examinations: 0.5, 1, 1.5 and 2 year(s);
- The combined use of MM and CBE: whether mammogram or CBE is given for every one or every two exams.

Using the model described earlier, we generate a cohort of women and their natural histories of disease, and assess how the screening strategies interact

Table 1. Balance sheet for two alternative screening strategies: annual MM and CBE screening and biennial MM and annual CBE. In both cases screening starts at 40 years of age and stops at age 79. Values are increments compared to no screening for a cohort of 10000 breast cancer women.

	Screening Strategy	
	MM/1,CBE/1	MM/2,CBE/1
Additional number of MM per woman	33	17
Additional number of CBE per woman	33	33
Additional number of false positives per woman	5.2	4.3
Additional years of life saved per woman	0.144	0.124
Additional women detected in preclinical state	867	810
Women treated unnecessarily	55	51

with the disease process and the survival after diagnosis. The quantities of interest are estimated using 100,000 Monte Carlo replicates, for each of the screening strategies.

In summary, we simulate a birth cohort of 100,000 women and follow them through the years. A fraction of them will develop breast cancer according to the age-specific incidence of pre-clinical breast cancer. For those women, we generate the natural histories of their disease, which include their ages at the onset of the preclinical disease, the pre-clinical durations (via tumor growth rates), and ages at the clinical onset of the disease. When a screening strategy is provided to a woman during a pre-clinical disease state, the probability that her cancer will be detected by this screening strategy is generated using the equations in Section 2.3, based on her age and tumor size at the time of the exam. If the diagnosis is missed during the exam, her breast cancer may be detected at her next scheduled exam or it may clinically manifest before the next exam depending on the sojourn time of her preclinical disease state. Once a woman is diagnosed to have breast cancer, we obtain her tumor size and age at the time of detection. The information is then used to predict the woman's survival and quality-adjusted survival after the detection using models developed in Section 2.2. The expected cost is estimated based on the average cost of screening exams from the 100,000 women for each screening strategy in the simulation.

A balance sheet is a summary of the expected benefits and harms of an intervention. Its goal is to inform decision makers, and enable them to weigh benefits and harms according to their individual values [MS99, BIG99]. Table 1 is a balance sheet for evaluating two alternative screening strategies, based on the model of this chapter. We consider annual MM and CBE screening (denoted by MM/1, CBS/1) and biennial MM and annual CBE (MM/2, CBE/1). Differences between the two columns can inform decision makers about whether annual or biennial MM are to be preferred once annual CBE is planned. Elmore and colleagues [EBM98] collected data on a retrospective cohort study of breast cancer screening and diagnostic evaluations among 2400 women who were 40 to 69 years old at study entry. False positive results occurred in 6.5% of the mammograms, an estimate that was used here to translate the estimated number of additional tests into estimated false positives. In addition we assume that positive CBE's would be followed by a mammography, that 10% of CBE are false positive, and the two tests are independent of each other. Then the overall false positive number per woman for the 1st strategy is: $(0.065+0.1-0.1*0.065)*33 = 5.2$; and the overall false positive number per woman for the 2nd strategy is $0.065*17+0.1*33-0.1*0.065*17 = 4.3$.

In Section 2.1, three model specification are discussed for the distribution of sojourn times in the preclinical state of the disease. It is of interest to investigate how these different models may impact the QALYs and expected cost of each screening strategy reported by [SP05]. We find that the analyses are fairly robust for the three model assumptions. The marginal QALYS is slightly higher (about 1-2%) for the lognormal model than for the exponential model for a given screening strategy. The relative marginal costs and QALYS among the screening strategies under evaluation are similar for the three model choices.

4 Discussion

Much attention has been focused on the early detection capabilities of new breast cancer screening technologies, including advances in mammography and MRI. The importance of clinical breast examination in breast cancer screening programs seems to be unclear. Even though some recent studies have indicated that regular CBE in addition to MM can be important in the early detection of breast cancer, few studies have investigated the optimal use of both mammography and clinical breast exam to reduce the mortality of breast cancer while balancing the associated burdens and costs to women and to the health care system.

Developing early detection guidelines and making public health policy requires careful consideration of the long-term benefits, costs, and feasibility associated with the screening strategies. In Shen and Parmigiani [SP05], we explore the trade-off between the QALYS and costs related to each screening strategy among several combinations of starting ages of screening, frequencies of screening, and the use of two screening modalities. The study indicates that starting from 40 years of age, a biennial mammogram is often cost-effective for women who undergo annual clinical breast exams. Given the cost to women who are already receiving care for other health issues or regular check-ups in a clinic, an annual CBE as part of their routine examination should not add much burden. Our analyses also indicate that CBE alone cannot replace regular mammography in screening practice, but can be used complementarily or alternatively in a screening program.

The decision analysis methodology and simulation techniques developed for this study can be directly applied to investigate other screening strategies, and even to other chronic diseases with certain modifications to the models. We have modeled screening sensitivity for MM and CBE, respectively, through age and tumor size at diagnosis. We have also introduced random variations for the parameters to incorporate uncertainty of data input and population heterogeneity. We have considered various models and parameters, and have derived them based on data from the large randomized breast cancer screening trials of the HIP [Sha97], CNBSS [MTB97], and the Nijmegen Trial [PVH93]. We have performed sensitivity analyses to assess the robustness of the patterns of benefit and cost with the alternative models.

Our study has several limitations. The cost of a biopsy following a CBE or MM that is positive for breast cancer has not been considered in the analysis. Moreover, we have not included the potential costs of false-positive exams, such as the anxiety, fear and discomfort that are associated with a biopsy. In fact, it is often difficult to convert these factors into dollar amounts [EBM98]. In addition, we have not included important cost components, which are the costs of follow-up procedures undertaken after the detection of breast cancer. This is in part due to the great variation in treatment protocols and in the cost of treating breast cancer that has existed over the years. Finally, we have used a hypothetical birth cohort of women with 100% compliance in the simulations for each screening strategy. In reality, it is rare to have 100% compliance for any screening program, and a real cohort would be dynamic, which would include changes in the cohort due to migration.

Acknowledgements

The authors thank Professor Marvin Zelen for his encouragement. This work was partially supported by the National Institutes of Health Grants R01 CA-79466 (YS), and by the NCI under the Johns Hopkins SPORE in Breast cancer P50CA88843 (GP).

References

[WTHR03] Weir HK, Thun MJ, Hankey BF, Ries LA, Lowe HL, Wingo PA, Jemal A, Ward E, Anderson RN, Edwards B.: Annual report to the nation on the status of cancer, 1975-2000, featuring the uses of surveillance data for cancer prevention and control. *J Natl Cancer Inst*, **95**(17):1276–1299, 2003

[SEER98] National Cancer Institute: The surveillance, epidemiology, and end results (seer) program. `www.seer.cancer.gov`, 1998.

[IARC99] Research on Cancer (IARC) IA. The cancer-mondial website. `www.dep.iarc.fr`, 1999.

[FE03] Fletcher SW; Elmore JG. Mammographic screening for breast cancer. *N Engl J Med*, 348(17):1672–1680, 2003.

[Wan03] Wang L. Mammography and beyond: Building better breast cancer screening tests. *J Natl Cancer Inst*, 94(18):1346–1347, 2002.

[Ber9] Berry DA. Benefits and risks of screening mammography for women in their forties: A statistical appraisal. *J Natl Cancer Inst*, 90:1431–1439, 1998.

[GBC97] Gordis L; Berry D; Chu S; et al. Breast cancer screening for women ages 40-49. *J Natl Cancer Inst*, 89:1015–1026, 1997.

[CTF01] Canadian Task Force on Preventive Health Care. Preventive health care, 2001 update: screening mammography among women aged 40–49 years at average risk of breast cancer. *CMAJ*, 164(4):469–476, 2001.

[Par02] Parmigiani G. Measuring uncertainty in complex decision analyses models. *Statistical Methods in Medical Research*, 11(6):513–37, 2002.

[VBH95] van Oortmarseen G; Boer R; Habbema J. Modeling issues in cancer screening. *Statistical Methods in Medical Research*, 4:33–54, 1995.

[PARM02] Parmigiani G. *Modeling in Medical Decision Making.* Wiley, Chichester, 2002.

[Par93] Parmigiani G. On optimal screening ages. *J Amer Stat Assoc*, 88:622–628, 1993.

[Dek00] De Koning H. Breast cancer screening; cost-effectiveness in practice. *Eur J Radiol*, 33:32–7, 2000.

[LR95] Lindfors KK; Rosenquist CJ. The cost-effectiveness of mammographic screening strategies. *J Am Med Assoc*, 274:881–884, 1995.

[Bro92] Brown ML. Sensitivity analysis in the cost-effectiveness of breast cancer screening. *Cancer*, 69 (7 Suppl):1963–1967, 1992.

[MF92] Mushlin AI; Fintor L. Is screening for breast cancer cost-effective? *Cancer*, 69 (7 Suppl):1957–1962, 1992.

[VVD93] van Ineveld BM; van Oortmarssen GJ; de Koning HJ; Boer R; van der Maas PJ. How cost-effective is breast cancer screening in different EC countries? *European Journal of Cancer*, 29:1663–1668, 1993.

[BF93] Brown ML; Fintor L. Cost-effectiveness of breast cancer screening: preliminary results of a systematic review of the literature. *Breast Cancer Res Treat*, 25:113–118, 1993.

[Eli91] Elixhauser A. Costs of breast cancer and the cost-effectiveness of breast cancer screening. *Int J Technol Assess Health Care*, 7:604–615, 1991.

[Cla92] Clark RA. Economic issues in screening mammography. *Am J of Roentgenology*, 158:527–534, 1992.

[PK93] Parmigiani G; Kamlet M. Cost-utility analysis of alternative strategies in screening for breast cancer. In C Gatsonis; J Hodges; RE Kass; N Singpurwalla, eds., *Case Studies in Bayesian Statistics*, 390–402. Springer, New York, 1993.

[EHM89] Eddy DM; Hasselblad V; McGivney W; Hendee W. The value of mammography screening in women under age 50 years. *J Am Med Assoc*, 259:1512–1519, 1989.

[CGV93] Carter R; Glasziou P; van Oortmarssen G; de Koning H; Stevenson C; Salkeld G; Boer R. Cost-effectiveness of mammographic screening in Australia. *Austr J Pub Health*, 17:42–50, 1993.

[SKP97] Saltzmann P; Kerlikowske K; Phillips K. Cost-effectiveness of extending screening mammography guidelines to include women 40 to 49 years of age. *Ann Intern Med*, 127:955–965, 1997.

[BBE01] Burnside E; Belkora J; Esserman L. The impact of alternative practices on the cost and quality of mammographic screening in the United States. *Clinical Breast Cancer*, 2(2):145–152, 2001.

[Fet01] Fett M. Computer modelling of the Swedish two country trial of mammographic screening and trade offs between participation screening interval. *J Med Screen*, 8(1):39–45, 2001.

[YRP03] Yasmeen S; Romano P; Pettinger M; Chlebowski R; Robbins J; Lane D; Hendrix S. Frequency and predictive value of a mammographic recommendation for short-interval follow-up. *J Natl Cancer Inst*, 95(6):429–436, 2003.

[KSR03] Kerlikowske K; Smith-Bindman R; Sickles E. Short-interval follow-up mammography: Are we doing the right thing? *J Natl Cancer Inst*, 95(6):418–419, 2003.

[BHF99] Barton MB; Harris R; Fletcher SW. Does this patient have breast cancer? The screening clinical breast examination: Should it be done? How? *J Am Mec Assoc*, 282:1270–80, 1999.

[BMB99] Baines CJ; Miller AB; Bassett AA. Physical examination. its role as a single screening modality in the canadian national breast screening study. *Cancer*, 63:1816–22, 1989.

[BLT00] Bobo J; Lee N; Thames SF. Findings from 752081 clinical breast examinations reported to a national screening program from 1995 through 1998. *J Natl Cancer Inst*, 92:971–6, 2000.

[SZ01] Shen Y; Zelen M. Screening sensitivity and sojourn time from breast cancer early detection clinical trials: mammograms and physical examinations. *J Clin Oncol*, 19:3490–9, 2001.

[SP05] Shen Y; Parmigiani G. A model-based comparison of breast cancer screening strategies: Mammograms and clinical breast examinations. *Cancer Epidemiology, Biomarkers and Prevention*, in press, 2005.

[MSL79] Moolgavkar SH; Stevens RG; Lee JAH. Effect of age on incidence of breast cancer in females. *J Natl Cancer Inst*, 62:493–501, 1979.

[PS01] Parmigiani G; Skates S. Estimating the age of onset of detectable asymptomatic cancer. *Mathematical and Computer Modeling*, 33:1347–1360, 2001.

[AGL78] Albert A; Gertman P; Louis T. Screening for the early detection of cancer: I. the temporal natural history of a progressive disease state. *Mathematical Biosciences*, 40:1–59, 1978.

[DW84] Day NE; Walter SD. Simplified models of screening for chronic disease: Estimation procedures from mass screening programmes. *Biometrics*, 40:1–13, 1984.

[BDM86] Brookmeyer R; Day NE; Moss S. Case-control studies for estimation of the natural history of preclinical disease from screening data. *Statistics in Medicine*, 5:127–138, 1986.

[ES97] Etzioni RD; Shen Y. Estimating asymptomatic duration in cancer: the AIDS connection. *Stat in Med*, 16:627–644, 1997.

[SPV97] Straatman H; Peer PG; Verbeek AL. Estimating lead time and sensitivity in a screening program without estimating the incidence in the screened group. *Biometrics*, 53:217–229, 1997.

[SZ99] Shen Y; Zelen M. Parametric estimation procedures for screening programmes: Stable and nonstable disease models for multimodality case finding. *Biometrika*, 86:503–515, 1999.

[SGH86] Spratt JS; Greenberg RA; Heuser LS. Geometry, growth rates, and duration of cancer and carcinoma in situ of the breast before detection by screening. *Cancer Research*, 46:970–974, 1986.

[PVH93] Peer P; van Dijck JAAM; Hendriks J; Holland R; Verbeek ALM. Age-dependent growth rate of primary breast cancer. *Cancer*, 71:3547–3551, 1993.

[PVS96] Peer P; Verbeek A; Straatman H; Hendriks J; Holland R. Age-specific sensitivities of mammographic screening for breast cancer. *Breast Cancer Research and Treatment*, 38:153–160, 1996.

[NIH02] National Cancer Institute: Chemoprevention of Estrogen Receptor (ER) Negative Breast Cancer Preclinical Studies. NIH homepage. `http://grants1.nih.gov/grants/guide/rfa-files/RFA-CA-03-005.html`, 2002.

[WWT85] Wood W; Weiss R; Tormey D; Holland J; Henry P; Leone L; et al. A randomized trial of CMF versus CMFVP as adjuvant chemotherapy in women with node-positive stage ii breast cancer: a CALGB study. *World J Surg*, 9:714–718, 1985.

[PNK96] Perloff M; Norton L; Korzun A; Wood W; Carey R; Gottlieb A; et al. Postsurgical adjuvant chemotherapy of stage ii breast carcinoma with or without crossover to a non-cross-resistant regimen: a cancer and leukemia group b study. *J Clin Oncol*, 14:1589–98, 1996.

[WBK94] Wood W; Budman D; Korzun A; Cooper M; Younger J; Hart R; et al. Dose and dose intensity of adjuvant chemotherapy for stage ii, node-positive breast carcinoma. *N Engl J Med*, 330:1253–1259, 1994.

[PBW99] Parmigiani G; Berry DA; Winer EP; Tebaldi C; Iglehart JD; Prosnitz L. Is axillary lymph node dissection indicated for early stage breast cancer—a decision analysis. *J Clin Oncol*, 17(5):1465–1473, 1999.

[SWZ01] Shen Y; Wu D; Zelen M. Testing the independence of two diagnostic tests. *Biometrics*, 57:1009–1017, 2001.

[KGB98] Kerlikowske K; Grady D; Barclay J; Frankel SD; Ominsky SH; Sickles EA; Ernster V. Variability and accuracy in mammographic interpretation using the american college of radiology breast imaging reporting and data system. *J Natl Cancer Inst*, 90(23):1801–1809, 1998.

[PRE02] PREVENTION. Cancer tests worth paying for. http://www.prevention.com/cda/ feature/0,1204,876,00.html, 2002.

[MS99] Matchar DB; Samsa GP. Using outcomes data to identify best medical practice: the role of policy models. *Hepatology*, Jun;29(6 Suppl):36S–39S, 1999.

[BIG99] Barratt A; Irwig L; Glasziou P; et al. Users' guides to the medical literature: XVII. how to use guidelines and recommendations about screening. evidence-based medicine working group. *J Am Mec Assoc*, 281(21):2029–2034, 1999.

[EBM98] Elmore JG; Barton MB; Moceri VM; Polk S; Arena PJ; Fletcher SW. Ten-year risk of false positive screening mammograms and clinical breast examinations. *N Engl J Med*, 338:1089–1096, 1998.

[Sha97] Shapiro S. Periodic screening for breast cancer: The HIP randomized controlled trial. *Monogr Natl Cancer Inst*, 22:27–30, 1997.

[MTB97] Miller A; To T; Baines C; Wall C. The Canadian national breast screening study: Update on breast cancer mortality. *Monogr Natl Cancer Inst*, 22:37–41, 1997.

Sequential Analysis of Quality of Life Rasch Measurements

Veronique Sebille[1] and Mounir Mesbah[2]

[1] Laboratoire de Biostatistiques, Faculté de Pharmacie, Université de Nantes, 1 rue Gaston Veil, BP 53508, 44035 Nantes Cedex 1, France. veronique.sebille@univ-nantes.fr

[2] Laboratoire de Statistique Théorique et Appliquée (LSTA), Université Pierre et Marie Curie - Paris VI, Boîte 158, - Bureau 8A25 - Plateau A. 175 rue du Chevaleret, 75013 Paris, France mesbah@ccr.jussieu.fr

Summary. Early stopping of clinical trials either in case of beneficial or deleterious effect of treatment on quality of life (QoL) is an important issue. QoL is usually evaluated using self-assessment questionnaires and responses to the items are combined into scores assumed to be normally distributed (which is rarely the case). An alternative is to use item response theory (IRT) models such as the Rasch model for binary items which takes into account the categorical nature of the items.

Sequential analysis and mixed Rasch models (MRM) were combined in the context of phaseII non-comparative trials. The statistical properties of the Sequential Probability Ratio Test (SPRT) and of the Triangular Test (TT) were compared using MRM and traditional average scores methods (ASM) by means of simulations.

The type I error of the SPRT and TT was correctly maintained for both methods. While remaining a bit underpowered, MRM displayed higher power than the ASM for both sequential tests. Both methods allowed substantial reductions in average sample numbers as compared with fixed sample designs (about 60%).

The use of IRT models in sequential analysis of QoL endpoints is promising and should provide a more powerful method to detect therapeutic effects than the traditional ASM.

Key words: Quality of life; Item Response Theory; Rasch models; Sequential Probability Ratio Test; Triangular Test; Clinical Trials

1 Introduction

Clinical trials usually focus on endpoints that traditionally are biomedical measures such as disease progression or survival for cancer trials, survival or hospitalization for heart failure trials. However, such endpoints do not reflect patient's perception of his or her well-being and satisfaction with therapy. Health-Related Quality of Life (QoL) which refers to "the extent to which one's usual or expected physical, emotional and social well-being are affected by a medical condition or its treatment" is an important health outcome (Cella and Bonomi, 1995; Fairclough, 2002).

Non-comparative phase II trials, which are commonly designed to evaluate therapeutic efficacy as well as further investigation of the side-effects and potential risks associated with therapy, often use QoL endpoints. Early stopping of such trials either in case of beneficial or deleterious effect of the treatment on QoL is an important matter (Cannistra, 2004). Ethical concerns and economic reasons for the use of early stopping rules include the fact that patients are recruited sequentially in a trial and that data from early recruited patients are available for analysis while later patients are still being included in the trial. Such a framework offers the possibility of using the emerging evidence to stop the study as soon as the treatment effect on QoL becomes clear. Early stopping of a trial can occur either for efficacy (when the trial seems to show clear treatment advantage), safety (when the trial seems to show clear treatment harm) or futility reasons (when the trial no longer has much chance of showing any treatment benefit). However, it is well-known that multiple looks at data result in inflation of the type I error α and in the risk of over-interpretation of interim results. Thus, specific early termination procedures have been developed to allow for repeated statistical analyses on accumulating data and for stopping a trial as soon as the information is sufficient to conclude. Among the sequential methods that have been developed over the last few decades (Pocock, 1977; O'Brien and Fleming, 1979; Lan and De Mets, 1983), the Sequential Probability Ratio Test (SPRT) and the Triangular Test (TT), which were initially developed by Wald (Wald, 1947) and Anderson (Anderson, 1960) and later extended by Whitehead to allow for sequential analyses on groups of patients (Whitehead and Jones, 1979; Whitehead and Stratton, 1983) have some of the interesting following features. They allow for: (i) early stopping under H_0 or under H_1, (ii) the analysis of quantitative, qualitative or censored endpoints, (iii) type I and II errors to be correctly maintained at their desired planning phase values, (iv) substantial sample size reductions as compared with the single-stage design (of about 30% reductions can often be achieved).

Patient's QoL is usually evaluated using self-assessment questionnaires which consist of a set of questions often called items (which can be dichotomous or polytomous) which are frequently combined to give scores for scales or subscales. The common practice is to work on average scores which are generally assumed to be normally distributed. However, these average scores are rarely normally distributed and usually do not satisfy a number of basic measurement properties including sufficiency, unidimensionality, or reliability.

More important, these scores are often used, knowingly or not, as a reduction of a bigger amount of data (each score is a "sufficient statistic" for a given set of observed categorical items, and then is used as a surrogate for this set of items), without indroducing clearly the mechanism of such reduction in the likelihood.

In Educational Sciences framework, or more generally in psychometry or sociometry, models relating a set of observed items to a hidden latent concept are called measurement models. Otherwise, models relating concepts (directly observed or latent) are called analysis models. Item Response Theory (IRT), which was first mostly developed in educational testing, takes into account the multiplicity and categorical nature of the items by introducing an underlying response model (Fisher and Molenaar, 1995) relating those items to a latent parameter having the nice property to be interpreted as the true individual QoL. In this framework, the probability of response of a patient on an item depends upon two different parameters: the "ability level" of the person (which reflects his/her current QoL) and the "difficulty" of the item (which reflects somehow the capacity of that specific item in discriminating between good and bad QoL). IRT models are specific generalized linear models which were more developed from a "measurement" point of view than from an "analysis" one. However, an equivalent modeling framework could be repeated measures logistic regression since IRT modeling deals with repeated items aimed at measuring an unobserved latent trait. IRT modeling, as a tool for scientific measurement, is not quite well established in the clinical trial framework despite a number of advantages offered by IRT to analyze clinical trial data including: helpful solutions to missing data problems, the possibility to determine whether items are biased against certain subgroups, an appropriate tool for dealing with ceiling and floor effects (Holman et al., 2003a). Moreover, it has been suggested that IRT modeling offers a more accurate measurement of health status and thus should be more powerful to detect treatment effects (McHorney et al., 1997; Kosinski et al., 2003). Hence, IRT modeling could be an interesting alternative to traditional sequential analysis of QoL endpoints based only on average scores. Thus, we tried to evaluate the benefit of combining sequential analysis and IRT methodologies in the context of phase II non-comparative trials. We performed sequential analysis of QoL endpoints (obtained from the observed data) using IRT modeling and we compared the use of IRT modeling methods with the traditional use of average scores methods.

2 Methods

2.1 IRT models

The basic assumption for IRT models is the unidimensionality property stating that all items of a questionnaire should measure the same underlying concept (e.g., QoL) often called latent trait and noted θ. Another important assumption of IRT models, which is closely related to the former, is the concept of local independence meaning that items should be conditionally independent given the latent trait θ. It can be expressed mathematically by writing the joint probability of a response pattern given the latent trait θ as a product of marginal probabilities. Let X_{ij} be the answer for subject i to item j and let θ_i be the unobserved latent variable (also called the ability, in our context, we call it the QoL) for subject i (i = 1, ..., N; j = 1, ..., k).

$$P\left(X_{i1}=x_{i1},X_{i2}=x_{i2},...,X_{ik}=x_{ik}/\theta_i\right)=\prod_{j=1}^{k}P\left(X_{ij}=x_{ij}/\theta_i\right)$$

where (X_{i1}, X_{i2}, ..., X_{ik}) are a set of items (either dichotomous or polytomous). In other words, the person's ability or the person's QoL should be the only variable affecting individual item response. For any person i, or more accurately for any given θ_i, the correponding response values $X_{ij}$to the various items j (j=1 to k) are independent as they were choosen randomly.

2.2 The Rasch Model

For binary items, one of the most commonly used IRT model is the Rasch model, sometimes called the one parameter logistic model (Rasch, 1960). The Rasch model specifies the conditional probability of a patient's response x_{ij} given the latent variable θ_ind the item parameters β_j:

$$P\left(X_{ij}=x_{ij}/\theta_i,\beta_j\right)=f\left(x_{ij}/\theta_i;\beta_j\right)=\frac{e^{(\theta_i-\beta_j)\,x_{ij}}}{1+e^{(\theta_i-\beta_j)}},$$

where β_j is called the difficulty parameter for item j (j = 1, ..., k). Contrasting with other IRT models, in the Rasch model, a subject's total score, $S_i = \sum_{j=1}^{k} X_{ij}$ is a sufficient statistic for a specific latent trait or ability θ_i. Thus, when the total score of a questionnaire with binary items is used as a measure of QoL, it is "knowingly or not" assumed that the Rasch model is the true underlying model.

2.3 Estimation of the parameters

Several methods are available for estimating the parameters (the θs and βs) in the Rasch model (Hamon, and Mesbah, 2002) including: joint maximum likelihood (JML), conditional maximum likelihood (CML), and marginal maximum likelihood (MML). JML is used when person and item parameters are considered as unknown fixed parameters. However, this method gives asymptotically biased and inconsistent estimates (Haberman, 1977). The second method CML consists in maximizing the conditional likelihood given the total score in order to obtain the items parameters estimates. The person parameters are then estimated by maximizing the likelihood using the previous items parameters estimates. This method has been shown to give consistent and asymptotically normally distributed estimates of item parameters (Andersen, 1970). The last method MML is used when the Rasch model is interpreted as a mixed model with θ as a random effect having distribution h(θ, ζ) with unknown parameters ζ. The distribution h is often assumed to belong to some family distribution (often Gaussian) and its parameters are jointly estimated with the item parameters. As with the CML method, the MML estimators for the item parameters are asymptotically efficient (Thissen, 1982). Furthermore, since MML does not presume existence of a sufficient statistic (unlike CML), it is applicable to virtually any type of IRT model.

2.4 Sequential Analysis

Traditional Sequential Analysis

In the traditional framework of sequential analysis (Wald, 1947; Whitehead, 1997; Jennison and Turnbull, 1999), θ_i is assumed to be observed (not to be a latent value) and the observed score S_i is used as a "surrogate" of the true latent trait θ_i. In that setting, we generally assume that $\theta_1, \theta_2, \ldots, \theta_N$ are N independent variables following distribution f(θ_1), f(θ_2), ..., f(θ_N) with unknown individual parameters φ_i and φ_i (i = 1, ..., N). We shall assume that those individual parameters are the same, i.e., that $\forall$i (i = 1, ..., N), $\varphi_i = \varphi$ (parameter of interest) and $\varphi_i = \varphi$ (vector of nuisance parameters), and that the trial involves the comparison of the two following hypotheses: H_0: $\varphi < 0$ against H_1: $\varphi > 0$, In that classical setting, the decision is based on the likelihood of the data, i.e. on:

$$L(\theta_1, \theta_2, ..., \theta_N / \varphi, \varphi) = f_{\varphi,\varphi}(\theta_1) \cdot f_{\varphi,\varphi}(\theta_2) ... f_{\varphi,\varphi}(\theta_N)$$

Values $\varphi_0 < \varphi_1$ are chosen and the following continuation region is used for the sequential test for suitable values of $B_{\alpha,\beta} < 1 < A_{\alpha,\beta}$ (Wald, 1947):

$$\frac{L(\theta_1, \theta_2, ..., \theta_N / \varphi_1, \hat{\varphi}(\varphi_1))}{L(\theta_1, \theta_2, ..., \theta_N / \varphi_0, \hat{\varphi}(\varphi_0))} \in (B_{\alpha,\beta}, A_{\alpha,\beta})$$

where $\hat{\varphi}(\varphi_0)$ ($\hat{\varphi}(\varphi_1)$) denotes the maximum likelihood estimate of φ for $\varphi = \varphi_0$ ($\varphi = \varphi_1$).

If the terminal value of the likelihood ratio is below $B_{\alpha,\beta}$, then H_0 is no rejected, if it is above $A_{\alpha,\beta}$, then H_0 is rejected. It is well-known that if φ_0 and φ_1 are assumed to be small (Whitehead, 1997), the log likelihood function $l(\varphi, \hat{\varphi}(\varphi))$ can be approximated using Taylor expansion up to quadratic terms in φ for $\varphi = \varphi_0$ or $\varphi = \varphi_1$. Thus, the continuation region can be simplified in the following way:

$$Z(S) \in \left(\frac{\log B}{\varphi_1 - \varphi_0} + \frac{1}{2}(\varphi_1 + \varphi_0) \cdot V(S), \frac{\log A}{\varphi_1 - \varphi_0} + \frac{1}{2}(\varphi_1 + \varphi_0) \cdot V(S) \right)$$

where the Z(S) statistic is the efficient score for φ depending on the observed scores S, and the V(S) statistic is Fisher's information for φ.

More precisely: $Z(S) = l_{\varphi}(0, \hat{\varphi}(0))$ *and* $V(S) = -\{l^{\varphi\varphi}(0, \hat{\varphi}(0))\}^{-1}$ where:

- $l_{\varphi}(0, \hat{\varphi}(0))$ denotes the first partial derivative of $l(\varphi, \varphi)$ with respect to φ, evaluated at $(0, \hat{\varphi}(0))$,

- the leading element of the inverse of the matrix of second derivatives is denoted by $l^{\varphi\varphi}(\varphi, \varphi)$, that is:

$$\{l^{\varphi\varphi}(\varphi, \varphi)\}^{-1} = l_{\varphi\varphi}(\varphi, \varphi) - \{l_{\varphi\varphi}(\varphi, \varphi)\}' \{l_{\varphi\varphi}(\varphi, \varphi)\}^{-1} l_{\varphi\varphi}(\varphi, \varphi)$$

where $l_{\varphi\varphi}(0, \hat{\varphi}(0))$ denotes the second partial derivative of $l(\varphi, \varphi)$ with respect to φ, evaluated at $(0, \hat{\varphi}(0))$, and $l_{\varphi\varphi}(0, \hat{\varphi}(0))$ denotes the mixed derivative.

Asymptotic distributional results have shown that for large samples and small φ, Z(S) follows a normal distribution: Z(S) $\sim$ N(φV(S), V(S)). More precisely, let a sequential study with up to K analyses produce the sequence of test statistics (Z_1(S), Z_2(S), ..., Z_K(S)). The sequence (Z_1(S), Z_2(S), ..., Z_K(S)) is multivariate normal with: Z_k(S) $\sim$ N(φV_k(S), V_k(S)) and Cov(Z_{k1}(S), Z_{k2}(S)) = V_{k1}(S) for k = 1, 2, ..., K and $1 \le k_1 \le k_2 \le K$ (Whitehead, 1997; Jennison and Turnbull, 1999).

Sequential Analysis based on Rasch measurements

We shall now be interested in the latent case, i.e., the case where θ_i is unobserved. Thus, the likelihood will be different, because the likelihood is traditionally a function of the observations, not of the unobserved variables. The following steps will be used in order to obtain the likelihood that we need for sequential testing.

1. The Rasch model specifies the conditional distribution of item response given the latent variable θ_i and item parameters β_j:

$$f(x_{ij}/\theta_i; \beta_j) = \frac{e^{(\theta_i - \beta_j)\, x_{ij}}}{1 + e^{(\theta_i - \beta_j)}} = f_{\beta_j}(x_{ij}/\theta_i)$$

2. We can then write:

$$f_{\beta_j}(x_{ij}/\theta_i) = \frac{f_{\beta_j,\varphi,\varphi}(x_{ij}, \theta_i)}{f_{\varphi,\varphi}(\theta_i)}$$

and get the joint distribution of the observed x_{ij} and the latent variable θ_i:

$$f_{\beta_j,\varphi,\varphi}(x_{ij}, \theta_i) = f_{\varphi,\varphi}(\theta_i) \cdot f_{\beta_j}(x_{ij}/\theta_i)$$

3. The likelihood is obtained after marginalizing over the unobserved latent variable θ_i:

$$f_{\beta_j,\varphi,\varphi}(x_{ij}, .) = \int f_{\varphi,\varphi}(\theta_i)\; f_{\beta_j}(x_{ij}/\theta_i)\; d\theta_i$$

4. Local independence of items allows us to derive the likelihood of a subject:

$$f_{\beta_1,\beta_2,...,\beta_k,\varphi,\varphi}(x_{i1}, x_{i2}, ..., x_{ik}) = \int f_{\varphi,\varphi}(\theta_i) \prod_j f_{\beta_j}(x_{ij}/\theta_i)\, d\theta_i$$

5. Finally, independence of subjects allows us to obtain the likelihood:

$$\prod_i f_{\beta_1,\beta_2,...,\beta_k,\varphi,\varphi}(x_{i1}, x_{i2}, ..., x_{ik}) = \prod_i \int f_{\varphi,\varphi}(\theta_i) \prod_j f_{\beta_j}(x_{ij}/\theta_i)\, d\theta_i$$

Using the notation $\eta = (\beta_1, \beta_2, \ldots, \beta_k, \varphi)$, we can write:

$$L(\theta_1, \theta_2, ..., \theta_N; \varphi, \eta(\varphi)) = \prod_i \int f_{\varphi,\varphi}(\theta_i) \prod_j f_{\beta_j}(x_{ij}/\theta_i)\, d\theta_i$$

or, more precisely:

$$L(\theta_1, \theta_2, ..., \theta_N; \varphi, \eta(\varphi)) = \prod_i \int f_{\varphi,\varphi}(\theta_i) \prod_j \frac{e^{(\theta_i - \beta_j)\, x_{ij}}}{1 + e^{(\theta_i - \beta_j)}}\, d\theta_i \qquad (1)$$

Estimation of parameters

We assumed that the latent trait θ followed a normal distribution $\sim$ N(μ, σ^2) and that we are testing: H_0: $\mu = \mu_0 = 0$ against H_1: $\mu > 0$. In this framework, the parameters of interest is μ and the vector of nuisance parameters is $\eta = (\beta_1, \beta_2, \ldots, \beta_k, \sigma)$.

>From (1), the log likelihood is:

$$l(\theta_1, \theta_2, ..., \theta_N; \mu, \eta(\mu)) = \sum_i \log\left\{ \int \frac{1}{\sigma\sqrt{2\pi}}\, e^{-\frac{1}{2\sigma^2}(\theta_i - \mu)^2} \prod_j \frac{e^{(\theta_i - \beta_j)\, x_{ij}}}{1 + e^{(\theta_i - \beta_j)}}\, d\theta_i \right\}$$

Let $\xi_i = (\theta_i - \mu)/\sigma$, then:

$$l(\xi_1, \xi_2, \ldots, \xi_N; \mu, \eta(\mu)) = \sum_i \log\left\{ \int \prod_j \frac{e^{(\sigma\xi_i + \mu - \beta_j)\, x_{ij}}}{1 + e^{(\sigma\xi_i + \mu - \beta_j)}} \cdot g(\xi_i)\, d\xi_i \right\}$$

where g is the density of the standard normal distribution.

Z and V statistics

The statistic Z which was previously defined and noted Z(S) will be depending this time on X, the responses to the items, which contain all the information on the items: Z(X) $= l_\mu(0, \hat{\eta}(0))$ where $\hat{\eta}(0)$ is the MLE of η under H_0 ($\mu = \mu_0 = 0$), and $\hat{\eta}(0) = \eta^* = (\beta_1^*, \beta_2^*, \ldots, \beta_k^*, \sigma^*)$, with $\beta^* = \hat{\beta}_1(0), \ldots, \sigma^* = \hat{\sigma}(0)$. Then, we can write: Z(X) $= l_\mu(0, \beta_1^0, \beta_2^0, \ldots, \beta_k^0, \sigma^0)$. We assumed that the β_1^0, β_2^0, ..., β_k^0 were known and we computed the MLE of σ under the null hypothesis in order to further estimate the Z(X) and V(X) statistics. More details are given in Appendix 1. Estimation of the statistics Z(X) and V(X) was done by maximising the marginal likelihood, obtained from integrating out the random effects. Numerical integration methods had to be used because it is not possible to provide an analytical solution. We used the well-known adaptive Gauss-Hermite quadrature to obtain numerical approximations (Pinheiro and Bates, 1995).

2.5 The Sequential Probability Ratio Test and the Triangular Test

The statistics Z and V were noted Z(S) and V(S) in the case of traditional sequential analysis based on sufficient scores and Z(X) and V(X) in the case of a joint sequential and Rasch analysis based directly on observed items. However, for the ease of the general presentation of the tests we shall use the notations Z and V here. The SPRT and the TT tests use a sequential plan defined by two perpendicular axes, the horizontal axis corresponds to Fisher's information V, and the vertical axis corresponds to the efficient score Z which represents the benefit as compared with H_0. The TT appears on figure 1.1. For a one-sided test, the boundaries of the test, delineate a continuation region (situated between these lines), from the regions of non rejection of H_0 (situated beneath the bottom line) and of rejection of H_0 (situated above the top line). The boundaries depend on the statistical hypotheses (values of the expected treatment benefit, α and β and on the number of subjects included between two analyses. They can be adapted at each analysis when this number varies from one analysis to the other, using the "Christmas tree" correction (Siegmund, 1979). The expressions of the boundaries for one-sided tests (Sébille and Bellissant, 2001) are given in Appendix 2. At each analysis, the values of the two statistics Z and V are computed and Z is plotted against V, thus forming a sample path as the trial goes on. The trial is continued as long as the sample path remains in the continuation region. A conclusion is reached as soon as the sample path crosses one of the boundaries of the test: non rejection of H_0 if the sample path crosses the lower boundary, and rejection of H_0 if it crosses the upper boundary.

2.6 Study framework

We simulated 1000 non-comparative clinical trials with patient's item responses generated according to a Rasch model. The latent trait θ_i was assumed to follow a normal distribution with mean μ and variance σ^2= 1 and the trial we considered involved the comparison of the two hypotheses: H_0: $\mu = \mu_0 = 0$ against H_1: $\mu \quad > 0$. The minimum clinically relevant difference (a difference worth detecting) often computed as an effect size (ratio of the minimum clinically relevant difference to the standard deviation) is often measured as $\frac{\mu-\mu_0}{\sigma}$ in clinical trial practice. Since $\mu_0 = 0$ and the standard

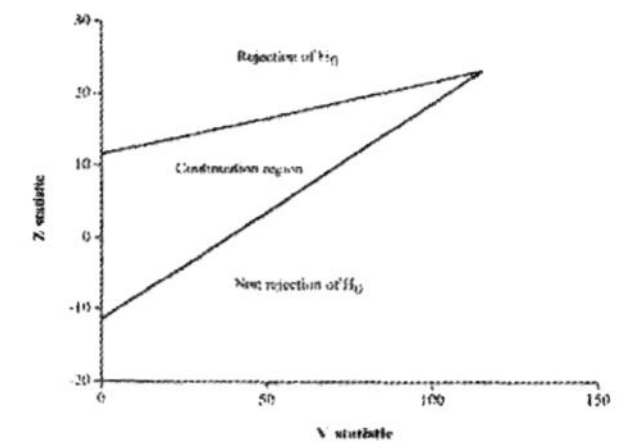

Fig. 1. Stopping boundaries based on the Triangular Test (TT) for $\alpha = \beta = 0.05$ with an effect size (ES) of 0.5.

deviation of θ is equal to one, the effect size will be equal to μ in our case. In practice, effect sizes of interest seen in published research range from 0.2 (small) to 0.8 (large) but the magnitude of effect size primarily depends on the subject matter. Indeed, in medical research effect sizes of 0.5 or 0.6 may be considered as large effect sizes. To our knowledge, there are no well-known conventional values for the effect size that could be most appropriate for QoL endpoints since they closely depend on the medical context under consideration. However, it seems that effect sizes ranging from 0.4 to 0.6 could be of interest (Lacasse et al., 1996). We assumed that the items under consideration formed part of a calibrated item bank, meaning that items parameters were assumed to be known (Holman et al., 2003b). The items parameters were uniformly distributed in the interval (-2, 2) and $\sum_j \beta_j = 0$. The average score methods simply used the sum of item scores S for each patient, assuming a normal distribution and the Z(S) and V(S) statistics were computed within the well-known framework of normally distributed endpoints (Whitehead, 1997).

We compared in the context of sequential analysis of QoL endpoints the use of Rasch modelling methods with traditional average scores methods. The statistical properties of the SPRT and of the TT were studied in the setting of one-sided non-comparative trials. We studied the type I error (α), power (1-β), average sample number (ASN) and 90th percentile (P90) of the number of patients required to reach a conclusion using simulations. The sequential tests were compared with the traditional method using the SPRT or TT based on the averages of patient's scores. We investigated scales with 10 or 20 items, 3 $\neq$ expected effect sizes (0.4, 0.5 and 0.6), and sequential analyses were performed every 20 included patients. The SAS PROC NLMIXED allowed Quasi-Newton procedures to maximise the likelihood and adaptive Gaussian quadrature was used to integrate out the random effects. The sequential tests were all programmed in C++ language.

3 Results

Table 1.1 shows the type I error for different values of the effect size, number of items and nominal power for the TT using either the average scores or the Rasch modelling method. The significance level was close to the target value of 0.05 for the average scores method but slightly increased when a 10 items scale was used as compared with a 20 items scale. The significance level was always lower than the target value of 0.05 for the Rasch modeling method for all effect sizes, number of items used, and nominal power values. Moreover, the significance level seemed to decrease as the effect size increased.

Table 1. Type I error for the Triangular Test using either the average scores method or the Rasch modelling method for different values of the effect size, number of items and power (nominal $\alpha = 0.05$). Data are $\widehat{\alpha}$ (standard errors).

Effect size	**Nb of items**	Average scores		**Rasch model**	
Power		***0.90***	***0.95***	***0.90***	***0.95***
0.4	10	0.056 (0.007)	0.062 (0.008)	0.033 (0.006)	0.035 (0.006)
0.4	20	0.049 (0.007)	0.049 (0.007)	0.036 (0.006)	0.034 (0.006)
0.5	10	0.057 (0.007)	0.055 (0.007)	0.027 (0.005)	0.033 (0.006)
0.5	20	0.050 (0.007)	0.044 (0.006)	0.020 (0.004)	0.033 (0.006)
0.6	10	0.052 (0.007)	0.053 (0.007)	0.020 (0.004)	0.028 (0.005)
0.6	20	0.049 (0.007)	0.046 (0.007)	0.014 (0.004)	0.018 (0.004)

Table 1.2 shows the power for different values of the effect size, number of items and nominal power for the TT using either the average scores or the Rasch modelling method. The TT was underpowered especially when using the averages scores method as compared with the Rasch modelling method. For instance, as compared with the target power value of 0.95, there were decreases in power of approximately 12% and 7% with 10 and 20 items, respectively for the averages scores method. By contrast, the decrease in power was of about only 5% for the Rasch modelling method, whatever the number of items used. Moreover, the power seemed to decrease as the effect size increased.

Table 1.3 shows the ASN of the number of patients required to reach a conclusion under H_0 and H_1 for different values of the effect size, number of items and nominal power for the TT using either the average scores or

Table 2. Power for the Triangular Test using either the average scores method or the Rasch modelling method for different values of the effect size, number of items and power (nominal $\alpha = 0.05$). Data are $\hat{\beta}$ (standard errors).

Effect size	**Nb of items**	Average scores		**Rasch model**	
Power		***0.90***	***0.95***	***0.90***	***0.95***
0.4	10	0.723 (0.014)	0.821 (0.012)	0.848 (0.011)	0.927 (0.008)
0.4	20	0.812 (0.012)	0.869 (0.011)	0.852 (0.011)	0.919 (0.009)
0.5	10	0.769 (0.013)	0.837 (0.012)	0.831 (0.012)	0.907 (0.009)
0.5	20	0.825 (0.012)	0.902 (0.009)	0.812 (0.012)	0.910 (0.009)
0.6	10	0.782 (0.013)	0.846 (0.011)	0.807 (0.012)	0.898 (0.010)
0.6	20	0.834 (0.012)	0.891 (0.010)	0.772 (0.013)	0.874 (0.010)

the Rasch modelling method. We also computed for comparison purposes the approximate number of patients required by a single-stage design (SSD) using IRT modelling from the results published in a recent paper (Holman et al., 2003a). As expected, the ASNs all decreased as the expected effect sizes increased whatever the method used. The ASNs under H_0 and H_1 were always much smaller for both methods based either on averages scores or Rasch modelling than for the SSD for whatever values of effect size, number of items or nominal power considered. The decreases in the ASNs were a bit larger for the averages scores method followed by the Rasch modelling method. For instance, under H_0 (H_1) as compared with the SSD, there were decreases of approximately 70% (65%) and 60% (55%) in sample size for the averages scores and the Rasch modelling method, respectively.

Table 1.4 shows the P90 of the number of patients required to reach a conclusion under H_0 and H_1 for different values of the effect size, number of items and nominal power for the TT using either the average scores or the Rasch modelling method. In most cases, the P90 values of the sample size distribution under H_0 and H_1were of the same order of magnitude for the average scores and the Rasch modelling method. Moreover, the P90 always remained lower for both methods based either on averages scores or Rasch modelling than for the SSD whatever values of effect size or number of items considered.

The operating characteristic (OC) function (figure 1.2), which is the probability of accepting H_0, was computed for the SPRT using either the average scores or the Rasch modelling method under H_0 (where it should be equal

Table 3. ASN required to reach a conclusion under H_0 for the Triangular Test using either the average scores method or the Rasch modelling method for different values of the effect size, number of items and power (nominal $\alpha = 0.05$).(* When using IRT: approximate number of subjects required in a single-stage design (SSD)).

Effect size	Nb of items.	IRT*		Average scores		Rasch model	
Power		*0.90*	*0.95*	*0.90* H_0 / H_1	*0.95* H_0 / H_1	*0.90* H_0 / H_1	*0.95* H_0 / H_1
0.4	10	~ 75	~ 125	34.24 / 42.58	41.50 / 50.70	49.70 / 62.62	60.78 / 71.44
0.4	20	~ 70	~ 120	34.30 / 40.62	42.42 / 46.86	42.10 / 52.50	51.44 / 59.16
0.5	10	~ 60	~ 100	24.74 / 29.10	29.04 / 33.84	33.72 / 43.10	41.34 / 48.60
0.5	20	~ 55	~ 95	24.46 / 28.50	29.02 / 33.06	29.68 / 37.16	35.02 / 42.52
0.6	10	~ 45	~ 80	21.24 / 22.78	23.16 / 25.60	26.24 / 32.46	30.38 / 36.88
0.6	20	~ 45	~ 80	21.20 / 22.78	22.96 / 24.98	23.38 / 27.64	26.92 / 31.52

Table 4. P90 of the number of patients required to reach a conclusion under H_0 for the Triangular Test using either the average scores method or the Rasch modelling method for different values of the effect size, number of items and power (nominal $\alpha = 0.05$).(* When using IRT: approximate number of subjects required in a single-stage design (SSD)).

Effect size	Nb of items	*IRT		Average scores		Rasch model	
Power		*0.90*	*0.95*	*0.90* H_0 / H_1	*0.95* H_0 / H_1	*0.90* H_0 / H_1	*0.95* H_0 / H_1
0.4	10	~ 75	~ 125	60 / 60	60 / 80	80 / 100	100 / 100
0.4	20	~ 70	~ 120	60 / 60	60 / 80	60 / 80	80 / 100
0.5	10	~ 60	~ 100	40 / 40	40 / 60	60 / 60	60 / 80
0.5	20	~ 55	~ 95	40 / 40	40 / 60	40 / 60	60 / 60
0.6	10	~ 45	~ 80	20 / 40	40 / 40	40 / 40	40 / 60
0.6	20	~ 45	~ 80	20 / 40	40 / 40	40 / 40	40 / 40

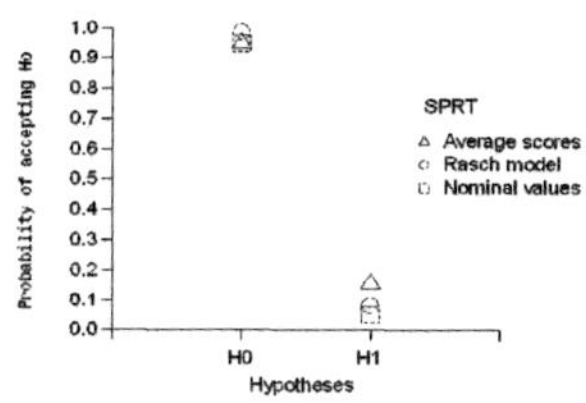

Fig. 2. Operating characteristic (OC) function (probability of accepting H_0) computed for the Sequential Probability Ratio Test (SPRT) using either the average scores or the Rasch modelling method under H_0 and under H_1 with an effect size of 0.5 and a nominal power of 0.95.

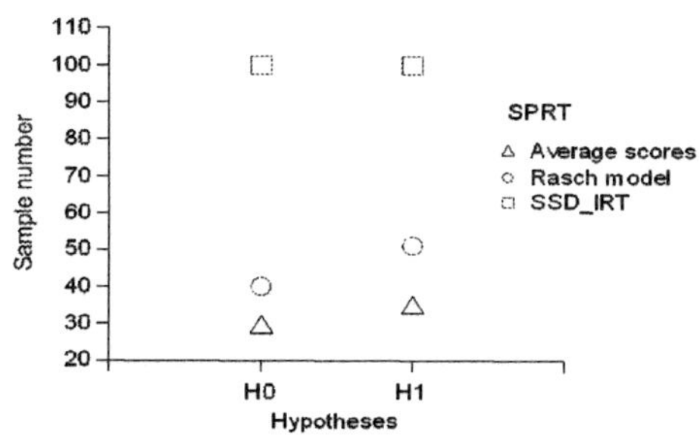

Fig. 3. Average Sample Number (ASN) under H_0 and H_1 (effect size of 0.5) for the Sequential Probability Ratio Test (SPRT) using the average scores or the Rasch modelling method and approximate sample size required by the SSD using IRT modelling (SSD_IRT).

to 0.95) and under H_1 with an effect size of 0.5 and a nominal power of 0.95 (where it should be equal to 0.05). As observed with the TT, we can see that the OC functions of the SPRT are quite similar under H_0 for both methods whereas under H_1, the Rasch modelling method seems more accurate, that is closer to the nominal value of 0.05 than the average scores method which is higher. Figure 1.3 shows the ASNs under H_0 and H_1 (effect size of 0.5) for the SPRT using the average scores or the Rasch modelling method as well as the approximate sample size required by the SSD using IRT modelling. As with the TT, the ASNs were always much lower using either the average scores or the Rasch modelling method as compared with the sample size required by the SSD. Moreover, we observed that the ASN of the SPRT was a bit higher using the Rasch model as compared with the average scores method.

Table 5. Distributions of the Z(S), V(S), Z(X), and V(X) statistics under H_0 estimated with the average scores (A) or the Rasch modeling (R) method. *: Number of Patients is Cumulated number of included patients since the beginning of the trial. Data are: $\overline{Z}(S), \overline{V}(S), \overline{Z}(X)$, and $\overline{V}(X)$: sample means; (Var): variance of $\hat{Z}(S)$ *or* $\hat{Z}(X)$ §: p (Kolmogorov-Smirnov)=0.005.

		Method A		*Method R*	
*Number of patients**	**Number of items**	$\overline{Z}(S)$ (Var)	$\overline{V}(S)$	$\overline{Z}(X)$ (Var)	$\overline{V}(X)$
40	10	0.087§ (38.897)	39.514	0.070 (23.158)	27.457
	20	-0.007 (39.135)	39.511	0.015 (31.456)	33.362
60	10	0.076 (61.188)	59.491	-0.056 (35.628)	39.923
	20	-0.060 (56.238)	59.532	0.102 (44.488)	48.598
100	10	-0.179 (104.283)	99.479	-0.262 (62.286)	65.186
	20	-0.381 (96.579)	99.517	-0.247 (76.248)	79.193

4 Discussion

We evaluated the benefit of combining sequential analysis and IRT methodologies in the context of phase II non-comparative clinical trials with QoL endpoints. We studied and compared the statistical properties of the SPRT and of the TT using either a Rasch modeling method or the traditional average scores method. Simulation studies showed that: (i) the type I error α was correctly maintained but seemed to be lower for the Rasch modeling method as compared with the average scores method, (ii) both methods seemed to be underpowered, especially the average scores method, the power being higher when using the Rasch modeling method, (iii) as expected using sequential analysis, both methods allowed substantial reductions in ASNs as compared with the SSD, the average scores method allowing smaller ASNs than the Rasch modeling method.

The fact that the Rasch modeling method seemed to be more conservative than the average scores method in terms of significance level might be partly explained by looking at the distributions of the Z(S), V(S), Z(X), and V(X) statistics under H_0 (table 1.5) under different conditions. According to asymptotic distributional results, we might expect the sequences of test statistics (Z_1(S), Z_2(S), ..., Z_K(S)) and (Z_1(X), Z_2(X), ..., Z_K(X)) to be multivariate normal with: $Z_k(S) \sim N(0, V_k(S))$ and $Z_k(X) \sim N(0, V_k(X))$, respectively, under H_0 for k = 1, 2, ..., K analyses (Whitehead, 1997; Jennison and Turnbull, 1999). The normality assumption was not rejected using a Kolmogorov-Smirnov test, except for the average scores method with a 10-items scale when Z(S) was estimated on only 40 patients (corresponding to the second interim analysis). Moreover, the variance of $\hat{Z}(S)$ and of $\hat{Z}(X)$ were quite close to $\overline{V}(S)$ and $\overline{V}(X)$, respectively, in most cases. However, the variance of $\hat{Z}$ was always lower when the estimation was performed using the Rasch modeling method ($\hat{Z}(X)$) as compared with the average scores method ($\hat{Z}(S)$, $p < 0.001$, for all cases), suggesting that the estimator of Z using Rasch modeling might be more efficient. The same feature was observed under H_1 (data not shown) except for the normality assumption which did not hold when a 20-items scale was used for both methods. This might explain why the SPRT and TT were underpowered, especially when using the average scores method. However, a more thoughtful theoretical study of the distributions of the statistics Z(S), Z(X) and V(S), V(X) which were obtained using both methods would be worth investigating.

Several limitations to our study are worth being mentioned. Firstly, we assumed all items parameters to be known which is unrealistic (at least for most scales). An option could be to investigate 2-stage estimation (Andersen, 1977) using item parameters estimates as known constants. However, problems with small sample sizes might occur especially in the context of sequential analysis of clinical trials where interim analyses are often performed on less than 50 patients and further work is needed. Secondly, some further sensitivity analyses could be worthwhile such as investigating the effects on the results of: (i) changing the number of items (either <10 or >20), (ii) looking at smaller or larger effects sizes than the ones investigated, and (iii) evaluating the potential effects of changing the frequency of the sequential analyses ($\neq$ 20 patients). Other types of investigations could also be interesting, such as: applying these combined methodologies to comparative clinical trials (phase III trials), evaluating the impact on the statistical properties of the sequential tests of the amount of missing data (often encountered in practice) and missing data mechanisms (missing completely at random, missing at random, non ignorable missing data). In addition, other group sequential methods could also be investigated such as spending functions (Lan and De Mets, 1983), and Bayesian sequential methods (Grossman et al., 1994) for instance. Finally, we only worked on binary items and polytomous items more frequently appear in health-related QoL scales used in clinical trial practice. Other IRT models such as the Partial Credit Model or the Rating Scale Model (Andrich, 1978; Masters, 1982) would certainly be more appropriate in this context and are currently being investigated (work in progress).

5 Conclusion

Item response theory usually provides more accurate assessment of health status as compared with summation methods (McHorney et al., 1997; Kosinski et al., 2003). The use of IRT methods in the context of sequential analysis of QoL endpoints seems to be promising and might provide a more powerful method to detect therapeutic effects than the traditional summation method. Even though the number of subjects required to reach a conclusion seemed to be a bit higher using IRT (one more sequential analysis was needed), the trade-off between small ASN versus a satisfying precision of the estimation of treatment effect is an open question.

Finally, there are a number of challenges for medical statisticians using IRT that may be worth to mention: IRT was originally developed in educational research using samples of thousands or even ten thousands. Such large sample sizes are very rarely (almost never) attained in medical research where medical interventions are often assessed using less than 200 patients. The problem is even more crucial in the sequential analysis framework where the first interim analysis is often performed on fewer patients. Moreover, IRT and associated estimation procedures are conceptually more difficult than the summation methods often used in medical research. Perhaps one of the biggest challenges for medical statisticians will be to explain these methods well enough so that clinical researchers will accept them and use them. As in all clinical research but maybe even more in this context, there is a real need for good communication and collaboration between clinicians and statisticians.

6 References

Andersen, E. B. (1970) Asymptotic properties of conditional maximum likelihood estimators. *J. R. Statist. Soc. B*, **32**, 283-301.

Andersen, E. B. (1977) Estimating the parameters of the latent population distribution. *Psychometrika*, **42**, 357-374.

Anderson, T. W. (1960) A modification of the sequential probability ratio test to reduce the sample size. *Ann. Math. Stat.*, **31**, 165-197.

Andrich, D. (1978) A rating formulation for ordered response categories. *Psychometrika*, **43**, 561-573.

Cannistra, S. A. (2004) The ethics of early stopping rules: who is protecting whom? *J. Clin. Oncol.*, **22**, 1542-1545.

Cella, D. F. and Bonomi, A. E. (1995) Measuring quality of life: 1995 update. *Oncology*, **9**, 47-60.

Fairclough, D. L. (2002) *Design and analysis of quality of life studies in clinical trials*. Boca Raton: Chapman & Hall/CRC.

Fisher, G.H. and Molenaar, I.W. (1995) *Rasch Models, Foundations, Recent Developments, and Applications*. New-York: Springer-Verlag.

Grossman, J., Parmar, M. K., Spiegelhalter, D. J., Freedman, L. S. (1994) A unified method for monitoring and analysing controlled trials. *Statist. Med.*, **13**, 1815-1826.

Haberman, S. J. (1977) Maximum likelihood estimates in exponential response models. *Ann. Statist.*, **5**, 815-841.

Hamon, A. and Mesbah, M. (2002) Questionnaire reliability under the Rasch model. In Mesbah, M., Cole, B. F., Lee, M. L. T. (eds.) *Statistical Methods for Quality of Life Studies: Design, Measurements and Analysis.* Amsterdam: Kluwer.

Holman, R., Glas, C. A., and de Haan, R. J. (2003a) Power analysis in randomized clinical trials based on item response theory. *Control. Clin. Trials*, **24**, 390-410.

Holman, R., Lindeboom, R., Glas, C. A. W., Vermeulen M., and de Haan, R. J. (2003b) Constructing an item bank using item response theory: the AMC linear disability score project. *Health. Serv. Out. Res. Meth.*, **4**, 19-33.

Jennison, C. and Turnbull, B. W. (1999) *Group Sequential Methods with Applications to Clinical Trials.* Boca Raton: Chapman & Hall/CRC.

Kosinski, M., Bjorner, J. B., Ware, J. E. Jr, Batenhorst, A., and Cady R. K. (2003) The responsiveness of headache impact scales scored using 'classical' and 'modern' psychometric methods: a re-analysis of three clinical trials. *Qual. Life. Res.*, **12**, 903-912.

Lacasse, Y., Wong, E., Guyatt, G. H., King, D., Cook, D. J., and Goldstein R. S. (1996) Meta-analysis of respiratory rehabilitation in chronic obstructive pulmonary disease. *Lancet*, **348**, 115-1119.

Lan, K. K. G. and De Mets, D. L. (1983) Discrete sequential boundaries for clinical trials. *Biometrika*, **70**, 659-663.

Masters, G. N. (1982) A Rasch model for partial credit scoring. *Psychometrika*, **47**, 149-174.

McHorney, C. A., Haley, S. M., and Ware, J.E. Jr. (1997) Evaluation of the MOS SF-36 Physical Functioning Scale (PF-10): II. Comparison of relative precision using Likert and Rasch scoring methods. *J. Clin. Epidemiol.*, **50**, 451-461.

O'Brien, P. C. and Fleming, T. R. (1979) A multiple testing procedure for clinical trials. *Biometrics*, **35**, 549-556.

Pinheiro, J. C. and Bates, D. M. (1995) Approximations to the Log-likelihood Function in the Nonlinear Mixed-effects Model. *J. Comput. Graph. Statist.*, **4**, 12-35.

Pocock, S. J. (1977) Group sequential methods in the design and analysis of clinical trials. *Biometrika*, **64**, 191-199.

Rasch, G. (1960) *Probabilistisc models for some intelligence and attainment tests.* Copenhagen, D.K.:Nielsen & Lydiche. [Expanded edition, 1980, Chicago: The University of Chicago Press].

Sébille, V. and Bellissant, E. (2001) Comparison of the two-sided single triangular test to the double triangular test. *Control. Clin. Trials*, **22**, 503-514.

Siegmund, D. (1979) Corrected diffusion approximations in certain random walk problems *Adv. Appl. Probab.*, **11**, 701-719.

Thissen, D. (1982) Marginal maximum likelihood estimation for the one-parameter logistic model. *Psychometrika*, **47**, 175-186.

Wald, A. (1947). *Sequential Analysis.* New York, U.S.A.: Wiley.

Whitehead, J. and Jones, D. R. (1979) The analysis of sequential clinical trials. *Biometrika*, **66**, 443-452.

Whitehead, J. and Stratton, I. (1983) Group sequential clinical trials with triangular continuation regions. *Biometrics*, **39**, 227-236.

Whitehead, J. (1997) *The Design and Analysis of Sequential Clinical Trials*, revised 2nd edition. Chichester, U.K.:Wiley.

7 Appendix 1

7.1 1. MLE of σ under $H_0(\mu = \mu_0 = 0)$

The first derivative of the log likelihood with respect to σ is:

$$\mathrm{l}_\sigma(\xi_1, \xi_2, \ldots, \xi_n; \mu, \sigma) = \frac{\partial l(\xi_1,\xi_2,\ldots,\xi_n;\mu,\sigma)}{\partial\sigma} = \sum_i \frac{\partial}{\partial\sigma} \log\left\{\int \prod_j \left[\frac{e^{(\sigma\xi_i+\mu-\beta_j)\,x_{ij}}}{1+e^{(\sigma\xi_i+\mu-\beta_j)}}\right] \cdot g(\xi_i)\, d\xi_i\right\}$$

We have to solve, for σ, the following equation to get $\hat{\sigma}(0) = \sigma^*$:

$$\mathrm{l}_\sigma(\xi_1, \xi_2, \ldots, \xi_n; 0, \sigma) = \sum_i \frac{\partial}{\partial\sigma} \log\left\{\int \prod_j \left[\frac{e^{(\sigma\xi_i-\beta_j)\,x_{ij}}}{1+e^{(\sigma\xi_i-\beta_j)}}\right] \cdot g(\xi_i)\, d\xi_i\right\} = 0$$

7.2 2. Efficient score: Z(X) statistic under $H_0(\mu = \mu_0 = 0)$

The first derivative of the log likelihood with respect to μ is:

The first derivative of the log likelihood with respect to μ is:

$$\mathrm{l}_\mu(\xi_1, \xi_2, \ldots, \xi_n; \mu, \sigma) = \frac{\partial l(\xi_1,\xi_2,\ldots,\xi_n;\mu,\sigma)}{\partial\mu} = \sum_i \frac{\partial}{\partial\mu} \log\left\{\int \prod_j \frac{e^{(\sigma\xi_i+\mu-\beta_j)\,x_{ij}}}{1+e^{(\sigma\xi_i+\mu-\beta_j)}} \cdot g(\xi_i)\, d\xi_i\right\}$$

$$\mathrm{Z(X)} = \mathrm{l}_\mu(\xi_1, \xi_2, \ldots, \xi_n; 0, \sigma^*) = \sum_i \frac{\partial}{\partial\mu} \log\left\{\int \prod_j \frac{e^{(\sigma^*\xi_i-\beta_j)\,x_{ij}}}{1+e^{(\sigma^*\xi_i-\beta_j)}} \cdot g(\xi_i)\, d\xi_i\right\}$$

7.3 3. Fisher's information: V(X) statistic under $H_0(\mu = \mu_0 = 0)$

Fisher's information V(X) will be: $V(X) = -\{\, l^{\mu\mu}(0,\sigma^*) \,\}^{-1}$ with:

$$\{\, l^{\mu\mu}(0,\sigma^*) \,\}^{-1} = l_{\mu\mu}(0,\sigma^*) - \{\, l_{\mu\sigma}(0,\sigma^*) \,\}' \{\, l_{\sigma\sigma}(0,\sigma^*) \,\}^{-1} l_{\mu\sigma}(0,\sigma^*)$$

8 Appendix 2

8.1 Stopping boundaries for the one-sided SPRT and TT

The stopping boundaries, allowing to detect an effect size (ES) with working significance level α and power 1-β (with $\beta = \alpha$), are:

Z = -a + bV (lower boundary) and Z = a + bV (upper boundary) for the one-sided SPRT,

Z = -a + 3cV (lower boundary) and Z = a + cV (upper boundary) for the one-sided TT,

with $a = a' - 0.583\sqrt{I}$, $b = \frac{1}{2} \cdot ES$, $c = \frac{1}{4} \cdot ES$ *and* $I = V_i - V_{i-1}$ where V_i is the information available at inspection i ($V_0 = 0$) for both tests, and

$a' = \frac{1}{ES} \log\left(\frac{1-\alpha}{\alpha}\right)$ for the one-sided SPRT, and $a' = \frac{2}{ES} \log\left(\frac{1}{2\alpha}\right)$ for the one-sided TT.

The correction $0.583\sqrt{I}$ is used to adjust for the discrete monitoring of the data (Siegmund, 1979). When $\beta \neq \alpha$, a corrected value of the effect size ES must be used to compute the equations of the boundaries. In this case, the boundaries of the tests are computed from an exact formula.

Three Types of Hazard Functions Curves Described

Sidorovich G.I., Shamansky S.V., Pop V.P., Rukavicin O.A.

Burdenko Main Military Clinical Hospital, Moscow, Russia name@email.address

Summary. Not doubts that measures of short-term treatment effects (remission or response rates) are presenting great interest to provide more efficient treatments. However, for all diseases with unfavorable prognosis, to which pertains hemoblastosis, life expectancy is the most important feature. The irrevocable decision about the choice between two different treatment options is usually based on survival functions comparison. Unfortunately, this analysis is not able to reveal critical periods in disease course with distinct maximum mortality rates. Clearly, this information is very important for clinicians efforts to distinguish time intervals when patients should be specially carefully monitored. A retrospective study of the overall survival function among patients with multiple myeloma (MM), acute nonlymphoblastic leukemia (ANLL) and chronic myeloprolypherative disorders (CMPD), treated in our hospital, was performed. These data were complemented with results for the hazard function estimations for each form of hemoblastosis. We found different types of hazard function curves, and we expect that it would be better for treatment results evaluation to use together both survival and hazard function analysis.

1 Patients and method

163 MM patients (120 male and 43 female), 125 ANLL patients (102 male and 23 female) and 106 patients with CMPD (79 male and 27 female) were registered. Age of MM and CMPD patients has demonstrated typically elderly patients predominance. Three fourth of those patients were more than 50 years old. However, ANLL patients age was atypical as three fourth of those patients were less than 50 years old [OLS99]. MM patient age ranged from 33 to 83, with a median of 66. ANLL patient age ranged from 18 to 86, with a median of 34. CMPD patient age varied from 26 to 84 with a median of 60. One may notice that all cohorts had an abnormally high male rate. Female fraction varied from 17,5% in ANLL patients to 26,4% in MM. This is a consequence of peculiar properties of contingent supervised in military hospital. Among patients with ANLL M0 leukemia subvariant was diagnosed in 3 patients, M1 - in 12, M2 - in 28, M3 - in 15, M4 - in 20, M4eos - in 5, M5

- in 6 and M7 - in 1 patient. Both lymphoid and myeloid antigen coexpression was revealed in 2 ANLL patients. In the CMPD structure, idiopathic myelofibrosis dominated (n = 71), then came polycythemia vera (n = 29) and essential trombocythemia was on the third place (n = 6). Diagnosis was defined according to standard criteria for each hemoblastosis form. All patients admitted to the hospital were included, without any exclusion. Patients were treated according to the standard options accepted in our clinic. Particularity of treatment methods have non been registered. Follow-up period for each patients was the time interval between the date of disease morphologic verification and the date of death or the date of last contact (according to the data obtained in January 2004). For analysis of the survival function, Kaplan-Meier method was used [KM58]. Overall survival was computed from diagnosis to death or last follow-up. STATISTICA 5.5 (StatSoft) software were used.

2 Results

In MM patients 75th survival percentile was 14,5 months (95% CI 8,8 - 20,2), median - 34,6 months (95% CI 27,5 - 41,7), 25th percentile - 68,0 months (95% CI 49, - 87,4). 12-month survival in MM patients was 0,78 (95% CI 0,72 - 0,85), 24-month survival - 0,63 (95% CI 0,55 - 0,71), 36-month survival - 0,49 (95% CI 0,41 - 0,58), 48-month survival - 0,37 (95% CI 0,29 - 0,46). 5-year survival was 0,29 (95% CI 0,21 - 0,38), 10-year survival - 0,10 (95% CI 0,04 - 0,17). MM survival curve shown on Fig. 1. Mean follow-up of MM patients who where alive was 38,6 months.

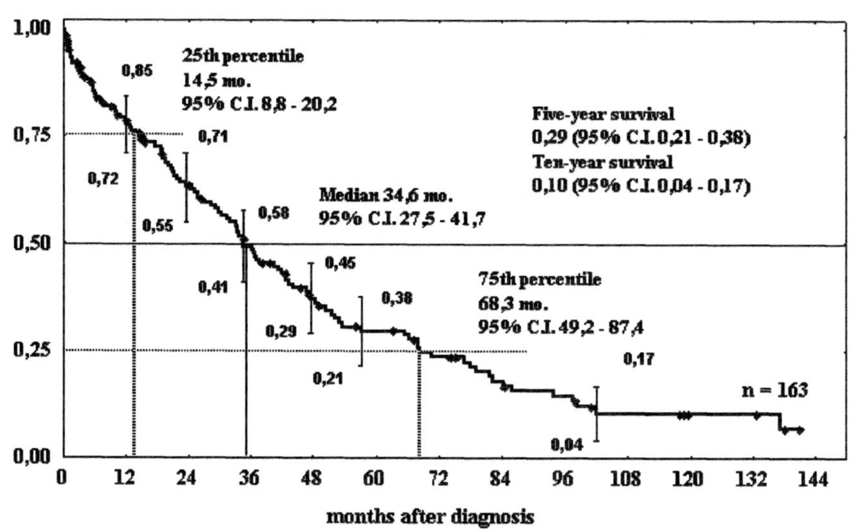

Figure 1. Overall survival curve of multiple myeloma patients

Fig 2. Estimated survival function of ANLL patients. 25th survival percentile was 2,9 months (95% CI 1,14 - 4,58), median - 14,6 months (95% CI 7,7 - 21,5). 12-month survival in ANLL patients was 0,54 (95% CI 0,44 - 0,64), 24-month survival - 0,38 (95% CI 0,27 - 0,49), 36-month survival - 0,32 (95% CI 0,21 - 0,43). 5-year survival was 0,27 (95% CI 0,13 - 0,40). Mean follow-up of ANLL patients who where alive was 17,8 months.

In CMPD patients 75th survival percentile was 65,0 months (95% CI 46,0 - 84,0), median - 142,0 months (95% CI 103,3 - 180,7), 25th percentile - 225,0 months (95% CI 200,5 - 249,5). 12-month survival in CMPD patients was 0,95 (95% CI 0,91 - 1,00), 24-month survival - 0,93 (95% CI 0,88 - 0,98), 36-month survival - 0,88 (95% CI 0,81 - 0,95), 48-month survival - 0,83 (95% CI 0,74 - 0,91). 5-year survival was 0,80 (95% CI 0,71 - 0,89), 10-year survival - 0,57 (95% CI 0,45 - 0,69). Mean follow-up of CMPD patients who where alive was 62,0 months. CMPD survival curve shown on Fig. 3.

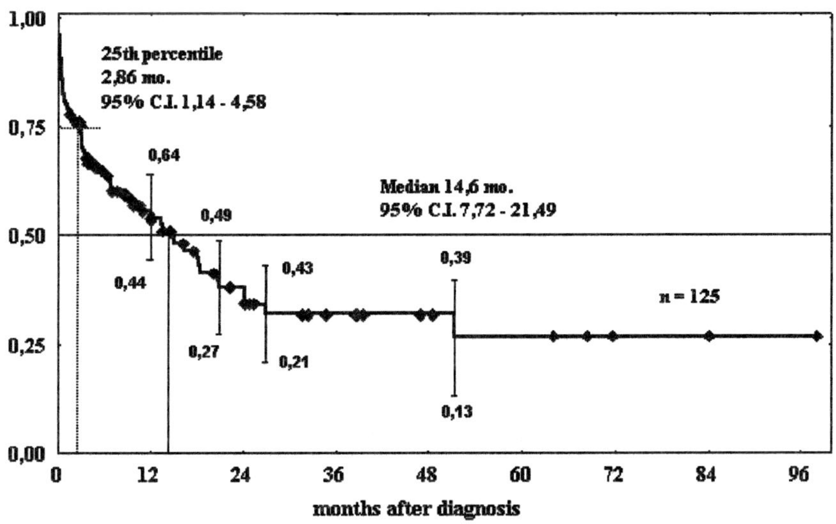

Figure 2. Overall survival curve for acute nonlymphoblastic leukemia patients

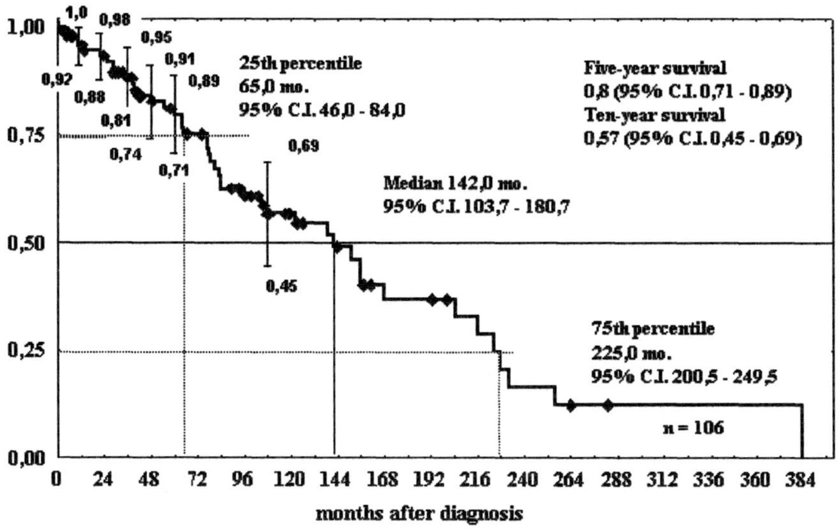

Figure 3. Overall survival curve of chronic myeloprolyphetative disorders patients

Average probability of death within the month in MM patients was 0,02; in ANLL - 0,05 and in CMPD - 0,005. It was shown that in MM the hazard function remained comparatively constant for the whole follow-up period (Fig. 4).

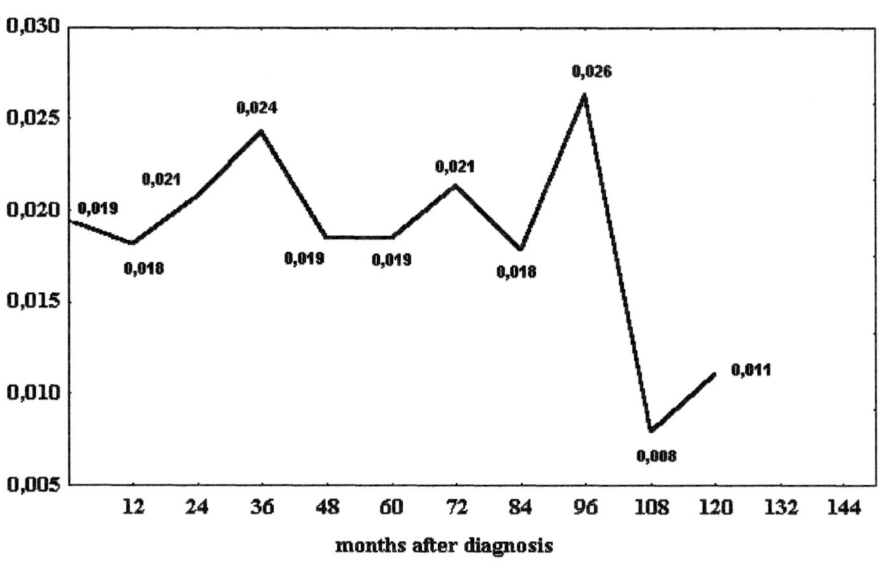

Figure 4. Hazard function of death in patients with multiple myeloma

ANLL course was described by "U"-formed hazard function curve. Mortality rate was maximal at the initial period (death probability within month was 0,05) with tenfold reduction to 36th month. Then it increased again (Fig. 5). In CMPD, hazard function showed linear death probability increase from 0,003 (during the first 12 months) to 0,02 at time of observation cessation (Fig. 6).

Thereby, three types of hazard function curves in different hemoblastosis have been estimated. MM course was characterized by comparatively constant death risk. In ANLL, mortality rate is the highest at first months after diagnostics, it decreased to minimum by 36 month and then increased. In CMPD patients death hazard constantly increases during the disease. We suggest that hazard function is important characteristic, describing the disease course. We recommend using this function with survival function in hematology and oncology practice for evaluation of treatment results.

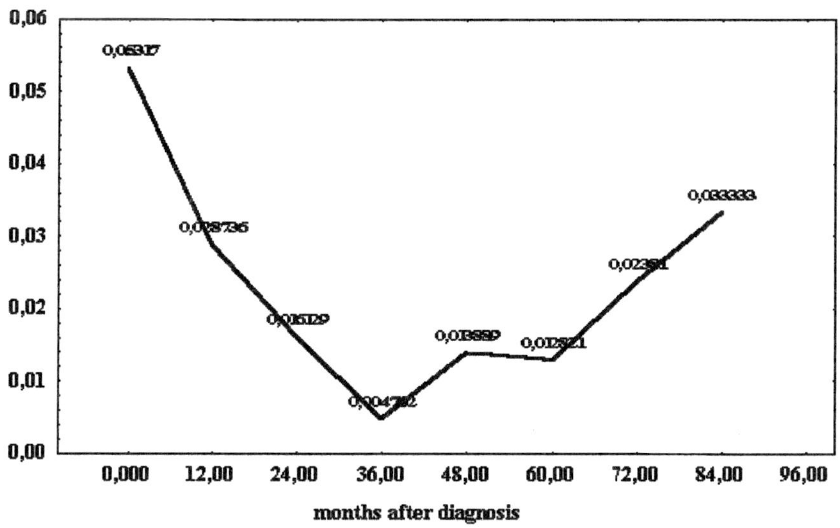

Figure 5. Hazard function of death in patients with acute nonlymphoblastic leukemia

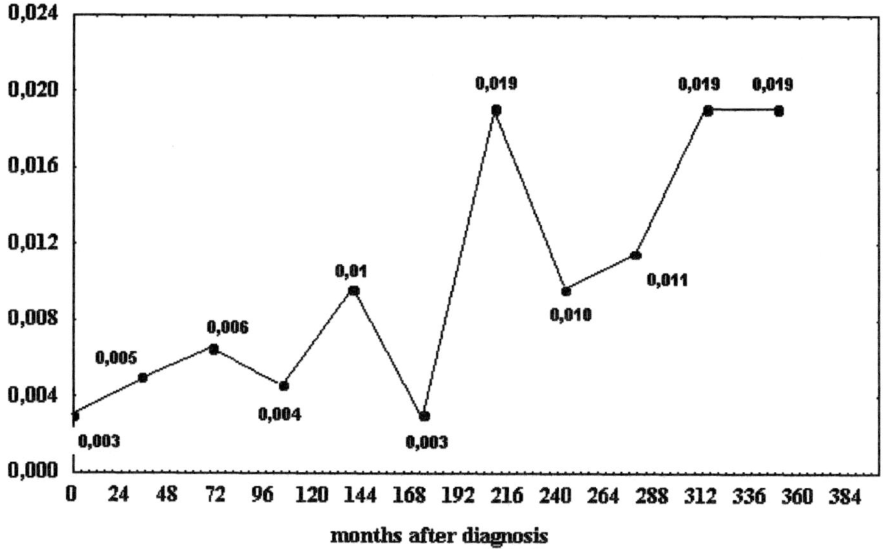

Figure 2. Hazard function of death in patients with chronic mycloprolyphetative disorders

References

[KM58] Kaplan, E.L., Meier, P. : Non-parametric estimation from incomplete observation. J. Am. Stat. Assoc., **53**, 457–481 (1958)

[OLS99] Olsen, J.H. : Epidemiology. In : Degos L., et al. (eds) The Textbook of malignant hematology. Martin Dunitz, London (1999)

On the Analysis of Fuzzy Life Times and Quality of Life Data

Reinhard Viertl

Vienna University of Technology, 1040 Wien, Austria R.Viertl@tuwien.ac.at

Summary. Life times, health data, and general quality of life data are often not adequately represented by precise numbers or classes. Such data are called non-precise or fuzzy, because their quantitative characterization is possible by so-called non-precise numbers. To analyze such data a more general concept than fuzzy numbers from the theory of fuzzy sets is necessary. A suitable concept are so-called non-precise numbers. Generalized methods to analyze such data are available, and basic methods for that are described in the paper.

1 Introduction

Life times of systems are frequently not adequately characterized by precise time values. Therefore precise numbers are not always suitable to describe life times. Generally all results of measurements of continuous quantities are not precise numbers. For details compare [Vie02]. Even more uncertainty is connected with recovering times from illness.

Quality of life is a complex task and several approaches to measure it are possible. There are different methodological difficulties, for example the necessity of aggregating variables. But at the beginning of the analysis process, data quality considerations are indispensable in order to avoid unrealistic results of analyses.

There are different kinds of uncertainty in life time data: Variability, errors, and imprecision. It is important to note that imprecision is the kind of uncertainty inherent in single measurement results. Imprecision should not be confused with errors. Errors can be modeled with probability distributions, but imprecision cannot be modeled adequately in this way, because imprecision is another kind of uncertainty.

The best up-to-date description of imprecision is - in case of one-dimensional quantities - by so-called *non-precise numbers* which are special fuzzy subsets of the set $\mathbb{R}$ of real numbers. Therefore such data are also called *fuzzy data.*

In the paper generalized methods for the description and analysis of fuzzy data are explained.

2 Fuzzy data

Quantitative data which cannot be characterized by precise numbers have to be characterized mathematically in a suitable way. This leads to so-called *non-precise numbers* which are defined by *characterizing functions*.

Definition 1. *A characterizing function* $\xi(\cdot)$ *is a real function* $\xi : \mathbb{R} \longrightarrow [0,1]$ *for which the so-called* δ*-cuts* $C_\delta\left[\xi(\cdot)\right]$,

$$C_\delta\left[\xi(\cdot)\right] := \{x \in \mathbb{R} : \xi(x) \geq \delta\}$$

are non-empty finite unions of bounded closed intervals, i. e.

$$C_\delta\left[\xi(\cdot)\right] = \bigcup_{j=1}^{n_\delta} \left[a_{j,\delta}, b_{j,\delta}\right] \qquad \forall \quad \delta \in (0,1].$$

Remark 1. Characterizing functions are special membership functions from the theory of fuzzy sets. But non-precise numbers are more general than so-called fuzzy numbers. This is necessary to characterize fuzzy data, for example data obtained from color intensity pictures, especially X-ray data. Examples of characterizing functions are given in Figure 1.

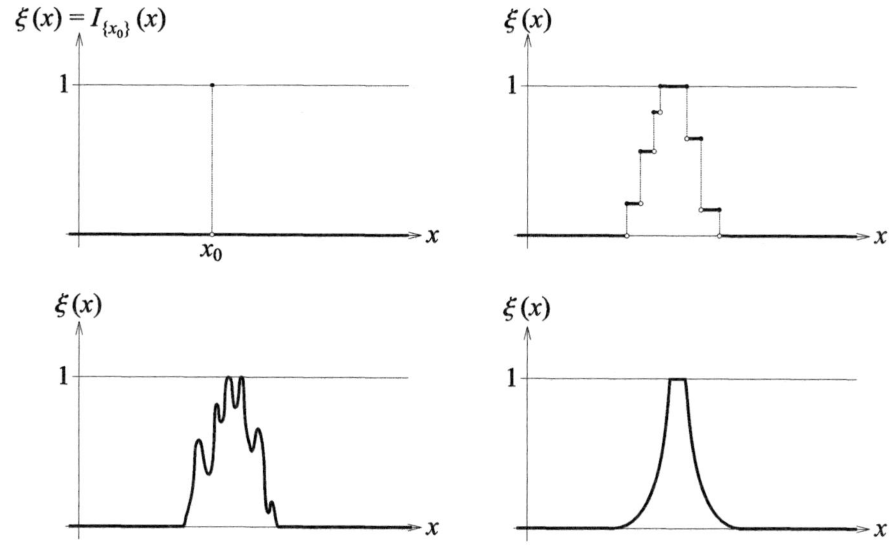

Fig. 1. Characterizing functions

Remark 2. The one-point indicator function $I_{\{x_0\}}(\cdot)$ is the characterizing function of a precise data point $x_0 \in \mathbb{R}$. Therefore the analysis of fuzzy data contains standard statistical procedures as special case.

A main problem is the determination of the characterizing function of a fuzzy data point $x^\star$. In case of color intensities this is possible in the following way:

Let $h(\cdot)$ be the light intensity of a fuzzy light point on a screen. Then the characterizing function $\xi(\cdot)$ is given by its values

$$\xi(x) = \frac{h(x)}{\max\limits_{x \in \mathbb{R}} h(x)} \qquad \forall \quad x \in \mathbb{R}.$$

For life time data $t^\star$ the characterizing function $\xi(\cdot)$ can be obtained from measurements of characteristic quantities which describe the degree of fulfilment of its objectives. Let $f(\cdot)$ be the function describing the degree of fulfilment of the characteristic quantity depending on the time t, then the characterizing function $\xi(\cdot)$ of the fuzzy life time is given by its values

$$\xi(t) = \frac{\frac{d}{dt} f(t)}{\max\limits_{t \geq 0} \frac{d}{dt} f(t)} \qquad \forall \quad t \in \mathbb{R}.$$

For $t \leq 0$ the value of $\xi(t) \equiv 0$.

In case of quality of life data, which are fuzzy by nature, these data usually contain subjectivity. Therefore also the corresponding characterizing functions are subjective to a certain degree. But still they contain important information for statistical analysis.

3 Empirical reliability functions for fuzzy life times

Real life time data consist of a finite sample $x_1^\star, \cdots, x_n^\star$ of non-precise numbers with corresponding characterizing functions $\xi_1(\cdot), \cdots, \xi_n(\cdot)$.

The standard empirical reliability function $\hat{R}_n(\cdot)$ for precise data $x_1, \cdots, x_n$, defined by its values

$$\hat{R}_n(x) = \frac{1}{n} \sum_{i=1}^{n} I_{(x,\infty)}(x_i) \qquad \forall \quad x \geq 0,$$

is generalized in the following way:

Assuming all characterizing functions $\xi_i(\cdot)$ to be integrable, the generalized empirical reliability function $\hat{R}_n^\star(\cdot)$ is defined by

$$\hat{R}_n^\star(x) = \frac{1}{n} \sum_{i=1}^{n} \frac{\int_x^\infty \xi_i(t)dt}{\int_0^\infty \xi_i(t)dt} \qquad \forall \quad x \geq 0.$$

Remark 3. The generalized estimate $\hat{R}^\star(\cdot)$ is a continuous function which is more realistic in case of continuous underlying life time distributions.

An example of fuzzy life times and the corresponding generalized empirical reliability function is given in Figure 2.

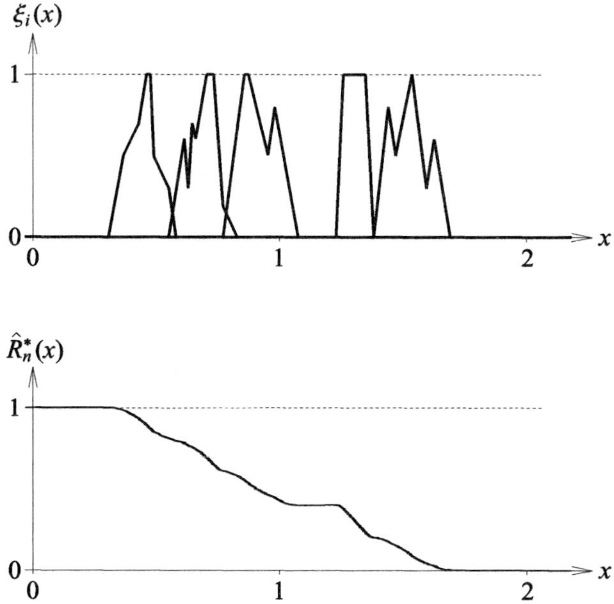

Fig. 2. Generalized empirical reliability function

4 Generalized classical statistical inference for fuzzy data

Statistical estimation of derived indicators of quality of life, based on fuzzy data is possible in the following way: Let $x_1, \cdots, x_n$ be classical pseudo-precise data and

$$I = f\left(x_1, \cdots, x_n\ ; w_1, \cdots, w_n\right)$$

be an indicator based on data $x_1, \cdots, x_n$ and weights $w_1, \cdots, w_n$. For standard data the indicator is a real number.

In case of fuzzy data $x_1^\star, \cdots, x_n^\star$ with corresponding characterizing functions $\xi_1(\cdot), \cdots, \xi_n(\cdot)$ the resulting value of the indicator becomes non-precise. In order to obtain the characterizing function $\eta(\cdot)$ of the non-precise value

$$I^\star = f\left(x_1^\star, \cdots, x_n^\star\ ; w_1, \cdots, w_n\right),$$

the so-called *extension principle* from fuzzy set theory is applied. The values $\eta(x)$ for all $x \in \mathbb{R}$ of the characterizing function are defined by

$$\eta(x) = \begin{cases} \sup\Big\{ \min\{\xi_1(x_1), \cdots \xi_n(x_n)\}\Big\} & \text{if } f(x_1, \cdots, x_n; w_1, \cdots, w_n) = x \\ 0 & \text{if} \nexists\ (x_1, \cdots, x_n) : f(x_1, \cdots, x_n; w_1, \cdots, w_n) = x \end{cases} .$$

The resulting non-precise value $I^\star$ of the indicator is a non-precise number in the sense of section 2.

Remark 4. Although the generalized indicators are non-precise they represent valuable information concerning the considered topic.

Moreover different statistical inference procedures can be generalized to the situation of non-precise date. These methods are described in the book [Vie96].

Recent work on the generalization of the concept of p-values is published in the paper [FV04].

5 Generalized Bayesian inference in case of fuzzy information

The generalization of Bayes' theorem by application of the extension principle from fuzzy set theory is not reasonable, because it is not keeping the sequential updating procedure of Bayesian inference. Therefore another method was developed which takes care of imprecision of a-priori distributions and fuzziness of data as well. This is presented in the paper [VH04].

It is important to note that a more general concept of probability distributions, so-called *fuzzy probability distributions*, is necessary to describe imprecision of a-priori distributions.

A fuzzy probability distribution $P^\star$, defined on a sigma field $\mathcal{A}$ of subsets of an observation space M, is defined in the following way:

Definition 2. *A* fuzzy number *is defined by a specialized characterizing function from section 2, for which all δ-cuts are non-empty compact intervals.*

Definition 3. *A fuzzy probability distribution $P^\star$ assigns to every event $A \in \mathcal{A}$ a fuzzy number whose support is a subset of $[0,1]$ and which obeys:*

1. $P^\star(\emptyset) = 0$ *and* $P^\star(M) = 1$,
   i.e. the extreme events have precise probabilities
2. *For any sequence $A_1, A_2, \cdots$ of pairwise disjoint events from $\mathcal{A}$, and for all δ-cuts $C_\delta\Big[P^\star(A_i)\Big] = \Big[\underline{P}_\delta(A_i), \overline{P}_\delta(A_i)\Big], \delta \in (0,1]$, the following has to be valid:*

$$\overline{P}_\delta\Big(\bigcup_{i=1}^{\infty} A_i\Big) \leq \sum_{i=1}^{\infty} \overline{P}_\delta(A_i)$$

and

$$\underline{P}_\delta\Big(\bigcup_{i=1}^{\infty} A_i\Big) \geq \sum_{i=1}^{\infty} \underline{P}_\delta(A_i)$$

3. *Fuzzy monotony: For $A \subseteq B$ it follows $P^\star(A)$ is fuzzy smaller than $P^\star(B)$, i. e. for the δ-cuts $C_\delta\left[P^\star(A)\right] = \left[\underline{P}_\delta(A), \overline{P}_\delta(A)\right]$ and $C_\delta\left[P^\star(B)\right] = \left[\underline{P}_\delta(B), \overline{P}_\delta(B)\right]$ the following has to be fulfilled:*

$$\underline{P}_\delta(A) \leq \underline{P}_\delta(B) \quad \textit{and} \quad \overline{P}_\delta(A) \leq \overline{P}_\delta(B) \qquad \forall \quad \delta \in (0,1]$$

More details can be found in the paper [TH04].

Remark 5. Special fuzzy probability distributions are obtained by so-called *fuzzy density functions*, which are also explained in [TH04].

6 Conclusion

Especially for quality of life data non-precise numbers are a more realistic description of quantitative data than precise real numbers. Generalized statistical analysis methods for this kind of data are available and provide valuable information in order to support well founded decisions.

References

[FV04] Filzmoser, P., Viertl, R.: Testing hypotheses with fuzzy data: The fuzzy p-value. Metrika, **59** (2004)

[OK95] Onisawa, T., Kacprzyk, J. (Eds.): Reliability and Safety Analyses under Fuzziness. Physica-Verlag, Heidelberg (1995)

[TH04] Trutschnig, W., Hareter, D.: Fuzzy Probability Distributions. In: Lopéz-Diaz, M. et al. (eds.) Soft Methodology and Random Information Systems. Springer-Verlag, Berlin (2004)

[Vie96] Viertl, R.: Statistical Methods for Non-Precise Data. CRC Press, Boca Raton, Florida (1996)

[Vie99] Viertl, R.: Nonprecise Data. in: Encyclopedia of Statistical Sciences - Update, Volume 3. Wiley, New York (1999)

[Vie02] Viertl, R.: On the Description and Analysis of Measurements of Continuous Quantities. Kybernetika, **38** (2002)

[VH04] Viertl, R., Hareter, D.: Fuzzy information and imprecise probability. ZAMM-Journal of Applied Mathematics and Mechanics, **84**, No. 10-11 (2004)

Statistical Inference for Two-Sample and Regression Models with Heterogeneity Effect: A Collected-Sample Perspective

Hong-Dar Isaac Wu

School of Public Health, China Medical University, 91 Hsueh-Shih Rd., Taichung 404, TAIWAN. honda@mail.cmu.edu.tw

Summary. Heterogeneity effect is an important issue in the analysis of clinical trials, survival data, and epidemiological cohort studies. This article reviews the works of inference for heterogeneity effect from a series of works by Hsieh who used the empirical process approach, and relevant works by Bagdonavičius, Nikulin, and coworkers. This includes two-sample models and Cox-type relative risk regression models. Heterogeneity property over the covariate space as well as non-constancy property are discussed for several models. In survival analysis, the log-relative risk as a function of time and of the covariates are plotted to present the heterogeneity property of Hsieh's and Bagdonavičius and Nikulin's hazards regression models.

Key words: Heterogeneity, two-sample problem, location-scale model, transformation model, Cox model, Hsieh model, Bagdonavičius-Nikulin model

1 Introduction

This article reviews the works of inference for heterogeneity effect from a series of works, mainly by Hsieh, and relevant works by Bagdonavičius, Nikulin, and coworkers. Before starting our discussion, the meaning of 'heterogeneity' is briefly defined. Concerning a measure of 'effect' and a set of subpopulations indexed by a variable $\mathcal{W}$, if the effect is fixed over $\mathcal{W}$, we say there is a homogeneity effect; otherwise, there is heterogeneity. In some situations, heterogeneity can be dealt with by stratified analysis, or by random effect analysis. By this, however, it is often assumed that there is *unobserved* or *unmeasured* factors according to which the effect heterogeneity exists and will be averaged or eliminated. The work of Hsieh proposed another possibility that the heterogeneity is a result of *observable* variables, and the impact of these variable 'should be' (and 'can be') estimated. A simple example of location-scale model can be used for an illustrative purpose. Consider the case of logistic regression: Let $X \sim F_X(x)$=Logistic$(a_1, b_1) = \frac{e^{a_1+b_1x}}{1+e^{a_1+b_1x}}$ and

$Y \sim G_Y(x)$=Logistic$(a_2, b_2) = \frac{e^{a_2+b_2x}}{1+e^{a_2+b_2x}}$, $F(\cdot)$ and $G(\cdot)$ are cdfs. For ordinary 2×2 table analysis, a possibly unknown cutoff value x_0 is assumed such that the *odds* of the first (F) and second (G) groups are

$$\text{odds}_F(x_0) = \frac{\Pr(X < x_0)}{1 - \Pr(X < x_0)} \text{ and } \text{odds}_G(x_0) = \frac{\Pr(Y < x_0)}{1 - \Pr(Y < x_0)},$$

respectively, which results in the *odds ratio* (OR) of G-group versus F-group:

$$\text{OR} = \frac{e^{a_2+b_2x_0}}{e^{a_1+b_1x_0}}.$$

The above two-sample problem can be simplified when the two distributions have an identical 'dispersion' (or 'scale'), that is $b_1 = b_2$. In that case, OR=$e^{a_2-a_1}$. The situation of 'identical dispersion' can be extended to the *ordinary logistic regression* if a_j is suitably modeled by a set of covariates $\mathbf{z} = (1, z_1, \ldots, z_p)^T$, for example,

$$a_j = \beta_0 + \beta_1 z_{1j} + \ldots + \beta_p z_{pj}.$$

However, when the dispersions are not identical (referred to as a case of 'heterogeneity'), the odds ratio is (with two-sample setting):

$$\text{OR} = \text{e}^{(a_2-a_1)+x_0(b_2-b_1)}.$$

Without loss of generality, we can set $a_1 = 0, a_2 = a$, and $b_1 = 1, b_2 = b$. Then OR=$\text{e}^{a+x_0(b-1)}$, which depends on the location difference a, as well as on the scale parameter b and the cutoff value x_0. The phenomenon of heterogeneity becomes more apparent if the dispersion parameter $b(> 0)$ is further expressed as a regression setting $e^{\gamma^T\mathbf{z}}$ through the same set of covariates $\mathbf{z}$.

There are, of course, other indices to be used as a measure of effect. If the variable indexes different locations (0 vs. a) will also index different scales (1 vs. b), the heterogeneity effect is said to be 'from the observable variable itself'. It is particularly important when the variable (say X) is continuous, and heterogeneity effect cannot be stratified out even by grouping the X-variable, because the 'effect' of X is to be estimated. This point will become more clear in a later context concerning a regression model with heterogeneity. To make reliable inference, the heterogeneity parameter needs to be estimated explicitly. This is very different from the other heterogeneity models in which the variable resulting in individual or cluster heterogeneity is *not* observed. So the heterogeneity discussed in this paper is not of the same type and not at the same level with, for example, the random effect models.

2 Two-Sample Models

Two-sample problem plays important role in the development of statistical inference. In clinical trials or epidemiological cohort studies, for example, data collected prospectively according to two treatments or retrospectively to diseased and healthy groups are analyzed to assess the effect of a treatment or the association between an exposure and the disease of concern. In what follows, we briefly refer to the measure of interest as 'treatment effect' or simply 'effect'. If the two-sample relation is described by a location-scale model and the goal is to make inference about the treatment effect, it is necessary to estimate with precision both of the location and scale parameters simultaneously. Ignoring the scale parameter (dispersion) leads to biased effect estimate. In this section, we introduce Hsieh's work on two-sample problems through the empirical process approach (EPA).

2.1 Two-sample location-scale model

The two-sample location-scale model studied in Hsieh [HSI95, HSI96a] assumes two distributions, say $F(\cdot)$ and $G(\cdot)$, satisfying

$$G(x) = F(\frac{x-\mu}{\sigma}) \quad \text{or} \quad G^{-1}(t) = \mu + \sigma F^{-1}(t),\ 0 < t < 1,$$

and two sets of samples $X_1, \ldots, X_m \sim F, Y_1, \ldots, Y_n \sim G$. Let $\mathbf{u} = (u_1, \ldots, u_J)^T$ be a set of grid (or cutoff) points, $0 < u_1 < \ldots < u_J < 1$, and J depends on n: $J = J(n)$. The EPA of Hsieh builds up the following regression-type setting for the specified points $u_1, \ldots, u_J$:

$$G_n^{-1}(\mathbf{u}) \cong \mu + \sigma F_m^{-1}(\mathbf{u}) + \sigma \mathcal{D} \mathcal{K}_{m,n}(\mathbf{u}), \tag{1}$$

where $\mathcal{D}$ =diag$(\ldots, 1/f(F^{-1}(u_j)), \ldots)$, $G_n^{-1}(\cdot)$ and $F_m^{-1}(\cdot)$ are the *empirical quantile processes* (Csörgő [CS83]). The process $\mathcal{K}_{m,n}(u)$ is different for complete (no-censoring) and censored data problems. For *complete* two-sample data, $\mathcal{K}_{m,n}(u)$ is a linear combination of two independent Brownian bridge process pertaining to the strong approximations of the two quantile processes respectively; for *right censored* data, $\mathcal{K}_{m,n}(u)$ is a combination of two independent generalized Kiefer processes.

To estimate $\theta = (\mu, \sigma)^T$, equation (1) is treated (at $\mathbf{u}$) as a regression setting and least squares method is used. However, the covariance matrix of $\mathcal{D}\mathcal{K}_{m,n}(\mathbf{u})$ may involve unknown μ and σ, a generalized least squares (GLS) estimate is then adopted. Let $\widetilde{\mathbf{X}}_{J\times 2} = (\mathbf{1}_{J\times 1}, F_m^{-1}(\mathbf{u}))$, where $\mathbf{1}_{J\times 1} = (1, \ldots, 1)^T$ is a $J \times 1$ column vector; further define $\Sigma_e = \mathcal{D}\Sigma_{\mathcal{K}}\mathcal{D}$, where $\Sigma_{\mathcal{K}}$ is the covariance matrix of the $\mathcal{K}(\cdot)$-process. Then we have the GLS estimate of θ:

$$\hat{\theta}_{\text{GLS}} = (\widetilde{\mathbf{X}}^T \widehat{\Sigma}_e^{-1} \widetilde{\mathbf{X}})^{-1} \widetilde{\mathbf{X}}^T \widehat{\Sigma}_e^{-1} \{G_n^{-1}(\mathbf{u})\}.$$

For which if a reweighted procedure is needed, Hsieh suggested a 'one-step' iteration only. Further, $f(F^{-1}(u_k))$ can be substituted by its *kernel-smoothed* estimate. The estimation has the same spirit of minimum chi-square method and, as a companion result, a testing statistic for overall model checking is rendered. In addition to the convenience of implementation, Hsieh's GLS estimate for the location-scale model has an important feature: It achieves the semiparametric Fisher information bound (Bickel et al. [BKRW93]) for large samples, and is thus asymptotically efficient.

2.2 Two-sample transformation model

Now suppose that the two populations have relationship $FG^{-1}(u) = \Psi(\mu + \sigma\Psi^{-1}(u))(0 < u < 1)$ for a *specified* transformation Ψ. For *complete data*, Hsieh [HSI95] proposed an EPA estimation procedure based on a strong approximation of the empirical receiver's operating characteristic (ROC) curve $F_m(G_n^{-1}(u))$ to the true curve $F(G^{-1}(u))$:

$$F_m(G_n^{-1}(u)) \cong F(G^{-1}(u)) + \mathcal{K}_{m,n}(u). \tag{2}$$

Here $\mathcal{K}_{m,n}(u)$ is a combination of two independent Brownian bridges. See also Hsieh [HSI96b] for the problem of ROC curve estimation. According to (2), for a set of points $\mathbf{u} = (u_1, \ldots, u_J)^T$,

$$\sqrt{n}\{F_m G_n^{-1}(\mathbf{u}) - FG^{-1}(\mathbf{u})\} \longrightarrow_D N(0, \Sigma_{\mathcal{K}}), \tag{3}$$

where $\Sigma_{\mathcal{K}}$ is the covariance matrix of $\sqrt{n}\mathcal{K}_{m,n}(\cdot)$. The following asymptotic distribution can be obtained by δ-method and the derivative of a inverse function:

$$\sqrt{n}\{\Psi^{-1}(F_m G_n^{-1}(\mathbf{u})) - (\mu + \sigma\Psi^{-1}(\mathbf{u}))\} \longrightarrow_D N(0, \mathcal{C}\Sigma_{\mathcal{K}}\mathcal{C}), \tag{4}$$

where $\mathcal{C}$ =diag$(\ldots, 1/\psi(\mu+\sigma\Psi^{-1}(u_j)), \ldots)$, $\psi(\cdot)$ is the derivative of $\Psi(\cdot)$. The previous formula implies

$$\Psi^{-1}(F_m G_n^{-1}(\mathbf{u})) = \mu + \sigma\Psi^{-1}(\mathbf{u}) + \varepsilon, \tag{5}$$

in which the covariance of ε is $\sigma_\varepsilon^2 = (1/n)\mathcal{C}\Sigma_{\mathcal{K}}\mathcal{C} \equiv \Sigma_e$. In view of this, a regression setting is built up. The case of $\Psi = \Phi$, the cumulative standard normal distribution, is studied in Hsieh [HSI96b]. For *censored data*, a similar setting was derived in Hsieh [HSI96c]:

$$\Psi^{-1}(\widehat{S}_{1,m}(\widehat{S}_{0,n}^{-1}(\mathbf{u}))) = \mu + \sigma\Psi^{-1}(\mathbf{u}) + \mathcal{K}_{m,n}(\mathbf{u}), \tag{6}$$

where $\widehat{S}_{1,m}$ and $\widehat{S}_{0,n}$ are Kaplan-Meier survival estimators for the two true survivor functions, and $\mathcal{K}_{m,n}(\mathbf{u})$ is again a combination of two independent generalized Kiefer processes. For unified exposition, we still denote the covariance of $\mathcal{K}_{m,n}(\cdot)$ in (6) as Σ_e. Note that the regression settings of (5) and (6) lead to the following least squares type estimation: For complete data, let $\widehat{ROC}(\mathbf{u}) = F_m G_n^{-1}(\mathbf{u})$; for right censored data, $\widehat{ROC}(\mathbf{u}) = \widehat{S}_{1,m}(\widehat{S}_{0,n}^{-1}(\mathbf{u}))$. Further define $\mathcal{D}(\mathbf{u}) = \Psi^{-1}(\widehat{ROC}(\mathbf{u})) - (\mu + \sigma\Psi^{-1}(\mathbf{u}))$. Then, because of the *normality* property of ε and $\mathcal{K}_{m,n}(\cdot)$, the (log-) likelihood comprises the *quadratic form* $\{\mathcal{D}(\mathbf{u})\}^T \Sigma_e^{-1}\{\mathcal{D}(\mathbf{u})\}$ plus a *remainder term*. Also note that the information of $\theta = (\mu, \sigma)^T$ contained in the remainder is asymptotically negligible compared to that contained in the quadratic term (Hsieh [HSI95, HSI96c]), taking derivatives of the quadratic term results in the estimating equation

$$\{\frac{\partial \mathcal{D}(\mathbf{u})}{\partial \theta}\}^T \Sigma_e^{-1} \mathcal{D}(\mathbf{u}) = 0. \tag{7}$$

This equation is convenient to use because, like the situation in linear regression with normal errors, a generalized least squares (GLS) estimate can be obtained by

$$\hat{\theta}_{\text{GLS}} = \{(\frac{\partial \mathcal{D}(\mathbf{u})}{\partial \theta})^T \widehat{\Sigma}_e^{-1}(\frac{\partial \mathcal{D}(\mathbf{u})}{\partial \theta})\}^{-1}(\frac{\partial \mathcal{D}(\mathbf{u})}{\partial \theta})^T \widehat{\Sigma}_e^{-1}\{\Psi^{-1}(\widehat{ROC}(\mathbf{u}))\}, \tag{8}$$

where $\widehat{\Sigma}_e$ is a consistent estimator of Σ_e.

The above estimation procedure has the following merits: it combines the estimation and hypothesis testing problems in a unified quadratic form, which is asymptotically chi-square distributed. This resembles the spirit of minimum chi-square inference. To elucidate, note that the quantity $\Delta = \{\mathcal{D}_\theta(\mathbf{u})\}^T \widehat{\Sigma}_e^{-1}\{\mathcal{D}_\theta(\mathbf{u})\} \sim \chi^2_{2J}$. The quadratic term Δ can be decomposed as

$$\Delta = \{\mathcal{D}_{\hat{\theta}}(\mathbf{u})\}^T \widehat{\Sigma}_e^{-1}\{\mathcal{D}_{\hat{\theta}}(\mathbf{u})\} + (\hat{\theta}-\theta)^T\{(\frac{\partial \mathcal{D}(\mathbf{u})}{\partial \theta})^T \widehat{\Sigma}_e^{-1}(\frac{\partial \mathcal{D}(\mathbf{u})}{\partial \theta})\}_{\hat{\theta}}(\hat{\theta}-\theta) + o_p(1),$$

where $\mathbf{Q_g} \equiv \{\mathcal{D}_{\hat{\theta}}(\mathbf{u})\}^T \widehat{\Sigma}_e^{-1}\{\mathcal{D}_{\hat{\theta}}(\mathbf{u})\} \sim \chi^2_{2J-2}$ is used as a statistic for testing the *global* model goodness-of-fit; and $\mathbf{Q_l} \equiv (\hat{\theta} - \theta)^T\{(\frac{\partial \mathcal{D}(\mathbf{u})}{\partial \theta})^T \widehat{\Sigma}_e^{-1}(\frac{\partial \mathcal{D}(\mathbf{u})}{\partial \theta})\}_{\hat{\theta}}(\hat{\theta} - \theta) \sim \chi^2_2$ can be used to test for a *local* hypothesis such as $H_0 : \theta = \theta_0$ vs. $H_a : \theta \neq \theta_0$ (for some specified θ_0) if under the validity of the global model. This issue will also be explored in the following discussion on hazards regression model.

3 Hazards Regression

The two-sample transformation model can be viewed as a special case of the linear transformation model: $H(T) = -\beta\mathbf{z} + \sigma\varepsilon$, or, after reparameterization, $\sigma H(T) = -\beta\mathbf{z} + \varepsilon$. Taking $H(t) = \log \Lambda(t)$ and $F_\varepsilon(t) = 1 - e^{-e^t}$ results in the 'σ-proportional hazards model' (Hsieh [HSI96c]):

$$\Lambda_1(t) = \{\Lambda_0(t)\}^\sigma \mu. \tag{9}$$

When μ and σ are further expressed as μ =exp($\beta^T\mathbf{z}$) and σ =exp($\gamma^T\mathbf{x}$) for two sets of p- and q-vectors $\mathbf{z}$ and $\mathbf{x}$, model (9) evolves into

$$\Lambda(t; \mathbf{z}, \mathbf{x}) = \{\Lambda_0(t)\}^{e^{\gamma^T\mathbf{x}}} e^{\beta^T\mathbf{z}}, \tag{10}$$

in terms of the cumulative hazard; or (when $\mathbf{z}$ and $\mathbf{x}$ are time-fixed)

$$\lambda(t;\mathbf{z},\mathbf{x}) = \lambda_0(t)\{\Lambda_0(t)\}^{e^{\gamma^T\mathbf{x}}-1}e^{\beta^T\mathbf{z}+\gamma^T\mathbf{x}}, \tag{11}$$

in terms of hazard function. When $\gamma = 0$, (11) reduces to the Cox's proportional hazards (PH) model (Cox [COX72]). The covariates in model (11) can be made time dependent:

$$\lambda(t;\mathbf{z},\mathbf{x}) = \lambda_0(t)\{\Lambda_0(t)\}^{e^{\gamma^T\mathbf{x}(t)}-1}e^{\beta^T\mathbf{z}(t)+\gamma^T\mathbf{x}(t)}. \tag{12}$$

The specific transformation model (12) are termed in Hsieh [HSI01] as the heteroscedastic hazards regression model, and will be called hereafter the *Hsieh model.* Different specific models corresponding to different transforms are listed in Hsieh [HSI95, page 741].

The heterogeneity property investigated in this paper can be explored through model (11) for $\mathbf{x} = \mathbf{z}$: Taking the one-dimensional case as an example, the log-relative risk (≡logRR(t)) between strata z_j versus z_i ($z_j - z_i = 1$) is

$$\log\{\text{RR}(t)\} = (e^{\gamma z_j} - e^{\gamma z_i})\log\Lambda_0(t) + (\beta + \gamma). \tag{13}$$

This one-dimensional case illustrates the ordinary interpretation that: for a multiple regression setting, the coefficient of Z corresponds to a unit-change of 'log-hazard' in Z while the other covariates remains fixed. If logRR(t) is the 'effect' of concern, (13) implies the effect is not only time-dependent, but also depends on the z-value. The 'time-dependence' is preferred to be called as *nonconstancy,* and the dependence on z-value to be as *heterogeneity,* which has the same meaning explained in the logistic regression example introduced in Section 1. The coexistence of nonconstancy and heterogeneity can be viewed as an 'interaction' between the heteroscedasticity component and the underlying hazard. Figure 1 gives examples of (13) in which the nonconstancy and heterogeneity properties of the Hsieh model is explored by plots of logRR(t) for the spectrum of $z_j = -5, -4, \ldots, 5$, and $\gamma = 0.1$ (Fig.1(a)), 0.2 (Fig.1(b)), 0.3 (Fig.1(c)), and 0.5 (Fig.1(d)); $\beta = 1$ for all cases. When the heteroscedasticity parameter is small ($\gamma = 0.1$, Fig.1(a)), the log-relative risk basically looks like a constant in time and they also coincide for much of the time $t \in (0, 2)$. From Figures 1(a) to (d), only $z = 5$ is plotted by a solid line to present the trend of log(RR)-plot in z. For larger γ, time-dependence of logRR(t) is clearer; moreover, for a fixed t, the 'effect' is different for different z's, which reveals larger heterogeneity. The selection of covariate vectors $\mathbf{x}$ and $\mathbf{z}$ is quite flexible: they can have a shared subset of variables (Wu, Hsieh, and Chen [WHC02]).

Estimation procedures proposed in Hsieh [HSI01] starts with a construction of the estimating equations. Let $(T_i, \delta_i, \mathbf{Z}_i, \mathbf{X}_i)$ be independent samples of failure time, censoring indicator, and covariate vectors, $i = 1, \ldots, n$, where (without loss of generality) $T_1 < \ldots < T_n$ be failure or right-censored times.

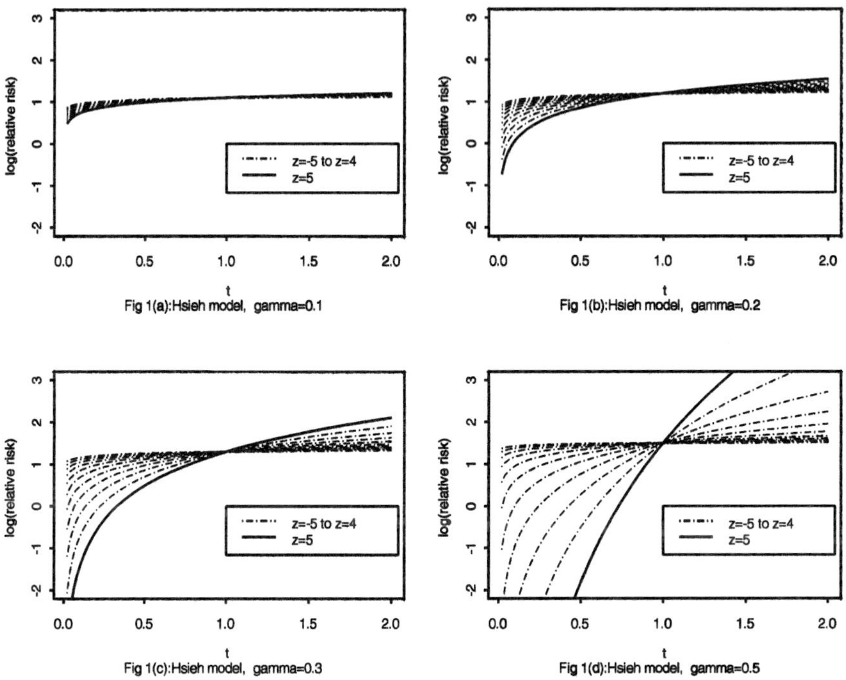

Fig. 1. Heterogeneous log-relative risks in the Hsieh model.

We denote $N_i(t) = \mathbf{1}_{\{T_i \le t, \delta_i = 1\}}$ and $Y_i(t)$ to be the counting process and at-risk indicator of individual i. Further let $\mathbf{V}_i(t) = \mathbf{X}_i(t)\{1 + e^{\gamma^T \mathbf{X}_i} \log \Lambda_0(t)\}$,$\theta = (\beta, \gamma)^T$, and

$$S_K(t; \Lambda_0, \theta) = (1/n) \sum_{i=1}^{n} Y_i(t) K_i(t) e^{\beta^T \mathbf{Z}_i + \gamma^T \mathbf{X}_i} \{\Lambda_0(t)\}^{e^{\gamma^T \mathbf{X}_i} - 1},$$

where $K_i(t) = 1, \mathbf{Z}_i(t)$, or $\mathbf{V}_i(t)$. The estimating equation processes constructed in Hsieh [HSI01] are

$$M_1(t) = \sum \int_0^t \{\frac{dN_i(u)}{S_1(u; \Lambda_0, \theta)} - \lambda_0(u) du\}, \tag{14}$$

$$M_2(t) = \sum \int_0^t \{\mathbf{Z}_i - \frac{S_{\mathbf{Z}}(u; \Lambda_0, \theta)}{S_1(u; \Lambda_0, \theta)}\} dN_i(u), \tag{15}$$

$$M_3(t) = \sum \int_0^t \{\mathbf{V}_i - \frac{S_{\mathbf{V}}(u; \Lambda_0, \theta)}{S_1(u; \Lambda_0, \theta)}\} dN_i(u), \tag{16}$$

where $t \in (0, u_J)$, for a maximal truncation time u_J (defined below). Setting $M_1(t)$=0 leads to

$$\Lambda_0(t) = \sum \int_0^t \frac{dN_i(u)}{\sum Y_i(u) e^{\beta^T \mathbf{Z}_i + \gamma^T \mathbf{X}_i} \{\Lambda_0(u)\}^{e^{\gamma^T \mathbf{X}_i} - 1}}. \tag{17}$$

Define the elements of the 3× 3 matrix of *covariation process* A as

$$A_{ik} = \lim(1/n) \sum \int E\{dM_i(u)\}\{dM_k(u)\} du.$$

Under several regularity conditions, Hsieh [HSI01] have the following property of $(M_1, M_2, M_3)(t)^T$. (I) The process M_1 is orthogonal to M_2 and M_3 in the sense that the covariation process $< M_1, M_2 >_t = < M_1, M_3 >_t = 0$. (II) The system of martingales $M(t) = (M_1, M_2, M_3)^T(t)$ converge weakly to $W(t) = (W_1, W_2, W_3)^T(t)$, which is a system of Gaussian processes with *independent increments.* The components of covariance of $W(t)$ are $A_{ik}(t), i, k = 1, 2, 3$. Moreover, by (I), $< W_1, W_2 >_t = < W_1, W_3 >_t = 0$. Conventional notations about the counting process model can be found in Andersen et al. [ABGK93]

The above limiting process $W(t)$ leads to the construction of an *approximated likelihood*: Let $G(t)$ be a stochastic process and $\Delta^{(J)} G(\mathbf{u}) = (G(u_1) - G(u_0), G(u_2) - G(u_1), \ldots, G(u_J) - G(u_{J-1}))^T$, where $G(u_0) = G(0) = 0$ and $G(u_J) = G(t_n)$. Also, denotes $\Delta_i G = G(u_i) - G(u_{i-1})$. By choosing suitable cutoff points $\mathbf{u} = (u_1, u_2, \ldots, u_J)^T$, an approximation of the likelihood due to the *independent-increment* property of $W(t)$ can be obtained: log-likelihood$\cong -\frac{1}{2}\sum_1^J (\Delta_i M)^T (\Delta_i A)^{-1} (\Delta_i M) \equiv L^{(J)}$. Note that, by the orthogonality of W_1 and $(W_2, W_3)^T$, $(\Delta_i A)^{-1} = \text{diag}((\Delta_i A_{11})^{-1}, (\Delta_i A_{(11)})^{-1})$, where $A_{(11)}$ is the submatrix of A deleting the first column and the first row. One can make statistical inference for θ and Λ_0 based on the approximated likelihood. Before doing this, Hsieh [HSI01] also introduces a piecewise-constant approximation to $\Lambda_0(t)$: $\Lambda_0^{(J)}(t) = \int_0^t \sum_1^J \alpha_i 1_{\{u_{i-1} < u \le u_i\}} du$, where $0 < \alpha_i < \infty$ and $J = O(n^{\frac{1}{3}})$. The score functions of parameters θ and $\alpha = (\alpha_1, \ldots, \alpha_J)^T$ are: $E_\beta = \frac{\partial}{\partial \beta} L^{(J)}$, $E_\gamma = \frac{\partial}{\partial \gamma} L^{(J)}$, and $E_\alpha = \frac{\partial}{\partial \alpha} L^{(J)}$. The estimated parameters of interest have the asymptotics:

$$\sqrt{n}(\hat{\theta} - \theta) \longrightarrow_D N(0, \{\lim[\Sigma_{\theta\theta} - \Sigma_{\theta\alpha} \Sigma_{\alpha\alpha}^{-1} \Sigma_{\theta\alpha}]\}^{-1}), \tag{18}$$

where $\Sigma_{\{.\}}$'s are the information matrices pertaining to the associated parameters. In addition, $\widehat{\Lambda_0^{(J)}}$ can have a $\sqrt{n}$- weak convergence. If the heteroscedasticity part, $\gamma^T \mathbf{x}$, is neglected, the estimate of β will be biased. (Wu [WU04a]).

The quadratic $L^{(J)}$ (omitting '$-\frac{1}{2}$') can be further decomposed as: $L^{(J)} = L_1^{(J)} + L_2^{(J)}$, where

$$L_1^{(J)} = \sum (\Delta_i M_1)(\Delta_i A_{11})^{-1} (\Delta_i M_1),$$

and

$$L_2^{(J)} = \sum(\Delta_i M_2, \Delta_i M_3)(\Delta_i A_{(11)})^{-1}(\Delta_i M_2, \Delta_i M_3)^T.$$

The part $L_2^{(J)}$ contains much of the information of the parameters of interest β and γ, and can be viewed as a projection of $L^{(J)}$ onto the space of $(\beta^T\mathbf{z}, \gamma^T\mathbf{x})$. Similar to the expression in Section 2.2, let $L_2^{(J)} = \mathbf{Q_g} + \mathbf{Q_l}$, where

$$\mathbf{Q_g} = \sum(\Delta_i\widehat{M}_2, \Delta_i\widehat{M}_3)(\Delta_i\widehat{A}_{(11)})^{-1}(\Delta_i\widehat{M}_2, \Delta_i\widehat{M}_3)^T \sim \chi^2_{(p+q)(J-1)}$$

can be used as a test statistic for *global* model validity (i.e. the Hsieh model), and

$$\mathbf{Q_l} = \sum\{\Delta_i(M_2-\widehat{M}_2, M_3-\widehat{M}_3)\}(\Delta_i\widehat{A}_{(11)})^{-1}\{\Delta_i(M_2-\widehat{M}_2, M_3-\widehat{M}_3)\}^T \sim \chi^2_{p+q}$$

is used to test for a *local* hypothesis: $H_0 : \theta = \theta_0$. For examples, the proportional hazards (PH) assumption can be checked under the nested family of the PH model ($H_0 : \gamma = 0$) within the Hsieh model ($H_a : \gamma \neq 0$) (Wu et al. [WHC02]); or, the equal-distribution null hypothesis ($H_0 : \beta = \gamma = 0$) can also be tested within the Hsieh model ($H_a : \beta \neq 0$ or $\gamma \neq 0$) (Wu [WU04b]).

4 Non-proportional Hazards Model

Non-proportional hazards modeling has been widely studied in the past decades. In this section, the Bagdonavičius and Nikulin's [BN99] generalized proportional hazards model and its variant are introduced, which also can deal with nonconstancy as well as heterogeneity. Before the discussion, several nonproportional hazards models are reviewed.

First consider a parametric model like the Weibull regression taking the cumulative hazard function as (see, for example, Gore, Pocock, and Kerr [GPK84, page 185] for a survivor-function expression):

$$\Lambda(t) = t^b e^{\beta_0+\beta_1 z_1+\ldots+\beta_p z_p} \equiv t^b e^{\beta^T\mathbf{z}},$$

which can be viewed as a Weibull-class with universal shape parameter b, but with different scales indexed by $\beta^T\mathbf{z}$. By this, the Hsieh model of Section 3 have a parametric (Weibull) regression model as a special case:

$$\Lambda(t) = t^{e^{\gamma^T\mathbf{x}}} e^{\beta^T\mathbf{z}}, \text{ for } \gamma^T\mathbf{x} = \gamma_0 + \gamma_1 x_1 + \ldots + \gamma_q x_q.$$

There are also several nonproportional hazards model studied in Gore at al. [GPK84] Among them, a Cox-type model with 'varying-proportionality hazards' is of interest:

$$\lambda(t;\mathbf{z}) = \lambda_0(t)e^{\theta(\mathbf{z},t)}, \tag{19}$$

where $\theta(\mathbf{z}, t)$ is a smooth function of covariate $\mathbf{z}$ and time variable t. Model (19) is capable of modeling heterogeneity *plus* nonconstancy. Because the time dependence is described simultaneously by $\lambda_0(t)$ and $\theta(\cdot, t)$, the form of $\theta(\mathbf{z},t)$ should be 'pre-specified' to make the setting *identifiable.* For modeling only the heterogeneity over the covariate space, a very flexible partly linear setting can be imposed on the relative risk function: $\lambda(t;\mathbf{x},\mathbf{z}) = \lambda_0(t)\exp\{\beta^T\mathbf{x} + g(\mathbf{z})\}$; see Sasieni [SAS92a], Nielsen, Linton, and Bickel [NLB98], and Heller [HEL01]. When $\mathbf{z}$ is categorical, the partly linear model reduces to the stratified proportional hazards model, by which the heterogeneity effect over $\mathbf{z}$ is further *stratified out* by putting a number of unknown baseline hazards. Moreover, a *continuously stratified* Cox model introduced in Sasieni [SAS92b]) and the 'time-varying' coefficients Cox model studied by Murphy and Sen [MS91], Murphy [MUR93] and Martinussen, Scheike, and Skovgaard [MSS01] (among others) are important works on the Cox-type relative risk modeling for heterogeneity and/or nonconstancy.

An alternative Cox-type model (other than the Hsieh model) which can deal with heterogeneity and nonconstancy together is the generalized PH model proposed by Bagdonavičius and Nikulin [BN99], see also Bagdonavičius and Nikulin [BN02] for more related works and models. The generalized proportional hazards model has the following form, in terms of hazard function,

$$\lambda(t;\mathbf{z}) = \lambda_0(t)g\{\mathbf{z}(t), \Lambda_{\mathbf{z}}, \theta\}, \tag{20}$$

where $g(\cdot)$ is a positive function, $\mathbf{z}(t)$ is a set of time-dependent covariate, and $\Lambda_{\mathbf{z}}$ is the corresponding cumulative hazard. A special form of (20) derived in Bagdonavičius, Hafdi, and Nikulin [BHN04] is

$$\lambda(t;\mathbf{z}) = \lambda_0(t)e^{\beta^T\mathbf{z}}\{1 + e^{(\beta+\gamma)^T\mathbf{z}}\Lambda_0(t)\}^{e^{-\gamma^T\mathbf{z}}-1}, \tag{21}$$

where $\Lambda_0(t) = \int_0^t \lambda_0(u)du$ is the baseline cumulative hazard function. Model (21) is called hereafter the *Bagdonavičius-Nikulin model.* Similar to the Hsieh model, the Bagdonavičius-Nikulin model also gives cross-effect for the cumulative hazards, but *not necessarily* for the hazard functions. When the log-relative risk is the effect of concern, the main differences between the Hsieh and the Bagdonavičius-Nikulin models are: (i) the former assumes the relative risk between groups to be possibly very large when t approaches 0, the model design of the latter relaxes this assumption; and (ii) relative risk of the Hsieh model is increasing or decreasing according to the relative direction of the heteroscedasticity part $\gamma^T x$, the Bagdonavičius-Nikulin model has more complex situation which depends on the configurations of β and γ in (21). Figure 2 gives a similar illustration to Figure 1 and corresponds to part of the configurations listed in Bagdonavičius et al. [BNLZ05]. When γ is large (0.5 or -0.5), the dependence of log-relative risks (LRR) on time is more significant. Also for larger γ, difference of LRRs in different z's is apparent, showing *heterogeneity* effect over the covariate-space. In Figure 2, the LRR also has an order in z (for $z = -5,\ldots,5$), so only $z = 5$ is plotted by a solid line; for all cases, $\beta = 1$ and $0 < t < 2$. The estimation of the Bagdonavičius-Nikulin model depends on iteratively solving the score equations derived from the 'partial likelihood' and a Breslow-type estimating equation for the baseline

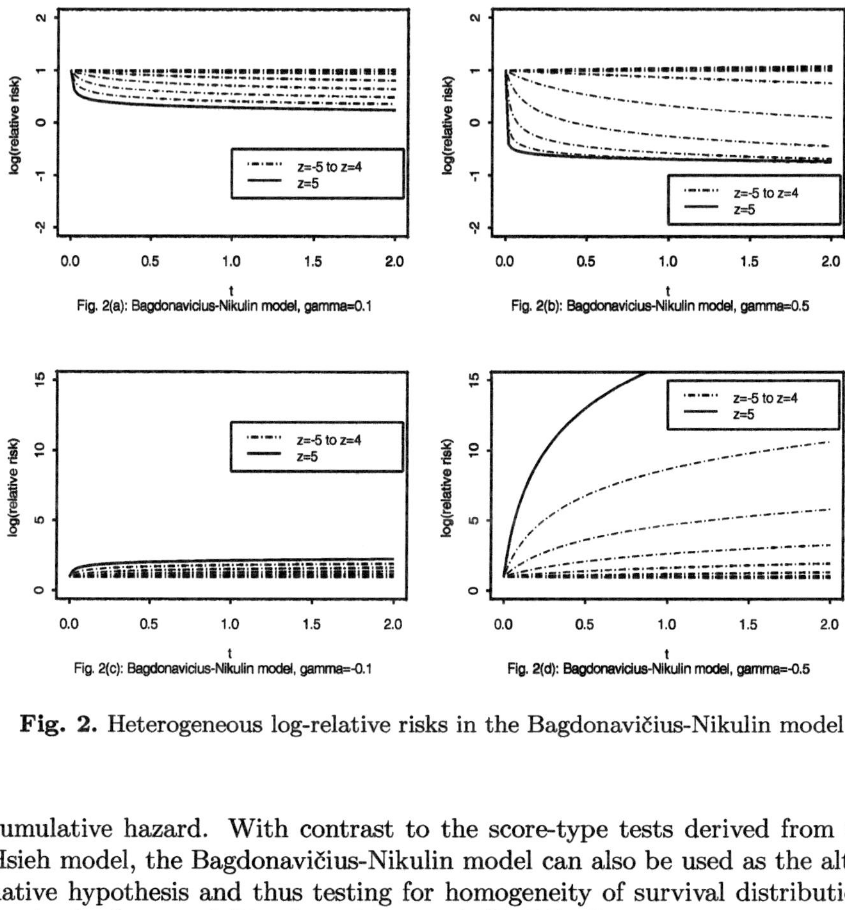

Fig. 2. Heterogeneous log-relative risks in the Bagdonavičius-Nikulin model.

cumulative hazard. With contrast to the score-type tests derived from the Hsieh model, the Bagdonavičius-Nikulin model can also be used as the alternative hypothesis and thus testing for homogeneity of survival distributions can be implemented (Bagdonavičius et al. [BNLZ05]).

5 Extensions and Brief Discussion

There are numerous papers targeting at heterogeneity and associated statistical inference. This paper is not intended to review the entire development, but only to focus on heterogeneity effect that can be modeled through *collected samples.*

The early works of Hsieh [HSI95, HSI96a, HSI96b, HSI96c] make statistical inferences on location-shift and scale-change parameters through EPA for two-sample models with complete or right-censored data. In Hsieh [HSI01], however, hazards regression is discussed and estimation and testing problems are solved with slightly different manner, though with similar spirit to the EPA method. The Hsieh model and the later introduced Bagdonavičius-Nikulin model allow for cross-effect modeling. Cross-effect or nonproportional hazards problem still is an important issue for future researches. In actual data analysis of follow-up data, *multiple cross-effect* may exist and extensions to these two models are of interests. First, consider the Hsieh model equipped with varying coefficients (in terms of hazard function):

$$\lambda(t; \mathbf{z}, \mathbf{x}) = \lambda_0(t)\{\Lambda_0(t)\}^{e^{\gamma(t)^T \mathbf{x}}-1} e^{\beta(t)^T \mathbf{z}+\gamma(t)^T \mathbf{x}}, \tag{22}$$

where $\beta(t)$ and $\gamma(t)$ are two sets of varying coefficients. Model (22) can be viewed as dealing with the 'time-heteroscedasticity interaction'. In practice, the same set of time-partition for a piecewise-constant approximation can be applied to $\lambda_0(t)$, $\beta(t)$ and $\gamma(t)$, and estimating equations similar to (15) and (16) while taking account the approximation can be used (Wu and Hsieh [WH04]). By suitably choosing the cutoff points which constitute the piecewise-constant intervals, multiple crossings among cumulative hazards according to different groups can be modeled. Second, a variant of model (21) is recently proposed by Bagdonavičius and Nikulin [BN04] by which at least two crossings can be properly captured:

$$\lambda(t; \mathbf{z}) = \lambda_0(t) e^{\beta^T \mathbf{z}(t)}\{1 + \gamma^T \mathbf{z}(t)\Lambda_0(t) + \delta^T \mathbf{z}(t)\Lambda_0^2(t)\}, \tag{23}$$

where γ and δ are two sets of parameters to be estimated.

Diagnostics for a hazards regression model is important for model validity and goodness-of-fit problem. By simply plotting the estimated (log-) relative risks versus time or the important covariates, time-constancy and effect homogeneity/heterogeneity can be checked. For example, Valsecchi, Silvestri, and Sasieni [VSS96] used the plot to check nonconstancy as well as proportionality for various explanatory variables of ovarian cancer patient's survival. Further applications of plotting the (estimated) relative risk as a model *discrimination* and *diagnostics* tool for the Hsieh and the Bagdonavičius-Nikulin models is interesting for future researches.

ACKNOWLEDGMENT

This research is partly supported by Grant NSC93-2118-M-039-001 of Taiwan's National Science Council. The relevant works of the author were mostly inspired by Professors Fushing Hsieh and Mikhail Nikulin; and also by Professor Marvin Zelen in a conversation during his visit to Taiwan.

References

[ABGK93] Andersen, P. K., Borgan, Ø., Gill, R. D., and Keiding, N.: Statistical Models Based on Counting Processes. Springer-Verlag, New York (1993)

[BHN04] Bagdonavičius, V., Hafdi, M. A., and Nikulin, M.: Analysis of survival data with cross-effects of survival functions. Biostatistics, **5** 415-425 (2004)

[BN99] Bagdonavičius, V. and Nikulin, M.: Generalized proportional hazards model based on modified partial likelihood. Lifetime Data Analysis, **5** 329-350 (1999)

[BN02] Bagdonavičius, V. and Nikulin, M.: Accelerated Life Models. Chapman and Hall/CRC, London (2002)

[BN04] Bagdonavičius, V. and Nikulin, M.: Semiparametric analysis of survival data with multiple crossing of survival functions. Preprint 0304,IFR "Public Health", University Victor Segalen Bordeaux 2, France (2004)

[BNLZ05] Bagdonavičius, V., Nikulin, M., Levuliene, R., and Zdorova, O.: Tests for homogeneity of survival distributions against non-location alternatives and statistical analysis of chemo and radio therapy data of the Gastrointestinal Tumor Study Group. Lifetime Data Analysis, to appear (2005)

[BKRW93] Bickel, P. J., Klaasen, C. A., Ritov, Y., and Wellner, J. A.: Efficient and Adaptive Inference in Semiparametric Models., Johns Hopkins University Press, Baltimore (1993)

[COX72] Cox, D. R.: Regression models and life-tables (with discussion). Journal of the Royal Statistical Society, Series B, **34** 187-220 (1972)

[CS83] Csörgő, M.: Quantile Processes with Statistical Applications. SIAM, Philadelphia (1983)

[GPK84] Gore, S. M., Pocock, S. J., and Kerr, G. R.: Regression models and non-proportional hazards in the analysis of breast cancer survival. Applied Statistics, **33** 176-195 (1984)

[HEL01] Heller, G.: The Cox proportional hazards model with a partly linear relative risk function. Lifetime Data Analysis, **7** 255-277 (2001)

[HSI95] Hsieh, F.: The empirical process approach for semiparametric two-sample models with heterogeneous treatment effect. Journal of the Royal Statistical Society, Series B, **57** 735-748 (1995)

[HSI96a] Hsieh, F.: Empirical process approach in a two-sample location-scale model with censored data. The Annals of Statistics, **24** 2705-2719 (1996a)

[HSI96b] Hsieh, F.: Nonparametric and semiparametric estimation of the receiver operating characteristic curve. The Annals of Statistics, **24** 25-40 (1996b)

[HSI96c] Hsieh, F.: A transformation model for two survival curves: an empirical process approach. Biometrika, **83** 519-528 (1996c)

[HSI01] Hsieh, F.: On heteroscedastic Cox's regression models and its applications. Journal of the Royal Statistical Society, Series B, **63** 63-79 (2001)

[MSS01] Martinussen, T., Scheike, T. H., and Skovgaard, Ib. M.: Efficient estimation of fixed and time-varying covariate effects in multiplicative intensity models. Scandinavian Journal of Statistics, **28** 58-74 (2001)

[MUR93] Murphy, S. A.: Testing for a time-dependent coefficient in Cox's regression model. Scandinavian Journal of Statistics, **20** 35-50 (1993)

[MS91] Murphy, S. A. and Sen, P. K.: Time dependent coefficients in a Cox-type regression model. Stochastic Processes and Their Applications, **39** 153-180 (1991)

[NLB98] Nielsen, J. P., Linton, O., and Bickel, P. J.: On a semiparametric survival model with flexible covariate effect. The Annals of Statistics, **26** 215-241 (1998)

[SAS92a] Sasieni, P.: Non-orthogonal projections and their application to calculating the information in a partly linear Cox model. Scandinavian Journal of Statistics, **19** 215-233 (1992a)

[SAS92b] Sasieni, P.: Information bounds for the conditional hazard ratio in a nested family of regression models. Journal of the Royal Statistical Society, Series B, **54** 627-635 (1992b)

[VSS96] Valsecchi, M. G., Silvestri, D., and Sasieni, P.: Evaluation of long-term survival: use of diagnostics and robust estimators with Cox's proportional hazards model. Statistics in Medicine, **15** 2763-2780 (1996)

[WU04a] Wu, H.-D. I.: Effect of model misspecification when omitting heterogeneity. In: Nikulin, M. S., Balakrishnan, N., Limnios, N., and Mesbah, M. (eds) Parametric and Semiparametric models with applications to reliability, survival analysis, and quality of life, 239-250. Birkhauser, Boston (2004a)

[WU04b] Wu, H.-D. I.: A partial score test for difference among heterogeneous populations. Preprint, China Medical University, TAIWAN. (Under revision in Lifetime Data Analysis.) (2004b)

[WH04] Wu, H.-D. I. and Hsieh, F.: Heterogeneity and varying effect in hazards regression. Preprint, China Medical University, TAIWAN (2004)

[WHC02] Wu, H.-D. I., Hsieh, F., and Chen, C.-H.: Validation of a heteroscedastic hazards regression model. Lifetime Data Analysis, **8** 21-34 (2002)

Failure Distributions Associated With General Compound Renewal Damage Processes

S. Zacks[1]

Department of Mathematical Sciences
Binghamton University
shelly@math.binghamton.edu

Key words: Failure Distributions, Reliability, Hazard Functions, Compound Renewal, Damage Processes

1 Introduction

In a recent paper [Z04] distributions of failure times due to random cumulative damage process were investigated. In particular the previous study was focused on random damage processes driven by non-homogeneous compound Poisson process, with a Weibull intensity function and exponential damage in each occurrence. The present paper generalizes the previous results to damage processes, which are driven by compound renewal processes and general damage distributions. The failure time is the first instant at which the cumulative damage crosses the system's threshold. In Section 2 we present the cumulative damage process as a general compound renewal process. The density function of the associated failure distribution is given, as well as its moments. In Section 3 we consider the special case of a homogeneous compound Poisson damage process (CCDP) with exponentially distributed jumps. The density of failure times, when the intensity λ of the CCDP is random (doubly stochastic Poisson process) is derived for λ having a Gamma distribution. The hazard function for this doubly stochastic case is illustrated in Figure 2. The results of Section 3 are extended in Section 4 to Erlang damage size distributions. The reader is referred to the book of Bogdanoff and Kozin [BK85] for illustrations of random cumulative damage processes. They used a discrete homogeneous Markov Chain to model the extent of damage and failure times (phase-type distributions). The reader is referred also to the book of Bogdanovicius and Nikulin [BN02], and the papers of Wilson [W00], Kahle and Wendt [KW00], Aalen and Gjessing [AG03].

2 The General Compound Renewal Damage Process, and The Associated Failure Distribution

Consider a renewal process, with renewal epochs $0 < \tau_1 < \tau_2 < \ldots < \tau_n < \ldots$. Let $T_i = \tau_i - \tau_0$ $(i = 1, 2, \ldots, \tau_0 \equiv 0)$ be the interarrival times. $T_1, T_2, \ldots$ is a sequence of independent identically distributed (i.i.d.) random variables, having a common distribution F. We assume in this paper that F is absolutely continuous, with density function f, and $F(0) = 0$. Let $N(t)$ denote the number of arrivals in the time interval $(0, t]$, with $N(0) \equiv 0$, i.e.,

$$N(t) = \max\{n \geq 0 : \tau_n \leq t\}. \tag{1}$$

It is well known [K97] that

$$P\{N(t) = n\} = F^{(n)}(t) - F^{(n+1)}(t), \quad n = 0, 1, \ldots \tag{2}$$

where $F^{(0)}(t) = 1$, all $t \geq 0$, $F^{(1)}(t) = F(t)$, and for $n \geq 2$

$$F^{(n)}(t) = \int_0^t f(x) F^{(n-1)}(t - x) dx. \tag{3}$$

That is, $F^{(n)}$ is the n-fold convolution of F. Let $f^{(n)}$ denote the corresponding n-fold convolution of the density f.

We model the cumulative damage (C.D.) process by the compound renewal process

$$Y(t) = \sum_{n=0}^{N(t)} Y_n, \tag{4}$$

where $Y_0 \equiv 0$, $Y_1, Y_2, \ldots$ are i.i.d. positive random variables having a common absolutely continuous distribution G, with density g, and $\{Y_n, n \geq 1\}$ is independent of $\{N(t), t \geq 0\}$. The distribution function of $Y(t)$, at time t, is

$$D(y; t) = \sum_{n=0}^{\infty} (F^{(n)}(t) - F^{(n+1)}(t)) G^{(n)}(y), \tag{5}$$

where $G^{(n)}$ is the n-fold convolution of G.

Notice that D has a jump point (atom) at $y = 0$, and $D(0; t) = 1 - F(t) \equiv \overline{F}(t)$. The density of $Y(t)$ on $(0, \infty)$ is

$$d(y; t) = \sum_{n=1}^{\infty} (F^{(n)}(t) - F^{(n)}(t)) g^{(n)}(t), \tag{6}$$

where $g^{(n)}$ is the n-fold convolution of g. Let β, $0 < \beta < \infty$, be a threshold value such that the system fails as soon as $Y(t) \geq \beta$. Thus, we define the stopping time

$$T(\beta) = \inf\{t \geq 0 : Y(t) \geq \beta\}. \tag{7}$$

Since $Y(t) \uparrow \infty$ a.s. as $t \to \infty$, $P\{T(\beta) < \infty\} = 1$ for all $0 < \beta < \infty$. Moreover,

$$P\{T(\beta) > t\} = D(\beta; t), \quad 0 \le t < \infty. \tag{8}$$

This is the reliability function of the system, which is obviously decreasing function of t, with $\lim_{t\to\infty} D(\beta; t) = 0$. The density function of $T(\beta)$ as obtained from (5), is given by

$$p_T(t; \beta) = f(t)\overline{G}(\beta) + \sum_{n=2}^{\infty} f^{(n)}(t)(G^{(n-1)}(\beta) - G^{(n)}(\beta)). \tag{9}$$

Let $S_n = \sum_{i=0}^{n} Y_i$, $n \ge 0$, and let

$$N^*(t) = \max\{n \ge 0 : S_n \le t\}. \tag{10}$$

$\{N^*(t), t \ge 0\}$ is the renewal process associated with $\{Y_0, Y_1, Y_2, \ldots\}$. Accordingly, the density of $T(\beta)$ can be written as

$$p_T(t; \beta) = \sum_{n=1}^{\infty} f^{(n)}(t) P\{N^*(\beta) = n - 1\}. \tag{11}$$

Theorem 1 *If the interarrival time T_1 has a moment of order m, μ_m $(m \ge 1)$ then*

$$E\{T^{(m)}(\beta)\} = \sum_{n=1}^{\infty} M_{n,m} P\{N^*(\beta) = n - 1\}, \tag{12}$$

where

$$M_{n,m} = E\left\{\left(\sum_{i=1}^{n} T_i\right)^m\right\}. \tag{13}$$

Proof. According to (11),

$$\begin{aligned} E\{T^{(m)}(\beta)\} &= \int_0^\infty t^m p_T(t; \beta) dt \\ &= \sum_{n=0}^{\infty} \left(\int_0^\infty t^m f^{(n)}(t) dt\right) P\{N^*(\beta) = n - 1\}. \end{aligned}$$

Moreover,

$$\int_0^\infty t^m f^{(n)}(t) dt = M_{n,m}.$$

Thus, if $E\{T_1\} = \mu_1$ then

$$\begin{aligned} E\{T(\beta)\} &= \mu_1 \sum_{n=1}^{\infty} n P\{N^*(\beta) = n - 1\} \\ &= \mu_1(1 + E\{N^*(\beta)\}). \end{aligned} \tag{14}$$

If $\mu_2 = E\{T_1^2\}$ then

$$E\{T^2(\beta)\} = \mu_2 \sum_{n=1}^{\infty} nP\{N^*(\beta) = n-1\} + \mu_1^2 \sum_{n=1}^{\infty} n(n-1)P\{N^*(\beta) = n-1\}$$
$$= \mu_2(1 + E\{N^*(\beta)\} + \mu_1^2(E\{(N^*(\beta))^2\} + E\{N^*(\beta)\}). \tag{15}$$

>From (14)-(15) we obtain the following formula for the variance of $T(\beta)$, namely

$$V\{T(\beta)\} = \sigma^2(1 + E\{N^*(\beta)\}) + \mu_1^2 V\{N^*(\beta)\}, \tag{16}$$

where $\sigma^2 = V\{T_1\}$.

In a similar fashion one can obtain formulae for higher moments of $T(\beta)$.

3 Compound Poisson With Exponential Damage

The process $\{Y(t), t \geq 0\}$ is a homogeneous compound Poisson process if $F(t) = 1 - e^{-\lambda t}$, $t \geq 0$.

In this case the damage distribution is

$$D(y; t) = \sum_{n=0}^{\infty} p(n; \lambda t) G^{(n)}(y), \tag{17}$$

where $p(n; \lambda t) = e^{-\lambda t} \dfrac{(\lambda t)^n}{n!}$ is the probability function of the Poisson distribution with mean λt. We consider the special case where $Y_1, Y_2, \ldots$ have a common exponential distribution, i.e., $G(y) = 1 - e^{-\mu y}$, $y \geq 0$. In this case $G^{(n)}(y)$ is the cdf of the Erlang distribution, and we have

$$G^{(n)}(y) = 1 - P(n-1; \mu y), \quad n \geq 1, \tag{18}$$

where $P(\cdot; \mu y)$ is the cdf of the Poisson distribution with mean μy. According to (8), (17) and (18) we obtain that the reliability function $R(t; \lambda, \zeta) = P(T(\beta) > t)$, where $\zeta = \mu\beta$, is

Theorem 2 *For a Compound Poisson damage process, with $G(y) = 1 - e^{-\mu y}$, the reliability function is*

$$R(t; \lambda, \zeta) = \sum_{j=0}^{\infty} p(j; \zeta) P(j; \lambda t). \tag{19}$$

Proof.

$$R(t; \lambda, \zeta) = \sum_{n=0}^{\infty} p(n; \lambda t)(1 - P(n-1; \zeta))$$
$$= 1 - \sum_{n=1}^{\infty} p(n; \lambda t) P(n-1; \zeta)$$

By changing order of summation we obtain

$$R(t;\lambda,\zeta) = 1 - \sum_{j=0}^{\infty} p(j;\zeta) \sum_{n=j+1}^{\infty} p(n;\lambda t).$$

This implies (19).

>From the definition of $T(\beta)$, it is obvious that $\zeta \mapsto R(t;\lambda;\zeta)$ is an increasing function. This follows also from (19), by applying Karlin's Lemma [K57]. The density of the failure times is, in this special case,

$$f_T(t;\lambda,\zeta) = \lambda \sum_{n=0}^{\infty} p(n;\zeta)p(n;\lambda t), \quad t \geq 0. \tag{20}$$

Indeed,

$$\frac{\partial}{\partial t} P(j;\lambda t) = -\lambda p(j;\lambda t)$$

and

$$f_T(t;\lambda,\zeta) = -\frac{\partial}{\partial t} R(t;\lambda,\zeta).$$

The expected value and variance of $T(\beta)$ are, in this special case,

$$E\{T(\beta)\} = \frac{1+\zeta}{\lambda}, \tag{21}$$

and

$$V\{T(\beta)\} = \frac{1+2\zeta}{\lambda^2}. \tag{22}$$

Indeed, in the present case $\mu_1 = \frac{1}{\lambda}$ and $E\{N^*(\beta)\} = \mu\beta = \zeta$. Similarly $V\{N^*(\beta)\} = \zeta$ and $\sigma^2 = \frac{1}{\lambda^2}$. Equation (22) follows immediately from (16).

The hazard function corresponding to (19)-(20) is

$$h(t;\lambda,\zeta) = \lambda \frac{\sum_{j=0}^{\infty} p(j;\zeta)p(j;\lambda t)}{\sum_{j=0}^{\infty} p(j;\zeta)P(j;\lambda t)}. \tag{23}$$

As proven in Zacks [Z04], $h(0;\lambda,\zeta) = \lambda e^{-\zeta}$ and $\lim_{t\to\infty} h(t;\lambda,\zeta) = \lambda$. Moreover, one can show that $t \mapsto h(t;\lambda,\zeta)$ is strictly increasing.

In Figure 1 we present the hazard function (23).

Wilson [W00] discusses the double stochastic Poisson process, in which the intensity parameter λ, of the Compound Poisson process is integrated with respect to some Lebesgue density. If we consider λ a gamma random variable with shape parameter ν and scale parameter $1/\lambda^*$, we obtain that the reliability function is

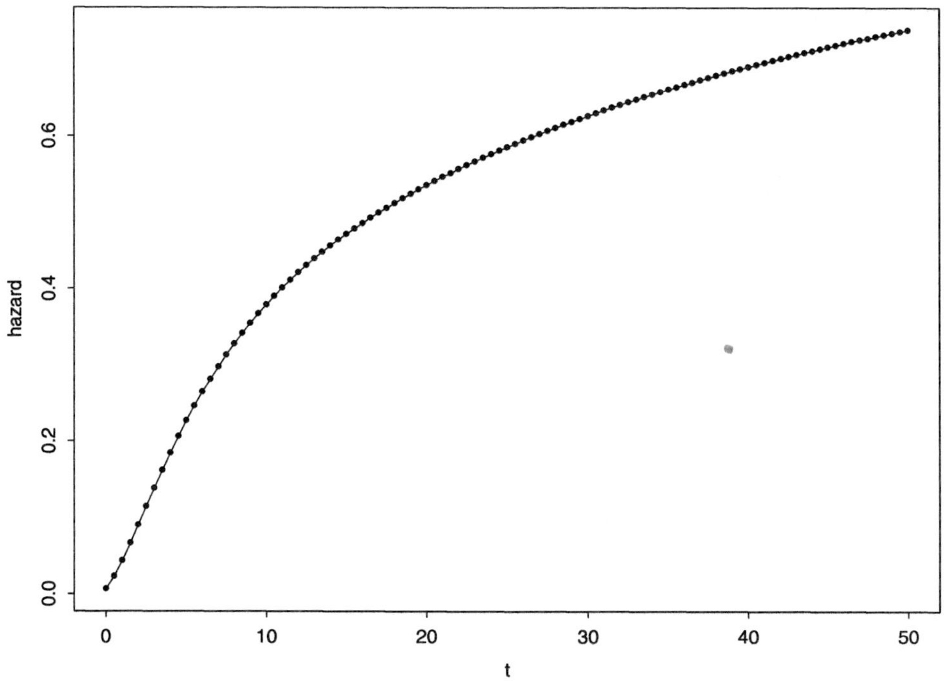

Fig. 1. Hazard Function, $\lambda = 1$, $\zeta = 5$.

$$R^*(t;\lambda^*,\zeta,\nu) = \sum_{j=0}^{\infty} p(j;\zeta) NB\left(j;\frac{t}{\lambda^*+t},\nu\right), \tag{24}$$

where $NB(j;\psi,\nu)$ is the c.d.f. of the negative-binomial distribution

$$NB(j;\psi,\nu) = \sum_{i=0}^{j} \frac{\Gamma(\nu+i)}{i!\Gamma(\nu)} \psi^i (1-\psi)^\nu. \tag{25}$$

Notice that in (24) $\psi = t/(\lambda^*+t)$. Also, $R^*(t;\lambda^*,\zeta,\nu)$ is an increasing function of t, from $\frac{\nu}{\lambda^*}e^{-\beta}$ to ν/λ^*.

The density function corresponding to (24) is

$$f_T^*(t;\lambda^*,\zeta,\nu) = \sum_{j=0}^{\infty} p(j;\zeta) \frac{\nu+j}{t+\lambda^*} nb\left(j;\frac{t}{t+\lambda^*},\nu\right) \tag{26}$$

where $nb\left(j;\frac{t}{t+\lambda^*},\nu\right)$ is the probability function corresponding to $NB\left(j;\frac{t}{t+\lambda^*},\nu\right)$.

The density function (26) is equivalent to

$$\begin{aligned} f_T^*(t;\lambda^*,\zeta,\nu) &= \frac{\nu}{t+\lambda^*}\sum_{j=0}^{\infty} p(j;\zeta)nb\left(j;\frac{t}{t+\lambda^*},\nu\right) \\ &+ \frac{\zeta}{t+\lambda^*}\sum_{j=0}^{\infty} p(j;\zeta)nb\left(j+1;\frac{t}{t+\lambda^*},\nu\right). \end{aligned} \tag{27}$$

The expected value and variance of the failure time under this double stochastic model is

$$E\{T(\beta)\} = \frac{(1+\zeta)\lambda^*}{\nu-1}, \quad \nu > 1 \tag{28}$$

and

$$V\{T(\beta)\} = \frac{\lambda^{*2}}{(\nu-1)(\nu-2)}\left(1+2\zeta+\frac{(1+\zeta)^2}{\nu-1}\right), \quad \nu > 2. \tag{29}$$

In Figure 2 we present the hazard function for the case of $\lambda^* = 1$, $\zeta = 5$ and $\nu = 1$.

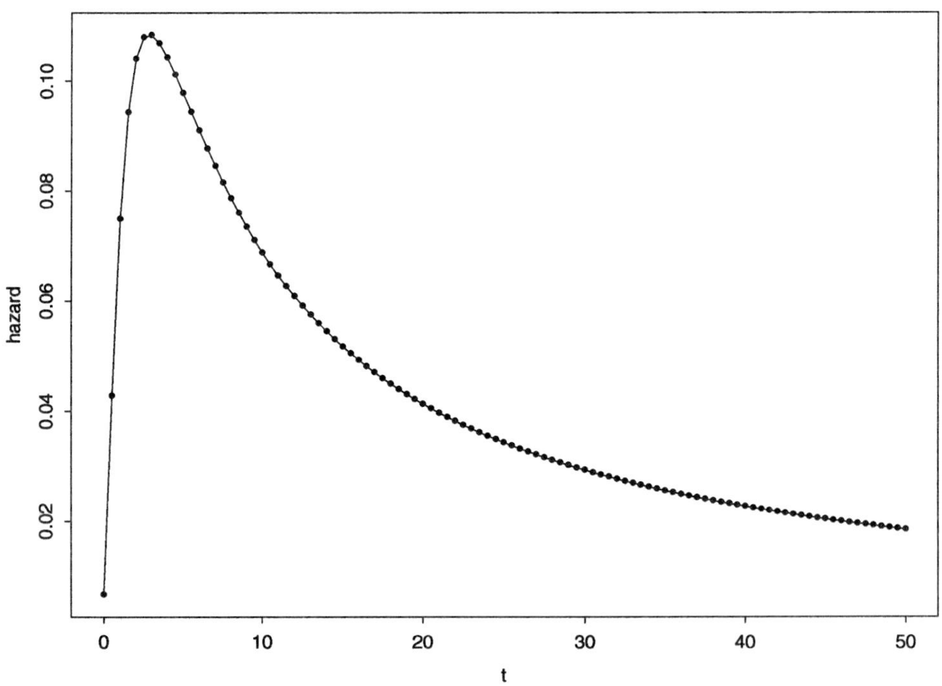

Fig. 2. Hazard Function for $\lambda^* = 1$, $\zeta = 5$, $\nu = 1$.

4 Compound Poisson With Erlang Damage

Suppose now that the amount of damage in each occurrence is a random variable having an Erlang (μ, k) distribution, $k \geq 2$. In this case the reliability function is

$$R(t; \lambda, \zeta, k) = \sum_{j=0}^{\infty} p(j; \lambda t)(1 - P(jk - 1; \zeta)), \tag{30}$$

where $\zeta = \mu\beta$. Changing the order of summation yields

$$R(t; \lambda, \zeta, k) = \sum_{j=0}^{\infty} p(j; \zeta) P\left(\left[\frac{j}{k}\right]; \lambda t\right), \tag{31}$$

where $[\alpha]$ is the maximal integer not exceeding α.

The corresponding pdf of $T(\beta)$ is

$$p_T(t; \lambda, \zeta, k) = \lambda \sum_{j=0}^{\infty} p(j; \zeta) p\left(\left[\frac{j}{k}\right]; \lambda t\right). \tag{32}$$

Define the probability weights

$$\Delta_j(k; \zeta) = P((j+1)k - 1; \zeta) - P(jk - 1; \zeta), \tag{33}$$

$j = 0, 1, \ldots$ where $P(-1; \zeta) \equiv 0$.

Notice that $\Delta_j(k, \zeta) > 0$ for all $j = 0, 1, \ldots$ and $\sum_{j=0}^{\infty} \Delta_j(k; \zeta) = 1$. One can write

$$R(t; \lambda, \zeta, k) = \sum_{j=0}^{\infty} \Delta_j(k; \zeta) P(j; \lambda t) \tag{34}$$

and

$$p_T(t; \lambda, \zeta, k) = \lambda \sum_{j=0}^{\infty} \Delta_j(k; \zeta) p(j; \lambda t). \tag{35}$$

The hazard function is the ratio of (35) over (34). As in Zacks [Z04] one can prove that

$$\lim_{t \to \infty} h(t; \lambda, \zeta, k) = \lambda$$

$$\lim_{t \to 0} h(t; \lambda, \zeta, k) = \lambda \Delta_0(k; \zeta) = \lambda P(k - 1; \zeta).$$

The moments of $T(\beta)$ are given by

$$E\{T^m(\beta)\} = \frac{1}{\lambda^m} \sum_{j=0}^{\infty} \Delta_j(k; \zeta) \prod_{i=1}^{m} (j + i). \tag{36}$$

Thus,

$$E\{T(\beta)\} = \frac{1}{\lambda}\left(1 + \sum_{j=1}^{\infty} j\Delta_j(k;\zeta)\right) \tag{37}$$

and

$$E\{T^2(\beta)\} = \frac{1}{\lambda^2}\left(2 + 3\sum_{j=1}^{\infty} j\Delta_j(k;\zeta) + \sum_{j=1}^{\infty} j^2\Delta_j(k;\zeta)\right). \tag{38}$$

Thus

$$V\{T(\beta)\} = \frac{1}{\lambda^2}\left(1 + \sum_{j=1}^{\infty} j\Delta_j(k;\zeta) + \left(\sum_{j=1}^{\infty} j^2\Delta_j(k;\zeta) - \left(\sum_{j=1}^{\infty} j\Delta_j(k;\zeta)\right)^2\right)\right). \tag{39}$$

Moreover, the increase of $h(t;\lambda,\zeta,k)$ from $\lambda P(k-1;\zeta)$ to λ is monotone.

In the double stochastic case, where λ is distributed like Gamma(λ^*,ν), the reliability function and $p_T^*(t;\lambda^*,\zeta,k,\nu)$ is as in (24) and (26) in which $p(j;\zeta)$ is replaced by $\Delta_j(k;\zeta)$.

References

[AG03] Aalen, O.O. and Gjessing, H.K.: A look behind survival data: underlying processes and quasi-stationarity. In: Bo H. Lindquist & K.A. Doksum (eds) Mathematical and Statistical Methods in Reliability. World Scientific, New Jersey (2003).

[BK85] Bogdanoff, J.L. and Kozim, F.: Probabilistic Models of Cumulative Damage. John Wiley, New York (1985).

[BN02] Bagdonavicius, V. and M. Nikulin: Accelerated Life Models: Modeling and Statistical Analysis. Chapman & Hall/CRC, Boca Raton (2002).

[K97] Kao, Edward, P.C.: An Introduction to Stochastic Processes. Duxbury Press, New York (1997).

[KW00] Kahle, W. and Wendt, H.: Statistical analysis of damage processes. In N. Limuios and M. Nikulin (eds) Recent Advances in Reliability Theory, Methodology, Practice and Inference. Birkhäuser, Boston (2000).

[K57] Karlin, S.: Polya type distributions, II, Ann. Math. Statist., **28**, 281-308 (1957).

[W00] Wilson, S.P.: Failure Models Indexed by Time and Usage. In N. Limuios and M. Nikulin (eds) Recent Advances in Reliability Theory, Methodology, Practice and Inference. Birkhäuser, Boston 2000).

[Z04] Zacks, S.: Distributions of Failure Times Associated With Non-Homogeneous Compound Poisson Damage Processes. In Anirban

Das Gupta (ed) A Fetstschrift for Herman Rubin. IMS-Lecture Notes-Monograph Series (2004).

Index